Springer Series in
MATERIALS PROCESSING

Springer
Berlin
Heidelberg
New York
Barcelona
Hong Kong
London
Milan
Paris
Tokyo

Springer Series in

MATERIALS PROCESSING

Series Editors: W. Michaeli H. Warlimont E. Weber

This series is focused on the science and application of materials processing as an essential part of progress in the materials field. It addresses researchers and process engineers alike. The scope of the series includes all classes of materials – metals, inorganic non-metallic materials, polymers and composites – in the form of bulk materials, thin films and layered structures as well as micro- and nanostructured forms. All aspects of materials processing from fundamental understanding to innovative strategies and methods of practical process implementation and control are covered. It is the aim of the series to provide comprehensive information on the science and application of leading-edge and well-established processing technologies.

Low-Pressure Synthetic Diamond. Manufacturing and Applications
Editors: B. Dischler and C. Wild

Purification Process and Characterization of Ultra High Purity Metals
Application of Basic Science to Metallurgical Processing
Editors: Y. Waseda and M. Isshiki

Supercritical Fluid Science and Technology
Editors: Y. Arai, T. Sako, and Y. Takebayashi

Epitaxy. Physical Foundation and Technical Implementation
By M.A. Herman and W. Richter

Series homepage – http://www.springer.de/phys/books/ssmp/

Yoshio Waseda   Minoru Isshiki (Eds.)

Purification Process and Characterization of Ultra High Purity Metals

Application of Basic Science to Metallurgical Processing

With 151 Figures and 56 Tables

Springer

Professor Dr. Yoshio Waseda
Professor Dr. Minoru Isshiki
Tohoku University
Institute of Multidisciplinary Research for Advanced Materials
Katahira 2-choume 1-1
Aoba-ku Sendai 980-8577, Japan
E-mail: waseda@tagen.tohoku.ac.jp
isshiki@tagen.tohoku.ac.jp

Series Editors:

Professor Dr. Walter Michaeli
Institut für Kunststoffverarbeitung, Pontstrasse 49
52062 Aachen, Germany

Professor Dr. Hans Warlimont
Institut für Festkörper- und Werkstofforschung e.V., Helmholtzstrasse 20
01069 Dresden, Germany

Professor Dr. Eicke Weber
University of California, Materials Science and Mineral Engineering
587 Evans Hall, Berkeley, CA 94720-1760, USA

ISSN 1434-9795
ISBN 3-540-41322-7 Springer-Verlag Berlin Heidelberg New York

Library of Congress Cataloging-in-Publication Data. Purification process and characterization of ultra high purity metals: Application of basic science to metallurgical processing/ Yoshio Waseda, Minoru Isshiki (eds.). p. cm. – (Springer series in materials processing, ISSN 1434-9795) Includes bibliographical references and index. ISBN 3540413227 (hc.: alk. paper) 1. Metals–Refining. I. Waseda, Yoshio. II. Isshiki, Minoru, 1948– III. Series. TN 673.P87 2001 669'.028–dc21 00-069259

Springer-Verlag Berlin Heidelberg New York
a member of BertelsmannSpringer Science+Business Media GmbH

http://www.springer.de

Typesetting: Data conversion by Frank Herweg, Leutershausen
Cover concept: eStudio Calamar Steinen
Cover production: *design & production* GmbH, Heidelberg

Printed on acid-free paper SPIN: 10747256 57/3141/ba - 5 4 3 2 1 0

Preface

Further development of high technology is based on the availability of adequate specialty materials. In this respect, various metals have been rediscovered and recognized as strategic materials of future technology, provided that they are purified to the required – often extremely high – levels. The preparation, characterization and application of ultra-high-purity metals therefore constitutes a relatively new and dynamically growing domain in modern materials science and engineering. Fundamental research has revealed that even trace amounts of impurities may significantly affect the physical and chemical properties of metals. Substantial enhancement of important properties is manifested when impurity concentrations are reduced to extremely low levels. This also implies that pure metals can serve as new functional materials, whose intrinsic properties, however, need to be further clarified. High-purity metals are also required in practical use as constituent elements of compound semiconductors and as structural materials for ultra-large-scale integration (ULSI) units. One of the current topics in the compound semiconductor field is focused on β-$FeSi_2$, a material for new opto-electronic devices in the infrared region. This, for example, directly represents a demand for semiconductor-grade (high-purity) iron.

Research has been continued to develop existing purification processes and to establish new purification methods, as well as to characterize pure metals. The frontiers of the present knowledge of theory and practice need further exploration. Emerging demands, created by sophisticated applications, may be met by the evolution of concepts and methodologies. This book offers an up-to-date survey and critical evaluation, starting with an extended introductory treatise on the fundamentals, followed by a description of the practically significant and relatively new methods of purification of transition, refractory and rare earth metals, and ending with their characterization. Physical and chemical properties, advanced applications, adequate purification techniques, quality assessment and characterization of pure metals are approached from the basic science, as well as the applied engineering points of view.

Chapter 1 starts with a general description of the principles underlying metal purification and process development. Due to its determining role in material qualification, the topic of purity evaluation – with special emphasis on the use of electrical resistivity – is discussed in detail. In order to provide the basis for process development, this chapter summarizes the currently applied and recently proposed chemical–metallurgical purification methods, and the recorded purity levels of the most important transition, refractory and rare earth metals. Special technologies using thermal plasma, mass separation and laser are described in Chap. 2. Furthermore, Chap. 3 discusses how lattice defects and impurities influence electrical and mechanical properties and surface stability, which are of increasing practical importance in advanced applications. Many references, illustrations and tables provide adequate help and guidance for specialists, nonspecialists or college students who wish to become acquainted with the processing and characterization of ultra-high-purity metals.

Since the properties of high-grade materials depend largely upon the behavior of defects, in correlation with the impurity content, further research on pure metals and impurity effects is required for producing "tailored materials". This demand is expected to become increasingly important in the future, with the progress of higher technologies. We hope that this book will be a step in promoting this progress.

The contributors were selected for their long and continuing expertise in their respective fields. We would like to express our gratitude to all those involved in the work of manuscript preparation and publishing. In particular, we should indicate the significant contributions of Dr. T. Uda and Mr. Y. Ito, for their assistance in preparing the electronic typesetting of this volume.

Sendai, Japan
August 2001

Yoshio Waseda
Minoru Isshiki

Contents

Part III Characterization

List of Contributors

Dr. David Fort
School of Metallurgy and Materials
The University of Birmingham
Edgbaston, Brimingham B15 2TT
United Kingdom
fortd@novell5.bham.ac.uk

Professor Eiji Hashimoto
Hiroshima Synchrotron
Radiation Center
Hiroshima University
Kagamiyama 2-313
Higashi-hiroshima
Hiroshima 739-8526, Japan
ehasimot@sci.hiroshima-u.ac.jp

Dr. Lutz Hollang
Technische Universität Dresden
Institut für Physikalische
Metallkunde
Mommsenstraße 13
01069 Dresden, Germany
hollang@physik.tu-dresden.de

Professor Minoru Isshiki
Institute of Multidisciplinary
Research for Advanced Materials
Tohoku University
Katahira 2-1-1, Aoba-ku
Sendai 980-8577, Japan
isshiki@tagen.tohoku.ac.jp

Professor Dr. K.T. Jacob
Department of Metallurgy
Indian Institute of Science
Bangalore 560012, India
katob@mrc.iisc.ernet.in

Dr. Tamas Kekesi
Institute of Multidisciplinary
Research for Advanced Materials
Tohoku University
Katahira-2-choume 1-1, Aoba-ku
Sendai 980-8577, Japan
kekesi@iamp.tohoku.ac.jp
University of Miskolc
Miskolc-Egyetemvaros 3515
Hungary
fekkt@uni-miskolc.hu

Professor Dr. Alfred Köthe
Institut für Festkörper und
Werkstofforschung, Dresden
Postfach 270 016
01171 Dresden, Germany
a.koethe@ifw-dresden.de

Dr. Kouji Mimura
Institute of Multidisciplinary
Research for Advanced Materials
Tohoku University
Katahira-2-choume 1-1, Aoba-ku
Sendai 980-8577, Japan
mimura@tagen.tohoku.ac.jp

Professor Kiyoshi Miyake
Department of Science and Technology, Saitama University
Shimo-Okubo 255, Urawa
Saitama 338-8570, Japan
kmiyake@d-kiki.ees.saitama-u.ac.jp

Professor Toru H. Okabe
Institute of Industrial Science
The University of Tokyo
4-6-1 Komaba, Meguro-ku
Tokyo 153-8505, Japan
okabe@iis.u-tokyo.ac.jp

Professor Shunichi Sato
Institute of Multidisciplinary Research for Advanced Materials
Tohoku University
Katahira-2-choume 1-1, Aoba-ku
Sendai 980-8577, Japan
shun@tagen.tohoku.ac.jp

Professor Dr. Alfred Seeger
Max-Planck-Institut für Metallforschung
Heisenbergerstraße 1
70569 Stuttgart, Germany
seeger@physix.mpi-stuttgart.mpg.de

Dr. Shigeru Suzuki
Advanced Technology Research Laboratories
Nippon Steel Corporation
20-1 Shintomi, Futtu
Chiba, 293-8511, Japan
ssuzuki@re.nsc.co.jp

Professor Yoshitake Ueda
Department of Clinical Radiology
Faculty of Health Sciences
Hiroshima International University
555-36, Gakuendai, Kurose-cho, Kamo-gun
Hiroshima 724-0695, Japan
y-ueda@hs.hirokoku-u.ac.jp

Professor Yoshio Waseda
Institute of Multidisciplinary Research for Advanced Materials
Tohoku University
Katahira 2-1-1, Aoba-ku
Sendai 980-8577, Japan
waseda@tagen.tohoku.ac.jp

Part I

Fundamentals and the Present Status of Purification of Metals

1 The Thermodynamics of Oxygen in Reactive Metals

T.H. Okabe, K.T. Jacob and Y. Waseda

1.1 Introduction

For some new applications of metals in functional devices, metals of high purity are required. In recent years, many high-purity metals have been produced commercially for use in electronics, but the demand for ultra-high-purity metals is increasing rapidly because of more stringent specifications for materials used in high-performance information devices.

The principle of removing trace impurities from metals is simple. Impurities in a metal can be removed by lowering the chemical potentials of the impurity elements outside the metal. Since the chemical potential of an impurity approaches minus infinity as the concentration tends to zero, there are practical limits to ultrapurification. It is even more difficult to eliminate impurities which have a strong affinity for the metals. High-purity reactive metals often contain more gaseous elements than metallic impurities, mainly because of contamination from the atmosphere during processing. For example, 99.9999% pure titanium (excluding carbon and gaseous elements) has been produced commercially as a target material [1–5]. However, this high-purity titanium is known to contain 50–500 mass ppm oxygen. The removal of metallic impurities below ppm level can be achieved by conventional purification processes such as electrolysis under a highly controlled atmosphere [4]. However, it is more difficult to reduce the amounts of gaseous impurities which form strong bonds with the metal. Commercially available high-purity rare earth metals also contain a similar quantity of oxygen, 50–500 mass ppm [6]. These reactive metals have an extremely strong affinity for oxygen and dissolve oxygen in large concentrations. Until recently, it was thought to be practically impossible to remove oxygen from these highly reactive metals to very low concentrations.

In recent years, a few deoxidation techniques have been developed for removing oxygen from titanium [7–9] and titanium aluminides [10]. Titanium containing less than 10 mass ppm oxygen, which is below the detection limit of conventional gas analysis, has been produced by an electrochemical technique [9]. The method has also been applied for oxygen removal from solid yttrium and rare earth (RE) metals; yttrium metal containing 5700 mass ppm oxygen, and rare earth metals (RE = La, Pr, Nd, Gd, Tb, Dy, Er) containing more than 2000 ppm, have been deoxidized below 100 ppm after a few hours of treatment [11–14].

Thermodynamic data for impurity elements in metals are essential for developing effective methods for removal of trace impurities in metals. As part of a systematic study on oxygen removal directly from reactive metals, a technique has, therefore, been developed for controlling extremely low oxygen chemical potentials, circa $1/2\mu_{O_2(\mathrm{Env.})} = -550\,\mathrm{kJ}$ at 1173 K ($p_{O_2} = 10^{-50}$ atm), which is far beyond the range of current solid oxide electrolyte sensors.

In the present chapter, the thermodynamics of trace impurities in metallic solid solutions is introduced briefly, and purification techniques are discussed with the aid of thermodynamic information. Oxygen in reactive metals is chosen as a representative impurity, not only because it is the most difficult element for removal, but also because the reactive metals are always subjected to contamination by oxygen at each step of processing. Some representative solid state deoxidation methods and their characteristic features are briefly reviewed in order to facilitate thermodynamic interpretation of the various deoxidation techniques.

1.2 Various Techniques for Removing Oxygen from Reactive Metals

Oxygen is the major impurity in reactive metals. The concentration of oxygen often increases during metal processing, as there are several sources of contamination. Several attempts to develope effective methods for reducing oxygen in reactive metals have been reported, and significant progress in solid state deoxidation has been achieved. In this section, some representative solid state deoxidation techniques which can be applied to reactive metals are presented.

1.2.1 Solid State Electrotransport

Solid state electrotransport (SSE) is based on slow migration of impurities in a metal rod during the passage of a large current at high temperatures. A schematic representation of oxygen removal by this technique is shown in Fig. 1.1, in comparison with other thermochemical deoxidation techniques. Under an electric field gradient, oxygen in solid solution migrates to one side, acting as if it is an ionic species with induced charge.

Among the various solid state purification techniques, SSE is effective not only for removing gaseous impurities but also other undesirable elements that may be present at low concentrations. This technique appears to be one of the most promising methods for producing ultra-high-purity metals [6]. The major drawback of this technique is its low productivity, imposed by restricted sample geometry and the requirement of a low initial impurity

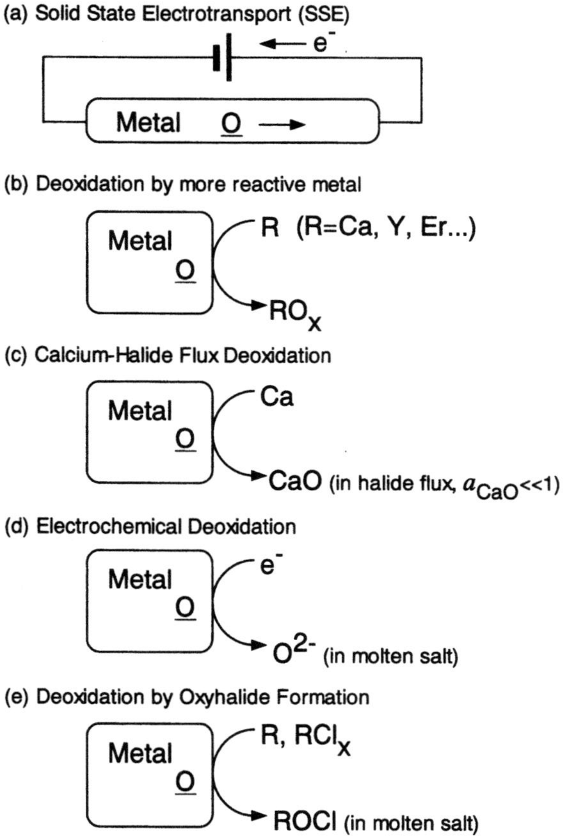

Fig. 1.1. A schematic illustration of some solid state purification methods for the deoxidation of reactive metals

concentration. It takes several tens or hundreds of hours to remove oxygen from one portion of the sample to another [15, 16], and the overall oxygen content of the sample usually increases slightly during processing in systems that do not have volatile oxide gases species. The deoxidation limit depends on the initial concentration, and only a small portion of the sample can be purified with this technique. For this reason, SSE is not suitable for large-scale commercial production, although this technique is quite useful for producing small amounts of ultra-high-purity metals. Details of SSE are presented in Chap. 5.

1.2.2 Deoxidation Using Metal/Metal Oxide Equilibrium

Deoxidation of a metal using "external gettering" by a more reactive deoxidant (R) placed in contact is widely used. For reactive metals, however, the number of candidate deoxidants is very limited. The overall deoxidation reaction may be expressed as

$$x\ \mathrm{O(Metal)} + \mathrm{R} = \mathrm{RO}_x\,. \tag{1.1}$$

Several efforts on deoxidation by reactive metals, such as calcium, have been made [7, 12, 17–27]. This technique has recently been applied to deoxidation molten reactive metals, and thermodynamic properties of deoxidant in molten metals have been investigated [28–30].

The deoxidation limit is governed by the oxygen chemical potential in the system, fixed by the $\mathrm{R/RO}_x$ equilibrium. The theoretical deoxidation limit at the desired temperature can be evaluated using information on the activity coefficient of oxygen in solid solution, γ_O or f_O, and the standard Gibbs energy of formation, $\Delta G^\circ_\mathrm{f}$, of the oxide ($\mathrm{RO}_x$). Calcium and some rare earth metals are found to be thermodynamically suitable for removing oxygen from Ti–O solid solutions down to ppm level [26, 27]. For example, titanium can be deoxidized to about the 500 mass ppm level when Ca/CaO equilibrium prevails at 1273 K. Although magnesium and sodium have strong affinities for oxygen, they do not act as deoxidants down to the oxygen level attained by calcium. For this reason, the current titanium production process, which employs magnesiothermic reduction [31], is susceptible to oxygen contamination [32].

1.2.3 Calcium Halide Flux Deoxidation

The oxygen level fixed by Ca/CaO equilibrium is suitable for commercial application; for example, removing oxygen in titanium scrap [25]. However, a lower deoxidation limit is preferred, as it can provide more flexibility in the subsequent remelting processes. Attempts to reduce the deoxidation limit by decreasing the activity of the CaO product were made in the calcium halide flux method [7, 8]. The principle of this method is shown schematically in Fig. 1.1c. Metal is deoxidized in calcium-saturated halide flux according to the following reaction:

$$\mathrm{O(Metal)} + \mathrm{Ca}(l, g) = \mathrm{CaO(Halide\ flux)}. \tag{1.2}$$

As a halide flux, liquid $CaCl_2$ is found to be suitable not only because it has large solubility for CaO [33] but also because the flux can be removed easily in the subsequent process. The deoxidation limit of the above reaction is governed by the activity of CaO in the flux. When the flux is saturated with CaO, the deoxidation limit becomes identical to that set by Ca/CaO equilibrium, discussed in the preceding section. This process is simple, and capable of producing high-purity titanium containing 20 mass ppm oxygen [7]. The method can be extended to other metals, such as niobium [34]. However, the method is not suitable for removing large amounts of oxygen, because the effectiveness of the process for oxygen removal decreases with increasing oxygen concentration in the flux. Some additional techniques that reduce the CaO activity are needed to achieve lower deoxidation limits.

1.2.4 Electrochemical Deoxidation

The electrochemical deoxidation method corresponds to a modification of the calcium-halide flux deoxidation method [9, 11, 14]. The main features are the production of the deoxidant calcium in situ by electrolysis of $CaCl_2$ flux, and the dissolution of the CaO deoxidation product into the flux for the subsequent electrolysis. In this method, represented by (1.3) and (1.4), the reactive metal placed at the cathode is deoxidized by calcium produced electrochemically:

$$\mathrm{O(Metal)} + \mathrm{Ca(Molten\ salt)} = \mathrm{O^{2-}(Molten\ salt)} + \mathrm{Ca^{2+}(Molten\ salt)}\,, \tag{1.3}$$

$$\mathrm{Ca^{2+}(Molten\ salt)} + 2\ \mathrm{e^-} = \mathrm{Ca(Cathode)}\,. \tag{1.4}$$

The overall reaction at the cathode can be visualized as

$$\mathrm{O(Metal)} + \mathrm{e^-} = \mathrm{O^{2-}(Molten\ salt)}\,. \tag{1.5}$$

This equation implies that oxygen in the reactive metals can be removed by applying an electrochemical potential, and that a metal deoxidant is not always necessary. At the carbon anode, O^{2-} ions are removed as CO (or CO_2) gas:

$$x\ \mathrm{O^{2-}(Molten\ salt)} + \mathrm{C} = \mathrm{CO}_x(g) + 2x\ \mathrm{e^-}\,. \tag{1.6}$$

In contrast to calcium halide flux deoxidation, the electrochemical method shown in Fig. 1.1d offers a number of advantages. First, the addition of metallic calcium as a deoxidant is not required, since the activity of calcium near the cathode can be increased electrochemically, even as high as unity, by controlling the applied voltage between the metal cathode and the carbon anode. Second, the CaO product is soluble in the ionic melt $CaCl_2$, and this facilitates continuous electrochemical removal of CaO. Therefore, the CaO activity in the melt is kept at a very low level and does not increase during the process. Third, in situ generation of calcium by electrolysis produces a deoxidant free of impurities that otherwise could contaminate the metal being deoxidized.

1.2.5 Deoxidation by Oxyhalide Formation

The formation of oxyhalides can theoretically be utilized to remove oxygen in reactive metals [35–37]. In some chloride-rich oxychloride melts, the activity of the oxide is considerably reduced. The deoxidation reaction can be expressed as

$$\begin{aligned}\mathrm{O(Metal)} &+ \frac{x-1}{x}\mathrm{R(Molten\ salt)} + \frac{1}{x}\mathrm{RCl}_x\mathrm{(Molten\ salt)}\\ &= \mathrm{ROCl(Molten\ salt\ or\ pure\ solid)}\,.\end{aligned} \tag{1.7}$$

Some rare earth metals form fairly stable oxyhalide compounds [35]. This reaction can, therefore, be used for deoxidation. From thermodynamic calculations, it appears that titanium can be deoxidized down to the ppm level using rare earth oxyhalide formation. However, the limit is comparable to that of the calcium halide flux method at 1200 K when the activity of RE-OCl is unity [13].

1.3 The Thermodynamics of Oxygen in Reactive Metals

To develop an efficient purification process, it is important to know the thermodynamic properties of the oxygen dissolved in metals. This information is essential for quantitative assessment of the deoxidation limit. Furthermore, information on oxygen chemical potentials in metal–oxygen (M–O) solid solutions can be utilized to optimize process parameters in the production of high-purity metals.

As shown in Fig. 1.2, a hypothetical partial pressure of oxygen gas, O_2, in equilibrium with a metal–oxygen solid solution, can be introduced. The chemical potential for diatomic oxygen gas in the environment, $\mu_{O_2(\mathrm{Env.})}$, and also that for the dissolved oxygen in metal, $\mu_{O(\mathrm{Metal})}$, can be expressed as follows:

$$\mu_{O_2(\mathrm{Env.})} = \mu^\circ_{O_2(\mathrm{Env.})} + RT \ln p_{O_2(\mathrm{Env.})}, \quad (1.8)$$

$$\mu_{O(\mathrm{Metal})} = \mu^\circ_{O(\mathrm{Metal})} + RT \ln a_{O(\mathrm{Metal})}, \quad (1.9)$$

where $\mu^\circ_{i(j)}$ is the standard chemical potential for species i in matrix j; $p_{O_2(\mathrm{Env.})}$ is the oxygen partial pressure around metal; and $a_{O(\mathrm{Metal})}$ is the activity of oxygen in metal, which is a function of oxygen concentration [O]. The deviation from this linear relation between the activity (a_O) and concentration of dissolved oxygen can be quantified using f_O, where f_O is the activity coefficient of oxygen in metal. When oxygen present in the environment

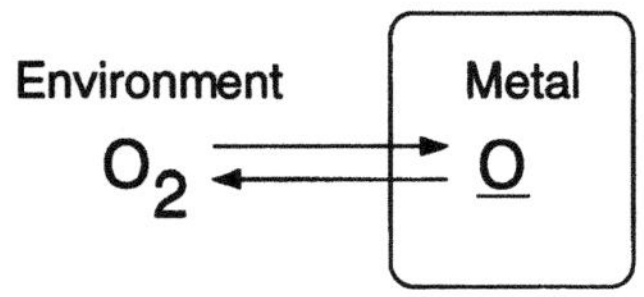

$$\mu_{O_2(\mathrm{Env.})} = \mu^\circ_{O_2(\mathrm{Env.})} + RT \ln p_{O_2(\mathrm{Env.})}$$

$$\mu_{O\,(\mathrm{Metal})} = \mu^\circ_{O\,(\mathrm{Metal})} + RT \ln a_{O\,(\mathrm{Metal})}$$

Fig. 1.2. Deoxidation of metal is thermodynamically feasible when $1/2\mu_{O_2(\mathrm{Env.})}$ is lower than $\mu_{O(\mathrm{Metal})}$

is in equilibrium with oxygen in metal, $1/2\mu_{O_2(Env.)}$ is equal to $\mu_{O(Metal)}$. Deoxidation of the metal is thermodynamically feasible when $1/2\mu_{O_2(Env.)}$ is lower than $\mu_{O(Metal)}$.

The extremely low oxygen chemical potentials in reactive metals, in the order of $\mu_{O_2(Env.)} = -1100\,kJ/mol$ at 1150 K, are far beyond the range of applicability of oxide solid electrolytes and conventional gas-equilibration techniques. However, they can be measured by an indirect technique involving equilibration of a metal–oxygen solution with reference solutions. As the reference material, titanium–oxygen (Ti–O) solid solutions can be used, since their thermodynamic properties are known [19,21,24,26,27]. The principle of oxygen chemical potential measurement for on yttrium–oxygen (Y–O) solid solution is explained. It serves as an example for the measurement of ultra-low-oxygen chemical potentials in solid solutions of reactive metals.

1.3.1 The Principle of Measurement of Ultra-Low-Oxygen Potentials

Oxygen chemical potential measurement is accomplished by an indirect technique involving equilibration of Y–O specimens with a Ti–O solid solution. Titanium can be chosen because the relationship between the oxygen concentration, [O], and the oxygen partial pressure, p_{O_2}, is well established in this material and can be extrapolated to extremely low levels of oxygen chemical potential in accordance with Henry's law.[1]

The principle of this technique is illustrated in Fig. 1.3, which gives the relationship between the square root of p_{O_2} and [O] in Ti–O and Y–O at constant temperature. The two lines were determined as follows. The coordinates of point A are p_1 and $[O]_A$. p_1 is the value of p_{O_2} in equilibrium with pure calcium and CaO (e.g. $p_{O_2} \sim 10^{-40}$ atm at 1273 K), and is available from the literature data [38]. $[O]_A$ is the concentration of oxygen in titanium (~ 500 mass ppm) at this p_{O_2}, and the value has been independently confirmed by several researchers [19,21,24,26].

On the assumption that Ti–O solutions exhibit Henrian behavior up to $[O]_A$, the relationship between $p_{O_2}^{1/2}$ (or a_O) and [O] in Ti–O solutions is represented by a straight line connecting point A and the origin. It is possible to measure the dissolved oxygen content [O] in titanium metal down to concentrations as low as 10 ppm by conventional techniques. At these values of [O] in titanium, the corresponding p_{O_2} is not measurable by direct methods. Nevertheless, such values can be accurately determined from the Ti–O line in Fig. 1.3. For example, when the concentration of oxygen in titanium is $[O]_C$, the equilibrium p_{O_2} is p_3.

[1] Henry's law: the partial pressure, or activity, of the solute is proportional to its concentration. It is empirically found if the concentration of the solute is sufficiently low at constant temperature.

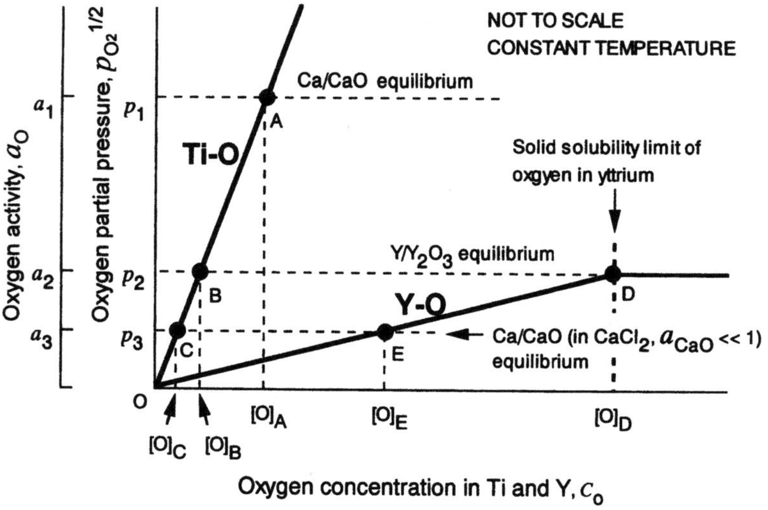

Fig. 1.3. The relationship between oxygen concentration and oxygen partial pressure in Ti–O and RE–O solid solutions (schematic)

If a Y–O solution is equilibrated with the same atmosphere in which $p_{O_2}^{1/2}$ is equal to p_3, and the oxygen concentration is measured as $[O]_E$, the thermodynamics of the Y–O solution can be evaluated. For the Y–O system, the relationship between p_{O_2} and [O] can be represented by a straight line connecting point E and the origin, when Y–O solution obeys Henry's law. The value of p_{O_2} at the solubility limit of oxygen in yttrium (p_2) can be given by point D. Points D and E in Fig. 1.3 fall on the same line through the origin, and this confirms that the Y–O solutions are Henrian across the entire composition range. Furthermore, [O] in titanium at Y/Y_2O_3 equilibrium gives the value of $[O]_B$ at p_2, which can also confirm the nature of the relationship between the oxygen concentration and the oxygen chemical potential in solid titanium.

1.3.2 Gibbs Energies of Solution of Oxygen in Reactive Metals

Using the thermodynamic properties of Ti–O and Y–O solid solutions, it is possible to extend the equilibrium method to other systems, the properties of which are unknown. An indirect technique involving equilibration of rare earth metal–oxygen (RE–O) specimens with Y–O and/or Ti–O solid solutions, which act as reference materials, is discussed further.

A diagrammatic summary of the data on the oxygen potential of the reference solutions is presented in Fig. 1.4, in comparison with data for metal–metal oxide couples [38]. More specifically, the chemical potential of oxygen at a fixed oxygen content is plotted as a function of temperature. Within the

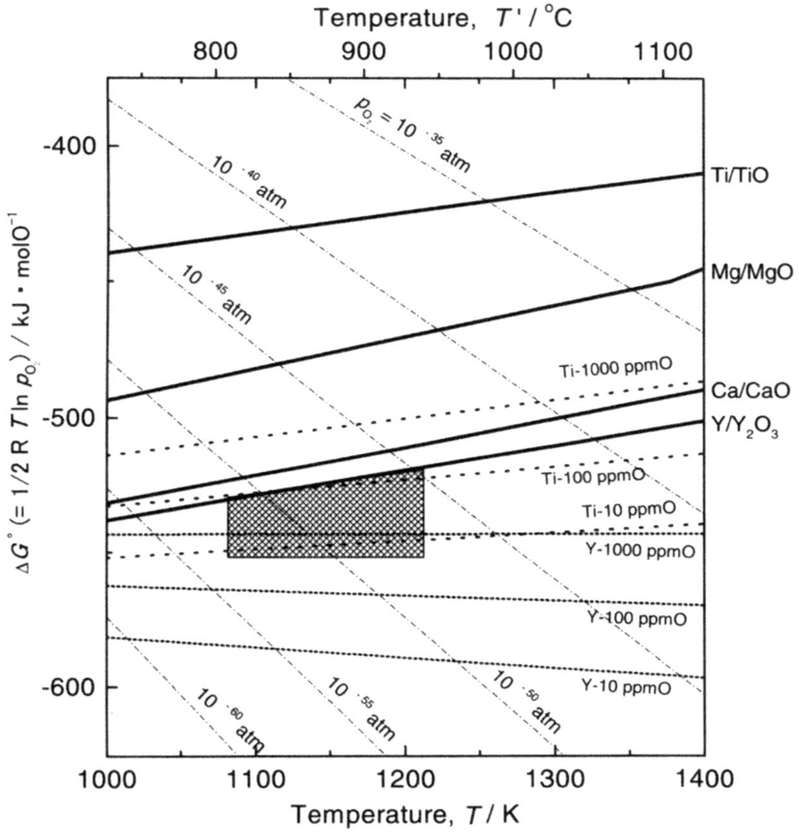

Fig. 1.4. Ellingham diagram for selected oxides and Ti–O and Y–O solid solutions. Range of oxygen chemical potentials encountered in Sect. 1.3 is depicted by the cross-hatched region

accuracy of measurement, the reference solutions were found to obey Sievert's law[2] [39, 40]. For a more quantitative discussion, some essential points of the thermodynamics of oxygen in solid solution are summarized.

The standard Gibbs energy change associated with the solution of oxygen in the rare earth metals, $\Delta G^\circ_{1,\mathrm{RE}}$, can be expressed as follows:

$$\tfrac{1}{2}\mathrm{O}_2(g) = \mathrm{O(RE)}, \tag{1.10}$$

$$\Delta G^\circ_{1,\mathrm{RE}} = \mu^\circ_{\mathrm{O(Metal)}} - \tfrac{1}{2}\mu^\circ_{\mathrm{O_2(gas.)}} = -\mathrm{R}T\ln\frac{a_{\mathrm{O(RE)}}}{p_{\mathrm{O_2}}^{1/2}} \tag{1.11}$$

[2] Sievert's law: the concentration of the dissolved oxygen is proportional to the square root of the partial pressure of oxygen in the gas phase. This is because bimolecular gases, such as oxygen, dissolve as dissociated atoms in a solid or in liquid metals.

$$= -RT \ln \frac{f_{O(RE)} \cdot [\text{mass\%O}]_{RE}}{p_{O_2}^{1/2}}, \tag{1.12}$$

where $[\text{mass\%O}]_{RE}$ is the equilibrium concentration of oxygen, in mass%, in the rare earth metal at temperature T (K) and oxygen partial pressure p_{O_2} (atm) and $f_{O(RE)}$ is the Henrian activity coefficient of oxygen, expressed relative to an infinitely dilute solution in which the activity of the solute is equal to its concentration in mass%. When Sievert's law is applied to dissolved oxygen, the value of the Henrian activity coefficient is unity. By measuring $[\text{mass\%O}]_{RE}$ at equilibrium under known values of p_{O_2} at various temperatures, the Gibbs energy change associated with the solution of oxygen, $\Delta G^\circ_{1,RE}$, can be evaluated as a function of T.

The oxygen partial pressure can be fixed using various methods. A two-phase mixture of a metal and its oxide controls the oxygen partial pressure. For example, the p_{O_2} established by Y/Y_2O_3 equilibrium may be expressed by the following equations:

$$\tfrac{2}{3}Y + \tfrac{1}{2}O_2(g) = \tfrac{1}{3}Y_2O_3, \tag{1.13}$$

$$\Delta G^\circ_{f,Y_2O_3} = -RT \ln \frac{a_{Y_2O_3}^{1/3}}{a_Y^{2/3} \cdot p_{O_2}^{1/2}}, \tag{1.14}$$

where $\Delta G^\circ_{f,Y_2O_3}$ is the standard Gibbs energy of formation of Y_2O_3, and a_i is the activity of species i. Among the different metal/metal-oxide buffers, the Y/Y_2O_3 equilibrium has the lowest p_{O_2}. The value of p_{O_2} can be calculated directly from the literature value for $\Delta G^\circ_{f,Y_2O_3}$ when both the metal and the oxide are at unit activity.

Another method for controlling oxygen partial pressure at very low values employs calcium-saturated $CaCl_2$ molten salt with a small amount of dissolved CaO. The p_{O_2} varies with the concentration of CaO dissolved in the fused salt. The equilibrium controlling the p_{O_2} can be represented as

$$Ca + \tfrac{1}{2}\, O_2(g) = CaO(CaCl_2), \tag{1.15}$$

$$\Delta G^\circ_{f,CaO} = -RT \ln \frac{a_{CaO}}{a_{Ca} \cdot p_{O_2}^{1/2}}. \tag{1.16}$$

The activity of calcium (a_{Ca}) in Ca-saturated $CaCl_2$ is close to unity because the solubility of chlorine in calcium metal is negligibly small [41]. When the activity of CaO (a_{CaO}) in $CaCl_2$ is high, p_{O_2} defined by (1.16) exceeds the value corresponding to the Y/Y_2O_3 equilibrium, because the standard Gibbs energy of formation of CaO is more positive than that of Y_2O_3 per mole of oxygen [27, 38]. By using $CaCl_2$ containing a low concentration of oxygen, where $a_{CaO} \ll 1$, p_{O_2} can be established at lower values. The value of p_{O_2} in the calcium-saturated $CaCl_2$ molten salt can be determined by measuring the oxygen content of the reference materials, such as titanium or yttrium metal in equilibrium with the salt. The Gibbs energy change associated with the dissolution of oxygen in yttrium and titanium can be expressed as

$$\tfrac{1}{2}\mathrm{O_2}(g) = \mathrm{O(Y)}, \tag{1.17}$$

$$\Delta G^\circ_{1,\mathrm{Y}} = -\mathrm{R}T \ln \frac{f_{\mathrm{O(Y)}} \cdot [\mathrm{mass\%O}]_{\mathrm{Y}}}{p_{\mathrm{O_2}}^{1/2}}; \tag{1.18}$$

$$\tfrac{1}{2}\mathrm{O_2}(g) = \mathrm{O(Ti)}, \tag{1.19}$$

$$\Delta G^\circ_{1,\mathrm{Ti}} = -\mathrm{R}T \ln \frac{f_{\mathrm{O(Ti)}} \cdot [\mathrm{mass\%O}]_{\mathrm{Ti}}}{p_{\mathrm{O_2}}^{1/2}}. \tag{1.20}$$

By combining (1.12) and (1.18), one obtains

$$\Delta G^\circ_{1,\mathrm{RE}} = \Delta G^\circ_{1,\mathrm{Y}} - \mathrm{R}T \ln \frac{f_{\mathrm{O(RE)}} \cdot [\mathrm{mass\%O}]_{\mathrm{RE}}}{f_{\mathrm{O(Y)}} \cdot [\mathrm{mass\%O}]_{\mathrm{Y}}}. \tag{1.21}$$

The standard Gibbs energy change for reaction (1.10), $\Delta G^\circ_{1,\mathrm{RE}}$, can be determined from the value of $\Delta G^\circ_{1,\mathrm{Y}}$ and the distribution coefficient of oxygen between the rare earth and yttrium metal, $K_{\mathrm{RE/Y}} = [\mathrm{mass\%}]_{\mathrm{RE}}/[\mathrm{mass\%}O]_{\mathrm{Y}}$, when Sievert's law is obeyed by the oxygen in RE and Y. Equation (1.21) can be rewritten as

$$\Delta G^\circ_{1,\mathrm{RE}} = \Delta G^\circ_{1,\mathrm{Y}} - \mathrm{R}T \ln K_{\mathrm{RE/Y}}. \tag{1.22}$$

The value of $\Delta G^\circ_{1,\mathrm{RE}}$ obtained can be cross-checked using the distribution coefficient of oxygen between the rare earth and titanium metal, $K_{\mathrm{RE/Ti}} = [\mathrm{mass\%O}]_{\mathrm{RE}}/[\mathrm{mass\%O}]_{\mathrm{Ti}}$,

$$\Delta G^\circ_{1,\mathrm{RE}} = \Delta G^\circ_{1,\mathrm{Ti}} - \mathrm{R}T \ln K_{\mathrm{RE/Ti}}. \tag{1.23}$$

1.4 The Removal of Oxygen from Reactive Metals

This section provides information about the removal of oxygen from its solid solution with some rare earth metals (RE = La, Pr, Nd, Gd, Tb, Dy, Er). As is well known, these reactive metals have an extremely strong affinity for oxygen. It is, therefore, very difficult to remove oxygen directly from these metals. Prior to the assessment of various deoxidation techniques, the thermodynamic properties of oxygen in rare earth metals, which have been experimentally determined in recent years, are briefly described.

1.4.1 The Thermodynamic Properties of Oxygen in Rare Earth Metals

A schematic diagram of the apparatus used for oxygen chemical potential measurements is shown in Fig. 1.5. Five to ten pieces of rare earth metals,

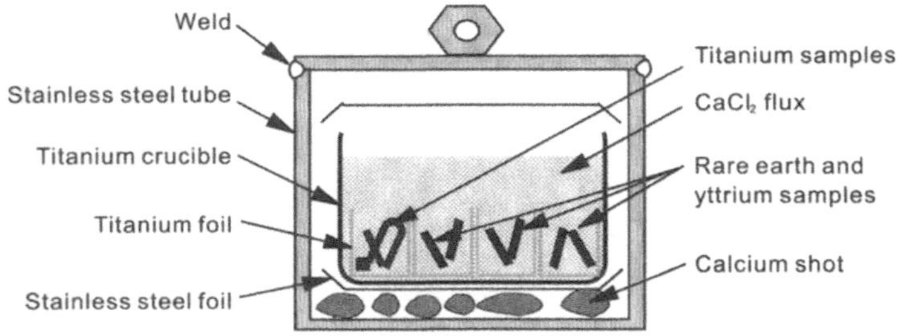

Fig. 1.5. The reaction vessel used for equilibration at ultra-low-oxygen chemical potentials

yttrium and titanium were placed in a titanium crucible filled with $CaCl_2$. To confirm equilibrium after experiments, several sets of metal samples with different initial oxygen concentrations were used: some samples had oxygen concentrations higher than equilibrium, with others had a lower oxygen content. The different metals were separated from each other by titanium foil. The titanium cup containing the metallic samples and $CaCl_2$ was sealed in an outer stainless steel vessel with an excess of calcium metal. To avoid contamination of the samples by impurities, mainly oxygen and nitrogen, in the metallic calcium, the calcium metal was physically isolated from the fused salt. Calcium was supplied to the $CaCl_2$ melt at unit activity via the vapor phase. The solubility of calcium metal in fused $CaCl_2$ at 1093 K is about 19 mol% [41]. The $CaCl_2$ acts as an oxygen reservoir in which the oxygen chemical potential is maintained at a constant value. The oxygen chemical potential is established by the Ca/CaO equilibrium, with CaO present in the fused salt at reduced activity. The maximum solubility of CaO in $CaCl_2$ at 1093 K is 18 mol% [33]. In another experimental setup, a mixture of yttrium metal turnings and yttria (Y_2O_3) powder were placed in the titanium cup along with $CaCl_2$, but without calcium metal, to set the oxygen chemical potential at a value corresponding to Y/Y_2O_3 equilibrium.

The sealed stainless steel vessel was heated in an electric furnace to a temperature between 1093 K and 1233 K. The holding time was between 690 ks (8 days) and 1210 ks (14 days), after which the vessel was quenched in water. The equilibration time was selected based on previous experimental results with the Ti–O [8, 26] and Y–O [27] systems.

The literature on the diffusion coefficients of oxygen in solid rare earth metals is limited. However, the values for oxygen in yttrium (Y), lanthanum (La) and gadolinium (Gd) have been reported to be larger than the value in titanium [42], as shown in Fig. 1.6. It is anticipated that the diffusivity of oxygen in other rare earth metals discussed in this section is close to that in La and Gd. On the right axis of Fig. 1.6, the equilibration time necessary to reach constant oxygen concentration within the accuracy of the chemical analysis is calculated. The equilibration time is calculated by assuming that

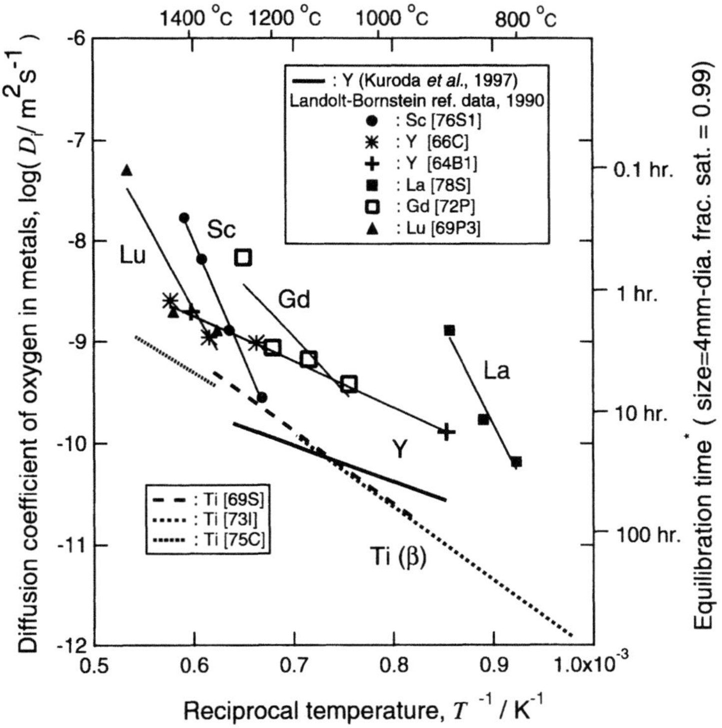

Fig. 1.6. The diffusion coefficient of oxygen in metals as a function of temperature [42]. The equilibrium time on the right vertical axis is calculated by assuming an infinitely long cylinder of 4 mm diameter [43, 49]

the metal sample is an infinitely long cylinder, of 4 mm diameter. The detailed procedure for this calculation is described by Darken and Gurry [43]. The actual time necessary for the equilibrium is expected to be much shorter than the calculated value, because the sample size is not infinite and oxygen diffusion along grain boundaries will accelarate the equibration process.

After quenching the reaction vessel to room temperature, the calcium-saturated fused salt in the titanium cup was mechanically removed, and the samples were collected. The concentrations of oxygen and nitrogen in the metal samples were determined by an inert gas fusion–infrared absorption method. For extraction of oxygen and nitrogen from the sample, 0.1 g of the metal sample enclosed in 1 g of oxygen-free platinum or nickel foil was dropped into a graphite crucible and heated to a temperature above 2000 K. By alloying the rare earth metal with platinum or nickel, the activity coefficient of the oxygen in the metal phase is enhanced significantly, thus facilitating easier extraction.

Table 1.1. The equilibrium oxygen concentration of titanium, yttrium and rare earth metal samples equilibrated with Y/Y_2O_3 [12]

	Oxygen concentration, C_O (mass ppm)					
		After equilibration with Y/Y_2O_3				
Sample	Initial	1213 K 551 ks	1173 K 518 ks	1123 K 807 ks	1114 K 892 ks	1093 K 1210 ks
Gd	200	480, 500	420, 450	290, 320	560, 710	240, 210
Tb	540	850, 890	1150, 1260	470, 470	840, 1240	660, 750
Dy	590	–	1360, 1360	870, 950	820, 920	720, 730
Er	1500	3390, 3320	7760, 9240	1750, 1800	1680, 1860	1720, 1650
Y	9700	18200, 18300	15000	11600, 11700	18000, 25200	15100, 12800
Ti-1	210	90	40, 80	240, 290	340, 350	160, 210
Ti-2	950	100, 220	160, 160	390, 410	540, 550	520, 530
Ti-3	670	80, 100	30, 240	400, 410	580, 320	470, 470

Representative oxygen concentrations in metal samples equilibrated with Y/Y_2O_3 via molten $CaCl_2$ are summarized in Table 1.1. In the experiments, the oxygen concentration of yttrium is found to be close to the solubility limit [46], and the oxygen concentration in the rare earth metals (Gd, Tb, Dy and Er) falls in the range from 200 to 2000 ppm. The nitrogen concentration of RE metal samples increased during the high-temperature experiment from the initial value by about 100–300 mass ppm. For titanium, the increase was 50–100 mass ppm. The calcium concentration in rare earth metals was unchanged from its initial value ($\sim$ 150 ppm), except for Gd, where the concentration increased to 500 ppm after equilibration. The distribution ratios of oxygen, $K_{\mathrm{RE/Ti}}$ and $K_{\mathrm{RE/Y}}$ [refer to (1.22) and (1.23)], obtained using Y/Y_2O_3 equilibrium, agree with the results from experiments at lower oxygen chemical potentials using Ca/CaO ($a_{\mathrm{CaO}} \ll 1$) equilibrium [12].

The temperature dependence of the distribution coefficients of oxygen between rare earth metals and yttrium, $K_{\mathrm{RE/Y}} = [\mathrm{mass\%O}]_{\mathrm{RE}}/[\mathrm{mass\%O}]_{\mathrm{Y}}$, is shown in Fig. 1.7. Values of $-RT \ln K_{\mathrm{RE/Y}}$, having the unit of Gibbs energy, are plotted along the y-axis. For the reaction

$$\mathrm{O(Y)} = \mathrm{O(RE)}, \tag{1.24}$$

$$\Delta G^\circ_{1,\mathrm{RE}} - \Delta G^\circ_{1,\mathrm{Y}} = -RT \ln K_{\mathrm{RE/Y}}. \tag{1.25}$$

Linear least-squares regression analysis of the data in Fig. 1.7 provides the following equations in the temperature range between 1093 K and 1223 K:

$$\Delta G^\circ_{1,\mathrm{Gd}} - \Delta G^\circ_{1,\mathrm{Y}} = -19,900 + 44.8\ T\ (\pm 5,000)/\mathrm{Jmol^{-1}}, \tag{1.26}$$

$$\Delta G^\circ_{1,\mathrm{Tb}} - \Delta G^\circ_{1,\mathrm{Y}} = 10,900 + 13.9\ T\ (\pm 4,200)/\mathrm{Jmol^{-1}}, \tag{1.27}$$

$$\Delta G^\circ_{1,\mathrm{Dy}} - \Delta G^\circ_{1,\mathrm{Y}} = 13,400 + 10.4\ T\ (\pm 3,000)/\mathrm{Jmol^{-1}}, \tag{1.28}$$

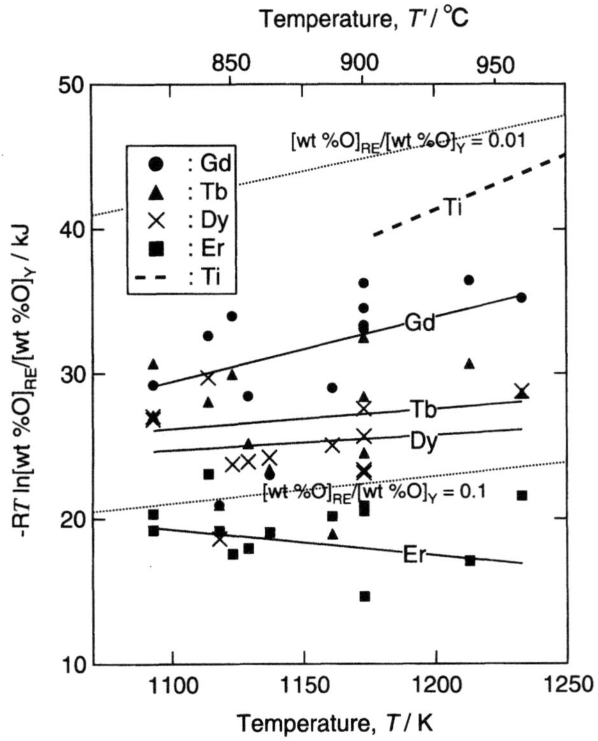

Fig. 1.7. The temperature dependence of the equilibrium distribution coefficients of oxygen between rare earth metals (RE = Gd, Tb, Dy, Er) and yttrium [12]

$$\Delta G^{\circ}_{1,\mathrm{Er}} - \Delta G^{\circ}_{1,\mathrm{Y}} = 38,400 - 17.4\ T\ (\pm 3,900)/\mathrm{Jmol}^{-1}. \qquad (1.29)$$

In view of the experimental uncertainties in the values shown in Fig. 1.7, these equations should not be extrapolated beyond the range of measurement. Therefore, the separate terms in (1.26)–(1.29) should not be interpreted in terms of the corresponding enthalpies and entropies.

The Gibbs energy change for the solution of oxygen in rare earth metals, $\Delta G^{\circ}_{1,\mathrm{RE}}$, can be determined from (1.26)–(1.29) and the value of $\Delta G^{\circ}_{1,\mathrm{Y}}$ from the literature [13]. The results are listed in Table 1.2, along with the reference data used for calculation. The expressions for $\Delta G^{\circ}_{1,\mathrm{RE}}$ are only valid within the investigated temperature range.

The relationship between the square root of the oxygen partial pressure, p_{O_2}, and the equilibrium oxygen concentration, $[\mathrm{mass\%O}]_{\mathrm{RE}}$, for the RE–O solid solutions at 1093 K is derived from the thermodynamic data summarized in Table 1.2. The results are shown in Fig. 1.8. The values of p_{O_2} were obtained from the equilibrium oxygen concentration in yttrium samples and known thermodynamic properties of the Y–O solid solution [27]. The dashed lines for the La–O, Pr–O and Nd–O systems are associated with large relative

Table 1.2. A summary of standard Gibbs energy changes for reactions involving oxygen and reactive metals

Reaction	Standard Gibbs energy change (J/mol)	Reference
$1/2O_2$ (g) = O(Gd, 1 mass%)[a]	$\Delta G^\circ_{1,\mathrm{Gd}} = -622\,000 + 109.8\,\mathrm{T}$ [1093–1223 K]	[12]
$1/2O_2$ (g) = O(Tb, 1 mass%)[a]	$\Delta G^\circ_{1,\mathrm{Tb}} = -591\,000 + 78.9\,\mathrm{T}$ [1093–1223 K]	[12]
$1/2O_2$ (g) = O(Dy, 1 mass%)[a]	$\Delta G^\circ_{1,\mathrm{Dy}} = -589\,000 + 75.4\,\mathrm{T}$ [1093–1223 K]	[12]
$1/2O_2$ (g) = O(Er, 1 mass%)[a]	$\Delta G^\circ_{1,\mathrm{Er}} = -564\,000 + 47.6\,\mathrm{T}$ [1093–1223 K]	[12]
$1/2O_2$ (g) = O(Y, 1 mass%)[a]	$\Delta G^\circ_{1,\mathrm{Y}} = -533\,000 + 12.4\,\mathrm{T}$ [1177–1438 K]	[27]
	$\Delta G^\circ_{1,\mathrm{Y}} = -602\,000 + 65.0\,\mathrm{T}^b$	Calculated from $\Delta G^\circ_{\mathrm{f},Y_2O_3}$ [38] and solubility data [46]
$1/2O_2$ (g) = O(Ti, 1 mass%)[a]	$\Delta G^\circ_{1,\mathrm{Ti}} = -583\,000 + 88.5\,\mathrm{T}^b$ [1173–1373 K]	[26]
	$\Delta G^\circ_{1,\mathrm{Ti}} = -562\,000 + 91.6\,\mathrm{T}$ [1323–1573 K]	[24]
2/3Gd (s) + $1/2O_2$(g) = $1/3Gd_2O_3$(s)	$\Delta G^\circ_{\mathrm{f},Gd_2O_3} = -605\,000 + 92.3\,\mathrm{T}$ [900–1300 K]	[38]
2/3Tb(s) + $1/2O_2$(g) = $1/3Tb_2O_3$(s)	$\Delta G^\circ_{\mathrm{f},Tb_2O_3} = -616\,000 + 90.5\,\mathrm{T}$ [900–1300 K]	[38]
2/3Dy(s) + $1/2O_2$(g) = $1/3Dy_2O_3$(s)	$\Delta G^\circ_{\mathrm{f},Dy_2O_3} = -614\,000 + 90.7\,\mathrm{T}$ [900–1300 K]	[38]
2/3Er(s) + $1/2O_2$(g) = $1/3Er_2O_3$(s)	$\Delta G^\circ_{\mathrm{f},Er_2O_3} = -629\,000 + 93.1\,\mathrm{T}$ [900–1300 K]	[38]
2/3Y(s) + $1/2O_2$(g) = $1/3Y_2O_3$(s)	$\Delta G^\circ_{\mathrm{f},Y_2O_3} = -631\,000 + 92.4\,\mathrm{T}$ [900–1300 K]	[38]
Ca(l) + 1/2 O_2(g) = CaO (s)	$\Delta G^\circ_{\mathrm{f},CaO} = -637\,000 + 106\,\mathrm{T}$ [1115–1300 K]	[38]

[a] Oxygen activity in the RE–O solid solution is taken relative to an infinitely dilute standard state in which activity is equal to its mass%. Dissolved oxygen is assumed to obey Sieverts' law up to saturation.

[b] Selected reference data for a Y–O solid solution.

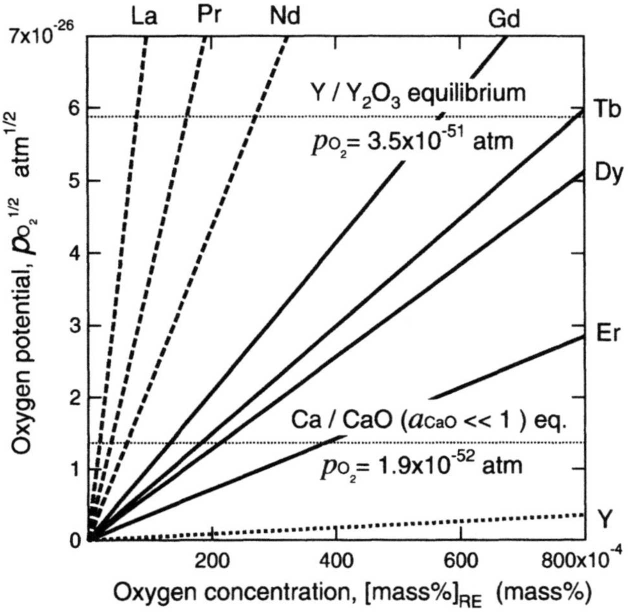

Fig. 1.8. The relationship between the square root of the oxygen partial pressure and the oxygen concentration in RE–O (RE = La, Pr, Nd, Gd, Tb, Dy, Er), and Y–O solid solutions at 1093 K [12, 13]

uncertainties in oxygen analysis (±10 ppm) [13]. The solid lines for the other RE–O (RE = Gd, Tb, Dy, Er) systems are more accurate, since the measured oxygen concentrations were higher and the data were obtained over a wider temperature range [12]. The affinity between oxygen and the rare earth metals increases with the atomic number. For all the systems investigated, the affinity between oxygen and lanthanide metals is weaker than that with yttrium metal. The oxygen potential of the Ti–O system is comparable to those of the lighter rare earth metals.

In Fig. 1.9, the Gibbs energy changes associated with the solution of diatomic oxygen gas, $\Delta G^{\circ}_{1,\mathrm{RE}}$, in the rare earth metals are compared as a function of temperature with those for titanium and yttrium. Also plotted are the Gibbs energies of formation of sesquioxides (RE_2O_3) of the corresponding rare earth metals, Y_2O_3, and CaO, ($= \Delta G^{\circ}_{\mathrm{f},i}$) per gram atom oxygen [38]. For the solution of oxygen in rare earth metals, the Gibbs energy change becomes progressively more negative with increasing atomic number. The stability of sesquioxides also exhibits the same general tendency, except that for Tb_2O_3, which is more stable than Dy_2O_3. The results in Fig. 1.9 show, for example, that rare earth metals containing 1 mass% oxygen can be deoxidized either by calcium or yttrium metal.

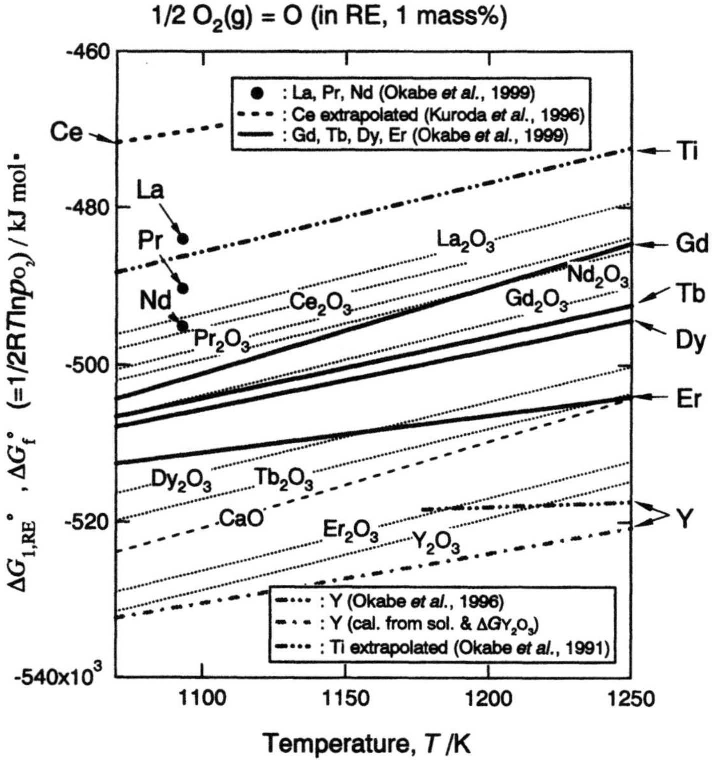

Fig. 1.9. Temperature dependence of the standard Gibbs energy change for oxygen dissolution, $\Delta G^{\circ}_{1,\mathrm{RE}}$, in rare earth metals. The standard Gibbs energies of formation of selected oxides, $\Delta G^{\circ}_{\mathrm{f},i}$, are also shown for comparison [12, 13, 38, 49]

Furthermore, it is seen that the oxygen chemical potentials for the RE–1 mass%O solid solutions ($\Delta G^{\circ}_{1,\mathrm{RE}}$), except for the Y–O system, are higher than those of their respective metal/oxide equilibria. This indicates that oxygen solubility in these metals (except yttrium) is less than 1 mass%. Literature values of oxygen solubility in four rare earth metals are summarized in Fig. 1.10 [45–49]. There is little information about the saturation solubility of oxygen in rare earth metals. The oxygen partial pressure at saturation (RE/RE_2O_3 two-phase field) can be calculated from the thermodynamic data of oxides [38, 44]. It is, therefore, possible to estimate the solubility limit for oxygen in rare earth metals as a function of temperature from the results shown in Fig. 1.9 by extending Sievert's law up to saturation. However, in view of the uncertainties in the data and possible deviations from Sievert's

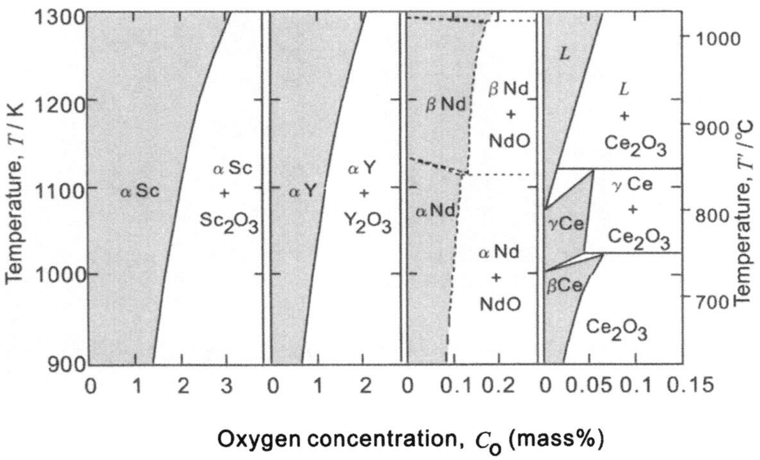

Fig. 1.10. The solvus at the metal-rich end of the RE–O systems [45–49]

law, especially at high solute concentrations, it may be better directly to measure the saturation solubility after equilibration with the oxides.

Information on the standard Gibbs energy change for oxygen dissolution, $\Delta G^{\circ}_{1,\mathrm{RE}}$, in rare earth metals, has an important application: estimation of the theoretical deoxidation limit. For example, ultra-low-oxygen chemical potentials ($p_{O_2} = 10^{-49}$ atm at 1223 K) can be attained by utilizing the electrochemical deoxidation technique. The electrochemical technique enables deoxidation of rare earth metals to levels below the current detection limit by chemical analysis (10 mass ppm O for a 0.1 g sample).

1.4.2 The Removal of Oxygen from Rare Earth Metals

From the thermodynamic data, it is estimated that the oxygen content of the rare earth metals can be brought down to the 10 ppm level. In this section, selected results are given for oxygen removal from reactive metals.

A schematic illustration of an apparatus used for electrochemical deoxidation [9, 11, 14] is shown in Fig. 1.11. Dehydrated $CaCl_2$ was contained in a metal crucible and set in a vertical stainless steel reaction tube, closed at one end. A cathode basket, made from strands of titanium wire, contained samples of rare earth metal, yttrium and titanium. The anode, cathode and thermocouple tubing were inserted into the reaction tube through a gas-tight water-cooled stainless steel lid.

After assembling the cell, the reaction tube was evacuated and checked for gas-tightness. The cell was then heated to about 600 K in vacuum to remove residual moisture. High-purity argon gas was then introduced into the reaction tube, and the temperature was increased to a preset value between 1073 K and 1176 K. After melting of the $CaCl_2$, the high-purity carbon anode was lowered into the molten salt, and pre-electrolysis was conducted to

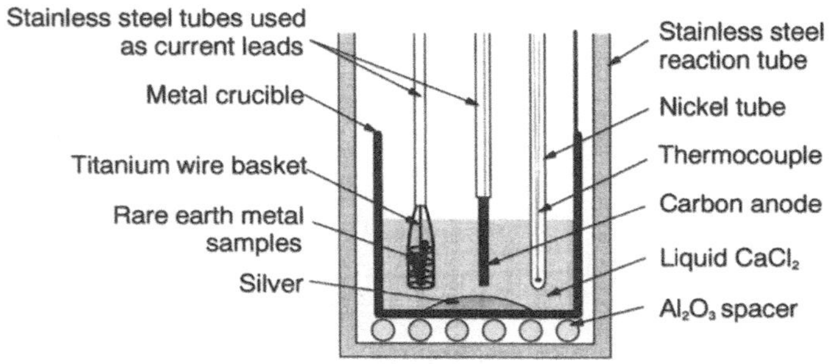

Fig. 1.11. The electrolysis cell used for electrochemical deoxidation of reactive metals

eliminate residual metallic and gaseous impurities in the molten salt. A dc potential of about 3.2 V was applied between the crucible and the graphite anode for about 4 ks, producing a current of about 2 A through the molten salt. About 50 g of silver was placed at the bottom of the crucible to absorb the calcium metal produced during pre-electrolysis.

After pre-electrolysis, the titanium cathode basket containing rare metal samples was lowered into the molten salt and positioned about 30 mm from the anode. Controlled current electrolysis was performed. After the reaction, the electrodes were removed from the melt, and the cell was allowed to cool in a stream of high-purity argon gas. In some cases, it was possible to use the same salt and anode to deoxidize additional metal samples. After deoxidation treatment, fused salt adhering to the surface of metal samples was removed mechanically. The samples were carefully polished and analyzed for oxygen, nitrogen and metallic trace elements.

Some representative oxygen and nitrogen concentrations of rare earth metal samples (RE = La, Pr, Nd) before and after electrochemical deoxidation are given in Table 1.3. The experimental results clearly indicate that lanthanum, praseodymium and neodymium can be deoxidized to 20–30 mass ppm by the electrochemical process in approximately 36 ks (10 h).

It is noteworthy that the nitrogen concentrations of the metal samples either remain unchanged or decrease by about 10–100 mass ppm during electrolysis. However, further quantitative discussion cannot be given, because there is no thermodynamic data on nitrogen solubility in rare earth metals or in the molten salt. The calcium concentration of solid La was found to be unchanged during deoxidation, but calcium in Pr and Nd increased by 350 and 150 mass ppm, respectively. The increase in calcium content of Pr and Nd after electrochemical deoxidation is substantially smaller than that after the equilibration experiment. This might be indicative of a smaller diffusivity of calcium in Pr and Nd compared to oxygen, and the possibility of kinetic control of calcium contamination in the lanthanide metals.

Table 1.3. The results of oxygen and nitrogen analysis of rare earth metal samples (RE = La, Pr, Nd) before and after electrochemical deoxidation [13]

Sample	C_O (mass ppm)			C_N (mass ppm)		
	Initial	After experiment		Initial	After experiment	
		1091 K 8 h	1073 K 10 h		1091 K 8 h	1073 K 10 h
La	2200	30, 50	60, 70	100	40, 50	10
Pr	2600	40, 50	20, 30	<10	<10	<10
Nd	1800	20, 40	40, 50	160	130	120, 130

Some representative results for oxygen concentration in rare earth (RE = Gd, Tb, Dy, Er), yttrium and titanium samples before and after deoxidation are listed in Table 1.4. Titanium and yttrium were used as reference materials in order to estimate the local partial pressure of oxygen at the cathode. As indicated by the results in Table 1.4, rare earth metals containing more than 2000 ppm of oxygen are deoxidized to below 10–50 mass ppm after an experiment at 1189 K for 36 ks (10 h). Yttrium, which has an approximately 15 times stronger affinity for oxygen than other rare earth elements at this temperature, is found to be deoxidized to 350 mass ppm using the same process. These results are consistent with the prevalence of local equilibrium at the cathode during the experiments, and with the Sievert's law behavior depicted in Fig. 1.8. The concentration of oxygen in titanium was reduced to 40 mass ppm after electrochemical deoxidation. This value is approximately seven times larger than the value expected from the Sievert's law constant for

Table 1.4. Oxygen and nitrogen in titanium, yttrium and rare earth metal samples (RE = Gd, Tb, Dy, Er) before and after electrochemical deoxidation [14]

Sample	C_O (mass ppm)			C_N (mass ppm)		
	Initial	After experiment		Initial	After experiment	
		1189 K 10 h	1179 K 10 h		1189 K 10 h	1179 K 10 h
Gd	2100	10, 50	30, 40	130	90, 100	20, 30
Tb	2600	20, 20	30, 50	380	330, 340	50, 60
Dy	2300	20, 30	50, 80	50	40, 50	20, 30
Er	2800	40, 40	20, 50	50	30, 40	10, 20
Y	11200	340, 350	500, 870	270	200, 200	170, 280
Ti	210	40, 50	50, 70	10	10	20, 20

oxygen in titanium, assuming that local equilibrium prevails at the cathode. As the diffusion coefficient of oxygen in titanium solid solution is smaller than the corresponding value for RE metals [42], the deoxidation may have been limited by oxygen transport in the solid titanium. The results shown in Table 1.4 were obtained by using the same solvent and anode. The results indicate that the molten salt is basically reusable to produce rare earth metals that contain less than 100 ppm O.

From the values of the residual oxygen concentrations in rare earth metal samples, one can determine the oxygen chemical potential at the cathode surface, assuming local equilibrium. The calculated oxygen partial pressure is approximately 10^{-50} atm at 1173 K, using data on the Gibbs energy of

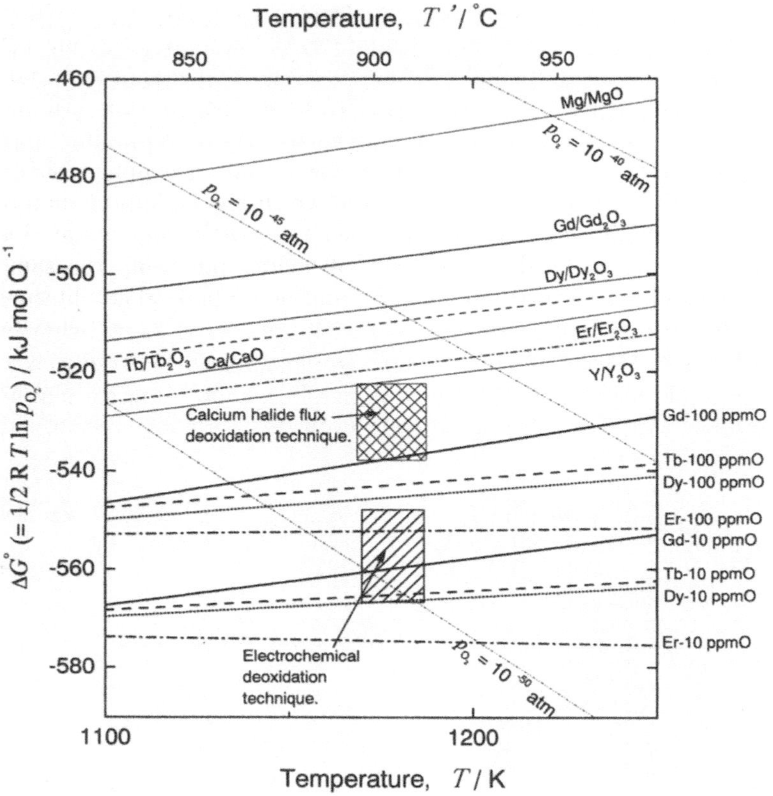

Fig. 1.12. An Ellingham diagram showing the oxygen chemical potentials of RE–O (RE = Gd, Tb, Dy, Er) solid solutions containing 10 and 100 mass ppm of oxygen [12] and selected M/M_xO_y couples [38]. The ranges of the oxygen chemical potentials established in different deoxidation processes involving $CaCl_2$ are also indicated

solution of oxygen discussed in the previous section. The range of oxygen chemical potentials attained at the cathode during electrochemical deoxidation is shown in Fig. 1.12. The oxygen chemical potentials of RE-O solutions (RE = Gd, Tb, Dy, Er) containing 10 and 100 mass ppm oxygen, which are calculated from data listed in Table 1.2, are also plotted for reference. The oxygen chemical potentials attained by electrochemical deoxidation are seen to be far below these for Y/Y_2O_3 equilibrium. Thus the ultra-low-oxygen chemical potentials obtained in the electrochemical deoxidation process can be used to deoxidize a variety of reactive metals.

It may also be added that the nitrogen concentration in the samples is found to be almost independent of the parameters of the deoxidation treatment. The nitrogen content generally decreased by about 10–30 mass ppm during deoxidation treatment. When the molten salt was reused, there was a more significant decrease in nitrogen content. It is noteworthy that the nitrogen concentration was indeed decreased, albeit by a small margin, by the electrochemical technique [14]. Further studies are required to explore electrochemical denitridation.

1.4.3 Comparison with Other Direct Oxygen Removal Techniques

Table 1.5 gives the results of electrochemical deoxidation of rare earth metals (RE = La, Pr, Nd) compared to data obtained by other deoxidation treatments reported in the literature [13, 15, 35, 37, 47, 50–55]. Kamihira et al. [51] investigated the deoxidation of rare earth metals by calcium metal at 1143 K, resulting in the formation of essentially pure CaO. They obtained metals containing between 130 and 370 ppm oxygen. From the thermodynamic information ($\Delta G^\circ_{1,\mathrm{RE}}$ in Table 1.2) and the standard Gibbs energy of formation of CaO ($\Delta G^\circ_{\mathrm{f,CaO}} = -521\,\mathrm{kJ/mol}$) [38]), it is possible to estimate the theoretical deoxidation limit. Calculated equilibrium oxygen concentrations in La, Pr, and Nd corresponding to Ca/CaO equilibrium at 1093 K are 180, 340 and 530 mass ppm, respectively. The measured values are considered to be in fair agreement with the theoretical values, considering the temperature difference between the experiment and the theoretical estimation, and the errors associated with the experimental data.

The formation of oxyhalides of the lanthanide elements can be utilized for removing oxygen dissolved in lanthanide metals [35, 37]. Corbett et al. reported oxygen concentrations ranging from 160 to 420 ppm using this technique [35]. In a chloride-rich oxychloride melt, the activity of the oxides can be reduced. The usefulness and validity of this process for oxygen removal, however, is restricted with increasing starting oxygen concentration in the metal, because the deoxidation limit depends on the oxide ion concentration in the oxyhalide flux.

In Table 1.6, the degree of deoxidation achieved by various solid state purification methods for metals (RE = Gd, Tb, Dy, Er, Y) is summarized

Table 1.5. A summary of the results of deoxidation of RE–O (RE = La, Pr, Nd) solid solutions by different processes

Sample	Method	Reference	Temperature (K)	Holding time (h)	Oxygen concentration (mass ppm)	
					Initial	After exp.
La	(Zone refining)	Fort et al. [50]	1193	–	34	20
	Oxyhalide	Corbett et al. [35]	1123	192	2095	416
	Ca/CaO	Kamihira et al. [51]	1143	18	1890	370
	Y/Y_2O_3	Okabe et al. [13]	1093	336	400	50
	Elec. Deox[a]	Okabe et al. [13]	1073	8	2200	30
Pr	SSE[b]	Muirhead and Jones [52]	1083	520	536	114
	Ca/CaO	Kamihira et al. [51]	1143	18	1300	128
	Ca/CaO	Ogawa et al. [53]	1163	12	2000	200
	Elec. Deox[a]	Okabe et al. [13]	1073	10	2600	20
Nd	SSE[b]	Mattocks et al. [54]	1113–1161	789	45	14
	Oxyhalide	Corbett et al. [35]	1063	257	1595	155
	SSE[b]	Fort et al. [55]	1093–1136	500	40	20
	Ca/CaO	Kamihira et al. [51]	1143	18	830	205
	Oxyhalide	Sano et al. [37]	1723	–	3000	200
	Elec. Deox[a]	Okabe et al. [13]	1073	8	1850	20

[a] Electrochemical deoxidation.

[b] Solid state electrotransport.

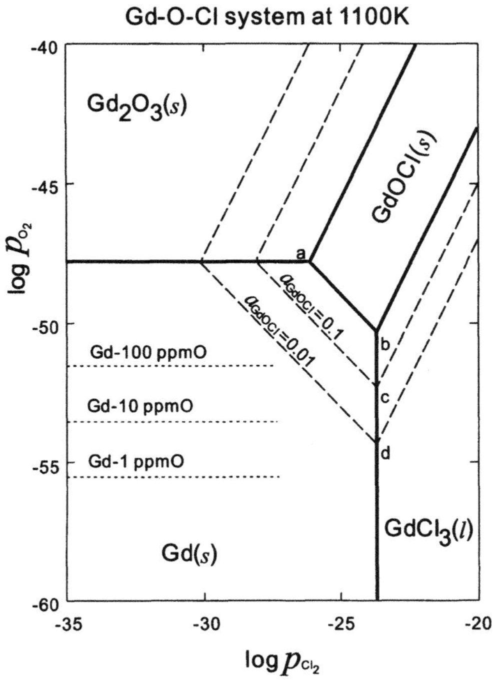

Fig. 1.13. A chemical potential diagram for the Gd–O–Cl system at 1100 K

[12, 14, 15, 50, 55–64]. Solid state electrotransport (SSE) is very effective for removal of various impurities, including oxygen, from rare earth metals. However, this method requires several hundreds of hours for processing, as indicated in the table. Furthermore, SSE imposes certain restrictions on sample geometry, since large currents have to flow through the sample in ultra-high vacuum for deoxidation to take place. The deoxidation limit depends on the initial oxygen content, and only one end of the sample is deoxidized. The electrochemical method has several advantages compared to SSE for deoxidation of high-oxygen-content samples and for recycling off-grade metals, because there are no restrictions on the sample shape and the initial oxygen level. Diffusion of oxygen may become a key factor when large samples are used. Zone refining can handle large sections [64], although the refining efficiency in each pass is limited by the distribution ratio of oxygen between solid and liquid.

It is interesting to discuss deoxidation by oxyhalide formation from a thermodynamic perspective. The stability field diagram for the Gd–O–Cl system at 1100 K, displayed in Fig. 1.13 as an example, is helpful for understanding the physico-chemical rationale of the oxychloride process. The thermodynamic data for Gd_2O_3, $GdCl_3$ and GdOCl from compiled data [38] and for Gd–O solid solutions from Table 1.2 [12] were used to construct this diagram. When gadolinium containing a high concentration of oxygen is reacted with pure $GdCl_3$, oxygen in the metal is transferred to the molten

Table 1.6. A summary of the results of deoxidation RE–O (RE = Gd, Tb, Dy, Er, Y) solid solutions by various processes

Sample	Method	Reference	Temperature (K)	Holding time (h)	Oxygen concentration (mass ppm) Initial	After deoxidation
Gd	SSE[a]	Peterson and Schmidt [56]	1 518	150	50	6
	SSE[a]	Jordan and Jones [57]	1 373	310	511	11
	(Zone refining)	Fort et al. [64]			56	25
	Oxyhalide	Corbett et al. [35]	1 123	240	758	49
	SSE[a]	Fort et al. [55]	1 303	1 272	56	3
	Ca/CaO	Kamihira et al. [51]	1 143	18	2 780	2 260
	Ca halide flux	Okabe et al. [12]	1 233	72	200	100
	Elec. Deox[b]	Hirota et al. [14]	1 189	10	2 200	10
Tb	SSE[a]	Jordan and Jones [58]	1 323	350	381	26
	Ca/CaO	Kamihira et al. [51]	1 143	18	1 360	1 520
	Ca halide flux	Okabe et al. [12]	1 118	192	540	140
	Elec. Deox[b]	Hirota et al. [14]	1 189	10	2 500	20
Dy	Ca/CaO	Kamihira et al. [51]	1 143	18	4 020	3 750
	SSE[a]	Fort et al. [59]	1 303	750	33	13
	Ca halide flux	Okabe et al. [12]	1 137	168	590	140
	Elec. Deox[b]	Hirota et al. [14]	1 189	10	2 300	20
Er	Ca halide flux	Okabe et al. [12]	1 118	192	1 500	170
	Elec. Deox[b]	Hirota et al. [14]	1 189	10	2 900	20
Y	SSE[a]	Williams and Huffine [60]	1 643	200	3 379	340
	SSE[a]	Carlson et al. [61]	1 448	190	782	90
	SSE[a]	Volkov et al. [62]	1 433	500	613	40
	Oxyhalide	Corbett et al. [35]	1 004	336	3 379	342
	Elec. Deox	Okabe et al. [11]	1 223	8	5 700	50
	SSE[a]	Kuroda et al. [63]	1 373	48	5 100	1 800
	Ca halide flux	Okabe et al. [12]	1 118	191	9 700	1 150
	Elec. Deox[b]	Hirota et al. [14]	1 189	10	11 100	340

[a] Solid state electrotransport.
[b] Electrochemical deoxidation using $CaCl_2$ molten salt.

salt until the salt is saturated and GdOCl forms as a separate phase [see (1.7) and Fig. 1.1e for reference]. The partial pressures of O_2 and Cl_2 are defined by point "b" in Fig. 1.13, when the metal coexists with oxide-saturated $GdCl_3$ and GdOCl. The corresponding equilibrium oxygen concentration in the metal is just above 100 mass ppm. If the initial oxygen content of the metal is relatively low and a large amount of the salt phase is present, then there may not be enough oxygen to form GdOCl as a separate phase, and the activity of GdOCl in the fused salt may be less than unity. The lines corresponding to reduced activities (0.1 and 0.01) of GdOCl are also shown in Fig. 1.13. The equilibrium partial pressures for metal/fused-salt equilibrium are then defined by points "c" and "d", respectively. For each order of magnitude reduction in the activity of GdOCl, the equilibrium oxygen concentration in metal is also reduced by the same order. The diagram shows that it is theoretically possible to produce gadolinium containing 10 mass ppm oxygen by oxyhalide formation when high-purity $GdCl_3$ with low a_{GdOCl} is available. Thermodynamic analysis using chemical potential diagrams can be used for other purification processes in a similar manner. The availability of thermodynamic data is, therefore, very important in order to optimize the purification process.

1.5 Purity Evaluation of Deoxidized Reactive Metals

So far, theoretical aspects for oxygen removal from reactive metals using various deoxidation techniques have been discussed. An electrochemical deoxidation method, which is capable of removing oxygen down to the 20 mass ppm level, appears to be most effective for oxygen removal among the pyrometallurgical deoxidation techniques. The quality of the obtained metal can be evaluated with the aid of the residual resistivity ratio (RRR) and the Vickers hardness (Hv). The following section gives results on high-purity titanium as an example.

1.5.1 The Removal of Oxygen from Titanium

Table 1.7 summarizes some representative results of titanium deoxidation by various solid state deoxidation methods. As shown in this table, titanium can be deoxidized to 470 mass ppm under Ca/CaO equilibrium at 1273 K [26]. This equilibrium value is fairly consistent with those reported by other researchers [19,21,24]. The temperature dependence of the deoxidation limit and the thermodynamics of Ti–O solid solutions are discussed in detail in the reference [26]. Fisher and Seagle also confirmed the feasibility of this process using calcium vapor; they obtained titanium with oxygen ranging from 230 to 400 ppm [25]. Deoxidation by metal/metal oxide equilibria shows that the deoxidation limit is around the 100 ppm level even with yttrium, the strongest deoxidant among the reactive metals [12,27].

Table 1.7. Some representative results of titanium deoxidation by various solid state deoxidation techniques

Method[a]	Oxygen (mass ppm)		Nitrogen[b] (mass ppm)		Conditions and reference (Exp.#)
	Init.	After deox.	Init.	After deox.	
Ca–CaO	430→ 1430→	470 470	20 → 100 →	340 470	1273 K, 86 ks [26]
$Y–Y_2O_3$	200→ 1100→	120 150, 170	N.A N.A		1231 K, 260 ks [27] (#2U)
$Y–Y_2O_3$	210→ 950→ 670→	90 100, 220 80, 100	10 → 20 → 30 →	(30) (50) (60)	1213 K, 551 ks [12] (#U)
$Y–Y_2O_3$	210→ 950→ 670→	40, 80 160, 160 30, 240	10 → 20 → 30 →	(50) (120) (80)	1173 K, 518 ks [12] (#T)
$Ca–Y–Y_2O_3$	210→ 950→ 670→	130 60, 90 60, 90	10 → 20 → 30 →	(40) (50) (120)	1173 K, 604 ks [12] (#W)
$Ca–CaCl_2$	130→ 130→	16 17	10 → 10 →	41 34	1273 K, 90 ks [8] (#17B1) $RRR^{c}_{4.2}$ = 98, Hv^{d} = 85
$Ca–CaCl_2$	130→ 130→	35 22	10 → 10 →	49 45	1373 K, 40 ks [8] (#19B1) $RRR^{c}_{4.2}$ = 95, Hv^{d} = 92
$Ca–CaCl_2$	150→	62	10 →	86	1273 K, 86 ks [8] (#3C) $RRR^{c}_{4.2}$ = 122, Hv^{d} = 82
Electro.Deox. $(CaCl_2)^{e}$	900→ 1400→	<10 <10	40 → 20 →	20 20	1223 K, 22 ks [9] (#49, contaminated by carbon)
Electro.Deox. $(CaCl_2)^{e}$	110→ 900→	10 20	<10 → 30 →	<10 50	1223 K, 27 ks [65] (#69, C : 20→30, 50→80 ppm) $RRR^{c}_{4.2}$ = 155, Hv^{d} = 64
Electro.Deox. $(CaCl_2)^{e}$	210→	40, 50	10 →	10	1189 K, 36 ks [14] (#12)
(O in Gd: 2100 → 10, 50 ppm, O in Tb: 2600 → 20, 20 ppm, O in Dy: 2300 → 20, 30 ppm, O in Er: 2800 → 40, 40 ppm, O in Y: 11200 → 340, 380 ppm)					
Electro.Deox. $(MgCl_2)^{e}$	110→ 900→	100 60	<10 → 30 →	40 70	1173 K, 86 ks [65] (#57, C : 20→50, 50→120 ppm)

[a] "M–MOx": deoxidation by metal/metal oxide, "$Ca–CaCl_2$": calcium halide flux deoxidation, "Electro. Deox.": electrochemical deoxidation

[b] Values in parentheses are approximate.

[c] Residual resistivity ratio, $\rho_{298}/\rho_{4.2}$, determined by measurment at 298 K and 4.2 K

[d] Vickers microhardness (kgf. mm^{-2}), measured using 500 g load at room temperature

[e] Salt (in parentheses) shows molten salt used for electrochemical deoxidation. N.A: Not analyzed or not reported

The calcium halide flux deoxidation method is also effective for titanium, and it is possible to reduce oxygen in titanium down to about 20 ppm [7]. When deoxidant calcium is added directly into the flux, titanium is likely to be contaminated by impurities such as nitrogen which are present in the calcium, although deoxidation itself proceeds successfully. Such contamination is avoided by using calcium vapor. However, the nitrogen content increases slightly by about 20–80 ppm. It may be noted that the maximum residual resistivity ratio (RRR) attained after this treatment was 122, with the Vickers hardness at 82 [8].

As can be seen from Table 1.7, electrochemical deoxidation enables us to obtain titanium containing about 10 ppm O, which is close to the detection limit of conventional gas analysis. Nitrogen contamination is not so serious, but contamination by carbon should not be neglected [9, 65]. Carbon contamination can be avoided technically by separating the anode and the cathode using a titanium plate separator. The highest RRR attained by this process is 155 with Hv = 64 [65]. These values show that the purity of the deoxidized titanium is at the highest level among the high-purity titanium samples currently being produced.

It is worth noting that even when $MgCl_2$ is used as the electrolyte, titanium can be deoxidized down to the 100 ppm level [65] by electrochemical deoxidation. This suggests that electrochemical polarization might be useful for oxygen reduction during the Kroll process.

1.5.2 Purity Evaluation of Deoxidized Titanium

Major technological progress in the development of commercial processes for producing high-purity titanium for use in electronics has been achieved in recent years [2, 4]. The total metallic impurity in titanium is of the order of 1 mass ppm. Among the many trace impurities in high-purity titanium, oxygen is still recognized as the major one, present in amounts ranging from 50 to 500 ppm. Compared to other impurities, the influence of oxygen on the mechanical and electrical properties of high-purity titanium is profound, not only because of its relatively larger concentration but also because oxygen dissolves interstitially and distorts the lattice. Ouchi [66, 67] recently reported on the mechanical and electrical properties of 6 N grade titanium produced by a modified large-scale iodide process [2]. Their results show that the extent of deoxidation can be evaluated by either electrical resistivity or hardness measurements, especially for ultra-high-purity titanium.

In Table 1.8, the electrical resistivity of pure titanium, ρ_T, at different temperatures T, residual resistivity ratios (RRR) and hardnesses (Hv) and with different major impurities reported by different investigators is summarized [8, 65, 66, 68–71]. The values listed in Table 1.8 are determined by assuming Matthiessen's rule, which can be expressed as follows:

$$\rho_T = \rho_{i,T} + \rho_0 + \Delta_{T,C}, \qquad (1.30)$$

Table 1.8. The electrical resistivity and Vickers hardness of titanium

References	Electrical resistivity[a]						Hardness	Impurities			
	$\rho_{298,i}$ (nΩm)	$\rho_{273,i}$ (nΩm)	$\rho_{77,i}$ (nΩm)	$\rho_{4.2}$ (nΩm)	$1/\mathrm{RRR}_{4.2}$	$\mathrm{RRR}_{4.2}$	Hv ($\mathrm{kgf\,mm^{-2}}$)	O (mass ppm)	N	C	Fe
White and Woods [68]	437	390	44.5	19.7	0.0431	23					
Wasilewski [69]	461	413.6	50.9	13.1	0.0276	36					
Elssner et al. [70]	(430)						105	210	20	40	20
Okabe et al. [8]	440		40.0	4.6	0.0103	97	85	30	50		
Nakamura et al. [65]	440		40.0	2.9		155	64	<10	<10	30	
Komatsu et al. [71]			40					20	20	20	
Ouchi [66]	(428)		(43.69)				70	40	75	18	10

[a] Values in parentheses include residual resistivity.

$\rho_T = \rho_{T,i} + \rho_o$.

ρ_T, Resistivity at temperature T; $\rho_{T,i}$, ideal resistivity at T; ρ_0, residual resistivity ($= \rho_{4.2}$).

Table 1.9. The contribution of oxygen to the electrical resistivity of titanium

References	Oxygen contribution	Temperature	Resistivity	Oxygen range
	$\Delta\rho_T/C$ (nΩm/mol% O)	T (K)	ρ_T (nΩm)	C (mol%)
Ames and McQuillan [72]	113	293.0	470	1.500
Wasilewski [73]	123	273.0	427	0.130–10.0
Elssner et al. [70]	82	298.0	430	0.063–1.0
Baur and Lehr [74]	103	20.4	2.68	0.009–0.32
Komatsu and Ikeda [76]	105	77.0	44.5	0.890–3.2
	102	295.0	431	0.890–3.2
Okabe et al. [8]	88	4.2	4.6	0.009–0.15
Komatsu et al. [71]	96	77.0	40	0.006–0.13
Ouchi [66]	106	77.0	43.69	–0.036

where $\rho_{i,T}$ is the ideal resistivity arising from scattering of electrons by thermal vibration; ρ_0 is the residual resistivity at absolute zero (0 K), essentially affected by impurities and lattice defects; and $\Delta_{T,C}$ is the deviation from Matthiessen's rule (DMR). The residual resistivity, ρ_0, may be approximated by the resistivity at 4.2 K, $\rho_{4.2}$, because $\rho_{i,4.2} + \Delta_{T,C}$ in (1.30) is negligible in the case of titanium.

As seen in Table 1.8, the ideal resistivity of titanium at 298 K, $\rho_{298,i}$, is about 440 nΩ m. The detected discrepancy among the researchers is attributed partly to the difficulties involved in preparing high-purity metal in wire form. The metal is easily contaminated by gaseous impurities during the annealing procedures that are essential before resistivity measurement.

The influence of oxygen on electrical resistivity is exemplified by the results in Table 1.9 [8, 66, 70–74, 76]. One can deduce, using the data shown in the table, that the oxygen content has to be less than 150 mass ppm ($C = 0.044$ mol%) to obtain high-purity titanium with RRR ≥ 100.

Figure 1.14 provides the relationship between $\mathrm{RRR}_{4.2}$ ($= \rho_{298}/\rho_{4.2}$) and RRR_{77} ($= \rho_{298}/\rho_{77}$) for different titanium grades based on the following equation [8]:

$$\frac{1}{\mathrm{RRR}_{77}} = A \cdot \frac{1}{\mathrm{RRR}_{4.2}} + B, \tag{1.31}$$

where A and B are $(\rho_{i,298} - \rho_{i,77})/(\rho_{i,298} - \rho_{i,4.2})$ and $(\rho_{i,77} - \rho_{i,4.2})/(\rho_{i,298} - \rho_{i,4.2})$, respectively. The values A and B in (1.31) are constant, and independent of impurity concentration. In other words, the plot shows the relationship between the resistivities at 4.2 and 77 K for various impurity levels, normalized by the resistivity at 298 K. Figure 1.14b, which is an inverse plot

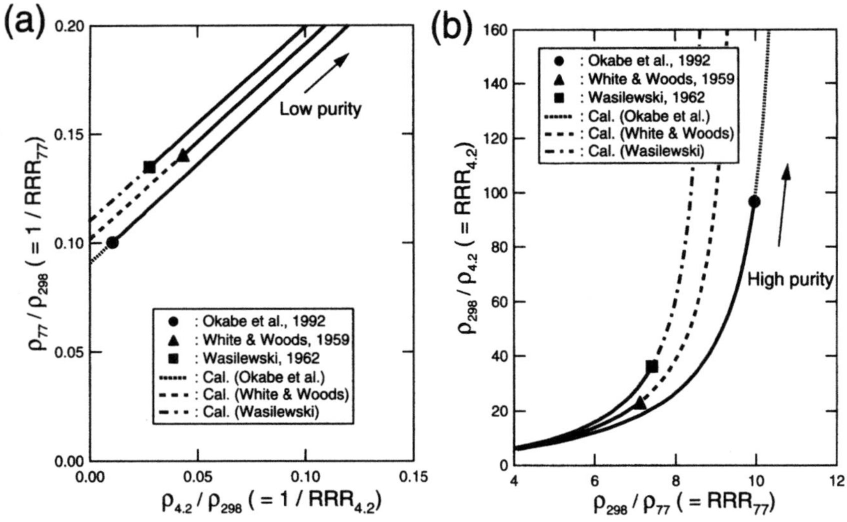

Fig. 1.14. (**a**) The relationship between titanium resistivity at 4.2 K and 77 K, normalized by the resistivity at 298 K. The data shown as markers are data for the sample used for the resistivity determination, and solid lines are for various impurity levels. (**b**) The relationship between RRR_{77} and $RRR_{4.2}$ for various impurity levels [8]

of Fig. 1.14a, illustrates that titanium with $RRR_{4.2} \geq 100$ has $RRR_{77} \geq 10$. On the contrary, when RRR_{77} is lower than 7, $RRR_{4.2}$ for this titanium is calculated to be lower than 20. This RRR_{77} value can, therefore, be used for purity evaluation of titanium, as it is easier and inexpensive to measure compared to $RRR_{4.2}$. For most commercially available titanium, the RRR_{77} value is sufficient for determination of purity because these samples contain at least 300 mass ppm of oxygen. Detailed discussion of different results reported in the literature is given elsewhere [8].

It is worth noting that high-purity titanium containing a low concentration of oxygen obtained by electrochemical deoxidation of 6N grade titanium produced by electrolytic refining [4] shows a fairly low Vickers microhardness at around 63 [8]. This low hardness value is also confirmed by Ouchi [66,67], using high-purity titanium produced by large-scale iodide processing [2]. Compared to electrical resistivity measurements, hardness is less accurate for purity evaluation. However, it can be used as a supplementary technique for purity evaluation, or as a quick approximate test.

1.6 Summary

In the present chapter, an attempt has been made to review various solid state deoxidation techniques, from a thermodynamic point of view. Thermodynamic information obtained from the measurement of ultra-low-oxygen chemical potentials in reactive metals is shown to be essential for estimating the deoxidation limits of the different processes.

Electrochemical deoxidation is found to be suitable for producing low-oxygen-containing reactive metals in a relatively short time compared to other purification methods. Evaluation of purity is discussed using titanium metal as an example.

The methods explained in this chapter are not specific to reactive metals. In principle, they can be generalized for oxygen removal from a variety of metals. Furthermore, it may be possible to remove other impurities besides oxygen.

Although solid state deoxidation is shown to be effective for purity control, a more convenient and effective anticontamination treatment of reactive metals needs to be developed before ultra-high-purity metals find wider application.

References

1. Y. Yoshimura, Y. Shimizu, Z. Imamura: In Titanium '92, Science and Technology, ed. by F.H. Froes, I.L. Caplan (TMS, Warrendale, 1993), 2459–2464
2. Y. Yoshimura: Titan·Japan (Bull Japan Titanium Soc.) **44**, 71–73 (1996)
3. Y. Shindo and H. Miyazaki, Proc. Fall Meeting of MMIJ (Mining Mate Process Inst Japan) Sapporo, 46–47 (1992)
4. Y. Shindo, S. Yamaguchi, S. Shimamoto: Titan·Japan (Bull Japan Titanium Soc) **44**, 74–76 (1996)
5. R.F.K. Ram: High purity titanium refining by iodide purification. In *Titanium extraction and processing* ed. by B. Mishra, G.J. Kipuros (TMS, Warrendale 1997) 3–22
6. K.A. Gschneider, Jr.: J. Alloys Compounds **193**, 1–6 (1993)
7. T.H. Okabe, R.O. Suzuki, T. Oishi, K. Ono: Tetsu-to-Hagane (J. Iron Steel Inst. Japan) **77**, 93–99 (1991)
8. T.H. Okabe, T. Oishi, K. Ono: J. Alloys Compounds **184**, 43–56 (1992)
9. T.H. Okabe, M. Nakamura, T. Oishi, K.Ono: Metall. Trans. B **24B**, 449–455 (1993)
10. T.H. Okabe, T. Oishi, K. Ono: Metall. Trans. B **23B**, 583–590 (1992)
11. T.H. Okabe, T.N. Deura, T. Oishi, K. Ono, D.R. Sadoway: J. Alloys Compounds **237**, 155–164 (1996)
12. T.H. Okabe, K. Hirota, E. Kasai, F. Saito, Y. Waseda, K.T. Jacob: J. Alloys Compounds **279**, 184–191 (1998)
13. T.H. Okabe, K. Hirota, Y. Waseda, K.T. Jacob: J. Mining Mat Process Inst Japan (Shigen-to-Sozai) **114**, 813–818 (1998)
14. K. Hirota, T.H. Okabe, F. Saito, Y. Waseda, K.T. Jacob: J. Alloys Compounds **282**, 101–108 (1999)
15. D. Fort: J. Less-Common Metals **134**, 45–65 (1987)
16. O.N. Carlson, F.A. Schmidt, R.R. Lichtenberg: Metall. Trans. A **6A**, 725–731 (1975)
17. W.C. Lilliendahl, M. Lakes, H.C. Rentschler: U.S. Patent 2,537,068 Jan. 9, 1951

18. W.C. Lilliendahl, D.M. Wroughton: J. Electrochem. Soc., **5**, 187–190 (1952)
19. O. Kubaschewski, W.A. Dench: J. Inst. Metals **82**, 87–91 (1953–54)
20. D.T. Peterson, R.G. Clark, W.A. Stensland: J. Less-Common Metals **30**, 169–172 (1973)
21. K. Ono, S. Miyazaki: J. Japan Inst. Metals **49**, 871 (1985)
22. R.L. Fisher: U.S. Patent 4923531, May 8, 1990 (U.K. Patent, 1989, GB 2224749A)
23. R.L. Fisher: U.S. Patent 5,022,935, June 11, 1991
24. H. Niiyama, Y. Tajima, F. Tsukihashi, N. Sano: J. Less-Common Metals **169**, 209–216 (1991)
25. R.L. Fisher, S.R. Seagle: In *Titanium science and technology* Proceedings of the 7th World Conference on Titanium, San Diego June 28–July 2, 1992, ed by F.H. Froes, I. Caplan (The Minerals, Metals & Materials Society 1993)
26. T.H. Okabe, R.O. Suzuki, T. Oishi, K. Ono: Mater. Trans. JIM **32**, 485–488 (1991)
27. T.H. Okabe, T.N. Deura, T. Oishi, K. Ono, D.R. Sadoway: Metall. Trans. B **27B**, 841–847 (1996)
28. Y. Kobayashi, F. Tsukihashi: Metall. Mater. Trans. B **28B**, 956–959 (1997)
29. Y. Kobayashi, F. Tsukihashi: Metall. Mater. Trans. B **29B**, 1037–1042 (1998)
30. Y. Kobayashi, F. Tsukihashi: Metall. Mater. Trans. B **30B**, 352–354 (1999)
31. W. Kroll, Trans. Electrochem. Soc.**78**, 35–47 (1940)
32. T. Noda: Titan·Japan (Bull. Japan Titanium Soc.) **44**, 85–89 (1994)
33. W.D. Threadgill: J. Electrochem. Soc. **112**, 632 (1965)
34. R.O. Suzuki, M. Aizawa, K. Ono: J. Alloys Compounds **288**, 173–182 (1999)
35. J.D. Corbett, J.D. Smith, E. Garcia: J. Less-Common Metals **115**, 343–355 (1986)
36. H. Sano, M. Tashiro, T. Fujisawa, C. Yamauchi: *Proc. Fall Meeting of MMIJ (Mining Mater. Process. Inst. Japan)*, Sapporo, Japan 99 (1997)
37. H. Sano, M. Tashiro, T. Fujisawa, C. Yamauchi: Mater. Trans. JIM **40**, 263–267 (1999)
38. I. Barin: *Thermochemical data of pure substances* (VCH, Weinheim, 1989)
39. A. Sieverts: Z. Phys. Chem. **77**, 591 (1911)
40. A. Sieverts: Z. Metalk. **21**, 37 (1929)
41. E.D. Eastman, D.D. Cubicciotti, C.D. Thurmond: *In Chemistry and metallurgy of miscellaneous materials*, ed. by L.L. Quill (McGraw-Hill, New York, paper 2 1950), 10
42. O. Madelung, (ed.): *Landolt-Bönstein numerical data and functional relationships in science and technology*, 26 (Diffusion in Solid Metals and Alloys, Springer-Verlag, Berlin, 1982)
43. L.S. Darken, R.W. Gurry: *Physical chemistry of metals* (McGraw-hill/Kogakusha, Tokyo/New York, 1953)
44. M.W. Chase, J.L. Curnutt, H. Prophet, R.A. McDonald, A.N. Syverrud: J. Phys. Chem. Ref. Data, *JANAF Thermochemical Tables Vol. 4 Supplement* (American Chemical Society and American Institute of Physics, 1975)
45. T.B. Massalski (ed.): *Binary alloy phase diagrams* (American Society for Metals, Metals Park, OH, 1990)
46. O.N. Carlson, R.R. Lichtenberg, J.C. Warner: J. Less-Common Metals **35**, 275–284 (1974)
47. H. Sano, T. Yoshida, S. Takai, K. Kuroda, T. Fujisawa, C. Yamauchi: J. Mining Mater. Processing Institute of Japan (Shigen-to-Sozai), **115**, 49–53 (1999)

48. H. Sano, T. Yoshida, K. Kuroda, T. Fujisawa and C. Yamauchi, J. Mining and Materials Process. Inst. Japan (Shigen-to-Sozai) **115**, 54–58 (1999)
49. K. Kuroda, S. Takai, T. Fujisawa, M. Okido, C. Yamauchi: J. Mining Mater. Process. Inst. Japan (Shigen-to-Sozai) **112**, 893–897 (1996)
50. D. Fort, D.W. Jones, B.J. Beaudry, K.A. Gschneidner, Jr.: J. Less-Common Metals **81**, 273–292 (1981)
51. K. Kamihira, Y. Hasegawa, O. Ogawa: Mater. Trans. JIM **34**, 243–247 (1993)
52. C.M. Muirhead, D.W. Jones: J. Less-Common Metals **50**, 73 (1976)
53. O. Ogawa, K. Kamihira, C. Song, A. Ando, R. Hasegawa: J. Alloys Compounds **193**, 17–19 (1993)
54. P.G. Mattocks, C.M. Muirhead, D.W. Jones, B.J. Beadry, K.A. Gchneidner, Jr.: J. Less-Common Metals **53**, 253–263 (1977)
55. D. Fort, B.J. Beaudry, K.A. Gschneidner, Jr.: J. Less-Common Metals **134**, 27–44 (1987)
56. D.T. Peterson, F.A. Schmidt: J. Less-Common Metals **29**, 321 (1972)
57. R.G. Jordan, D.W. Jones: J. Less-Common Metals **21**, 125 (1973)
58. R.G.Jordan, D.W.Jones, V.J. Hems: J. Less-Common Metals **42**, 101–110 (1975)
59. D. Fort, V.K. Pecharsky, K.A. Gchneidner, Jr.: J. Alloys Compounds **226**, 190–196 (1995)
60. M. Williams, C.L. Huffine: Nucl. Sci. Eng. **9**, 500 (1961)
61. O.N. Carlson, F.A. Schmidt, D.T. Peterson: J. Less-Common Metals **10**, 1–11 (1966)
62. V.T. Volkov, T.V. Nikiforova, A.M. Ionov, A.N. Pustovit, G.G. Sikharulidse: J. Less-Common Metals **79**, 199–205 (1981)
63. K. Kuroda, T. Yoshida, T. Fujisawa, C. Yamauchi: J. Mining Mater. Process. Inst. of Japan (Shigen-to-Sozai) **113**, 179–184 (1997)
64. D. Fort, J. Crystal Growth: **94**, 85–96 (1989)
65. M. Nakamura, T.H. Okabe, T. Oishi, K. Ono: In *Molten salt chemistry and technology 1993*, ed. by M.L. Saboungi, H. Kojima (The Electrochemical Society Inc., 1993) p. 529–540
66. C. Ouchi: Bull. Japan Inst. Metals (Materia Japan) **37**, 22–26 (1998)
67. C. Ouchi, A Kato: Titan·Japan (Bull. Japan Titanium Soc.) **44**, 81–83 (1996)
68. G.K. White, S.B. Woods: Phil. Trans. R. Soc. London, Ser. A **251**, 273–302 (1959)
69. J. Wasilewski: Trans. Metall. Soc. AIME **224**, 5–7 (1962)
70. G. Elssner, U. Krohn, O.Runao: Z. Metallkd. **67**, 311–317 (1976)
71. S. Komatsu, M. Ikeda, M. Yoshimura: In Ultra High Purity Base Metals, UHPM-94 (The Japan Institute of Kitakyushu, 1994) p. 452–457
72. S.L. Ames, A.D. McQuillan: Acta Metall. **4**, 619–626 (1956)
73. J. Wasilewski: Trans. Metall. Soc. AIME **224**, 8–12 (1962)
74. G. Baur, P. Lehr: J. Less-Common Metals **69**, 203–218 (1980)
75. B.N. Aleksandrov: Phys. Met. Metall. **61**, 10–25 (1986)
76. S. Komatsu, M. Ikeda: Bull. Japan Inst. Metals (Materia Japan) **33**, 44–47 (1994)

2 Principles of Metal Purification and Purity Evaluation

T. Kekesi and M. Isshiki

Techniques of metal refining and purification rely on the differences in physicochemical properties of the base metal and the impurities. In order to reach the desired ultra-high purity, a sequence of complementary refining steps is required. These steps fundamentally apply chemical, electrochemical and physical methods, usually in this order. A substance is considered to be practically pure if its properties are determined by the atomic-crystalline structure and the intrinsic defects of the crystal lattice, while the relative effect of impurities is negligible [1]. Concentrations of impurities as low as one part in a million of the matrix metal can significantly influence the physical properties. The effect of a certain impurity also depends on the state of occurrence. Electrical properties are influenced practically by substitutional impurities, homogeneously dissolved in the material, while mechanical properties are strongly dependent on interstitial ones. The influence of minute quantities of imp urities on the physical properties of the base metal can be applied for the purpose of indirect purity evaluation, thus providing comparative methods to accompany direct elemental analysis.

2.1 Methods of Purification

Impurity elements can be removed from the base metal by assuring that they are eventually transferred – in elemental or compound form – into a separate phase. The methods of purification can be classified according to the main property utilized in separation (Table 2.1).

The general scheme of purification is shown in Fig. 2.1. Aqueous precipitation techniques and fire refining are suitable for removing large quantities of different impurities, as applied in commercial hydrometallurgical and pyrometallurgical operations, and under appropriate conditions they can also be applied to fine purification. In order to achieve ultra-high purity, however, further ion-selective methods have to be applied, usually in aqueous media. The metal is extracted from the purified solution by electrolysis or hydrogen reduction of the crystallized and dehydrated salt. Remaining traces of impurities can be eliminated by refining electrolysis or by different physical methods, including vacuum melting, distillation, zone melting, hydrogen-plasma melting or, in special cases, electrotransport. The result of the procedure may equally depend on the primary (gross) and the secondary (fine) purification steps. The purified metal is annealed in ultra-high vacuum or hydrogen, and

Table 2.1. A classification of methods applied to metal purification

No.	Nature of separation	Property utilized	Method of separation	Section
1	Chemical	Reactivity	Selective reaction	2.1.1
2	Chemical	Solubility	Selective precipitation	2.1.2.1
3	Chemical	Electrode potential	Cementation	2.1.2.1
4	Chemical	Distribution of dissolved ions between phases	Ion exchange, solvent extraction	2.1.2.2
5	Electrochemical	Electrode potential	Electrowinning, Electrorefining	2.1.3
6	Physical	Distribution of atoms between solid and liquid phases	Zone melting, fractional crystallization	2.1.4
7	Physical	Vapor pressure	Distillation, sublimation	2.1.5
8	Physical	Ion mobility	Electrotransport	2.1.6

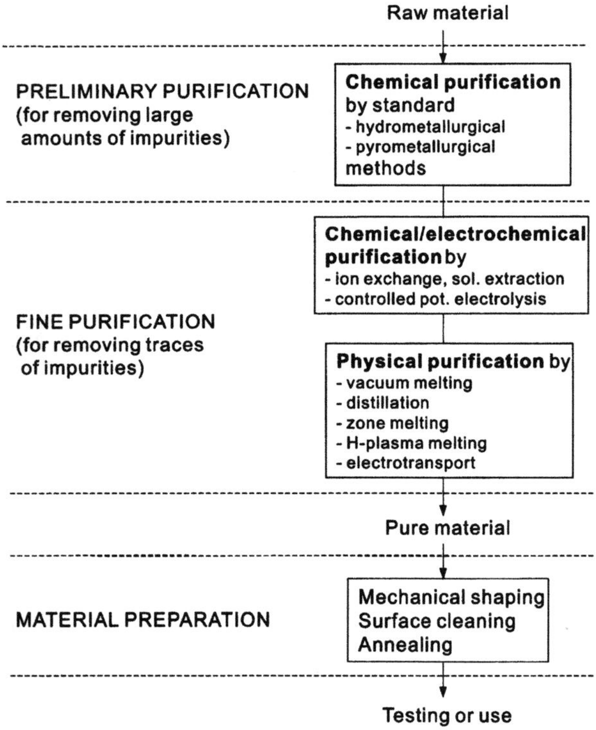

Fig. 2.1. A schematic procedure of ultra-high purification

finally brought to a form suitable for the application or the testing procedure in several steps of mechanical shaping and annealing. Concomitant superficial contamination is removed by repeated chemical or electrochemical polishing.

2.1.1 Pyrometallurgical Refining

Differences in the chemical properties of the base metal and the impurity permit a large number of chemical–metallurgical refining techniques. The principal chemical property is reactivity, or affinity toward added agents, such as oxygen, chlorine or sulfur. In the common fire refining process, the reagent (A) is added to the molten metal at an appropriate temperature. Conditions are favorable if the Gibbs free energy change in the reaction of one mole of the reagent with the impurity element (M),

$$x\mathrm{M} + \mathrm{A}_n = n\mathrm{M}_{x/n}\mathrm{A}, \tag{2.1}$$

is more negative than that in the reaction with the base metal (Me),

$$y\mathrm{Me} + \mathrm{A}_n = n\mathrm{Me}_{y/n}\mathrm{A}. \tag{2.2}$$

This can be expressed mathematically with the following relationship:

$$\Delta G^{\circ}_{n\mathrm{M}_{x/n}\mathrm{A}} + \mathbf{R}T \ln \frac{a^n_{\mathrm{M}_{x/n}\mathrm{A}}}{a^x_{\mathrm{M}} a_{\mathrm{A}_n}} < \Delta G^{\circ}_{n\mathrm{Me}_{y/n}\mathrm{A}} + \mathbf{R}T \ln \frac{a^n_{\mathrm{Me}_{y/n}\mathrm{A}}}{a^y_{\mathrm{Me}} a_{\mathrm{A}_n}}, \tag{2.3}$$

where ΔG° is the change of the Gibbs free energy accompanying the formation of the noted compounds when all the components of the reaction are in the normal state. The symbol a stands for the activities of the noted species in the respective phases, $\mathbf{R}$ is the gas constant and T is the thermodynamic temperature.

During the refining process the impurity element is selectively reacted; therefore its concentration (activity) in the molten metal phase is continuously decreasing. The resulting impurity compound is taken up by another phase, usually referred to as the slag or dross, which practically should not be miscible with the molten metal. The activity of the impurity compound in the produced phase is continuously increasing. Therefore, the two sides of (2.3) will become equal, terminating the selective removal of the impurity. The activity of the remaining impurity in equilibrium at this point and during any further operation will be:

$$a^x_{\mathrm{M}} = a^y_{\mathrm{Me}} \left(\frac{a_{\mathrm{M}_{x/n}\mathrm{A}}}{a_{\mathrm{Me}_{y/n}\mathrm{A}}} \right)^n \exp \left(\frac{\Delta G^{\circ}_{n\mathrm{M}_{x/n}\mathrm{A}} - \Delta G^{\circ}_{n\mathrm{Me}_{y/n}\mathrm{A}}}{\mathbf{R}T} \right). \tag{2.4}$$

After the period of selective reaction the base metal is simultaneously reacted, which involves a corresponding loss, but it also means a decrease in the $a_{\mathrm{Me}_x\mathrm{A}}/a_{\mathrm{Me}_y\mathrm{A}}$ ratio, reducing the residual concentration of the impurity element. This effect can be utilized advantageously if the amount of the secondary (slag) phase is reduced to a minimum before the period of simultaneous reaction. The refining efficiency can be further enhanced by introducing such agents in the slag that reduce the activity of M_xA by binding it in stable complex compounds. It is especially advantageous if the compound formed from the removed impurity is volatile, and its activity can be kept to a minimum by continuous removal (applying vacuuming or flushing) of the produced gas.

At high temperatures and in well-designed reactors, most systems may approach these equilibrium conditions within practical time intervals. Unless equilibrium is appropriately approached, kinetic conditions may dominate in determining the actual impurity concentration.

2.1.2 Hydrometallurgical Purification

High-purity metals are often obtained from aqueous solutions by hydrometallurgical methods, offering the possibility of eliminating a large number of impurities from a variety of base metals. Preliminary purification of the solution, usually by precipitation techniques, is capable of removing large quantities of the main impurities. The remaining portion may be eliminated by fine purification techniques (usually by ion exchange or solvent extraction).

The base metal of high purity can be extracted from the solution by the following alternative methods:

- evaporation to dryness, followed by hydrogen reduction of the compound (usually a chloride of the metal)
- precipitation of the metal (with NH_4OH as hydroxides, or with organic agents as oxalates, stearates or citrates) and calcination followed by hydrogen reduction of the oxide
- cathodic reduction by direct electrolysis of the purified solution

The procedure and the applied reagents used for extracting the base metal from the purified solution are selected with due care to avoid contamination. Metals are obtained from the solution phase generally in a loose, powdery form, requiring melting to produce bulk material in a compact shape.

2.1.2.1 Preliminary Purification by Precipitation. Impurity metals (M) of more noble character than the base metal can be removed from the solution in metallic form by cementation. Contamination of the solution is avoided by applying the base metal (Me) in high-purity powder or granule form as the precipitation agent:

$$n\mathrm{M}^{m+} + m\mathrm{Me} = n\mathrm{M} + m\mathrm{Me}^{n+}. \tag{2.5}$$

This process may continue as long as the electrode potential of the impurity element is sufficiently high:

$$E^\circ_\mathrm{M} + \frac{\mathbf{R}T}{m\mathbf{F}} \ln \frac{a_{\mathrm{M}^{m+}}}{a_\mathrm{M}} > E^\circ_\mathrm{Me} + \frac{\mathbf{R}T}{n\mathbf{F}} \ln \frac{a_{\mathrm{Me}^{n+}}}{a_\mathrm{Me}}, \tag{2.6}$$

where E°_{M} and E°_{Me} are the standard electrode potentials of the impurity and the base metal, respectively. The activities of the metal ions in the solution are indicated as $a_{\mathrm{M}^{m+}}$ and $a_{\mathrm{Me}^{n+}}$. The activities in the pure metallic phases, a_{M} and a_{Me}, are equal to unity, while the molar charge (**F**) is approximately 96 500 As mole^{-1} for univalent species. If the standard electrode potential of the impurity is at least $\sim$ 360 mV higher than that of the base metal, a 10^6 times lower concentration of this impurity may be reached under ideal conditions at room temperature and for ions of equal charge ($m = n$).

The precipitated impurity metal covers the added metal particles. This porous layer, however, does not prevent the progress of cementation. The kinetic conditions of the reaction are usually governed by material transport (diffusion) through the boundary layer of the solution around the particle, and through the porous deposit. Most often, the former step controls the rate of the process, expressed by the amount (n) of M^{m+} ions transported in unit time (t) across the boundary layer:

$$\frac{\mathrm{d}n}{\mathrm{d}t} = \mathcal{D} A \,\mathrm{grad}\, a_{\mathrm{M}^{m+}}, \tag{2.7}$$

where A is the interfacial area, $\mathcal{D}$ is the diffusion coefficient. As suggested by (2.7), efficient stirring of the solution, higher temperature and a larger specific surface area of the added metal particles will increase the rate of the process. Hydrogen gas co-deposition and evolution, which take place especially at higher acid concentrations, generally contribute to the loosening of the deposited metal layer.

Some dissolved metals may be partially eliminated from the solution by hydrolytic precipitation in the hydroxide form at an appropriate pH value. As metals are processed primarily in acid solutions, hydrolytic precipitation can be achieved by consuming much of the free acid:

$$\mathrm{M}^{m+} + m\mathrm{H_2O} = \underline{\mathrm{M(OH)}_m} + m\mathrm{H}^+. \tag{2.8}$$

At the beginning of precipitation – for example, in the cases of copper and zinc – some basic salts of a $x\mathrm{M}_y\mathrm{Y}_m \cdot \mathrm{M(OH)}_m$ general composition may be formed, which are decomposed to the hydroxide and the Y^{y-} anion as the pH is further increased.

Acid liberated during hydrolysis requires further addition of the neutralizing agent, which has to be chosen so that the salt produced will not contaminate the solution. This is practically achieved by applying the oxide or hydroxide of the base metal as the agent:

$$2n\mathrm{H}^+ + \mathrm{Me_2O}_n = 2\mathrm{Me}^{n+} + n\mathrm{H_2O}\,, \tag{2.9}$$

$$n\mathrm{H}^+ + \mathrm{Me(OH)}_n = \mathrm{Me}^{n+} + n\mathrm{H_2O}\,. \tag{2.10}$$

The solubility products of the most important metal hydroxides and basic salts are summarized in Table 2.2. Equilibrium metal ion activities, derived from the tabulated data, are plotted as functions of pH in Fig. 2.2. It is important to note that precipitates of amphoteric metals, which are capable of

Table 2.2. Solubility products of hydroxides at 18–25°C [2,3]

Compound	Solubility product,	
	Definition	Value
$Te(OH)_4$	$a_{Te^{4+}} \times a^4_{OH^-}$	0.7×10^{-54}
$Ti(OH)_4$	$a_{Ti^{4+}} \times a^4_{OH^-}$	1.0×10^{-53}
$Tl(OH)_3$	$a_{Tl^{3+}} \times a^3_{OH^-}$	1.4×10^{-53}
$Co(OH)_3$	$a_{Co^{3+}} \times a^3_{OH^-}$	3.2×10^{-45}
$Sb(OH)_3$	$a_{Sb^{3+}} \times a^3_{OH^-}$	4.0×10^{-42}
$Fe(OH)_3$	$a_{Fe^{3+}} \times a^3_{OH^-}$	1.6×10^{-39}
$Al(OH)_3$	$a_{Al^{3+}} \times a^3_{OH^-}$	3.2×10^{-34}
$Cr(OH)_3$	$a_{Cr^{3+}} \times a^3_{OH^-}$	1.6×10^{-30}
$Hg(OH)_2$	$a_{Hg^{2+}} \times a^2_{OH^-}$	1.0×10^{-26}
$Sn(OH)_2$	$a_{Sn^{2+}} \times a^2_{OH^-}$	5.0×10^{-26}
$Be(OH)_2$	$a_{Be^{2+}} \times a^2_{OH^-}$	5.0×10^{-22}
$Cr(OH)_2$	$a_{Cr^{2+}} \times a^2_{OH^-}$	2.0×10^{-20}
$Cu(OH)_2$	$a_{Cu^{2+}} \times a^2_{OH^-}$	4.8×10^{-20}
$Zn(OH)_2$	$a_{Zn^{2+}} \times a^2_{OH^-}$	3.5×10^{-17}
$Ni(OH)_2$	$a_{Ni^{2+}} \times a^2_{OH^-}$	6.3×10^{-16}
$Fe(OH)_2$	$a_{Fe^{2+}} \times a^2_{OH^-}$	7.9×10^{-16}
$Pb(OH)_2$	$a_{Pb^{2+}} \times a^2_{OH^-}$	1.0×10^{-15}
$Co(OH)_2$	$a_{Co^{2+}} \times a^2_{OH^-}$	1.3×10^{-15}
$Cd(OH)_2$	$a_{Cd^{2+}} \times a^2_{OH^-}$	4.5×10^{-15}
$Mn(OH)_2$	$a_{Mn^{2+}} \times a^2_{OH^-}$	1.6×10^{-13}
$Mg(OH)_2$	$a_{Mg^{2+}} \times a^2_{OH^-}$	7.1×10^{-12}
$Ag(OH)$	$a_{Ag^+} \times a_{OH^-}$	2.0×10^{-8}
$Ca(OH)_2$	$a_{Ca^{2+}} \times a^2_{OH^-}$	6.5×10^{-6}

forming dissolved hydroxo-complex anions, are re-dissolved on increasing the pH above the range of efficient precipitation. Examples are Cr(III), Al(III), Pb(II), Sn(II) and Zn(II). The appropriate pH ranges of hydrolytic precipitation for the most common (transition) metals are given in Table 2.3.

The removal of iron (III) by hydrolytic precipitation has an additional purifying effect. At lower pH, the hydrolytic reaction produces $Fe(OH)_3 \cdot xH_2O$, an extremely fine precipitate, capable of adsorbing a significant amount of metal ions (primarily As, Sb, Sn, Ge, Te, etc.). At high pH, Fe(III) hydroxide tends to form the $[Fe(H_2O)_2(OH)_4]^-$ type anionic complex species, which are

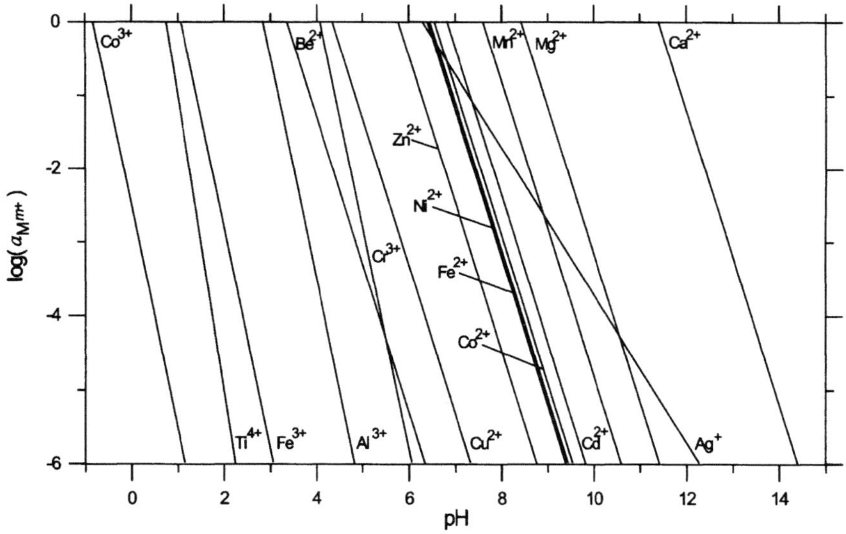

Fig. 2.2. Solubilities of metal hydroxide precipitates in the form of aquo-ions, as functions of pH

Table 2.3. Practical pH ranges of metal precipitation using NaOH [4]

Species	pH value		
	Precipitation onset	Precipitation end	Re-dissolution
Sn(II)	3.8	~ 4.3 (colloid)	9.5–10.5
Cu(II)	5.8	7.6	–
Fe(II)	7	8.9	–
Zn(II)	7.6	8.5–9.0	10.8–11.0
Ni(II)	7.8	9.3–9.9	–
Cd(II)	9.1	9.8–10.4	–
Pb(II)	6.5–7.0	9.5–10.3	–
Fe(III)	2.8	3.5	–
Al(III)	4.3	4.8	8.7
Cr(III)	5.5–5.8	6.5–6.8	8.3–9.2

especially efficient in adsorbing cations. The precipitated hydroxide forms a gel-type material with a high specific surface and a high content of retained solution. Coagulation of the precipitate retains the adsorbed impurities by occlusion [2].

Impurity metals can be efficiently removed from the solution by introducing anions that form insoluble compounds with the cations to be eliminated.

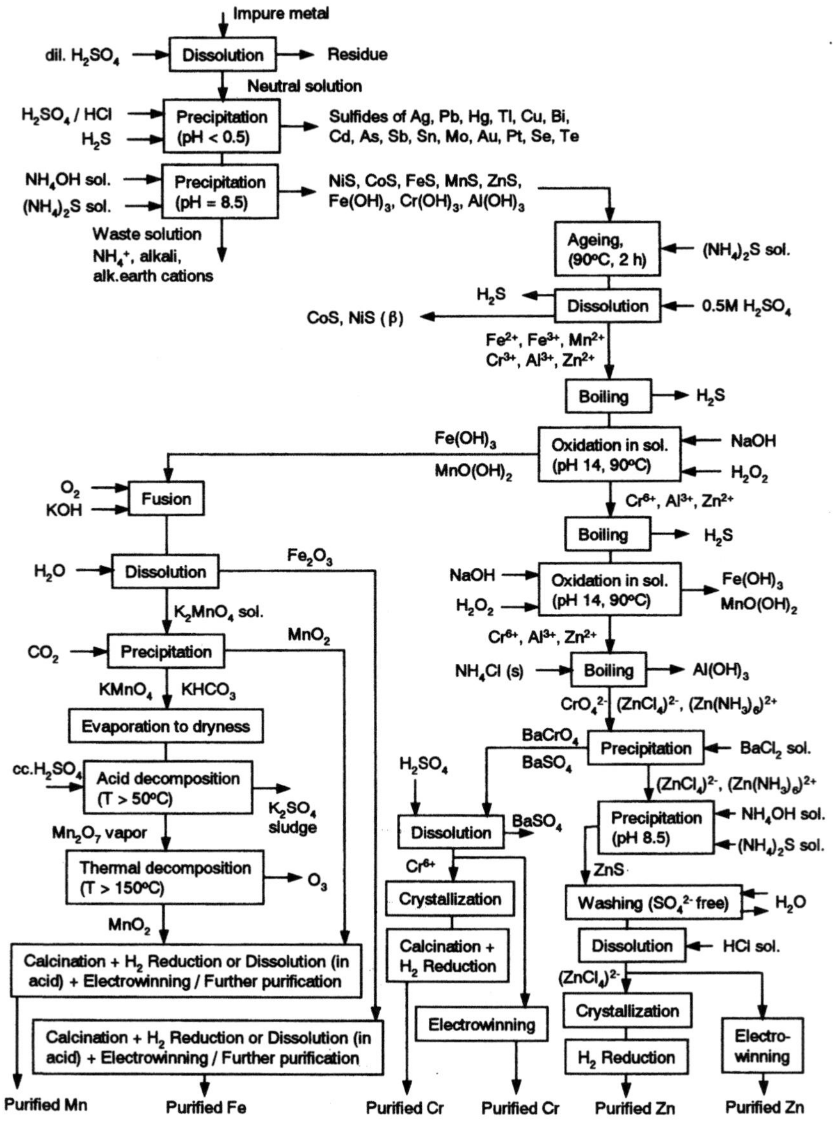

Fig. 2.3. The purification of Cr, Mn, Fe and Zn by aqueous processing

The most frequently precipitated compounds are sulfides, sulfates and carbonates, but some metals may also be precipitated as chlorides, oxychlorides, phosphates or oxalates. This method does not always require large changes in the pH value of the solution, and allows selective precipitation. Solubility products of the most common precipitated compounds are available in compilations [2–9].

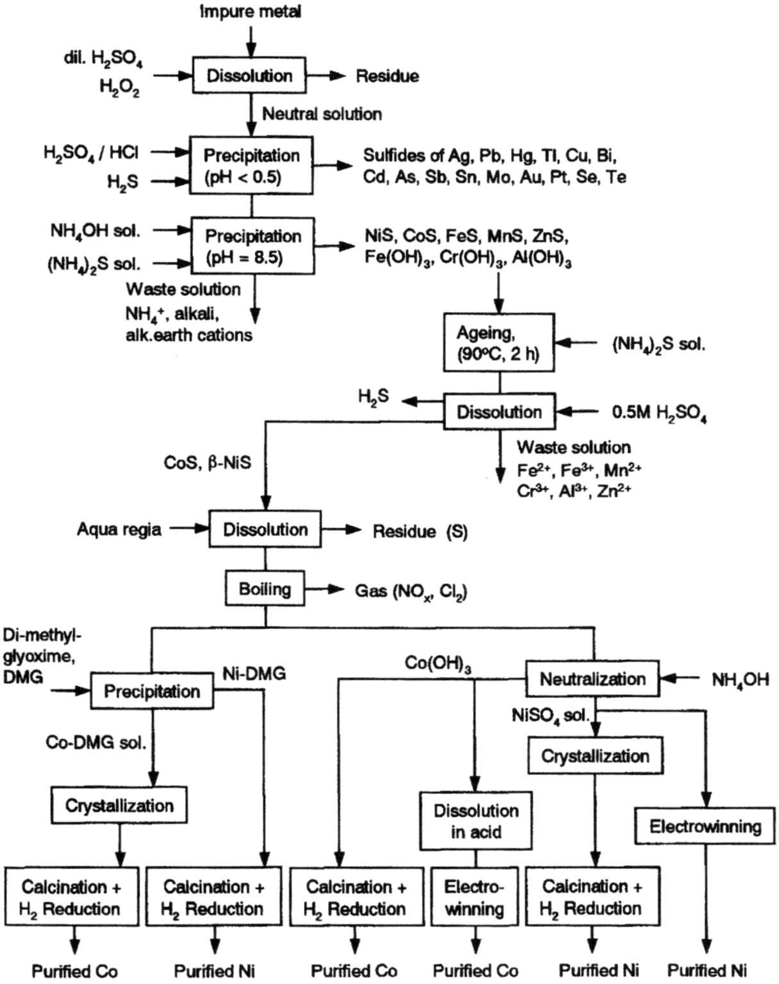

Fig. 2.4. The purification of Co and Ni by aqueous processing

Solution pre-treatment, based on the combination of different precipitation techniques, may eliminate significant amounts of impurities. Known solubility data and set chemical techniques [2–12] have been used as the basis for designing separation procedures for the practically important cations. Possible processes, treating multi-component aqueous systems, to obtain the purified solutions of a base metal (from Cr to Zn) are outlined in Figs. 2.3 and 2.4. Aqueous solution treatment is generally supplemented with additional steps of metal extraction. Boxes in the flowcharts stand for the principal steps, also including solid–liquid separation where it applies. The solution phase (filtrate) is depicted to leave the boxes always at the bottom, whereas residual solid material and additives are marked on the sides. In some cases the outlined aqueous purification processes seem rather complicated, involving a number of consecutive steps, but generally the operations have been compiled to involve only common reagents and standard techniques. The simplest, and safest methods were given preference as a rule: however, in some cases alternative routes had to be allowed. In all cases, the pure metal is extracted from the purified solution either by hydrogen reduction of the crystallized salt or by direct electrowinning. However, for the preparation of ultra-high-purity metals, these steps of bulk purification need to be followed by further (fine) purification of the solution.

2.1.2.2 Fine Purification by Ion Exchange or Solvent Extraction. Distribution between the aqueous solution and an organic phase is a frequently utilized chemical property for separating dissolved metal ions. These special organic materials, constituting the cation or anion exchange resins and liquid extractants, always contain counter-ions (C^+ cations or A^- anions), which may be exchanged for the corresponding aquated or complexed ions of the metal (Me^{n+} or $[MeX_y]^{n-y}$) in solution. The generalized reversible reactions for the cation and anion exchanges, respectively, are expressed as follows:

$$Me^{n+} + n(R \cdot C) \Leftrightarrow nR \cdot Me + nC^+ \tag{2.11}$$

or

$$[MeX_y]^{n-y} + n(R \cdot A) \Leftrightarrow nR \cdot MeY_y + nA^- . \tag{2.12}$$

Strength of sorption is practically characterized by the equilibrium distribution coefficient:

$$D = \frac{\overline{C_{Me}}}{C_{Me}} \tag{2.13}$$

which refers to the amounts of the metal ions in unit volumes of the organic (solid or liquid) and the contacted aqueous solution phases, respectively. The preference of the organic phase for one of two competing metals can be expressed by the selectivity coefficient, which is a ratio of the respective distribution coefficients.

The separation of metal ions in solution can be effected by applying selective sorption and/or elution. This is carried out by adjusting the chemistry of the aqueous solution in a way that favors selective uptake or release of one metal by the organic phase. In cases when a larger number of different elements have to be separated, as in the case of achieving ultra-high-purity, conditions of perfectly selective sorption and elution are impossible. Nevertheless, efficient purification is attainable by chromatographic separation, which is physically possible when ion-exchange columns are used. The column operations can be considered as a large number of repeated batch equilibration steps. The same eluent may desorb metals at different rates, forming elution fronts, which move at different velocities along the column. The separated metal is obtained in a certain volume fraction of the effluent.

The separating effect of ion exchange can be enhanced and supplemented by applying such aqueous media where the metal ions may be complexed, some of them forming anionic species. The stability of complexes may show great differences [5–7] and may depend on the concentration of the complexing anions in the solution, according to the following equilibrium:

$$\mathrm{Me}^{n+} + y\mathrm{Y}^{-} = [\mathrm{MeY}_y]^{n-y}\,. \tag{2.14}$$

Anionic complexes of the metal arise when $y > n$. The negatively charged complexes are subject to sorption of different strengths on anion exchange resins and are released from cation exchange resins. Anion exchange can more sharply differentiate between transition metals: therefore, ultra-high purification can be carried out primarily by this technique, which may be supplemented by a cation exchange step in some special cases.

Chloro-complexes of a large number of metals can be formed at appropriate concentrations of chloride ions in the solution, while other common inorganic acids are often ineffective. For the purpose of ultra-high-purity, it is best to avoid the introduction of foreign cations; therefore, the complexing chloride ions are supplied in the form of pure hydrochloric acid. The sorption of the elements by a strong-base anion exchange resin in HCl solutions is summarized in Fig. 2.5. Sorption functions may be determined directly by elution tests, using ion-exchange columns [13]. In cases where precipitation, unstable oxidation states or high values of the distribution coefficient cause practical difficulties, batch equilibration is required for the examination of ion exchange characteristics [14]. Results referring to this technique are shown by the dotted lines in Fig. 2.5.

Those impurities not forming anionic complex species can be directly separated by anion exchange if the base metal is strongly complexed and sorbed under the applied conditions. The great differences in sorption characteristics of the elements in hydrochloric media (Fig. 2.5) suggest that an optimum elution process, applying different concentrations of HCl for rinsing, may be compiled for the overall purification of the majority of base metals [15]. Proper adjustment of the HCl concentration in the eluent fractions requires elution experiments with bench-scale columns [16].

Cation exchange may be used in addition to anion exchange steps for the separation of cationic impurities from a base metal which can form only neutral complex species, even at high HCl concentrations.

Similarly to ion exchange, the method of solvent extraction is also based on the distribution of metals between the aqueous and the organic phases, the latter being liquid in this case. The reactions involved are similar to those applied in ion exchange, but the equilibration and the subsequent separation

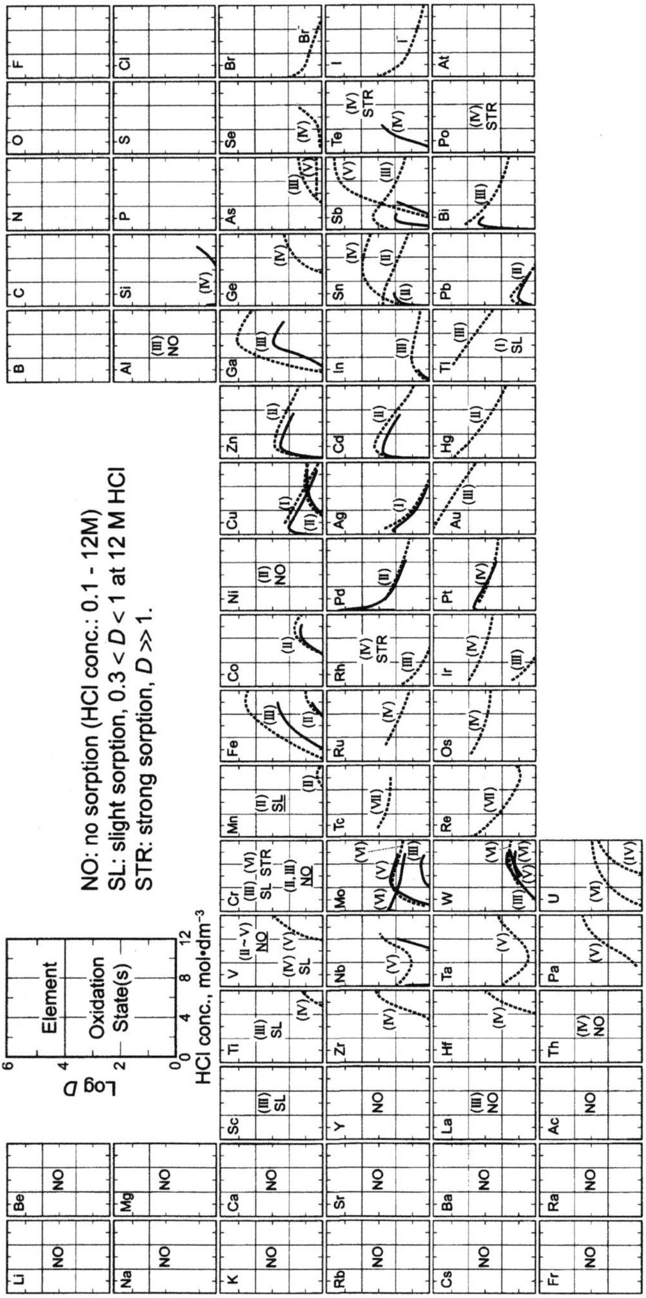

Fig. 2.5. The sorption of elements from hydrochloric acid solutions on strong base anion exchange resins (solid lines and underlined refer to equilibrating tests [14], while broken lines and simple notes are based on column elution results [13])

of the two phases imply more complicated technology. Handling of the often volatile and toxic organic extractants may cause difficulties. A further disadvantage is the lack of chromatographic (dynamic) operation. Therefore, solvent extraction is usually applied for special purposes. Highly selective (chelating) reagents have been developed for the extraction of some metals. Separation by solvent extraction may primarily find application in removing specific impurities, for which general purification procedures have proved to be less efficient [17].

2.1.3 Electrochemical Purification

Besides extracting the metal from the electrolyte solution, electrowinning may contribute significantly to the purpose of purification. The technique of electrowinning requires insoluble, inert anodes, usually made of a noble or passivated metal, but pure graphite may also be considered. If the starting cathode is made of the same metal as the one to be obtained from the solution, further processing requires no separation of the cathode layers. Alternatively, conditions have to be found for convenient mechanical separation of the produced deposit from the substrate.

Due to the different electrochemical properties, expressed by the electrode potentials, of the metals in the electrolyte solution, cathodic deposition may offer a significant purifying effect. The characteristic electrode potential of a reduction process,

$$Me^{n+} + ne^- = Me, \tag{2.15}$$

is composed of the equilibrium electrode potential and the overpotential. The former component – described by (2.6) – is determined by the free energy change in the electrode process of (2.15). The latter component is associated with the displacement of the system from equilibrium, in order to produce irreversible chemical changes, indicated by a net current on the electrode. A higher (more positive) equilibrium electrode potential means a stronger tendency of cathodic reduction for the metal, since the equilibrium state is sustained despite a higher deficiency of electrons in the cathode. The relative values of the electrode potentials determine the order of cathodic deposition of competing cations. In order to obtain selective cathodic deposition of a metal, it is necessary to ensure that its electrode potential will be sufficiently higher than that of the other metal ions in the solution.

A net current density (i) of the cathode relies on a continuous supply of ions carried by diffusion. This requires a lower effective concentration of the discharged ions at the cathode interface ($a_{\mathrm{Me}^{n+},\mathrm{eff}}$) than in the bulk solution ($a_{\mathrm{Me}^{n+},b}$), resulting in a shift of the electrode potential, defined as the diffusion overpotential (η_{diff}) [18,19]:

$$\eta_{\mathrm{diff}} = \frac{\mathbf{R}T}{n\mathbf{F}} \ln \frac{a^{n+}_{\mathrm{Me,eff}}}{a^{n+}_{\mathrm{Me},b}} \approx \frac{\mathbf{R}T}{n\mathbf{F}} \ln \left(1 - \frac{i}{i_l}\right), \tag{2.16}$$

where i_l is the limiting current of diffusion, associated with zero surface concentration of the ionic species taking part in the cathodic process.

The activation energy related steps of charge transfer and atom incorporation in the crystal lattice bring about the activation-type overpotential (η_{act}), which can be estimated at practical levels of current densities as follows:

$$\eta_{act} \approx \frac{\mathbf{R}T}{\beta\mathbf{F}} \ln \frac{i_o}{|i|}, \tag{2.17}$$

where i_o is the exchange current density at the zero net current equilibrium state, and β is the charge transfer factor [18–20].

Cathodic overvoltages, implying negative values, diminish the effective electrode potential:

$$E_{Me,eff} = E_{Me} + \eta_{diff} + \eta_{act}, \tag{2.18}$$

thereby facilitating the co-deposition of impurities. According to (2.16) and (2.17), higher current densities correspond to more negative cathode potentials, restricting the conditions of selective deposition. Therefore, high-purity cathodes may generally be obtained by electrowinning, on the condition that the potential is regulated by limiting the practical range of the current density.

In order to obtain the extracted metal at the highest purity, it is usual to deposit the more noble impurity metals at a higher initial cathode potential. The lower this initial potential is set, the more of these impurities will be eliminated, but more of the main metal will also be lost by co-deposition. After replacing the cathode with the high-purity starting sheet, the base metal is deposited at an appropriate cathode potential, which leaves the less noble impurities in the electrolyte solution.

Most metals can be electrodeposited from aqueous solutions on solid electrodes under proper conditions. The limitation is imposed by the possibility of hydrogen evolution:

$$H^+ + e^- = \tfrac{1}{2}H_2, \tag{2.19}$$

suppressing the desired cathodic process of (2.15). In order to achieve metal deposition from an aqueous medium, the effective electrode potential of the metal has to be higher than that of hydrogen:

$$E_{Me,eff} > E_{H_2,eff}. \tag{2.20}$$

The equilibrium electrode potential of hydrogen evolution depends on the activity of hydrogen ions (a_{H^+}) in the solution:

$$E_{H_2} = \frac{\mathbf{R}T}{\mathbf{F}} \ln \frac{a_{H^+}}{\sqrt{p_{H_2}}}, \tag{2.21}$$

where p_{H_2} is the partial pressure of hydrogen gas evolved at the electrode. The non equilibrium component of the effective hydrogen electrode potential,

Table 2.4. Tafel constants of the hydrogen overpotential at 20°C [2,18]

Cathode material	Electrolyte	$-a$	$-b$
Cadmium	1M H_2SO_4	1	0.180
Cobalt	1M HCl	0.62	0.140
Copper	0.5M H_2SO_4	0.8	0.115
Iron	1M HCl	0.7	0.125
Iron	1M H_2SO_4	0.67	0.116
Iron	2M NaOH	0.76	0.112
Mercury	0.5M H_2SO_4	1.42	0.113
Mercury	1M HCl	1.41	0.116
Zinc	0.5M H_2SO_4	1.24	0.118

the hydrogen overpotential, is essentially determined by process steps that require activation energy, in which the catalytic effect of the metal substrate also plays an important role.

According to experimental data, the hydrogen overpotential can be estimated by the Tafel semi-logarithmic relationship [18]:

$$\eta_{H_2} = a + b \log i, \qquad (2.22)$$

where the parameters a and b are dependent on the conditions and the material of the cathode, as is shown in Table 2.4 for a few practical examples. The absolute value of the hydrogen overpotential is usually increased by applying smoother cathode surfaces. Sufficiently negative potentials of hydrogen evolution – with a given metal as cathode material – can be attained by increasing the pH and by setting a lower temperature and higher current densities for electrolysis. In this way, metals with low standard electrode potentials, such as zinc or manganese, can be electrowon from neutral (slightly acidic) electrolytes of sufficiently high metal concentration. If all efforts fail to ensure an acceptable current efficiency (as in the cases of the most reactive metals, such as Al, Mg and Ti), electrowinning can be carried out from a molten salt electrolyte, depositing the metal either in the molten or in the solid state.

Metals produced by electrowinning may be further purified by placing them in electrorefining cells as anodes. The metal is anodically dissolved in the electrolyte, and is deposited at the cathode as highly pure material. Noble impurities, of significantly higher electrode potential, tend to remain undissolved and mixed with the fallen particles of the anode, constituting a slime by-product. Reactive elements of lower electrode potential may enter the electrolyte but they do not endanger cathode purity until their concentration – and electrode potential – increases to a critical level.

If the base metal in the solution has a relatively low electrode potential, it is advisable to separate the anode and cathode compartments in the cell with a porous diaphragm or even by an ion-exchange membrane. In the case of electrowinning with insoluble anodes, this separation also prevents the direct transport of the hydrogen ions produced at the anode,

$$H_2O - 2e^- = 2H^+ + \tfrac{1}{2}O_2, \tag{2.23}$$

across the cell to the cathode. In the case of electrorefining, separation of the anode and cathode compartments allows the circulation of the anolyte out of the cell for purification before it is fed to the cathode compartment.

When it is warranted by extreme purity requirements, the electrorefining step may be repeated. For this purpose, those types of electrolytes (e.g. chlorides, nitrates) are preferred whose basic components – causing trace occlusions in the cathode – are efficiently removed by the following melting step, carried out under vacuum or in a proper gas atmosphere.

2.1.4 Purification Methods Based on Solid–Liquid Equilibria

The relative positions of liquidus and solidus lines in the equilibrium phase diagrams indicate that the solubility of impurities may differ considerably in the molten and the solid metal phases. The portion of impurities rejected by the solidifying base metal is pushed along by a solidification front. If the direction and rate of solidification is controlled by proper temperature gradients, these kinds of impurities can be concentrated in a moving molten zone, which carries them to one end of the metal rod. Impurities whose solubility is higher in the solid phase are partially rejected by the molten zone; therefore, they are concentrated in an opposite sense. The equilibrium partition (or segregation) coefficient is the ratio of the equilibrium concentrations of the impurity in the solid and in the liquid phases:

$$k = \frac{c_{i(\mathrm{solid})}}{c_{i(\mathrm{liquid})}} . \tag{2.24}$$

For dilute solutions – as in the case of almost pure metals – the segregation coefficient is calculated as the ratio of the intercepts of the initial solidus and liquidus tangents with any horizontal line drawn from the temperature axis [21, 22].

Due to the slow diffusion transport of the segregated impurities from the boundary layer preceding the solidification front, the effective value of the partition coefficient [23],

$$k_{\mathrm{eff}} = \frac{k_{\mathrm{o}}}{k_{\mathrm{o}} + (1 - k_{\mathrm{o}}) \exp\left(-\nu_s \frac{\delta}{\mathcal{D}}\right)} , \tag{2.25}$$

at finite rates of solidification set by the travel speed of the molten zone, is lower than the equilibrium k_{o} value, where ν_s is the rate of solidification, δ is the thickness of the diffusion boundary layer and $\mathcal{D}$ is the diffusion coefficient. The efficiency of this purification method can be enhanced by applying relatively longer specimens, slower rates of zone travel and narrower molten zones, in which case the technique is considered as zone refining. The relative impurity concentration (c/c_o) along the length of the bar after n passes of the molten zone is given as [23]:

$$\frac{c}{c_o} = \left[1 - (1 - k_{\text{eff}}) \exp\left(-k_{\text{eff}} \frac{x}{L}\right)\right]^n , \tag{2.26}$$

where x is the distance from the head of the rod of total length L. The initial and the actual concentrations of the impurity in a section of the rod are c and c_o, respectively. The effect of the various values of the segregation coefficients on the concentration profile of the impurity as a result of one zone pass at the optimum speed of travel is demonstrated in Fig. 2.6.

As the impurity content of the liquid phase is continuously increasing during the operation, a single pass of the molten zone cannot yield any appreciable degree of purification. High purity can only be achieved by applying sufficiently pre-refined material and multiple (15–25) passes of the zone. Impurities with effective partition coefficients less than unity ($k_{\text{eff}} < 1$) move in the direction of zoning, and are concentrated at the end of the rod, while those with partition coefficients higher than unity ($k_{\text{eff}} > 1$) behave in the opposite manner. However, no segregation can be expected when the value of the partition coefficient does not differ significantly from unity ($k_{\text{eff}} \approx 1$). The purifying effect of zone melting is most often combined with that of

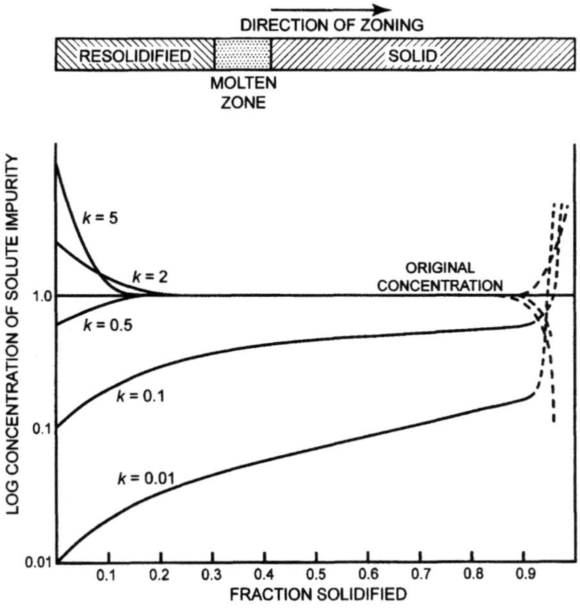

Fig. 2.6. The distribution of impurities of different segregation coefficients along a bar traversed by a molten zone [23]

melting in vacuum or reactive gas. For volatile metals, an inert or reactive gas atmosphere may be required. The metals ified by zone refining are Fe, Mg, Al, Cu, Mo, W, Co, Re, Ti, Zr, Pb, Zn, U, Sn, Bi, Sb, In, Be, Ga, Pt and Si [24].

The purity of the product is highly sensitive to the material of the container – as a source of contaminants – applied in horizontal arrangements. The use of a container is avoided by the floating zone melting of vertically arranged specimens. The molten zone is stabilized by the surface tension of the material and the electromagnetic forces of the induction coil. This technique has the added advantage of efficiently stirring the molten zone, which enhances the transport of impurities and results in higher effective segregation coefficients.

Zone refining is a slow and expensive technique. Therefore, it is usually applied as the final purification step in the preparation of ultra-high-purity metals, following large-scale purification by less expensive and more generally efficient methods.

Fractional crystallization is a purifying method based on the segregation phenomenon, just as zone refining is. However, fractional crystallization is capable of removing higher amounts of impurities and offers significantly higher levels of productivity, although the possible degree of purification is lower. This method is practically combined with continuous casting, where the molten metal is poured into a graphite mould, whose top section is heated while the bottom is cooled. The solidified material, forming the cast billet, leaves the mould at the bottom. During solidification, impurities with low partition coefficients ($k_{\text{eff}} < 1$) are essentially left behind in the molten phase. In order to reduce the thickness of the diffusion layer at the solidification front, the liquid pool of metal is intensively stirred. Continuous production of high-purity crystals is facilitated by a constant bleed of the molten metal, collecting the segregated impurities, from the top of the mould. The achieved purity depends on the rate of solidification and the speed of casting.

2.1.5 Purification by Selective Volatilization

2.1.5.1 Distillation of Metals. Sufficient differences in vapor pressure may allow purification techniques involving the preferential volatilization of either the base metal or the impurities. The former technique applies to relatively volatile metals, characterized by high vapor pressures at convenient temperatures, whereas the latter option may serve the purification of refractory metals. The vapor pressure of some components may reach a significant value even below the melting point, in which case sublimation can be carried out in a simple way. Although distillation requires a relatively sophisticated operation, it offers higher efficiencies.

The vapor pressure of a quasi-pure metal ($p^*_M e$) is determined by the equilibrium constant of the evaporation reaction:

$$\text{Me(condensed)} = \text{Me(vaporized)}\,, \tag{2.27}$$

$$\gamma_{\mathrm{Me}} \frac{p^*_{\mathrm{Me}}}{p^\circ} = \exp\left(-\frac{\Delta G^\circ}{\mathbf{R}T}\right) , \tag{2.28}$$

where p° is the standard pressure and γ_{Me} is the fugacity coefficient (which is close to unity at low vapor pressures). The vapor pressure of a substance is an exponential function of temperature. The numerical forms of these functions are readily available in compilations of thermochemical data [25, 26]. The vapor pressure of the impurity (M) also depends on its activity (a_{M}) in the condensed phase:

$$p_{\mathrm{M}} = a_{\mathrm{M}} p^*_{\mathrm{M}} \approx x_{\mathrm{M}} p^*_{\mathrm{M}} , \tag{2.29}$$

where x_{M} is the mole fraction of the impurity and p^*_{M} is the vapor pressure in the pure state.

Evaporation – which is a strongly endothermic process – usually requires heating. The process of evaporation is mostly restricted to the surface in contact with the gas atmosphere, since high metallostatic pressures, surface tension and heat of evaporation tend to prevent bubble formation in liquid metal baths. Therefore, the specific surface area of the liquid should be increased to a technical maximum. Due to the mechanism of surface evaporation, the concentration of the volatile component decreases at a thin boundary layer. This effect constitutes the driving force of diffusion. Distillation of metals consists of four consecutive process steps [27]:

- transport of volatile components (and heat) to the surface of the metal,
- evaporation,
- transport in the gas phase,
- condensation.

The rate-determining step is generally evaporation at low pressures ($< 10\,\mathrm{Pa}$) and moderate temperatures, whereas diffusion through the boundary layer in the melt is expected to be dominant at high temperatures. The transport process through the boundary layer in the gas phase may influence the rate of distillation only at high vapor pressures ($> 100\,\mathrm{Pa}$) [27].

If diffusion within the liquid phase or in the gas determines the rate of distillation, the major governing parameters are indicated by the following relationships, respectively:

$$\frac{\mathrm{d}n}{\mathrm{d}t} = -A\mathcal{D}\frac{\Delta c}{\delta_{\mathrm{D}}} , \tag{2.30}$$

$$\frac{\mathrm{d}n}{\mathrm{d}t} = -A\frac{\mathcal{D}P}{\mathbf{R}Tl} \ln\left[\frac{P - p_1}{P - p_2}\right] , \tag{2.31}$$

where n (moles) is the amount of the metal evaporated, t is the time in seconds, A (m^2) stands for the surface area of evaporation, $\mathcal{D}$ ($\mathrm{m}^2\mathrm{s}^{-1}$) is the diffusion coefficient, Δc (mole m^{-3}) indicates the concentration difference of the metal to be evaporated between the bulk and the surface of the melt, δ_D (m) is the thickness of the diffusion layer, and P (Pa) is the total pressure, constituted from the partial pressures of the inert gas and the metal vapor. The vapor pressure p_1 at the outer surface of the gas-phase boundary layer is slightly lower than the equilibrium vapor pressure, and p_2 is the corresponding vapor pressure – only slightly higher than the equilibrium value – at the condensation surface. The condenser–evaporator distance is denoted by l.

When the presence of an inert gas is negligible, the rate of evaporation from an open surface can be expressed as [27]:

$$\frac{\mathrm{d}n}{\mathrm{d}t} = 4.4 \times 10^{-2} A\varepsilon \left(p_{\mathrm{evap}} - p_{\mathrm{cond}}\right) \sqrt{\frac{1}{MT_{\mathrm{evap}}}}, \tag{2.32}$$

where p_{evap} and p_{cond} are the vapor pressures at the temperature of evaporation (T_{evap}) and at the temperatrure of condensation, respectively. The condensation coefficient, ε, expresses the ratio of atoms condensed to those impinging on the surface under equilibrium conditions. It is assumed by (2.32) that at the temperature of the condenser a marked degree of re-evaporation takes place, superimposing an opposite material transport to that of distillation.

At higher pressures of residual gases, evaporated particles may suffer frequent collisions, resulting in a proportional back-scattering in the direction of the surface of the melt. Therefore, the distillation rate is pressure-dependent, and p_{cond} in (2.32) can be replaced by p_1 – defined above – at higher residual gas pressures than a critical value in the 4–40 Pa range, depending on the atomic properties (effective scattering cross-section and mass) [27]. A number of common metals (Mg, Ca, Cr, Sr, Mn, Sb, Te, Cd, Zn, Si, Pd, Fe, V, Ag, Ni) have a vapor pressure of the critical order of magnitude at the melting point, or at slightly higher temperatures. This relatively high vapor pressure of the base metal may particularly suppress – by back-scattering – the evaporation of those impurities which have a lower atomic mass. Oxide films or any superficial contamination may also strongly hinder the process of di llation.

In vacuum, distillation rates of pure metals may be determined by the heat transfer to the surface boundary layer:

$$\frac{\mathrm{d}n}{A\mathrm{d}t} = \frac{\lambda \Delta T}{\Delta H_{\mathrm{evap}} \delta_T} = 4.4 \times 10^{-2} \varepsilon p_{\mathrm{evap}} \sqrt{\frac{1}{MT_{\mathrm{evap}}}}, \tag{2.33}$$

where ΔT is the temperature difference in the boundary layer of thermal conduction at the surface of the melt, ΔH_{evap} is the molar heat of evaporation, δ_T is the thickness of the boundary layer, and λ is the thermal conductivity in the melt.

Material transport to the interface can be enhanced by vigorous agitation of the melt. Besides increasing the evaporation rate, the application of a vacuum displaces the equilibrium in the favorable direction, allowing lower distillation temperatures. Efficient removal and condensation of the metal vapor greatly enhances the rate of evaporation. Beyond this aspect, condensation gains particular importance in the case when the base metal is distilled. For high efficiency of metal recovery from the vapor phase, relatively large drops are to be formed and the volume of the off-gas has to be low. Slower rates of gas cooling, moderate supercooling below the dew point – which should be as close to the boiling point as possible – and very low concentration of foreign seeds reduce the portion of metal entrapped in the gas stream leaving the condenser.

2.1.5.2 Reactive Distillation of Metal Compounds. In order to purify metals efficiently, distillation – in a broader sense – may be carried out in combination with a chemical reaction, forming a volatile and sufficiently unstable compound of the metal. Both the van Arkel–de Boer method, generally referred to as the "Iodide Process", and the Mond–Langer (or carbonyl) process, apply the reaction of the raw metal with a reactive gas (I_2 and CO, respectively). The gaseous compound formed is transported to another location, where the reverse reaction is induced by different thermodynamic conditions. The vaporized compound can be decomposed by significantly changing the temperature, pressure or concentration of the reagent. The van Arkel method uses an electrically heated filament of the metal in the center of a low-pressure (0.1–20 Pa) reaction vessel loaded with a thin layer of the raw metal along the molybdenum-lined walls. Decomposition of the carrier compound yields the metal and recovers the reactive gas:

$$m\mathrm{Me} + n\mathrm{X} \underset{\rightarrow T_1}{\overset{\leftarrow T_2}{=}} \mathrm{Me}_m\mathrm{X}_n \; (T_2 > T_1)\,. \tag{2.34}$$

Purification of the metal is possible on the condition that the impurities do not behave in the same manner as the base metal.

For practical processes, halides and carbonyls have proven suitable. The low thermal stability and restricted reactivity of iodides make them especially suitable. Van Arkel has summarized the criteria and favorable conditions [28] for this process of metal refining:

- the formation of volatile iodides of the metal,
- a higher melting point for the metals than the dissociation temperature of the iodides,
- a manageable temperature required for iodide evaporation,
- readiness of the iodide to decompose at elevated temperatures,
- a low vapor pressure of the metal, even at the decomposition temperature.

Although the most essential requirements are met by a number of metals – Ti, Zr, Hf, V, Nb, Ta, Cr, Cu, Ag, Fe, B, Ge, Th, Y, Ni, U, W and Si – this method has gained practical importance only in the purification of

Ti, Zr, Hf, Th, Cr and V. Even in these cases, the removal rates of some metallic impurities (e.g. Fe, Cu, Ca, S) are not satisfactory, and therefore the process is not perfectly suitable for overall purification. Depending on the raw material, metal grades of > 99.95% are commonly achieved [28].

A similar procedure is the reaction of the raw metal with the higher oxidation state ($n+$) halide vapor of the same metal at an appropriate temperature:

$$\frac{n-m}{m}\mathrm{Me(cond)} + \mathrm{MeX}_n(g) \underset{T_1\rightarrow}{\overset{\leftarrow T_2}{=}} \frac{n}{m}\mathrm{MeX}_m(g)\ (T_1 > T_2, n > m)\,. \qquad (2.35)$$

Due to the change in the total volume of the gas, lower pressures shift the equilibrium of the reaction to the right. The subhalide vapor of MeX_m leaves the reactor to be purified in a rectification column, and it finally enters the second reaction chamber to undergo disproportionation at a reduced temperature and increased pressure. The reverse reaction of (2.35) yields the purified metal and recovers the reactant halide MeX_n. The method has reached the semi-industrial level for aluminum and titanium purification with chloride compounds.

2.1.6 Purification by Electrotransport

Charged particles in a solid piece of metal may migrate under the effects of an electric field and collision with electrons in a direction determined by their effective charge. This phenomenon is applicable to the purification of metals if the velocity of migrating impurities is different or opposite in comparison with that of the base metal [2]. The longitudinal particle flux in a relatively long bar of metal placed along the imaginary x-axis is generally a result of complex driving forces caused by the gradients of concentration, temperature and electric potential.

A sufficient duration of electrotransport brings about concentration gradients resulting in equal magnitudes of diffusion fluxes in opposite directions:

$$-\mathcal{D}\frac{\mathrm{d}c}{\mathrm{d}x} + \frac{\mathcal{D}c\nu\mathbf{F}}{\mathbf{R}T}E = 0\,, \qquad (2.36)$$

where ν is the effective valence and E is the electric field strength. The resulting equilibrium concentration along the longitudinal (x) axis of the ion transport is as follows:

$$c = c_\mathrm{o}\exp\left(\frac{\nu\mathbf{F}E}{\mathbf{R}T}x\right)\,. \qquad (2.37)$$

Migration becomes observable only at very high current densities in the metal (10^3–10^5 A cm^{-2}), which enables only very thin wires to be purified by this method in practice. Due to the slow rate of migration, the operational costs are extraordinarily high. Therefore, the application of electrotransport to metal purification is warranted only in special cases of extremely valuable materials and ultra-high levels of purity. Electrotransport is mainly efficient in removing traces of oxygen, nitrogen, carbon or hydrogen, since the mobility of interstitial impurities is significantly higher than that of substitutional elements. This property makes this method a good complement to zone refining, which is efficient for dissolved impurities.

2.2 Purity Evaluation

The development of purification processes, assessment of the achieved purity and characterization of the product require appropriate analytical results. In most cases, the typical range of ultra-high purity is 6–7 N, implying a total impurity concentration of the order of 0.1–1 parts per million (ppm) in mass. Therefore the sensitivity of the analytical method has to be extremely high (usually of the order of 0.001–0.01 ppm), while it should not distort the results by appreciable contamination.

2.2.1 Direct Indication of Purity

In order to express the purity of the base metal matrix, a systematic analysis of the largest possible number of impurities is required. Atomic Absorption Spectrometry (AAS) and Inductively Coupled Plasma Atomic Emission Spectrometry (ICP AES) or ICP Mass Spectrometry are convenient methods of analyzing metallic impurity elements. The precision of the method is usually increased by appropriate matrix matching in the standard and blank samples, or by preliminary separation of the major components and pre-concentration of the impurities.

Probably the most advantageous direct method of providing a virtually complete analysis of high-purity metals is Spark Solid Source or Glow Discharge Mass Spectrometry (GDMS). The sensitivity is very high, and contamination is excluded by generating the ions directly on the appropriately prepared solid metal samples.

Nonmetallic light element impurities (C, N, O and H) – in gaseous form – may be analyzed by the conventional technique of gas chromatography, applying infrared or thermal conductivity detectors. Oxygen, nitrogen and hydrogen are determined after vacuum or inert gas fusion of the metal, while carbon and sulfur are typically determined using the high-temperature combustion method.

Another efficient method is activation followed by γ-*ray* spectrometry. Activation analysis, based on irradiation with γ-*rays*, neutrons or charged particles, is a sensitive analytic method to be applied to the non metallic elements [29]. Neutrons, which are readily available in nuclear reactors, are often utilized for activation purposes. However, for the analysis of light elements (C, N, O), photon irradiation, which implies the (γ, n) type reaction, may be particularly important. In order to suppress unwanted and more complex nuclear reactions, which produce different radionuclides, the energy of irradiation is of essential importance [30]. In certain cases, the (γ, n) reaction cannot be suitable, such as for analyzing oxygen, because the produced ^{15}O radioisotope has a relatively short half-life (2.03 min), and some coexisting metallic impurities may be activated to a disturbing level [31]. For the radio-activation analysis of oxygen, especially in reactive media, irradiation by charged particles (p or ^{3}He) may be the only way to achieve high precision and sensitivity [31, 32].

In the cases of iron, chromium, nickel and aluminum as the base metal, neutron activation analysis plays an important role in determining the residual impurities that remain after purification [33–36]. In the case of cobalt, however, neutron bombardment would induce strong activity from the matrix (^{59}Co) through an (n, γ) reaction. Therefore Proton Activation Analysis (PAA) – at an energy level lower than 11 MeV – is more suitable for the analysis of trace impurities in ultra-pure cobalt specimens [37].

Charged particle activation analysis (CPAA) is one of the most sensitive methods for analyzing light elements. The effect arising from the activated surface oxide layer can be easily eliminated by subsequent etching. In order to reduce the level of the background activity, and thereby to enhance the sensitivity and accuracy, it may be necessary to separate radionuclides – produced by the irradiation – other than the analyzed species. Applying proper chemical separation after irradiation and etching, the limit of accurate oxygen detection in an iron matrix may be in the ppb range [32]. On the other hand, radionuclides with different half-lives can be precisely distinguished and detected by repeating the *γ-ray* spectrometry several times during the cooling period. This method is capable of multi-element analysis of purified samples below the ppm level [37].

Sensitivity, accuracy and scope requirements make such modern instrumental methods favored for the chemical analysis of ultra-high-purity metal samples. Accurate and precise impurity analysis on a trace level requires sophisticated and – consequently – expensive and time-consuming procedures. Determination of the overall purity level essentially depends on the accuracy, precision and completeness of the analysis, which has to cover virtually all the appreciable components in the matrix. Typical analytical results are given in Chap. 3, referring to the tested purification processes of metals.

2.2.2 Indirect Indication of Purity

A global indication, based on the effects of all the impurities, has a great significance in providing evidence for material quality. There are a number of properties that are sensitive to the presence of even trace amounts of foreign elements in the base metal matrix. Mechanical properties, such as yield stress, hardness or recrystallization temperature after a standardized cold work are relatively straightforward to determine and are useful in comparing a series of samples. The most practical indication of purity is, however, offered by electrical resistivity, measured under specified conditions.

According to Matthiessen's rule, the resistivity of a bulk metal may be expressed as the sum of three principal contributions [38, 39]:

$$\rho_B = \rho_T + \rho_D + \rho_i \,, \tag{2.38}$$

where ρ_T is the temperature-dependent ideal resistivity component arising mainly from the thermal vibration of the lattice (phonon scattering), ρ_D expresses the effect of structural defects (vacancy, dislocation, grain boundary, etc.) disturbing the regularity of the lattice, and ρ_i is the contribution caused by the impurities in solid solution.

The residual resisitivity ratio, as a measure of purity, is usually expressed as the ratio of the resistivity at room temperature to that obtained at a temperature close to 0 K. To examine high-purity materials, a sufficiently low temperature has to be reached. In earlier practice, the more available liquid hydrogen was used to set the low-temperature point (20.4 K), but currently the temperature provided by liquid helium (4.2 K) has become generally accepted to express the residual resistivity ratio [40]:

$$RRR = \frac{\rho_{298}}{\rho_{4.2}} \approx \frac{R_{298}}{R_{4.2}} ts. \tag{2.39}$$

To a good approximation, this ratio can also be expressed using the directly measurable resistance of the sample at the two set temperatures.

In the cases of iron, cobalt and nickel, for example, the temperature-dependent component of resistivity can be approximated as

$$\rho_T = \rho_p(T) + \alpha T^2 \,, \tag{2.40}$$

where $\rho_p(T)$ is the phonon scattering term and α is a constant coefficient in the component expressing magnon scattering or the electron–electron scattering effects. Derived from resistivity measurements on different iron specimens (under an applied longitudinal magnetic field), Fig. 2.7 shows the observed change in resistivity as a function of temperature, in comparison with the estimated contribution of phonon scattering, determined by the Grueneisen formula [41]. It is clearly seen that this effect, which is important primarily at higher temperatures, vanishes below 10 K. The dependence of the resistivity on temperature is mainly determined by the quadratic term of (2.40) in the lowest temperature range [41]. This change is negligible for materials of higher residual resistivity (RRR $\lesssim$ 5000) at temperatures lower than 4.2 K.

In the case of magnetic metals, a magnetoresistivity term has to be allowed for, due to the spontaneous magnetic field in the material. However, this is eliminated by superimposing an external (longitudinal) magnetic field.

A further resistivity contribution may arise from the scattering of electrons by the surface of the metal, ρ_s, which has to be taken into consideration

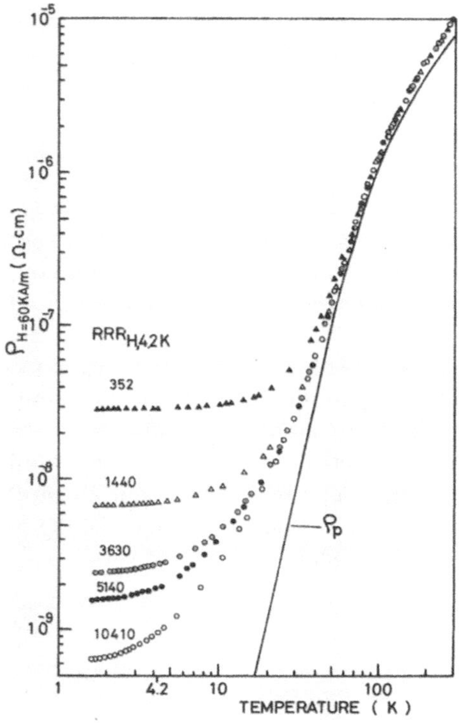

Fig. 2.7. The temperature dependence of the electrical resistivity of different iron specimens under an applied longitudinal magnetic field (60 kA/m) [41]. The specimens of 352 and 10 410 RRR values correspond to purity levels of 3N and 6N, respectively. The solid line indicates the calculated phonon scattering term, $\rho_D(T)$

if the mean free path of the electrons, λ, becomes commensurate with the diameter, d, of the specimen [42, 43]:

$$\rho_s \approx \rho_B C \frac{\lambda}{d} , \tag{2.41}$$

where the parameter C depends on the surface shape and only slightly on the λ/d ratio, and can usually be considered as a constant [43].

Due to the additivity of contributions, expressed in (2.38), (2.41), the resistivity of the sample can be reduced to that caused by the impurities:

ρ_T vanishes at temperatures close to 0 K,
ρ_D can be diminished by preliminary annealing,
ρ_s can be determined by varying the diameters of the samples, although for comparative assessment this factor can be eliminated by keeping the dimensions constant and generally as large as is technically possible.

The levels of purity may be expressed by the changes in the residual resistivity only if all the impurity content is homogeneously dissolved in the base metal matrix. The effects of segregation or precipitation may lead to false assessment of chemical purity. Therefore, the use of residual electric resistivity for such purposes requires appropriate preparation and measurement techniques, assuring the proper state of the impurities and excluding any further reactions, especially internal oxidation.

2.2.3 Comparison of Achieved Purity Levels

It is not always easy to compare the purity of different materials. Direct analytic results are most often incomplete, since not all the impurities are analyzed. For commercial purposes, therefore, the term "nominal purity" is frequently used, meaning that some selected impurities are analyzed, but others are not accounted for. This shortcoming is usually aggravated by the precision and detection limit factors, which depend strongly on the method and the operator.

The application of measured RRR values for the comparison of purity also implies some difficulty. The aim of preliminary annealing is not restricted to the reduction of structural defects, but it also serves to dissolve any precipitated fraction of the impurities. This may often require elevated temperatures and a reducing (usually hydrogen) atmosphere, followed by a high vacuum. Internal oxidation of the sample may cause part of the impurity content to be undetected by this method [43]. Therefore, normalization of the treatment preceding RRR testing is important if this indirect indication is to be used as a measure of material purity.

If the examined metal is of ultra-high purity, as indicated by measured RRR values of the 10^4 order of magnitude, and the diameter of the wire specimen is comparable to the mean free path of the electrons, practically in the $< 1\,\mathrm{mm}$ range, the size effect has to be taken into consideration. If the size effect is appreciable, the usual practice is to convert the directly measured RRR values into ones representing the bulk material, excluding the resistivity contribution due to the scattering of electrons at the surface [44]. At this high level of purity, the total surface area of the grain boundaries may also represent a remarkable contribution to the resistivity, which can only be excluded by growing single crystals. In this case, the crystallographic orientation also becomes important. Furthermore, examination of ferromagnetic metals by this method presupposes the application of a longitudinal magnetic field (of about 800 Oe), which serves the reproducible alignment of the magnetic moments and minimization of the resistivity contribution caused by electron scattering due to the spontaneous magnetic field in the material.

Impurities may have very different coefficients of influence on electrical resistivity. Therefore, exact impurity assessment is directly possible in the case of samples obtained by the same purification procedure, exhibiting the same

Table 2.5. Reported RRR values of high-purity metals [45]

Group	Me	$RRR_{4.2K}$	Purification method	Ref.
Ib	Cu	36 000 (bulk)	6N→ FZM. under red. H_2 atm.	[46]
		40 000 (single cr.)	6N→ FZM in He + internal oxidation	[47]
	Au	>10 000	Vacuum ann.	[48]
	Ag	7 000	RRR 140→ Ann. under red. O_2 atm.	[49]
		10 000 (single cr.)	Internal oxidation	[50]
IIa	Mg	3 200	Sublimation + ZM	[51]
	Ca	148	HP oxide→ Al-red. in vac. + vac. dist.	[52]
	Sr	64	ef	[52]
	Ba	1 100	ef	[52]
IIb	Zn	50 000 (single cr.)		[53]
		25 600	FZM in H_2 atm. (Bridgeman method)	[54]
	Cd	41 600 (single cr.)	Vac. Dist. + ZM	[55]
IIIa	Y	1 200	ET	[56]
	Nd	120 ($RRR_{1.2K}$)	ef	[57]
	La	260	ef	[58]
IIIb	Al	90 000 (bulk)	EW + segregation + ZM	[59]
	Ga	128 000 (bulk)		[60]
IVa	Ti	155	Deoxidataion (Ca hydroflux)	[61]
	Zr	600	EBFZM + ET	[62]
	Hf	15	Decomp. of iodide + arc melting + ET	[63]
IVb	Sn	68 000 (bulk)	ZM	[64]
	Pb	15 900 (bulk)	ef	[64]
Va	V	1 880	ER + arc melting + ET	[65]
	Nb	24 500	ER + EBFZM + UHV ann.	[66]
	Ta	8 000 (single cr.)	EBFZM + UHV ann.	[67]
VIa	Cr	667	FZM + H_2 ann.	[68]
		>700	ET (in Zr foil)	[69]
	Mo	1 000 000 (si. cr.)	SX + Reactive Dist. + EBFZM	[70]
	W	300 000 (single cr.)	EBFZM	[71]
VIII	Fe	20 210 (RRR_{0K})	AIEX + FZM + H_2 ann.	[72]
	Co	334	AIEX + EW + FZM + H_2 ann.	[73]
	Ni	7 000	AIEX + EW + H_2 ann. + EBFZM	[74]
	Pt	5 000		[75]
	Pd	>25 000 (bulk)	FZM in air (internal oxidation)	[76]

Ann., annealing; EB, electron beam; (F)ZM, (floating) zone melting; HP, high purity; UHV, ultra-high vacuum; SX, solvent extraction; Dist, distillation; ET, electrotransport; EW, electrowinning; ER, electrorefining; red., reduced/reduction; si., single; cr., crystal.

– or similar – relative impurity concentrations. In other cases, the use of low-temperature resistivities requires the analytic data of the major components too.

The purity levels achieved in different laboratories and for different metals may be compared by RRR results. Reported values [45] and the corresponding processes are listed in Table 2.5. Purity evaluation by the RRR results is, however, uncertain in some cases, due to unspecified or improper sample preparation. Purity is correctly indicated in those cases when the purified and formed material had been subjected to hydrogen annealing for a sufficient time before RRR testing, and the size effect had also been corrected for. Due to the large differences in the chemical properties of the base metals, the tabulated RRR values show great differences.

References

1. C.V. Kopeczky: Encyclopedia of materials science and engineering, Vol. 3 (Pergamon Press, Oxford 1986) p. 2148
2. Z. Horvath, A. Mihalik, K. Sziklavari: Elmeleti kohaszattan (Tankonyvkiado Publishing, Budapest 1986)
3. A.J. Monhemius: Trans. Inst. Min. Metall. C **86**, C202 (1977)
4. L. Hartinger: Handbook of effluent treatment and recycling for the metal finishing industry (ASM, Finishing Publ. 1994)
5. L.G. Sillen, A.E. Martell: Stability constants of metal–ion complexes, Spec. Publ. No. 17 (The Chemical Society, London, 1964)
6. L.G. Sillen, A.E. Martell: Stability constants of metal–ion complexes, Suppl., Spec. Publ. No. 25 (The Chemical Society, London, 1964)
7. E. Hogfeldt: Stability constants of metal–ion complexes, IUPAC, Chem. Dat. Series No. 21 (Pergamon Press, Oxford New York 1982)
8. D. Sandulescu: Manualul inginerului chimist, Vol. 1 (Editura Tehnica, Bucharest 1972)
9. L. Erdey, L. Mazor: Analitikai kezikonyv (Muszaki Konyvkiado, Budapest, 1974)
10. F. Habashi: Textbook of hydrometallurgy (Metallurgie Extractive Quebec, Enr., Sainte Foy Quebec 1993)
11. R.J. Meyer, E.H.E. Pietsch: Gmelins handbuch der anorganischen chemie (Verlag Chemie, Weinheim 1951)
12. U. Zwicker: Titan und Titanlegierungen (Springer, Berlin Heidelberg New York 1974)
13. K.A. Kraus, F. Nelson: International conference or peaceful uses of atomic energy Geneva Vol. 7 (1956)
14. T. Kekesi, M. Isshiki: Mater. Trans. JIM. **35**, 406 (1994)
15. T. Kekesi, M. Isshiki: Publ. Univ. Miskolc, Ser. B **39**, 155 (1995)
16. T. Kekesi, M. Isshiki: Hydrometallurgy **45**, 345–353 (1997)
17. O. Dimitrov, J. Bigot, M. Fedoroff: In Ultra high purity base metals, UHPM-94 (The Japan Institute of Metals, Kitakyushu, 1994) p. 51
18. T. Erdey-Gruz: Kinetics of electrode processes, (Wiley-Interscience, New York London Sydney Toronto 1972)

19. T. Kekesi: Acta Technica Acad. Sci. Hung. **105**, 153 (1993)
20. J.M. Bockris, A.K.N. Reddy: Modern electrochemistry (Plenum Press, New York 1974)
21. N.L. Parr: Zone refining and allied techniques (George Newnes, London 1960)
22. T. Lyman: Metals handbook (American Society for Metals, Cleveland, Ohio 1948)
23. W.G. Pfann: Zone melting (Wiley, New York 1958)
24. I. Meszaros, I. Molnar: Magyar Aluminium, **19** 46 (1982)
25. O. Kubaschewski, C.B. Alcock: Metallurgical thermochemistry (Pergamon, Oxford New York Toronto Sydney Paris Frankfurt 1979)
26. D.R. Stull, H. Prophet: JANAF thermochemical tables (National Bureau Standards Washington DC 1971)
27. O. Winkler, R. Bakish: Vacuum metallurgy (Elsevier, Amsterdam London New York 1971)
28. A.E. van Arkel: Reine Metalle (Springer, Berlin, 1939)
29. M. Isshiki, Y. Fukuda, K. Igaki: Trans. Japan Inst. Metals **27**, 449 (1986)
30. C. Engelmann: J. Radioanal. Chem. **6**, 399 (1970)
31. C. Engelmann: J. Radioanal. Chem. **6**, 227 (1970)
32. M. Isshiki et al.: Trans. ISIJ **23**, 796 (1983)
33. M. Isshiki, K. Igaki: Trans. Japan Inst. Metals. **18**, 412 (1977)
34. M. Isshiki et al.: Radioisotopes **28**, 349 (1979)
35. K. Igaki, M. Isshiki, K. Yakushiji: Trans. Japan Inst. Metals. **20**, 611 (1979)
36. M. Isshiki et al.: Radioisotopes **30**, 211 (1981)
37. M. Isshiki, Y. Fukuda, K. Igaki: J. Radioanal. Nucl. Chem., Articles, **82**/1, 135 (1984)
38. C. Kittel: Introduction to solid state physics (Wiley, New York 1953) p. 305
39. L.R. Weisberg, R.M. Josephs: Phys. Rev. **124**, 36 (1961)
40. A. Kurosaka et al.: In Advances in cryogenic engineering (materials), Vol. 36, ed. by R.P. Reed, F.R. Ficket (Plenum Press, New York 1990) p. 749
41. M. Isshiki, K. Igaki: Trans. JIM **19**, 431 (1978)
42. H. Mende, G. Thummes: Appl. Phys. **6**, 93 (1975)
43. J. Peterseim, G. Thummes, H. Mende: Z. Metallk. **70**, 266 (1979)
44. K. Mimura, Y. Ishikawa, M. Isshiki, M. Kato: Mater. Trans. JIM **38**, 714 (1997)
45. Y. Ishikawa, M. Isshiki: In Recent developments of bulk crystal growth, ed. by M. Isshiki (Research Signpost, India, 1998) p. 1
46. Y. Ishikawa, K. Mimura, M. Isshiki: Ann. Rep. IAMP, Tohoku Univ. **51**, 10 (1995)
47. Z.S. Basinski, J.S. Dugdale: Phys. Rev. B **32**, 2149 (1985)
48. C. Buchal, R.M. Mueller, M. Kubota, F. Pobell: Physica **108B**, 331 (1981)
49. J.F. Kos: Can. J. Phys. **51**, 1602 (1973)
50. A.C. Ehrlich, J.T. Schriempf: Solid State Communs **14**, 469 (1974)
51. T. Chaudron, G. Revel: Mem. Sci. Rev. Metall. **64**, 561 (1967)
52. J. Evers, G. Ohlinger, A Weiss, C. Probst, M. Schmidt, P. Schramel: J. Less-Common Metals **81**, 15 (1981)
53. J.H. Wernick, E.E. Thomas: Trans. Metal Soc. AIME **218**, 763 (1960)
54. M. Isshiki et al.: Bull. Japan Inst. Metals **48**, 1176 (1984)
55. J.E.A. Anderson, C. Hurd: Phys. Rev. B **12**, 501 (1975)
56. T.V. Nikroforova, V.T. Volkov: J. Less-Common Metals **115**, 91 (1986)

57. P.G. Mattocks, C.M. Muirhead, D.W. Jones: J. Less-Common Metals **153**, 253 (1977)
58. P.H. Pan et al.: Phys. Rev. B **21**, 2809 (1980)
59. E. Hashimoto, Y. Ueda: Mat. Trans. **35**, 262 (1994)
60. D. Novak, S. Meszaros, K. Vad, K. Botos: Z. Metallk. **78**, 478 (1987)
61. M. Nakamura, T.H. Okabe, T. Oishi, K. Ono: Proc. Int. Symp. Molten Salt Chemistry and Technology 529 (1993)
62. R.H. Zee, G.J.C. Carpenter, F.A. Schmidt: Scr. Metall. **18**, 489 (1984)
63. O.N. Carlson, F.A. Schmidt, J.C. Sever: Metall. Trans. **4**, 2407 (1973)
64. B.N. Aleksandrov, I.G. Dyakov: Sov.Phys. JETP. **16**, 603 (1963)
65. O.N. Carlson, F.A. Schmidt, J.C. Sever: Metall. Trans. **3**, 1279 (1972)
66. K. Schulze, M. Krehl: Nucl. Instrum. Methods Phys. Res. A **236**, 609 (1985)
67. R. Lachenmann, H. Schultz: Scr. Metall. **4**, 709 (1970)
68. M. Isshiki, K. Arakawa, K. Igaki: J. Less-Common Metals **96**, 157 (1984)
69. S. Hirano, O. Yoshinari, M. Koiwa: J. Less-Common Metals, **113**, 17 (1985)
70. K.H. Berthel, D. Elefant: Wissenschaftliche Berichte des ZFW Dresden, 1974
71. K.H. Berthel: Phys. Stat. Sol. **5**, 159, 399 (1964)
72. M. Isshiki, K. Igaki: Trans. JIM **19**, 431 (1978)
73. M. Isshiki, Y. Fukuda, K. Igaki: J. Less-Common Metals, **105**, 211 (1985)
74. S.V. Plyushcheva, L.V. Malyarevich, C.V. Kopetskii: Izv. Akad. Nauk. SSSR, Metall. **6**, 27 (1976)
75. J.J. Martin, P.H. Sidles, G.C. Danielson: J. Appl. Phys. **38**, 3075 (1967)
76. N.B. Sandesara, J.J. Vuillemin: Metall. Trans. B **8**, 693 (1977)

3 The Purification of Base Transition Metals

T. Kekesi and M. Isshiki

Possible methods of purification, outlined in Sect. 2.1, are applied in practice to purify the technically important 3d-type (first series) transition metals used in modern industry. This group includes the elements in the d-block of the fourth row of the periodic table, from titanium to zinc. Although zinc is not considered a transition metal by virtue of its chemical characteristics, it is also included in the discussion due to the similar practical significance and purification techniques. These metals are produced in large quantities at a commercial level of purity and are used as pure materials (Fe, Cu, Zn and Ti), or as components in various grades of alloys (V, Cr, Mn, Co, Ni and Zn).

Having discussed the principles underlying the methods of metal purification in the previous chapters, the attention is now focused on:

- the major industrial applications of high-purity grades
- starting materials for purification
- practically proven methods of separation
- achieved levels of purity

for each important transition metal, approaching the subject primarily from the angle of overall purification by chemical methods, supplemented by physical steps of specific effects.

In general, similarities in the properties of the base metal and the impurities usually make the elimination of certain elements especially difficult. Therefore, a combination of different chemical and physical purification methods may have the desired effect of overall purification. It is especially difficult to remove impurities from the most reactive members in the first transition series, such as Ti, V and Cr. The attainable levels of global purity, indicated by the measured RRR values, are relatively low in these cases.

3.1 The Purification of Titanium

The application of ultra-high-purity titanium in VLSI chips requires virtually complete elimination of: (i) alkali metals, which migrate into the gate insulation film and cause deterioration of MOS interface characteristics; (ii) radioactive elements, which cause microprocessor operational errors by α-particle emission; and (iii) heavy metals, especially Fe, Cr and Ni, which impair interface junctions [1]. Titanium is also used as a minor addition to low-activation structural alloys, to improve the metallurgical properties.

Commercial titanium sponge is usually obtained by the (Kroll) magnesium reduction process after removing relatively large residual amounts of the reductant metal and salt impurities. Eventually, no significant impurities are introduced by the reducing metal [2]. Minute amounts of chlorides may remain sequestered in pores, unaffected by the washing treatment. These residues are fully removed by vacuum melting. The iron impurity of the Kroll sponge can hardly be avoided because of the contamination from the reactor wall. The sodium reduction option (Hunter process) has the advantage of separating the reactor wall from the sponge with sodium chloride.

Electrolytic reduction of $TiCl_4$ is capable of producing purer sponge: however, this method has not yet been fully commercialized. As Ti(IV) chloride and oxide are virtually insoluble in molten electrolytes, lower titanium chlorides (mainly $TiCl_2$) [4, 5], are generated in the NaCl or LiCl–KCl eutectic electrolyte from the added $TiCl_4$ material. In order to prevent the reverse reaction with chlorine, the anode is isolated from the bulk of the electrolyte by a diaphragm. Titanium dendrites adhering to the steel cathode are cooled in an inert atmosphere and scraped off.

Electrolytic titanium contains less than 0.2% of impurities, of which Fe ($< 0.02\%$), Cl ($< 0.035\%$), O ($< 0.065\%$) and C ($< 0.011\%$) are significant. The total impurity content of the Kroll sponge may be 3–4 times higher, with Cl ($< 0.12\%$), Fe ($< 0.12\%$), O ($< 0.1\%$), Mg ($< 0.08\%$), Si ($< 0.04\%$) and C ($< 0.02\%$) as the main impurities. The Hunter sponge has lower Fe ($< 0.05\%$) and higher Na ($< 0.19\%$) concentrations [3].

Purification of titanium-containing solutions by standard aqueous chemical methods would imply a number of steps, and only the oxi-hydroxide compound could be obtained. The metal can be dissolved in boiling (sulfuric) acid with added hydrogen peroxide. After separating the cations forming precipitated sulfides at $pH < 0.5$, titanium is co-precipitated with iron, chromium and aluminum as hydroxides at neutral pH. Separation of Al by hot alkalic treatment can be followed by re-dissolution in H_2SO_4, and precipitation of Cr as fluoride. Titanium and iron can be finally precipitated by neutralization. Pure $TiO(OH)_2$ may be obtained after dissolving iron in the oxalate form, but this titanium compound cannot be reduced directly with hydrogen. Reduction of the purified TiO_2 material with hydrogen usually leads to a mixture of lower oxides, unless a pure hydrogen stream is applied at elevated temperatures. To extract the metal, it is necessary to convert titanium primarily into the chloride form by chlorination under reducing conditions at $\sim 1000°C$. Even then, hydrogen reduction can only be carried out at high temperatures (900–1100°C) and under a reduced pressure.

Titanium does not form strongly sorbed anionic complex species in HCl media; thus anion-exchange separation may be applied to eliminate those impurities from the solution which exhibit high anion-exchange distribution coefficients in the practical 0–9M HCl range (Fig. 2.5). Cationic impurities can be separated from the neutral complex species of Ti (which is possibly formed at increasing HCl concentrations) by applying an auxiliary step of cation exchange. The major difficulty in aqueous processing is, however, caused by the very large negative standard electrode potential (−1.75 V) of the reduction to the metallic state, which prevents electrodeposition from the solution. Electrorefining in molten salt systems combined with vacuum melting or the iodide/subchloride process may offer a viable method of fine purification [3].

Titanium sponge has been traditionally treated to reach ultra-high purity by the iodide (van Arkel–de Boer) process, described in Sect. 2.1.5. According to (2.34), iodine is reacted with crude titanium at a temperature either below 200°C or above 500°C. The reaction product is TiI_4 in the low temperature range, and the lower iodides (TiI_2, TiI_3) above 200°C, which can, however, be volatilized only above 500°C. The iodide vapor is decomposed on an electrically heated filament at approximately 1000°C, yielding high-purity titanium. The liberated iodine is recycled to the iodidization.

Thermal disproportionation by the sub-chloride or sub-bromide process, as described in Sect. 2.1.5, also yields high-purity metal. Passing $TiCl_4$ vapor over the crude titanium material at elevated temperatures (1000–1500°C depending on the gas phase composition and pressure) forms $TiCl_2$ vapor, which is decomposed below 1000°C at increased pressures according to (2.35).

Heavy metal impurities can be efficiently eliminated from titanium by electrorefining in a LiCl(NaCl)–KCl molten salt electrolyte at ~ 720°C un-

Table 3.1. Typical analyses of raw and electrorefined titaniums [5]

Element	Concentration (ppm)		Analytical
	Raw Ti	Cathode Ti	method
Na	–	<0.02	FAAS[a]
K	–	<0.02	FAAS
Fe	6–250	<0.1	FAAS
Ni	3–200	<0.1	FAAS
Cr	1	<0.1	FAAS
Sn	2	<0.1	FAAS
Mn	37	<0.1	FAAS
Cu	2	<0.1	FAAS
Al	2	<0.1	FAAS
U	0.0006	<0.00001	ICP-MS[b]
Th	<0.0005	<0.00001	ICP-MS
O	250	<100	IGF-GC[c]
H	–	<10	IGF-GC

[a] Flameless atomic absorption spectrophotometry.
[b] Inductively coupled plasma mass spectrometry.
[c] Inert gas fusion – Gas Chromatography.

der a protective Ar atmosphere [1]. The electrolyte contains both $TiCl_2$ and $TiCl_3$ compounds in solution. The cathodic reactions may imply the Ti^{3+}/Ti^{2+}, Ti^{2+}/Ti and alkali deposition processes at gradually increased cathodic polarization (approximately −1.0, −1.3 and −1.8 V versus Ag/0.1 mole-fraction AgCl electrode, respectively). Higher temperatures and lower current densities promote the growth of larger Ti crystals. The anodic dissolution potentials of Cr, Cu, Fe and Ni, the main heavy metal impurities, are approximately 0.4, 0.5, 0.6 and 1.0 V higher than that of titanium, offering the possibility of practically high refining efficiencies. Chemical analyses of the anode (raw) material and the purified Ti metal are given in Table 3.1. Due to the special difficulties in removing impurities from titanium, the highest reported RRR value is only 155 [6].

3.2 The Purification of Vanadium

High-purity vanadium has recently found application as a basic component in special V–Cr alloys constituting low-activation materials, characterized by a low yield of long-lived activation products [2]. Other uses with high purity requirements are vanadium-containing titanium alloys, nickel-based superalloys for the aerospace industry, and TiAl6V4 alloy for load-bearing components, compressor discs and blades in jet engines. Vanadium is also used as a cladding material for nuclear fuels in sodium-cooled fast breeder nuclear reactors and in the production of superconducting alloys.

Vanadium, like the other elements in the same group, is capable of dissolving large quantities of interstitial impurities, primarily oxygen, nitrogen, hydrogen and carbon, causing significant changes in the mechanical properties. The commonly practiced aluminothermic reduction results in a raw metal that contains as much as 10–15% of aluminum. In order to minimize the uptake of other impurities, the alumina-lined reaction vessel is sometimes substituted by a water-cooled copper crucible. The initial aluminum content of the V–Al material is reduced to 0.1–0.2% by vacuum annealing at 1800°C, and further to approximately 300 ppm by electron-beam melting. If the material is electron-beam melted directly, the aluminum content is typically reduced to $\sim$ 1000 ppm, and by re-melting to 100 ppm [1].

Preparation of a pure precursor compound may practically start from an available water-soluble salt, sodium vanadate ($Na_3VO_4 \cdot 12H_2O$). Vanadium is conveniently precipitated as ammonium meta-vanadate (NH_4VO_3), which is converted into the pentoxide by calcination. Reduction of V_2O_5, however, implies difficulties. On the other hand, anion-exchange purification in chloride solutions, as indicated by the equilibrium results in Fig. 2.5, can eliminate only those impurities that exhibit sufficiently high anion-exchange distribution coefficients ($D \gtrsim 50$–100) in the practical HCl range (0–9M). An additional cation-exchange step may be applied to separate cationic impurities from vanadium, which is converted into the neutral complex species of V(V) (formed at sufficiently high HCl concentrations). After separation, however, there is still the major difficulty of extracting the metal from the aqueous solution, which is virtually prevented by the largely negative standard electrode potential, just as in the case of titanium.

High-temperature vacuum treatment of vanadium results in the preferential evaporation of primarily Al, O, H, Ca, Mg, Mn, Cr and to a lesser extent Fe, among the practically important impurities. However, analytic results have indicated an appreciable removal of other impurity metals too. This may be caused by mechanical effects of evolved vanadium vapor and by the formation of volatile compounds, according to the reaction [4–7]:

$$x\mathrm{Me}(v) + \mathrm{O}(v) = \mathrm{Me}_x\mathrm{O}(g)\,. \tag{3.1}$$

Vanadium is usually melted by the consumable electrode vacuum arc or the electron-beam techniques. The former option requires the auxiliary step of electrode preparation by welding under a protective atmosphere, implying contamination of the metal with interstitial impurities and tungsten. The removal of oxygen (from 0.35% to 0.05% by arc melting, and to 0.014% by electron-beam melting) is efficient, while the concentrations of nitrogen and hydrogen remain unaffected [8]. The main metallic impurities (Si, Ti, Mo, Al and Fe) cannot be eliminated even by repeated vacuum melting. The oxygen content decreases mainly due to volatile oxide (VO) evaporation [9].

Similar-purity ($>$ 99.95%) vanadium can be produced by the van Arkel method [3, 10, 11]. The impure metal is reacted with iodine vapor at 800–900°C, forming volatile VI_2 and VI_3 compounds, which are subsequently decomposed and deposited on a hot filament at approximately 1300°C. This process is carried out in an evacuated and sealed tube, as described in Sects. 2.1.5 and 3.1, implying high expenses and complex equipment.

The purification of vanadium is also possible by converting the pentoxide into the volatile oxy-trichloride, by chlorinating with thionyl chloride [2]. The distillate is collected in a sealed container at a reflux temperature of 127°C. Although there are a number of other elements capable of reacting in a similar way, most of them form compounds of low vapor pressure.

Similarly to the case of titanium, only fused salt systems are viable for vanadium electrodeposition. The potential electrolyte consists of a KCl–LiCl eutectic, containing 2–5% vanadium in the form of VCl_2 at a temperature of 650–675°C [12, 13]. A purity of 99.995% can be attained in the vanadium crystals or flakes deposited at the cathode.

Repeatedly electrorefined vanadium may be further purified by electrotransport based on the principles given in Sect. 2.1.6 [14]. Specimens can be prepared from the double-electrorefined metal by, for example, arc melting in a gettered inert atmosphere and swaging into thin rods, which should be electropolished in a methanol–perchloric acid solution [15]. After lengthy degassing and recrystallizing at a temperature of $\sim$ 850°C and under a pressure of $\sim 10^{-4}$ Pa, further degassing at an elevated (1400°C) temperature for a few hours under 10^{-7} Pa pressure may be useful as a preparatory step.

Table 3.2. Concentrations of the main impurities in fused salt electrorefined and further purified vanadium samples [15]

Purification method	RRR	Impurity concentration (ppm)				
		O	C	N	H	Si
Electrorefining	–	150	60	10	15	< 50
DER	–	40	15	7	< 10	15
DER + EBZM	655	13	20	5	1	–
DER + AM	185	58	23	4	< 1	15
DER+AM + ET	511	11	< 2	1.3	–	15
DER+AM + ET	989	8	< 2	2	–	15
DER+AM + ET	1470	5.1	< 2	< 0.5	–	15

DER, double electrorefining; EBZM, electron-beam zone melting; AM, arc melting; ET, electrotransport.

Subsequently, the vacuum chamber has to be separated from the pump and filled up with purified helium to a pressure of 1–2 kPa. A steady state condition for impurity movement can be attained by applying a current density of 2150 A cm^{-2} at a temperature of 1650°C for 210–220 h. Helium g as suppresses the sublimation of vanadium during the prolonged heating period. As a result of repeated runs, the interstitially dissolved C, O and N atoms migrate to the negative end, leaving the anode section (about one third of the total length) significantly purer (Table 3.2). There is, however, no evidence for the migration of silicon [15]. Typical residual resistivity ratios along the rod, relative to the value (1100–1900) measured in the anodic section, are ~ 50% at the middle section and 2–3% at the cathode section. In order to reduce the flux of interstitial solutes from the cooler ends, previously purified pieces can be butt-welded to the anode end of the specimen.

Similar residual resistivity ratios have been obtained by electron-beam floating-zone melting of a double electrorefined vanadium starting material consolidated into rods by pressing under ultra-high vacuum (10^{-7}–10^{-8} Pa) conditions [16]. The residual resistivity results were obtained by suppressing the superconductivity with a magnetic field of varied strength, and extrapolating the measured values to zero field. Direct electron-beam floating-zone melting of the pressed dendritic rods produced vanadium single crystals with RRR values of 1000–1900, depending on the details of the procedure. Both the interstitial and the substitutional impurity contents are markedly diminished. The removal of Cr, Cu and Fe is efficient (> 90–99%), but the Si concentration is not changed. This method is less time-consuming than the electrotransport procedure, and it can practically combine purification with single crystal growth. Contamination during the pressing and degassing steps should, however, be avoided as much as possible.

3.3 The Purification of Chromium

Chromium finds its most frequent application in ferrous and non ferrous alloys, imparting special mechanical properties, and corrosion and oxidation resistance, to the material. The production of high-performance alloys relies on the availability of high-purity chromium. A special application – in combination with vanadium and titanium – is the production of low-activation structural materials [2]. Other uses of ultra-high-purity chromium, warranting the increased production costs, are in electronics, basically to provide the primary layer on ceramic boards for hybrid circuits [17], but it is also applied in the recently developed magneto-optical (MO) thin films. Pure chromium compounds are used in a variety of applications, including paints, inks and dyes, tanning, corrosion inhibition agents and catalysts.

The starting material of purification is usually electrolytic chromium obtained by the acid chrome-alum process or by aluminothermic reduction [4]. Typical analyses of crude and vacuum annealed chromium grades, primarily obtained by the pyro- and hydrometallurgical routes, are compared in Table 3.3.

The purification of chromium by standard chemical separation techniques in aqueous media is possible, as outlined in Fig. 2.3. The proposed procedure, however, consists of a number of precipitating and filtering steps, implying the probability of contamination, and imperfect separation. As indicated in Fig. 2.5, chromium does not form anionic complexes in the practical (0–9M) HCl concentration range in the divalent and the trivalent states: therefore, those impurities that can form such species may be separated by anion-exchange sorption. However, cationic or neutral complex species can be formed at increasing HCl concentrations, allowing separation of some impurities from the chromium solution in an additional cation-exchange step [18]. Electrolytic extraction of the metal from aqueous solutions is commonly practiced.

Electrolytic chromium is brittle, due to its high oxygen content, which may be efficiently lowered [3] by various methods:

- Annealing at temperatures up to 1500°C in a stream of dry hydrogen gas (to $\sim$ 50 ppm).
- Iodinating by the van Arkel process – described in Sect. 2.1.5 – at 900°C and decomposing the iodide vapor at 1200°C. This technique results in very low concentrations of gaseous impurities (below 10–20 ppm) and eliminates the majority of other impurities.
- Reacting the impurities (oxygen, sulfur, etc.) with calcium vapor at about 1000°C in a titanium-lined vessel, at a pressure of approximately 3 Pa.
- Vacuum melting, implying a simple operation but less efficiency.

Zone melting of chromium is difficult because of the high melting point, reactivity and vapor pressure (1000 Pa at 1890°C). The application of a protective gas of atmospheric pressure is required. The choice of hydrogen allows

Table 3.3. The typical compositions of various grades of crude chromium [3,4]

Grade	Impurity concentration (%)								
	Cr	C	O	Si	S	P	N	Al	Fe
Alth.[a]	99.4	0.05	0.1	0.1	0.01	0.01	0.02	0.1	–
El.[b]	99.1	0.02	0.5	0.01	0.03	0.01	0.05	0.01	0.1
El.+V.[c]	99.5	0.05	0.02	0.05	0.04	0.01	0.002	0.01	0.1

[a]Aluminothermic.[b]Electrolytic.[c]Vacuum treated electrolytic.

Table 3.4. The effect of zone-melting in a H_2 atmosphere on the purity of chromium [19]

Impurity	Concentration (ppm)		
	Starting material	Zone-refined material	
	(van Arkel)	Head end	Tail end
C	40–100	30	30
N	30–75	10	70
Sb	2.3	0.025	0.009
Fe	17.1	9.2	18
Cu	0.09	0.14	0.57
Ni	0.8	0.26	–
Co	0.02	0.02	0.02
V	3.4	–	–
Al	<0.01	–	–
W	<0.01	0.01	0.04

Sc, Mo, Hf, As, Ag, Au, Zn, Cd and Ga less than 0.01 ppm in all samples.

the simultaneous removal of some impurities (e.g. oxygen and sulfur) by reaction. The effect of ten molten zone passes – at a travel speed of $4.5\,\mathrm{cm\,h^{-1}}$ – on the purity of the iodide chromium starting material held in a water-cooled copper container within the quartz furnace tube is demonstrated in Table 3.4. Some impurities (such as tungsten) may be concentrated during the process due to the loss of chromium by distillation. The highest reported RRR values, obtained after floating-zone melting and hydrogen annealing or Zr treatment, are around 700 [20,21].

3.4 The Purification of Manganese

Manganese is primarily used in combination with other metals or nonmetals in alloys and compounds. In order to assure the carefully designed properties of the speciality alloys, the addition of manganese must not imply a considerable source of contamination. Therefore, the use of pure manganese has been increasing for the purpose of alloying. Beside the main fields of application, the electronic components industry has recently shown a growing demand for high-purity manganese; for example, for the production of Mn–Zn soft ferrites, high-magnetoresistance materials and magnetic semiconductors.

Relatively pure manganese metal is commercially produced by the hydro-electrometallurgical route. Electrolytic manganese usually has a nominal purity of 99.4–99.8%, and contains 0.2% O, 0.005% N and 0.01–0.02% S. The concentrations of metallic impurities (Fe, Cu, Co and Ni) are usually each below 10 ppm. The purity of the deposited manganese and the economy of the process depend on scrupulous solution purification and careful control of the operation. The elimination of metallic impurities may be enhanced by a repeated (refining) electrolysis.

The reactivity and negative electrode potential of manganese renders its refining by selective reaction and cathodic deposition rather difficult. It can be efficiently purified neither by zone melting nor by the van Arkel method. Its vapor pressure is higher than that of the major impurities (therefore, vacuum treatment cannot be useful), but not sufficiently high to allow its distillation under ordinary conditions. Therefore, extensive elimination of the impurities may rely upon the separation techniques in the solution phase. The conventional aqueous chemical separation technique (Fig. 2.3), however, implies a large number of steps and reagents, and – consequently – various sources of potential contamination.

According to Fig. 2.5, manganese is capable of forming anionic complexes in concentrated hydrochloric solutions. Impurity elements that form anionic chloro-complex species at lower Cl^- ion concentrations can be separated by preferential sorption in anion-exchange resins. Impurities that persist as aqua cations even at high Cl^- concentrations may be eliminated either by cation or anion exchange. Equilibrium distribution coefficients of manganese and

Table 3.5. A comparison of the anion-exchange distribution coefficients of different metals in HCl and $MnCl_2$ media [22]

Sorbed	Distribution coefficient, D	
elements	8.5M HCl solution	$MnCl_2$ sol., 8.5M Cl^-
Zn(II)	100	670
Mo(VI)	95	300
W(VI)	40	340
Cu(II)	9	22
Fe(III)	∼ 25 000	740
Zr(IV)	9	74
Co(II)	0.5	0.3
Ni(II)	0	0
Mn(II)	0	2

the main impurity metals in hydrochloric acid and neutral $MnCl_2$ solutions of 8.5M Cl^- concentration are given in Table 3.5. The valuable portion of the ion-exchange column effluent is highly purified, and can be subjected to subsequent metal extraction by evaporation to dryness and reduction with, for example, calcium or carbon. The obtained manganese metal may still contain approximately 1.5 ppm of Cu, 0.5 ppm of Zn, 3 ppm of Mo, 2 ppm of W, 6 ppm of Fe, 0.1 ppm of Co and 0.5 ppm of S [22]. The reduction procedure, however, may cause an appreciable contamination with carbon (approx. 700 ppm). This metal may be further purified by physical methods, such as zone refining or vacuum melting.

3.5 The Purification of Iron

The practical application of pure, commercial high-purity iron has been growing dynamically in recent years, involving massive demands by magnet technology and special powder metallurgy. Ultra-high-purity (semiconductor grade) iron has long been used in basic research, and it is promising also in electronic applications, in the new optoelectronic semiconductor (β-$FeSi_2$) material and as the basic component in hard magnetic films and in the (Fe–Tb–Co–Cr) thin film applied in the recently developed magneto-optical (MO) recording technology. The investigation of the intrinsic and limiting properties of chemically pure iron bears supreme importance in ferrous material development, requiring the production of well-defined samples as a first step. The starting material for purification is the conventionally produced technically pure iron. Higher purity may only be achieved by a sequence of a number of different methods.

Vacuum treatment of the liquid metal is practiced on a massive scale in the steel-making industry. A technical vacuum is effective in enhancing the reactions of degassing, decarburization or deoxidation by continuous removal of the evolved gases (N_2, H_2, CO and H_2O, respectively). The usual levels of nitrogen, hydrogen, carbon and oxygen are 300–400 ppm, 20–40 ppm, 1000–1500 ppm and 6000–7000 ppm, respectively. This treatment, however, has no or little effect on metallic impurities, of which Al, Si, Co, Cr, Mo, Sb, Sn, Ti and As have concentrations in the range 100–1000 ppm, and Cu, Mn, Ni and S are usually in the 1000–5000 ppm range in raw iron.

In order to produce an iron material of significantly higher purity, the method of electrorefining in aqueous media can be applied. The low-carbon soft iron starting material is anodically dissolved in the $FeCl_2$ or $FeSO_4$ solution – which contains supporting electrolytes and organic additives – and the Fe^{2+} ions are plated on the cathode. The brittle and coherent or powdery iron deposit can be separated mechanically from the starting cathode sheet. The resulting iron particles – after washing and drying – are subjected to vacuum melting or hydrogen + vacuum annealing. Hydrogen treatment serves the removal of absorbed oxygen. Electrolysis significantly reduces the concentrations of impurities, especially in the cases of Cu, Mn, Sn, C, O and S, leading to a purity level of 99.0–99.4% Fe. Higher purity can be achieved by further chemical and physical purification steps.

The method of reactive distillation can be implemented by reacting fine iron powder with carbon monoxide. The exothermic reaction of volatile carbonyl formation,

$$Fe + 5CO \Leftrightarrow Fe(CO)_5 , \tag{3.2}$$

can be reversed in a separate reactor by reducing the pressure and increasing the temperature. The process of carbonyl synthesis is significantly faster at higher temperatures, and – correspondingly – at high pressures (180–200°C, 15–25 MPa). Carbonyl vapor is removed from the water-cooled autoclave and decomposed in a special tower at standard pressure. A number of other metals are also capable of forming carbonyls, the most important examples being $Ni(CO)_4$ and $[Co(CO)_4]_2$. Differences in vapor pressures, however, allow their separation by fractional distillation.

Hydrometallurgical methods are potentially efficient in eliminating large amounts of various impurities. The conventional technique is based on precipitation processes. It has been shown in Fig. 2.3 that adjusting the pH value and introducing sulfide ions under controlled conditions results in the precipitation of dissolved metals as hydroxides or sulfides. According to (2.5), metal impurities of sufficiently high electrode potential may also be precipitated by contact reduction (cementation) with pure iron powder. The technique of overall solution purification is, however, made more straightforward and more efficient by the application of ion exchange.

It has been demonstrated [23, 24] that anion exchange is capable of eliminating impurity elements (e.g. Co, Ni, Cr, Mn and Cu) which cannot be removed from iron by physical methods, such as zone melting. Passing a reduced Fe(II) chloride solution of an appropriate HCl concentration through an anion-exchange resin bed results in the sorption of impurities that form exchangeable anionic species (Fig. 2.5). The iron-containing effluent can be oxidized and fed into a second column, where iron – in the Fe(III) form – is separated from the remaining impurities [25]. Iron is readily obtained from the purified solution by evaporating to dryness, reducing the $FeCl_3$ salt to the metallic state with hydrogen and melting in vacuum [24]. A combination of anion-exchange purification steps in two different valence states of iron can result in $RRR_{4.2\,K}$ values of approximately 5000 in the metal specimen [24, 26, 27].

The purifying effect of melting techniques in general is strongly limited in the case of iron, because of the high solvent power of the liquid metal [28]. Horizontal zone melting in a container involves a high risk of contamination. Water-cooled copper or silver vessels offer pseudo-containerless conditions, and this has been further improved by the introduction of the vertical floating zone technique, in which the molten zone of the iron specimen does not

Table 3.6. Equilibrium partition coefficients of impurities in iron [28]

Experimental data		Phase diagram data		Approximate values	
Impurity	k_o	Impurity	k_o	Impurity	k_o
O	0.033	B	0.05	Pb, Bi, Ag	< 0.1
S	0.05	Ti	0.27	As	0.1–0.2
P	0.17	N	0.28	Sb, Nb	0.3–0.4
C	0.2	H	0.32	Ge	0.4–0.5
Ta	0.43	Pd	0.55	Au, Zr	0.5–0.6
Ni	0.74	Rh	0.78	Be, Si, Sn, Cu	0.6–0.7
Mn	0.86			Mo	0.7–0.8
				V, Co	0.8–0.9
				Al,Cr,W,Ru,Pt	0.9–1.0
				Re	> 1.0

come into contact with any supporting material. Table 3.6 lists the equilibrium partition coefficients for a number of impurities in iron [28]. Unfortunately, carbon exhibits distribution coefficients close to unity if the residual concentration is below 10–15 ppm [29]. The removal of selectively oxidizable impurities (Ca, Mg, Al, Si, Zr, Ti and C) can be enhanced by applying a controlled oxygen concentration in the gas phase. The oxide phase is separated to the surface of the bar during the zone-melting procedure [30].

A powerful combination of vacuum melting and zone refining is achieved by the ultra-high-vacuum electron-beam floating-zone melting technique (UHV-EBFZM). Although ultra-high vacuum may result in iron losses, experimental proof [27] has shown that it is a powerful technique for purifying iron, although the efficiency may strongly depend on the composition of the starting material.

A comparison of the purifying effects of different procedures is shown in Table 3.7, for a few of the most important impurities in iron. The tabulated data demonstrate that Ni, Mo, Cu and Co are eliminated successfully by the anion-exchange route, yielding residual concentrations that are much lower than in the cases of electrorefining and zone melting [23].

For trace analysis, iron can be activated for gamma-ray spectroscopy by neutron irradiation. The application of thermal neutrons helps to avoid interference by Mn radionuclides produced in the $^{56}Fe(n,p)^{56}Mn$ reaction [31]. The effect of purification is directly evidenced by a significantly reduced background level. Comparative results of testing the anion-exchange and the melting methods are shown in Table 3.8.

The highest reported RRR value referring to conditions after H_2 annealing (indicating the total impurity level in the bulk material) is 20 210 [32].

Table 3.7. The efficiencies of iron purification by different common methods [23]

Impurity	Concentration (ppm)			
	Starting material	Electrorefining	Electrorefining + zone melting[a]	Anion exchange + electrowinning
Ni	160	75	50–190	1.5
Mo	19	2	6.3–14.5	0.7
Co	0.22	0.1	0.24–0.8	0.2
Cu	7.7	1.5	0.7–3.4	0.1
Mn	5.3	1	0.1–0.3	0.05
C	100	40	25–66	25

[a]The first value corresponds to the concentration in the head, and the second to that in the tail end of the bar.

Table 3.8. Neutron activation analytic results for different iron specimens [31]

Preparation	RRR	Concentration (ppm)						
		Ga	In	As	W	Na	Mn	Cu
Electrolytic (original)	80	59	50	5.2	1	0.65	1.7	28
AIEX Fe(II–III)		7.2	<0.02	<0.1	<0.08	0.2	0.05	<0.4
AIEX Fe(II–III) + ZM	6700	<0.26	<0.01	<0.6	<0.1	<0.1	0.1	<0.1

AIEX Fe(II–III), two-stage anion exchange in Fe(II) and Fe(III) form + evaporation + H_2 reduction; ZM, zone melting in vacuum.

3.6 The Purification of Cobalt

Cobalt is primarily used in materials for advanced applications, such as superalloys for aircraft turbine vanes and blades, alloys for powerful, high-coercive-force permanent magnets (Alnico and rare earth–cobalt types), hard metal alloys for cutting tool materials (Co–Cr–W–C and Co–Cr–W–Mo–Ni–Fe–C) and electrodeposited Ni–Co alloys for protective hard coatings. Cobalt is also used as a basic component in speciality materials, including alloys for dental and surgical implants or bone fracture fixations (Vitallium Co–Cr–Mo–Si–C, Co-Cr–Ni–Ti–Mo), low thermal expansion alloys (Fe–Ni–Co, Co–Fe–Cr), fine spring alloys (Co–Cr–Ni–Fe–Mo–Mn–C–Be) and, as a recently expanded application, magnetic recording thin film materials (Co–Ni, Co–Fe–Mg–C, Co–Mo–Zr–Nb, γ-Fe_2O_3–Co), and as a constituent in the (Fe–Tb–Co–Cr) thin film used for magneto-optical (MO) recording technology. A large portion of cobalt production is converted into different chemicals that are used for glasses, ceramics, dryers, paints, varnishes, catalysts, electroplating materials, electronics and solid state devices. Black cobalt oxide, Co_3O_4, has recently gained application as a selective coating material for high-temperature solar collectors. High-purity cobalt is also promising as a basic component in gate electrode materials (cobalt silicide) in ULSI technology. Most of these strategic uses require materials based on the production of pure cobalt.

The most important impurities in cobalt are Ni, Fe, Cu, Zn and Pb. Iron and nickel are not removed efficiently by zone melting [33]. Standard chemical separation methods, eliminating impurity metals from solutions – as shown in Fig. 2.4 – are based on the varying solubility and stability of their compounds. Iron can be eliminated after dissolving commercial cathode cobalt in hydrochloric acid, also by a solvent extraction step in the methyl–isobutyl ketone–0.8M HCl system. According to Fig. 2.5, an overall solution purification in acid chloride media is possible by the application of anion exchange resins. Anion-exchange chromatography is capable of eliminating almost all of the significant impurities during a 9M HCl rinsing step. Cobalt is subsequently eluted by applying 4–5M HCl as the eluent solution. However, as indicated by Fig. 2.5, satisfactory separation of copper may require a preliminary anion-exchange step retaining copper either in the divalent or in the monovalent state [33]. Controlled potential electrowinning with gradually increased current densities, as discussed in Sect. 2.1.3, can further enhance

Table 3.9. Proton activation analyses of cobalt specimens obtained after each step in the anion exchange (AIEX) – electrowinning (EW) – floating-zone melting (FZM) purification procedure [35]

Element	Radioisotope and	Concentration (ppm)			
	γ-ray energy (keV)	Raw	After AIEX	After EW	After FZM
Ti	^{48}V (983.5)	15	1.7	0.2	<0.08
Cr	^{52}Mn (744.2)	22	<0.4	<0.09	<0.09
Fe	^{56}Co (846.8)	21	14	3.8	3.3
Ni	^{61}Cu (656.0)	272	<4	<0.2	<0.2
Cu	^{63}Zn (962.1)	45	0.08	0.01	<0.006
Zn	^{66}Ga (833.7)	32	<2	<1	<0.3
Ga	^{69}Ge (1106.4)	<5	<0.9	<0.04	<0.01
Ge	^{72}As (834.0)	<4	<0.2	<0.4	<0.4
As	^{75}Se (264.7)	<1	<1	<4	<4
Se	^{82}Br (776.5)	<6	<2	<1	<0.7
Zr	^{90}Nb (1129.2)	<1	<0.2	<0.05	<0.03
Nb	^{93}Mo (684.7)	<0.01	<0.005	<0.003	<0.002
Mo	^{95}Tc (765.8)	<3	<1	<0.5	<0.7
Cd	^{111}In (254.4)	<30	<8	<2	<1
Sn	^{122}Sb (564)	<3	<2	<1	<0.6
Sb	^{121}Te (573.1)	<0.2	<0.02	<0.02	<0.02
Te	^{130}I (536.1)	<0.1	<0.09	<0.02	<0.02
Pb	^{206}Bi (803.1)	<200	<40	<80	<90

separation, and produce cathode cobalt of superior purity, characterized by residual resistivity ratios well over 200. However, the concentrations of interstitial impurities may remain in the 1–5 ppm range [34].

The steps in the anion-exchange–electrowinning–floating-zone melting process are characterized in terms of relative importance and efficiency by the corresponding increments in the measured residual resistivity ratios. Starting from an initial purity level of RRR $\sim$ 65 (measured on the crude metal reduced directly with hydrogen from the $CoCl_2$ raw material) it was increased to approximately 200 after a single-step anion exchange (followed by evaporation and hydrogen reduction). Application of electrowinning further increased the purity to a level indicated by an RRR value of $\sim$ 240. Application of the floating-zone melting method to the purified metal could raise the RRR value to $\sim$ 330 [35]. In order to assure the proper state of the impurities, as defined in Sect. 2.2.3, the RRR was measured after annealing the specimens in dry hydrogen at 1123 K for 214 h [36]. Proton activation analysis has shown that dry hydrogen annealing is also effective in removing C, N and O impurities [36]. The corresponding compositions are given in Table 3.9.

3.7 The Purification of Nickel

The traditional markets for nickel (stainless steel and other common ferrous and nonferrous alloys) represent little demand for high-purity metal. There is, however, a growing market for special-purpose materials, whose designed properties rely upon the quality of the raw materials used as components. Nickel is a key element in high-temperature stress- and corrosion-resistant superalloys applied in the aerospace industry, mainly as parts in gas turbine engines. Other special uses are alloys for nuclear reactors, electrodeposited Ni–Co materials for protective hard coatings, biomaterials and low-expansion alloys. The electronics industry accounts for approximately 6–8% of nickel consumption, which is a constantly growing share. A special nickel–iron alloy is widely used in lead frames, while a Cu–Ni–Sn alloy is used in terminals, clips and springs. Pure nickel is also required for the production of hydrogenation catalysts and other chemicals.

The starting material for fine purification is the commercially available electrolytic Ni, obtained by cathodic reduction from sulfate solutions, as discussed in Sect. 2.1.3, or carbonyl Ni, produced by the method described in Sects. 2.1.5 and 3.5. Elements that impose adverse effects in technical applications primarily include Sb, As, Bi, Co, Cu, Fe, Pb, P, S, Sn and Zn. Due to their closely similar chemical and physical properties, cobalt presents great difficulties in its removal from nickel by conventional methods. Fortunately, it only exerts slightly harmful effects on the properties of the base metal.

Conventional aqueous chemical techniques (Fig. 2.4) and standard hydrometallurgical methods are capable of purifying nickel to a significant degree. According to a practically proven method [37], commercial-grade nickel powder is dissolved in sulfuric acid. The dissolution is accelerated by the presence of sparged oxygen. While cobalt and iron go into solution, copper is not dissolved as long as metallic nickel is in contact with the solution. Iron is subsequently removed as ferric hydroxide, after oxidation with hydrogen peroxide or ammonium persulfate, by adjusting the pH value using ammonia. Cobalt ions can be removed by preferential electrolytic deposition on a high-speed rotating cathode, or by precipitating with 1-nitroso-2-naphtol at pH 2. Nickel is extracted as pure powder from the aqueous system with hydrogen gas under pressure at elevated ($\sim 180°C$) temperatures (in autoclaves). The use of catalysts for nucleation (ammonium carbonate) and for densification (nickel powder) is essential. The carbon and oxygen impurities of the produced nickel powder can be removed by wet and dry hydrogen treatment at 450°C.

The purification of electrolytic or carbonyl nickel material by zone melting can take place by selective vaporization and/or segregation. The equilibrium vapor pressures of the most important impurity elements in the pure state at the temperature of molten nickel are in the 0.1–1.0 Pa range for Ni, Si and Fe, in the 20–80 Pa range for Cu and Al, but ~ 0.1 MPa for Ca and > 0.5 MPa for Mg. These data suggest that only Mg and Ca can be efficiently removed from nickel by selective vaporization [38]. Zone melting in an atmosphere of dry hydrogen eliminates carbon, oxygen, arsenic, antimony, zinc, sodium and a number of other minor impurity elements. The removal of carbon can be assisted by preliminary oxidation [38]. Table 3.10 demonstrates, in cases of electrowon and carbonyl nickel starting materials, that zone refining under a hydrogen atmosphere in a refractory container efficiently removes non metallic impurities, while the concentrations of the most important metallic components – copper, iron, and especially cobalt – are virtually unaffected, or may even increase because of contamination from the container wall and the vaporization loss of nickel [39]. Therefore, zone melting of nickel has to be preceded by a different method of purification, one which is capable of eliminating cobalt, copper and iron.

Table 3.10. The effect of zone melting (22 passes, at 3.8 cm h^{-1} under a H_2 atmosphere, in a refractory container) on the concentration ranges of the most important impurities in nickel [38]

Material	Concentration (ppm)						
	Co	Fe	As	Cu	S	O	C
Cathode Ni							
Raw	5300	1000	350	140–220	<10	80	27
Zone-melted	4500–7300	1100–1100	~ 4	165–185	1.5–4.5	0–2	7
Carbonyl Ni							
Raw	6.9	93	0.04	37–60	14–40		52
Zone-melted	5.3–7.8	93–102	0.02–0.07	12	2.5–7.0	<1	9

The purification of nickel in chloride solutions by anion exchange is especially suitable for the purpose of eliminating these important impurities, as suggested by Fig. 2.5. Nickel does not form anionic chloro-complex species: thus it is entrained in the solution passing the ion-exchange column directly, while a large number of impurities (including Co, Cu and Fe) may be retained in the resin bed. High concentrations of chloride ions favor sorption – and thereby elimination – of these impurities. A parallel reduction in the solubility of nickel, however, suggests that an optimum concentration (of approximately 8.6 mole dm^{-3}) should be applied [40].

In order to assess the efficiency of the anion-exchange purification step, Table 3.11 shows neutron activation analytic results [41] of relevant materials. Three commercial nickel grades (99.99% Johnson–Matthey, electrolytic and Mond carbonyl nickels), the commercial-grade nickel oxide and chloride reagents, which were used as the starting materials for the the anion-exchange purification, are compared to the nickel obtained as a result of anion exchange purification of the chloride solution and the evaporation–H_2 reduction process steps. It is obvious that anion-exchange is especially effective in the removal of Fe, Co, Zn, Sb, Hg and Mo. Further purification by floating-zone melting in vacuum efficiently eliminates volatile elements, such as Na, K, Cr, Ag, Sb and Hg, in addition to the purifying effect of segregation. However, the two most important impurity elements are not affected by the vacuum-zone melting step.

Instead of H_2 reduction after evaporation, nickel may be electrodeposited from the purified ion-exchange effluent after separating the excess hydrochloric acid content. A starting concentration of 120 g Ni dm^{-3} may be reduced to a final value of 20 g dm^{-3} in a simple cell containing no diaphragm between the anode and cathode compartments. The required range of acidity is pH 1–3 and the temperature should be higher than 40°C, at a current density of approximately 10 A dm^{-2} to obtain a pure metallic deposit. Enhanced purification by the electrowinning step is possible by applying a preliminary electrolysis at lower current densities, which dominantly deposits elements of a more noble character than nickel. The inert anode is conveniently made of platinum, while the starting cathode is made of a high-purity nickel sheet [42]. The metal purified by the anion-exchange procedure, and electrodeposited directly from the chloride electrolyte, is characterized by significantly higher purity than that listed in Table 3.11, and even higher than that achieved by vacuum-zone melting of the H_2-reduced metal. The concentrations of Na, Cr, Co and Fe do not exceed the 0.05, 0.14, 0.1 and 2.3 ppm levels, respectively. The highest reported purity of nickel, obtained by the anion-exchange–electrolysis–H_2 annealing–EBFZM route, is characterized by an RRR value of 7000 [37].

Table 3.11. The neutron activation analysis of the commercial and purified nickel grades and reagent-grade nickel compounds [41]

Me	Concentration (ppm)					
	J-M (4N)	Electrolytic	Mond Ni	NiO	$NiCl_2$	AIEX Ni
Na	0.31	<0.2	0.11	–	600	7.7
K	<0.6	<21	<0.5	–	28	4.1
Sc	<0.0004	<0.004	<0.0009	0.004	<0.0004	0.0005
Cr	<0.3	<0.2	0.2	6.4	3.7	2.2
Fe	25	<32	130	460	62	3.2
Co	1.2	110	0.69	14	7.2	0.04
Zn	<0.2	72	<0.5	6.9	1.0	<0.06
Ga	<0.02	<0.6	<0.01	<3	<0.04	<0.05
As	<0.007	1	0.021	<5	<0.03	<0.02
Se	0.16	0.75	<0.1	<0.2	<0.03	<0.03
Rb	<0.1	<2	<0.4	<0.6	<0.2	0.04
Mo	0.5	<0.3	0.05	<1	0.5	<0.04
Ag	<0.03	<0.2	<0.1	<0.09	<0.04	0.02
Cd	<0.3	<0.9	<0.3	22	0.4	1.6
Sn	<5	<20	<16	<20	<4	<3
Sb	<0.004	<2	<0.002	130	2.1	0.05
Cs	<0.006	<0.06	<0.02	0.02	<0.007	<0.003
Ba	<2	<17	<5	<8	8.7	2.9
La	<0.001	<0.008	<0.0004	<0.06	<0.01	0.02
Ce	<0.02	<0.06	<0.04	0.1	–	<0.2
Nd	0.52	–	<0.4	0.3	0.23	2.6
Gd	<1	<0.02	<0.4	<200	<1	0.002
Sm	0.002	<0.001	<0.0003	<0.007	0.002	0.004
Eu	0.011	<0.01	<0.003	<0.01	0.01	<0.002
Tb	<0.02	<0.1	–	1.2	<0.01	<0.002
Yb	<0.005	<0.04	<0.002	<0.03	<0.002	0.0003
Lu	<0.005	<0.002	<0.0004	–	0.0002	<0.003
Hf	<0.004	<0.02	<0.01	0.03	<0.004	<0.004
Ta	0.3	<0.1	<0.04	<0.06	<0.02	<0.008
W	1.0	<0.09	0.011	<2	<0.008	0.03
Hg	<0.02	<0.1	0.09	2.4	1.5	<0.002
Th	<0.01	<0.1	<0.01	0.03	0.006	0.11
U	<0.04	<0.2	0.03	7.3	0.6	–

3.8 The Purification of Copper

The electrical and mechanical properties of copper strongly depend on the degree of purity. The variations in properties are linked to the chemical and physical lattice imperfections. High-purity copper is extremely ductile, exhibiting an especially high electrical (and thermal) conductivity. Mechanically detrimental impurities, which decrease the strength at high temperatures, are principally Pb, Bi, Sb, Se, Te and S, which form oxides or other precipitated compounds at the grain boundaries during heating. Appreciable amounts of oxygen make the metal susceptible to hydrogen embrittlement, which is caused by the precipitation of water vapor at grain boundaries, arising from the reaction of dissolved oxygen with incoming hydrogen.

Demands are extremely stringent in cryogenic uses, where the high electrical and thermal conductivity of copper is indispensable for its use as a low-temperature superconductor stabilizer. Recently developed applications in microelectronics [43] include extremely soft bonding wires, which replace the traditionally used gold wires, copper-to-ceramic bonding in high-thermal-conductivity substrates, interconnection of ULSI circuits, or pure copper alloys. Due to the low level of impurities and the immunity to hydrogen embrittlement, ultra-high-purity copper is widely used in high-vacuum technologies, such as materials for gaskets, high-energy vacuum tubes, or cavities in particle accelerators. The homogeneous grain structure of high-purity copper is an advantage in producing laser mirrors [44]. This material is also used for hollow conductors (for high-field magnets, induction coils and generator windings) cooled with water or liquid hydrogen, or for tubing of deep-water optical cables. Special copper grades are also necessary for the production of high-fidelity audiovisual cables [45,46].

Conventional refining, which is widely applied in industrial production technology, comprises the major steps of: (i) fire refining, removing large amounts of oxidizable impurities from the crude melt; (ii) electrolytic refining, eliminating the majority of residual impurities to the anode slime or to the electrolyte solution; and, finally, (iii) cathode melting and casting. The last step can be carried out under strictly controlled reducing conditions, so as to reach a standard OFC (oxygen-free, high-conductivity) grade, where the remaining oxygen content is typically in the 1–5 ppm range, without the presence of deoxidants [44]. The industrial refining procedure is hardly capable of achieving higher levels of purity than 99.99%.

The purification of copper in aqueous solutions by standard chemical precipitation techniques can be considered as a possibility for rejecting the larger part of the bulk impurities. The separation of copper from aqueous solutions by chemical precipitation is relatively simple. After dissolution in sulfuric acid, with the addition of hydrogen peroxide, Ag^+, Hg^+, Pb^{2+} and Tl^+ ions are precipitated (in the chloride form) on adding HCl. Copper can be extracted from the solution by reducing the divalent ions to the monovalent form by the addition of dissolved glucose or formaline, which will result in the precipitation of CuCl. Copper chloride is readily reduced with hydrogen, although the temperature has to be regulated under very close control, due to the low melting point of copper chloride [47].

The efficiency of commercially applied purification may be increased basically by repeating the electrorefining step in a specially purified $CuSO_4$–H_2SO_4 electrolyte (at, for example, 25°C and 55 A m^{-2} current density) under strictly controlled operating conditions and with the application of a diaphragm around the anode [48]. The electrodeposited copper is stripped from the stainless steel plates and melted in the presence of oxygen to remove sulfur arising from the entrapped sulfates of the electrolyte. The impurities that remain and are additionally picked up can be removed by subsequent electrolysis in purified $CuNO_3$ solution (at, for example, 25°C and 160 A m^{-2} current density). The nitrate impurities are eliminated by thermal decomposition during the final melting and casting, under a protective N_2 atmosphere.

Table 3.12. Typical analyses of commercial- and special-grade coppers [48, 49]

Element	Concentration (ppm)			
	Fire refined	Electrolytic	Oxygen-free	Special (5N)
Ag	10–40	10	10	<0.3
As	1–10	<5	3	<2
Bi	5–15	3	1	<0.1
Cd	1–10	1	Not detected	
Co	1–10	1	~ 0	
Fe	5–20	2	5	<0.7
Mn	<50	<0.5	0.5	
Ni	30–80	3	6	<1
O	350	350	1.7	<0.1
P	<10	<10	Not detected	
Pb	10–50	2	6	<1
S	15	10	25	<1
Sb	<5	<2	5	<1
Se	30–70	<10	2	<1
Si	<8	<8	4	<0.1
Sn	5–50	5	2	<1
Te	<10	<10	1	<2
Zn	<10	<10	Not detected	
Total	<550–800	<440	<72	<11.8
%IACS[a]	100	101.5	101.7	103.6

[a] Resistivity ratio of the International Annealed Copper Standard ($1.7241 \times 10^{-6} \Omega$ cm at 20°C) to the sample.

The composition of the corresponding copper product [49] is compared with that of common grades in Table 3.12.

Special copper purification processes that have recently been reported, as shown in Table 3.13, are principally based on electrolytic refining in acidic sulfate and/or nitrate solutions and zone melting. Since nitrogen, unlike sulfur, is practically insoluble in solid copper, possible nitrate inclusions from the final electrolysis step are readily eliminated by melting the cathodes. However, nitrate electrolysis has the disadvantage of poor reproducibility and extreme sensibility to the pH of the solution. The favorable cathodic process may be disturbed by the undesirable deposition of Cu_2O, which takes place at low acid concentrations according to the following reaction:

$$2Cu^{2+} + H_2O + 2e^- = Cu_2O + 2H^+, \tag{3.3}$$

characterized by an electrode potential of $E = 0.20 + 0.06\,\mathrm{pH\,V}$ in 1M Cu solutions. On the other hand, reduction of the nitrate ion can interfere with the cathode process at low pH:

$$\begin{aligned} &NO_3^- + 4H^+ + 3e^- = NO + 2H_2O, \\ &E = 0.95 - 0.08\mathrm{pH\ V}, \quad \text{if}\,\frac{[NO_3^-]}{p_{NO}} = 1. \end{aligned} \tag{3.4}$$

Kinetic inhibition of reaction (3.4), however, makes it possible to deposit copper despite the unfavorable electrode potential [50].

Continuous and close control is a general requisite for any electrorefining operation. The deposition of impurities may be avoided efficiently by a proper adjustment of the cathode potential. Separation of the anode with a diaphragm is useful in preventing the formed anode slime particles from being carried over to the cathode, in which case forced circulation of the electrolyte may also be applied to enhance material transport and to diminish the harmful reduction in the copper concentration in the cathodic boundary layer [51–53]. Constructing the cell with separate anodic and cathodic compartments can reduce the effects of the anodic process on the composition of the catholyte. A portion of the impure anolyte can be drained out continuously for chemical purification (e.g. precipitating Ag on Cu), followed by recirculation to the cathodic compartment. Further improvement of the chemical conditions in the electrolyte is possible by addition of reagents, such as free Cl^- or H_2 to decrease the Ag and S concentrations, but organic additives such as levelling agents should be avoided if ultra-high purity is desired. Further refining steps are generally necessary after electrolysis, including the conventional melting or zone refining under a protective (reactive) atmosphere or in vacuum.

In order to achieve an overall purification of copper in a virtually single-step operation, the method of anion-exchange chromatography in hydrochloric media has been proved efficient [54–56]. Proper selection of HCl concentrations and oxidizing–reducing conditions during the anion-exchange operation

Table 3.13. Conventional methods for the production of ultra-high-purity copper

	Electrolytic copper (3–4N) starting material					4N Cu starting material		5N Cu st. mat.	6N Cu st. mat.
PROCESS	El.R. (SO_4) Melt. (O_2) El.R. (NO_3) Melt (N_2) Cast (N_2)	El.R. (SO_4) Cathode pot. Control Cast. (Vac)	El.R. (SO_4) Rep. Rep. . . . Melt. (Vac) Z.R.	El.R. (SO_4) El.R. (SO_4) Z.R. (La doping)	El.R. (NO_3) Diaphragm Cl^-add.	El.R. (SO_4) Diaphragm Ag precip. Cathode pot. control El.R.(NO_3)	El.R. (SO_4) low c.d. Cl^-add. low el. pot. El.R.(NO_3) red.atm.(H_2) [Melt.(Vac.)]	El.R. (NO_3) [Melt.(V) Roll. Ann.(V)] [Z.R. (floating)] [Ann.(V)]	FZM under H_2 atm., reduced pressure electropolish
PRODUCT	5N – **As, Te** < 2 S, Se, Sb, Ni < 1 Pb, Sn, Fe < 0.5–1.0 Smart et al. (1941)	>5N RRR ∼ 800 **Fe** = 1 **Ag** = 0.45 **S, Si** = 0.2 Al, Ni, Cr, P Na = 0.01– 0.05 Kurosaka et al. (1990)	>5N RRR > 3000 Morita et al. (1984)	6N RRR ∼ 1000 **Ag** < 0.5 Fe, Pb, Te Se, As, Sb Bi, S, La < 0.1 Morikawa et al. (1991)	6N – **Ag** = 0.3 Ni < 0.07 S, Fe, Cr Sb, As, Bi Al, Si <0.05 Kawato et al. (1987)	– RRR > 10 000 Ag < 0.01 S < 0.05 Fe < 0.05 Ni < 0.01 Ogata et al. (1989)	7N – Ag = 0.01 Cd, Sb, Zn <0.02 S, Fe, Ni, Cr Mg, Na, K, Al Si, As, Bi, Pb <0.01 Toya et al. (1988)	RRR = 1000– 22 000 Ag ≈ 0.01 Fe = 0.04–0.4 BA, Ti, Na, In Ir, Pt, Co.. = 0.02–0.4 Bruning et al. (1985)	RRR_B = 36 000 Isshiki et al. (1999)

El.R., electrorefining; Melt., melting; Cast., casting; Rep., repeated; pot., potential; add., addition; red., reducing; atm., atmosphere; Z.R., zone refining; FZM, floating-zone melting; Roll., rolling; Ann., annealing; Vac., vacuum; RRR_B, bulk RRR value.

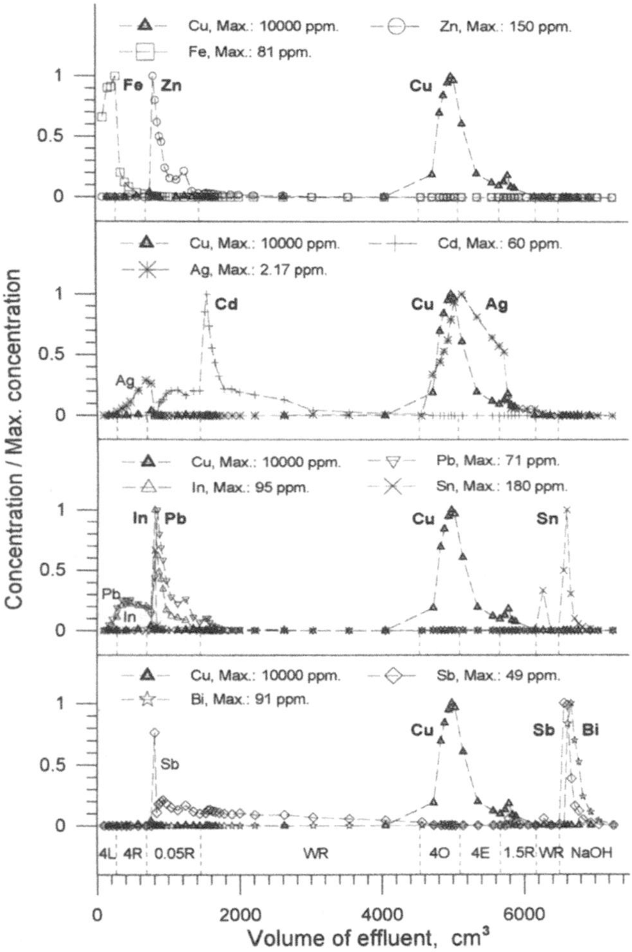

Fig. 3.1. Elution curves of the CuCl purification process [56]. Procedure: loading (4L) – 4M HCl + M_xCl_y, 4M HCl rinse (4R), 0.05M HCl rinse (0.05R); water rinse (WR); oxidation (4O) – 4M HCl + H_2O_2; Copper elution (4E) – 4M HCl, 1.5M HCl rinse (1.5R); water rinse (WR), 2M NaOH rinse (NaOH) (M_xCl_y – added chlorides of the tested elements)

may result in the efficient elimination of practically all of the impurities. Elution curves referring to the characteristic impurities are shown in Fig. 3.1. The operation starts under reducing conditions and the sorbed Cu(I) species are converted into Cu(II) for better elution by in situ oxidation in the column.

The ratio of elimination shows how many times the relative amount of an impurity decreased as a result of the procedure:

$$R_i = \frac{m_{i,\text{loaded}}/m_{m,\text{loaded}}}{m_{i,\text{collected}}/m_{m,\text{collected}}} = \frac{c_{i,o}}{c_{m,o}} \cdot \frac{\int_{V_s}^{V_e} c_m \, \mathrm{d}V}{\int_{V_s}^{V_e} c_i \, \mathrm{d}V} , \tag{3.5}$$

where V_s and V_e are the effluent volumes at the starting and ending points of the purified solution collection, respectively. The amounts of the main and the impurity elements collected in the purified solution are determined from the respective concentrations, c_m and c_i, of the effluent samples. The loaded amount of an element is obtained from the analysis, $c_{m,o}$ and $c_{i,o}$, of the loaded solution. The standard values refer to the effluent volume limited by the rising and declining half-height concentration points of the principal elution peak of the main element.

Optimization of the rinsing and elution steps of the process shown in Fig. 3.1 results in standard elimination ratios of the important and characteristic impurities (Fe, Bi, Sn, Zn, Cd, Sb, etc.) approaching or surpassing the value of 1000 at a Cu yield of approximately 70%. Virtually complete elimination of silver and lead, however, requires a second ion-exchange step. As suggested by Fig. 2.5, passing the primary effluent after dilution to ~ 1.5 M HCl concentration, through a second anion-exchange column removes the remaining portions of silver and lead impurities.

The production of copper from the purified effluent can be achieved by direct electrowinning [57]. The efficiency of purification is demonstrated by Table 3.14, in which the starting materials and the major steps of the process are identified [54]. A comparison of the third and fourth columns, which represent the same 6N grade Cu standard materials analyzed before and after melting, proves that the short melting at relatively low temperatures did not cause further purification. Thus any substantial reduction in the concentrations of the impurities can be attributed to the preceding anion-exchange–electrowinning purification procedure. Silicon, tungsten and thorium contaminants (of the order of 0.1–2.0 ppm), arising from the quartz ion-exchange column, and from the plasma furnace cathode, respectively, were omitted from the table of analytic results, as they can be avoided by the choice of different materials and melting methods [55].

Zone refining offers efficient removal of certain impurities, but the equilibrium partition coefficients are close to unity for a number of significant elements, such as Co, Fe and Zn [58, 59]. In these cases, the purification efficiencies are correspondingly low. Therefore, zone refining of copper may be included in the procedure to remove certain residual impurities from the material which had been subjected to an overall primary purification. During zone melting, the high thermal conductivity of copper is disadvantageous, because of diminishing temperature gradients in the copper bar, thereby ac-

Table 3.14. Typical concentrations of impurities in the copper samples (elements detected at least in one of the samples) [54]

Starting mat.	CuCl	6N Cu	6N Cu	CuCl
Purification	–	–	–	Anion exchange
Reduction	H_2-red. after evaporation	–	–	Electrowinning
Melting	H–Ar plasma	–	H–Ar plasma	H–Ar plasma
Element		Concentration (ppm)		
Na	0.07	0.04	0.08	<0.0045
Mg	<0.002–0.007	0.0009	0.0008	<0.0057
P	0.061–0.066	<0.003	0.45	0.06
S	0.22–0.29	0.01	0.006	<0.1
Cl	0.018–0.041	0.013	0.041	<0.024
K	0.023–0.024	0.0087	0.025	<0.007
Ca	0.043–0.058	<0.02	0.0059	<0.21
Ti	0.0026–0.0071	0.027	0.019	0.0047
V	0.003–0.026	<0.0005	0.00008	<0.004
Cr	0.062–0.097	<0.002	0.0006	<0.0054
Fe	0.57–0.73	0.0061	0.04	0.057
Mn	0.0035–0.0040	<0.001	0.00043	<0.0045
Ni	<0.002–0.054	<0.002	0.0018	0.0081
Co	0.0012–0.0013	<0.001	<0.00074	<0.004
Zn	<0.08–<0.09	<0.2	0.16	<0.064
As	0.0050–0.0059	0.0064	0.0032	0.0088
Se	0.17–0.41	<0.04	<0.03	0.0295
Br	<0.02–0.028	<0.04	<0.02	<0.017
Zr	0.0011–0.0012	<0.0017	0.011	<0.009
Nb	0.00072–0.0069	0.00083	0.0008	0.007
Mo	0.0048–0.0068	0.00052	<0.004	<0.018
Ag	3.50–3.63	0.113	0.013	0.025
Sn	0.015–0.023	0.024	<0.02	<0.053
Sb	0.0063–0.014	<0.008	0.0039	0.006
Te	0.12–0.15	<0.06	<0.05	<0.064
Pb	0.0071–0.011	0.0055	0.0035	<0.019

celerating unwanted solid state diffusion, and causing considerable longitudinal heat losses (which necessitate continuous variation of the heating power, depending on the position of the molten zone). The high density and low sur-

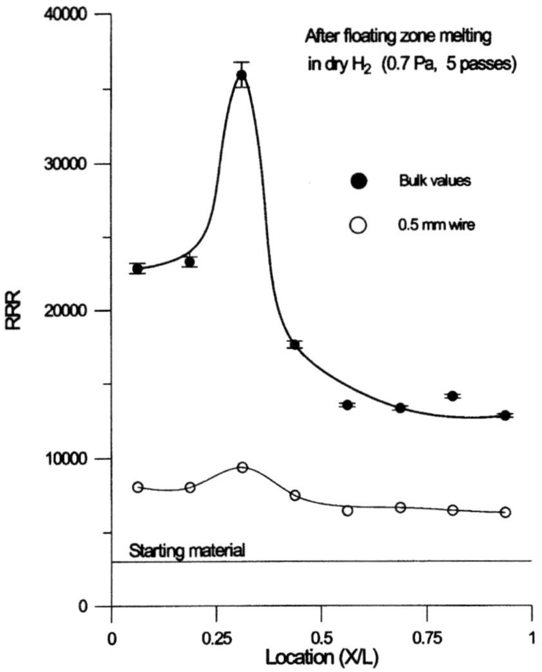

Fig. 3.2. The residual resistivity ratio profile of floating-zone-refined copper [61]. Specimens were cold worked and annealed in dry hydrogen at 650°C for 24 h. Solid marks, bulk value; open marks, measured values for 0.5 mm wire

face tension makes the zone of molten copper particularly unstable, which is disadvantageous for the application of the floating-zone technique. On the other hand, horizontal zone refining implies the danger of contamination caused by the crucible. Floating-zone melting of copper requires stabilization of the molten zone by the application of a specially designed conical induction coil [60].

Figure 3.2 shows the RRR profile of a copper specimen of 6N initial purity, floating-zone refined in a reduced pressure (0.7 Pa) hydrogen atmosphere [61]. The RRR values are plotted against the normalized distance (X/L) from the starting point along the length (L) of the specimen. Open marks indicate the measured values, while solid marks show the bulk values ($\mathrm{RRR_B}$) estimated by the size effect correction [62], according to the following formula:

$$\mathrm{RRR_B^{-1}} = \mathrm{RRR_W^{-1}} - (3.8 \times 10^{-8} \pm 4 \times 10^{-10})d^{-1}\,, \tag{3.6}$$

where $\mathrm{RRR_W}$ is the residual resistivity ratio measured on a wire specimen of diameter d (m). The error bars represent the error of the empirical coefficient in (3.6). The resistivity measurements were carried out after cold working (to form the 0.5 mm wire), followed by annealing in hydrogen at 650°C for 24 h.

Most of the metallic impurities have equilibrium partition coefficients lower than unity in copper. The only important exception is nickel, while the coefficients of cobalt and iron are close to unity. This explains the obtained RRR_B profile, which shows lower RRR values (related to the segregation of the majority of impurities) at the end portion. It is also notable that the RRR_B values measured at the end portion of the sample are still higher than the RRR_B of the starting material (indicated by an arrow). This is proof that the special melting technique under a reduced hydrogen pressure and dynamic gas flow is effective in removing the relatively high vapor pressure impurities. The highest RRR_B value, of 36 000, was obtained at around the $X/L = 1/3$ position along the specimen. This is the highest value as reported on polycrystalline copper which had been annealed under a strongly reducing (hydrogen) atmosphere, where the possibility of internal oxidation was entirely excluded. It has also been found that ultra-high-vacuum annealing fails to meet this criterion [61].

It has been demonstrated that dissolved hydrogen has only a negligible effect on the residual resistivity of the copper samples [61], whereas internal oxidation may be significant even in the case of applying ultra-high-vacuum. Therefore, annealing the samples in a hydrogen atmosphere at 650°C for 24 h is suitable as a pretreatment for RRR measurement, to indicate the material purity. This procedure is recommended as the standard heat treatment for purity evaluation purposes.

3.9 The Purification of Zinc

The largest demand for high-purity zinc (of the 99.99–99.995% grade) is generated by the production of pressure casting alloys. However, the preparation of ultra-high-purity zinc has been primarily promoted by recent developments in the semiconductor industry. One of the expanding applications is the production of zinc selenide single crystals, the direct transition-type wide-gap semiconducting compound, for blue light emitting diodes [63].

Pyrometallurgical methods can produce crude zinc that contains large amounts of impurities, the most important of which are Pb (0.7–3%), Fe (0.05–0.2%), Cd (0.3%) and also As in smaller amounts. The crude metal is commercially purified in high temperature distillation columns constructed of ceramic bonded SiC elements [4], making use of the relatively low boiling point (910°C) of zinc. Most of the commercially pure metal is currently produced by the hydrometallurgical–electrowinning route, applying sulfate solutions.

Standard aqueous precipitation techniques may be combined to purify the crude zinc starting material (Fig. 2.3), although the process is complicated, implying the danger of contamination. The conventional method of producing ultra-high-purity zinc consists of repeated vacuum distillation (10^{-3} Pa) of the 4–5N grade starting material, followed by zone refining in a pure argon

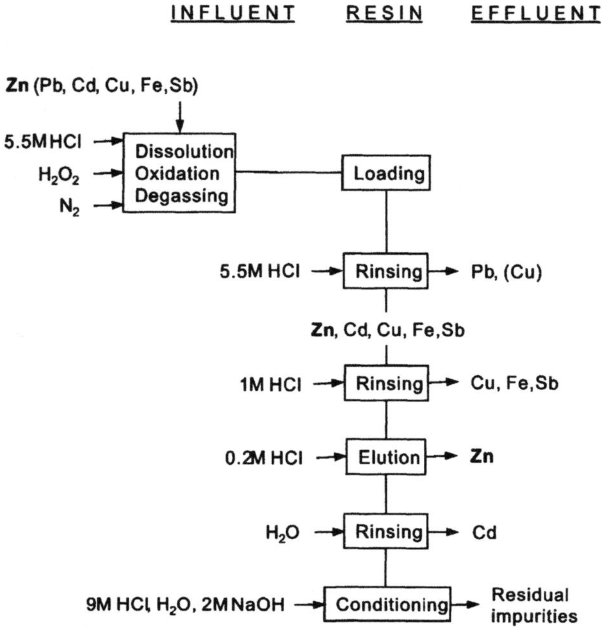

Fig. 3.3. A flow-sheet of the anion-exchange separation procedure for zinc purification in HCl solutions [65]

atmosphere [64]. Unfortunately, this procedure requires rather a sophisticated technique and offers only low productivity: furthermore, the final purity is strongly dependent on the quality of the starting material.

Similarly to the case of copper, anion exchange in hydrochloric acid media can be adapted to meet the principle objectives of zinc separation. On the basis of experimentally determined equilibrium distribution coefficients, as functions of Cl^- ion concentration (Fig. 2.5), a proper scheme of separation can be designed (Fig. 3.3) [65]. Practical elimination of the main impurity elements (Cd, Cu, Fe, Pb and Sb) from the $ZnCl_2$–HCl starting solution by anion exchange is demonstrated by Fig. 3.4. Contrarily to the copper purification procedure, in this case the starting solution had to be oxidized. In order to provide the optimum elution conditions, the hydrochloric acid concentration of the eluent has to be adjusted not only to allow desorption (Fig. 2.5) but also to avoid hydrolytic precipitation (Sect. 2.1.2.1). Optimization of the anion-exchange separat ion procedure requires repeated elution tests on the laboratory scale [56].

The purified solution of zinc chloride can be treated by conventional methods of metal extraction to produce an ultra-high-purity metal product. Two straightforward alternatives are direct electrowinning or complete evaporation to dryness followed by hydrogen reduction. The anhydrous $ZnCl_2$ salt, obtained in the latter case, is suitable for use as a high-purity dopant [66]

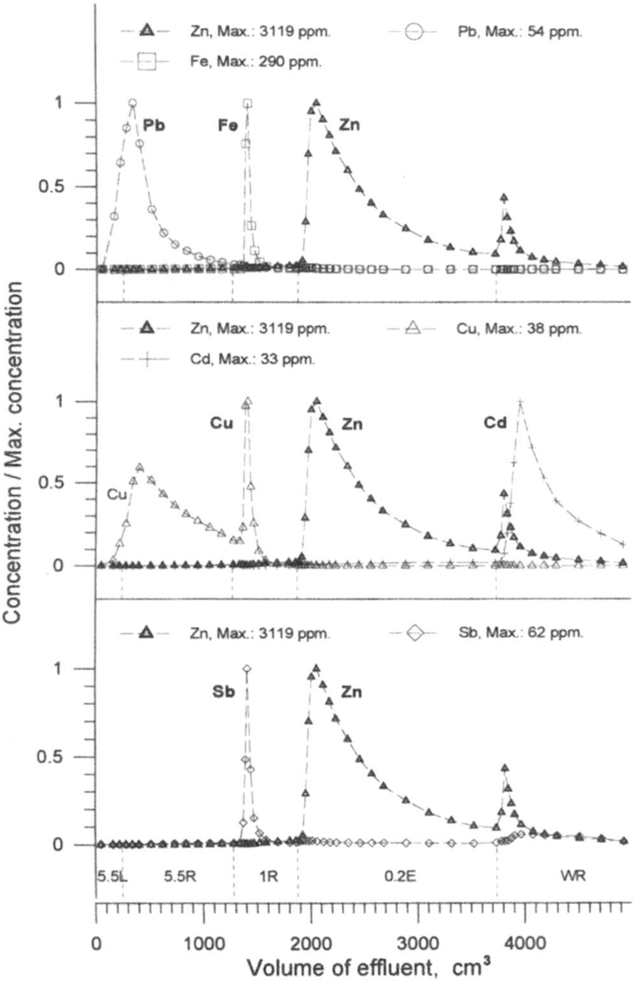

Fig. 3.4. Elution curves of the $ZnCl_2$ purification process [65]. Procedure: loading (5.5L) – 5.5M HCl + M_xCl_y + H_2O_2, 5.5M HCl rinse (5.5R), 1M HCl rinse (1R); zinc elution (0.2E) – 0.2M HCl; water rinse (WR). (M_xCl_y – added chlorides of the tested elements)

to incorporate controlled amounts of chlorine in the ZnSe single crystals. If sources of contamination can be excluded throughout the entire procedure, the final purity attainable is determined by the anion-exchange separation step.

The example of zinc purification also demonstrates that anion-exchange chromatography is an expedient way of separating elements that exhibit similar chemical properties. Overlap zone melting in an H_2 atmosphere of the pre-purified material has been proved to be capable of achieving purity levels denoted by an RRR value of 25 600, the highest measured on zinc [67].

References

1. Y. Shindo, H. Miyazaki, Y. Yamakoshi: In Ultra high purity base metals, UHPM-94 (The Japan Institute of Metals, Kitakyushu 1994) p. 413
2. D. Murphy, G.J. Butterworth: J. Nucl. Mater. **191–194** 1444 (1992)
3. Kirk–Othmer encyclopedia of chemical technology (Wiley, New York 1991)
4. B. Elvers: Ullmann's encyclopedia of industrial chemistry (VCH, Weinheim 1990)
5. U. Zwicker: Titan und Titanlegierungen (Springer, Berlin Heidelberg New York 1974)
6. M. Nakamura, T.H. Okabe, T. Oishi, K. Ono: Proc. Int. Symp. Molten Salt Chemistry and Technology (1993) p. 529
7. G. Hörz: Z. Metallk. **60**, 50 (1969)
8. P. Vitanyi: Magyar Aluminium **20**, 158 (1983)
9. H.J. Gottwald, K. Krone, J. Kruger: Metall. **23**, 1284 (1969)
10. O. Winkler, R. Bakish: Vacuum metallurgy (Elsevier, Amsterdam London New York 1971)
11. A.E. van Arkel: Reine Metalle (Springer, Berlin 1939)
12. T.A. Sullivan: J. Metall **17**, 45 (1965)
13. K.P.V. Lei, T.A. Sullivan: J. Less-Common Metals **14**, 145 (1968)
14. F.A. Schmidt, J.C. Warner: J. Less-Common Metals **13**, 493 (1967)
15. O.N. Carlson, F.A. Schmidt, D.G. Alexander: Metall Trans. **3**, 1249 (1972)
16. J. Bressers, R. Creten, G. van Holsbeke: J. Less-Common Metals **39**, 7 (1975)
17. S. Weinig: In Ultra high purity base metals, UHPM-94 (The Japan Institute of Metals, Kitakyushu, 1994) p. 156
18. Y.Nishimura: M. Eng. Thesis, Tohoku Univ. (1999)
19. J. Bigot: Mem. Sci. Rev. Metall **65**, 61 (1968)
20. M. Isshiki, K. Arakawa, K. Igaki: J. Less-Common Metals **96**, 157 (1984)
21. S. Hirano, O. Yoshinari, M. Koiwa: J. Less-Common Metals **113**, 17 (1985)
22. M. Nardin: Mem. Sci. Rev. Metaall. **67**, 431 (1970)
23. V.Q. Kinh: Mem. Sci. Rev. Metall. **65**, 49 (1968)
24. M. Isshiki, K. Igaki: Techn. Rep. Tohoku Univ. **44**, 331 (1979)
25. T. Kekesi, M. Uchikoshi, K. Mimura, Y. Zhu, M. Isshiki: In Proc. Second Int. Conf. on Processing Materials for Properties, PMP2000 (TMS, San Francisco, 2000) p. 979
26. S. Takaki, K. Igaki: Trans. Japan Inst. Metals. **17**, 353 (1976)
27. S. Takaki, H. Kimura: Scr. Metall. **10**, 1095 (1976)
28. R.L. Smith, E.J. Koepel: Mem. Sci. Rev. Metall. **65**, 27 (1968)
29. B.F. Oliver, F. Garofalo: AIME Trans. **233**, 1318 (1965)
30. B.F. Oliver, E.W. Troy: Mem. Sci. Rev. Metall. **65**, 39 (1968)
31. M. Isshiki et al.: Radioisotopes **28**, 3 (1979)
32. M. Isshiki, K. Igaki: Trans. JIM **19**, 431 (1978)
33. T. Kekesi, M. Uchikoshi, K. Mimura, M. Isshiki: Spring Meeting Japan Inst. Metals, No. 126, (Japan Inst. Metals, Yokohaama 2000) Abstracts, p. 328

34. M. Bourahla, B. Dubois: C.R. Acad. Sc. Paris **278C**, 679 (1974)
35. M. Isshiki, Y. Fukuda, K. Igaki: J. Less-Common Metals **105**, 211 (1985)
36. M. Isshiki, Y. Fukuda, K. Igaki: J. Radioanal. Nucl. Chem., Articles, **82/1**, 135 (1984)
37. N. Zubryckyj, W. Kunda, D.J.I. Evans: J. Electrochem. Soc. **116**, 542 (1969)
38. R.R. Soden, V.J. Albano: J. Electrochem. Soc. **115**, 766 (1968)
39. B. Dubois: Mem. Sci. Rev. Metall. **65**, 67 (1968)
40. M. Nardin, A.M. Wache, V.Q. Kinh: Mem. Sci. Rev. Metall. **66**, 312 (1969)
41. M. Isshiki et al.: Radioisotopes **30**, 19 (1981)
42. M. Dinu: Metalurgia (Bucharest) **35**, 599 (1983)
43. R. Lohmann: Metall. **11**, 1108 (1987)
44. H. Rajainmaki, A. Helenius, M. Kolehmainen: J. Metall. **45** (3), 68 (1993)
45. M. Kato: In Ultra high purity base metals, UHPM-94, (The Japan Institute of Metals, Kitakyushu 1994) p. 441
46. Y. Miyake: Bull. Japan Inst. Metals. **31**, 267 (1992)
47. T. Kekesi, K. Mimura, M. Isshiki: Mater. Trans. JIM **36**, 649 (1995)
48. J.S. Smart, A.A. Smith, A.J. Phillips: Trans. Amer. Min. Metall. Eng. **143**, 272 (1941)
49. P. Gregory, A.J. Bangay, T.L. Bird: Metallurgia **71**, 207 (1965)
50. H. Bruning, J. Le Hericy, K. Lucke: Ann. Chim. Fr. **10**, 521 (1985)
51. W.C. Cooper: J. Appl. Electrochem. **15**, 789 (1985)
52. V.A. Ettel, A.S. Gendron, B.V. Tilak: Metall. Trans., **6B**, 31 (1975)
53. T. Kekesi: Acta Techn. Acad. Sci. Hung. **105**, 153 (1993)
54. T. Kekesi, K. Mimura, Y. Ishikawa, M. Isshiki: Metall. Mats. Trans. B **28B**, 987 (1997)
55. M. Uchikoshi, T. Kekesi, Y. Ishikawa, K. Mimura, M. Isshiki: Mater. Trans. JIM **38**, 1083 (1997)
56. T. Kekesi, M. Isshiki: Hydrometall. **45**, 345 (1997)
57. T. Kekesi, M. Isshiki: J. Appl. Electrochem. **27**, 982 (1997)
58. E.D. Tolmie, D.A. Robins: J. Inst. Metall. **85**, 171 (1956–57)
59. J. Gerlach, F. Pawlek, D. Rogalla: Metall. **18**, 1158 (1964)
60. J. Le Hericy: C.R. Acad. Sci. Paris **251**, 1509 (1960)
61. Y. Ishikawa, K. Mimura, M. Isshiki: Mater. Trans. JIM **40**, 87 (1999)
62. K. Mimura, Y. Ishikawa, M. Isshiki, M. Kato: Mater. Trans. JIM **38**, 714 (1997)
63. M. Isshiki, T. Yoshida, T. Tomizono, S. Satoh, K. Igaki: J. Cryst. Growth, **73**, 221 (1985)
64. M. Isshiki, T. Tomizono, T. Yoshida, T. Okawa, K. Igaki: J. Japan Inst. Metals **48**, 1176 (1984)
65. K. Takahashi, T. Kekesi, Y. Aoki, M. Isshiki: Zinc & Lead '95, Int. Symp. Extr. Appl. of Zinc and Lead, Sendai, Japan, 22–24 May, 1995, p. 321
66. M. Isshiki: J. Cryst. Growth **86**, 615 (1988)
67. M. Isshiki et al.: Bull. Japan Inst. Metall. **48**, 1176 (1984)

4 Refractory Metals

A. Koethe

4.1 Introduction

The definition of "refractory metals" is not unanimous. Often, they are defined as those metals having a melting point in excess of 2200 K; that is, as comprising the elements Hf, Nb, Ta, Mo, W and Re [1]. Other authors (see, for example, [2]) include all the transition metals of group IVa to group VIa of the Periodic Table (i.e. also comprising the metals Ti, Zr, V and Cr with melting points between 1941 K and 2180 K) and the group VIIa element Re. This latter definition is followed here. Table 4.1 presents a number of relevant properties of the refractory metals [3]. The group IVa metals Ti, Zr and Hf have a hcp structure with a phase transition to bcc at high temperatures, whereas the group Va and VIa metals are bcc, and the group VIIa metal Re is hcp at all temperatures. The success of purification and the purity level reached until now is different for the various refractory metals. The group IVa metals and vanadium are the most difficult to purify, primarily because of the high reactivity of these metals and their lower melting temperatures in comparison with the other refractory metals. Taking the residual resistivity ratio, $\mathrm{RRR} = \rho_{(293\mathrm{K})}/\rho_{(\mathrm{T}\to 0\mathrm{K})}$, as a simple but illustrative indicator of purity, the highest degree of purity has been reached up to now for the refractory metals Nb, Mo and W.

The extractive metallurgy of refractory metals – for example, their production from ores or concentrates by thermic reduction with Ca, Si, Al, Mg or C – will not be dealt with here. Instead, it is assumed (and recommended) that ultra-purification should begin with starting materials (metals or compounds) of the highest purity commercially available. In this context, it is very important to identify the goal of the research work [4–7]:

- What degree of purity is necessary, for which property or application?
- What are the critical, most intolerable impurities to be removed?
- What are the possibilities and constraints of the various processing steps?

The properties of UHP refractory metals are influenced by both substitutional and interstitial impurity elements (especially in bcc metals, the interstitial impurities are often more influential than the substitutional ones). Since the possibilities and methods for removing both types of impurities are quite different, they will be treated separately: we shall discuss the removal of

Table 4.1. Refractory metals of groups IVa–VIIa of the Periodic Table [3]

Group	Metal	Rel. atomic mass	Structure	Melting point (K)	Young's modulus (GPa)	Density (g/cm^{-3})	Supercond. transition temperature (K)
IVa	Ti	47.867	hcp→bcc (1155 K)	1941	116	4.507	0.40
	Zr	91.224	hcp→bcc (1135 K)	2128	68	6.511	0.61
	Hf	178.49	hcp→bcc (2013 K)	2506	78	13.31	0.128
Va	V	50.942	bcc	2180	128	6.110	5.40
	Nb	92.906	bcc	2750	105	8.570	9.25
	Ta	180.948	bcc	3290	186	16.650	4.47
VIa	Cr	51.996	bcc	2180	279	7.140	No data
	Mo	95.941	bcc	2896	329	10.280	0.915
	W	183.84	bcc	3695	411	19.250	0.015
VIIa	Re	186.207	hcp	3459	463	21.020	1.70

metallic impurities in Sect. 4.2–4.5, and that of interstitial impurities (gases and carbon) in Sect. 4.6.

The high melting point of refractory metals allows the evaporation of volatile elements such as alkali metals, Al, Cr, Fe, Si and so on by vacuum or ultra-high-vacuum melting. Therefore vacuum-arc melting (VAM), electron-beam melting (EBM) and especially electron-beam floating-zone melting (EBFZM) play a very important role in the preparation of UHP refractory metals. However, trace impurities of other refractory metals cannot be removed by remelting, due to their low vapor pressure. Particular problems are posed by impurity/host metal pairs, in which the impurity element is especially difficult to remove due to its chemical and physical similarity with the host metal. Such combinations of interest here are Hf and Zr, Ta and Nb, or W and Mo. To purify these elements down to the ppm level, special and often complex processes have been developed, including combinations of various chemical or physical methods which yield high separation factors for the system under consideration. The processes applied include liquid–liquid extraction, halogenization, crystallization from solvents, electrolysis, distillation, sublimation, thermal reduction from the gas phase and reduction with hydrogen. As to purification by electrolysis, for the refractory metals the application of molten salt electrolysis is necessary, since the electrolysis of aqueous solutions (with the exception of Cr) is unsuitable due to the low overvoltage of hydrogen evolution.

In general, each purification method is restricted to the removal of a certain group of impurities. Therefore, it is often necessary to combine several methods in order to obtain a UHP metal. The total process, as well as each single step, has to be considered from two main aspects:

- the purification effect attainable,
- the possible detrimental influence of new impurity sources.

The elimination or control of impurity sources in the processing chain may be difficult and costly. Thus, if an impurity introduced in an early processing step (e.g. from the container or handling materials, from chemicals or reagents used, from the atmosphere etc.) can easily and quantitatively be removed in a later processing step, it is not necessary to take too much care to avoid this contamination.

It should be remembered that any efforts to produce UHP metal can only be successful if appropriate methods for characterizing the level of purity reached in intermediate steps of purification, as well as in the final UHP metal, are developed and applied in parallel to the refining processes. The continuous improvement of analytic tools by physical or chemical methods often requires efforts and costs that are similar to those of the purification processes themselves. Nevertheless, the present section concentrates on purification methods and will discuss neither the analytic techniques nor the – often new and surprising – properties of these UHP metals. However, most specialists working in this field are convinced that any successful research on UHP metals can only be carried out within the trinity of preparation, characterization and properties.

Finally, we would like to point to a fact that is often overlooked: during investigations or applications of UHP metals the possibility of re-contamination or deterioration of the crystal perfection – for example, by absorption of gases or carbon at elevated temperature, or by inadvertent deformation during mounting or handling of samples – should be taken into account.

In order to facilitate a rough estimate of the possible concentrations of substitutional impurity elements from measurements of the residual resistivity, resistivity contributions published in the literature for a number of elements are compiled in Table 4.2. Similar data for interstitital impurities are given in Table 4.13.

4.2 Group IVa Metals

4.2.1 Titanium

Recently, two processes have been published which are intended to provide high-purity titanium in commercial quantities.

By a combination of electrorefining and electron-beam melting [8], a 6N Ti (with respect to metallic impurities) was produced from sponge Ti. The

Table 4.2. The contribution of substitutional impurity elements to the electrical resistivity of refractory metals (in $\mu\Omega$ cm per at%)

Group	Metal	$\rho(273\,K)$ ($\mu\Omega$ cm)	ρ_{imp} ($\mu\Omega$ cm/at%)							
			Ti	Zr	Hf	V	Nb	Ta	Mo	W
IVa	Ti	40^{a}	×	1.8^{b}			1.8^{b}		16.3^{b}	
	Zr	42^{a}		×						
	Hf	30^{a}			×					
Va	V	18.9^{c}				×				
	Nb	13.3^{c}	1.4^{d}	$0.6\text{-}1.4^{d}$	0.11^{c}	0.54^{c}	×	0.22^{c}	0.21^{d}	0.55^{c}
				0.84^{c}	1.4^{d}			0.25^{d}	0.52^{e}	$0.4\text{-}1.1^{d}$
				0.53^{f}				0.11^{f}	0.19^{f}	0.33^{f}
	Ta	12.1^{c}		0.74^{f}			0.35^{c}	×	0.58^{f}	0.71^{f}
VIa	Cr	11.8^{c}								
	Mo	4.88^{g}							×	0.077^{c}
	W	4.84^{h}	1.68^{c}				1.18^{c}	1.11^{c}	0.2^{c}	×
VIIa	Re	18^{a}								

Reference: [a] [3]; [b] [67]; [c] [64]; [d] [5]; [e] [49] [f] [4]; [g] [65]; [h] [66]

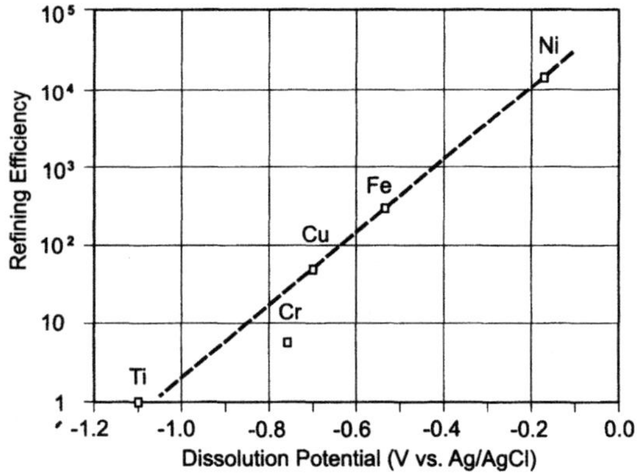

Fig. 4.1. The relationship between the dissolution potential and the refining efficiency of impurities in Ti [8]

electrorefining was carried out in an electrolyte that consisted of an eutectic mixture of LiCl, NaCl and KCl under an Ar atmosphere. The refining efficiency (the ratio of impurities in the raw material to impurities in the deposited material) as a function of the dissolution potential (measured in a molten KCl–NaCl electrolyte at 993 K) is shown in Fig. 4.1. The main impurities in the raw material (in mass ppm: Fe 6–250, Ni 3–200, Cr 1, Sn 2, Mn 37, Cu 2, Al 2) were all reduced to < 0.1 mass ppm. The content of interstitial impurities was 80–100 mass ppm O, < 10 mass ppm C and < 110 mass ppm H. The purified material had a RRR of 120.

In [9] it is reported that it was possible, by carefully controlling and continuously improving the electrorefining process (fused salt electrolysis and electron-beam melting), to produce Ti ingots that typically had 10–15 mass ppm of metallic impurities (Fe, Al, Zr, Cr and V). The mean concentrations of the interstitial impurities were < 300 mass ppm (220 ppm O, 20 ppm C, 10 ppm N and 4 ppm H). Since the vapor pressure of TiO is about the same as that of the metal, the ingot oxygen content after EBM was nearly identical to that prior to EBM. At CRNS-CECM in Vitry/France, titanium has been purified by the van Arkel process [10, 11]. Iodine reacts at about 473 K with impure Ti, forming the volatile iodide TiI_4, which is then thermally dissociated at a pure Ti wire heated to 1370–1470 K. Purification is related to

Table 4.3. An analysis of titanium (in mass ppm) [11]. MRC and iodide Ti, activation analysis; JM and iodide Ti, GD-MS and LECO (C,N,O,S)

Element	MRC titanium		JM titanium	
	Crude metal	Iodide metal	Crude metal	Iodide metal
Al	9	0.6	2.7	0.08
As			0.01	<0.01
Cr	0.3	0.02	0.69	0.04
Cu			0.175	0.07
Fe	3	2	9.45	1.9
Hf			0.05	<0.05
Mo			0.05	<0.05
Na			<0.01	<0.01
Ni	0.1	0.07	0.06	0.07
P			<0.01	<0.01
Sb			0.05	<0.05
Si			0.37	0.04
V	0.9	0.6	1.5	0.21
W			0.01	<0.05
Zr	1.9	0.4	0.58	0.67
C			16	18
N			14	8
O		20	299	68[a]
S			3.17	2

[a] Activation analysis: 30 ppm

Table 4.4. A typical analysis of crude Ti sponge, Ti deposited by the improved iodide process [12] and remelted Ti [13] (in mass ppm)

Element	Crude Ti sponge [12]	Deposited Ti [12]	Remelted Ti [13]
Fe	50	0.5	2
Ni	15	0.1	3
Cr	8	0.4	<3
Cu			<1
Al	7	1.1	<5
Si	3	0.4	
Mn			5
Na	<0.02	<0.02	
K	<0.02	<0.02	
U	<0.001	<0.001	
Th	<0.001	<0.001	
C	30	30	10
O	300	80	30
H	<10	<10	4
N	10	10	20

the vapor pressure differences of the iodides formed. The vessel is carefully sealed by ultra-high-vacuum techniques, degassed at 573 K and evacuated to $< 1.3 \times 10^{-6}$ Pa. Since the purification effect with respect to iron was low, a low-iron starting material has been used. Table 4.3 gives the results obtained with MRC Ti and with electrolytic Ti from Johnson and Matthey. With JM Ti, a metal with RRR = 200 could be obtained. The major difficulties reported were due to: (a) the instability of TiI_4, which transforms at 623 K into TiI_2, whose vapor pressure is low, thus causing the thermal decomposition to stop; and (b) fast etching of the cold part of the Ti deposition wire by the iodine gas, causing failure of the wire after a few hours of operation. After 4 days of continuous operation, the 70 cm pin may reach a mass of 1 kg [9].

In [12], the van Arkel process has been modified to produce 99.999% Ti on the 50–100 kg per batch scale by the so-called "improved iodide process", making use of the lower-Ti iodides. Unlike the conventional process, where the deposition occurs on a hot Ti filament, here a high-purity Ti tube of 1 mm wall thickness is used as the substrate, thus significantly enlarging the deposition surface area. The Ti tube is internally heated by a graphite heater. The crude Ti sponge of 99.98% purity (compacted by compression into a tubular form) surrounds the tube. TiI_4 is introduced into the reactor, which has previously been evacuated to 1–10 Pa and heated to 1150 K. The TiI_4 reacts with the crude Ti, forming lower-Ti iodides (TiI_2, TiI_3) which are decomposed at the tube surface, yielding a high-purity Ti deposit on its outer wall. The deposition rate is about 1 μm/min^{-1}. The tube is then remelted in an Ar plasma arc.

Table 4.4 shows typical values of the chemical composition of the crude and the deposited Ti, illustrating especially the removal of Fe, Ni and Cr.

In [13], the afore-mentioned material was further purified by skull melting in a 250 kW electron-beam furnace. The analysis of this UHP-Ti is also given in the last column of Table 4.4. The low concentrations of the interstitial impurities O, N and C obtained, which are in the range of 10–30 mass ppm, should be especially noticed.

4.2.2 Zirconium and Hafnium

Unlike the pairs of similar elements Nb and Ta, and Mo and W, the elements Zr and Hf have identical outer shell electron arrangements ([Kr]$4d^2 5s^2$ for Zr and [Xe]$4f^{14} 5d^2 6s^2$ for Hf), giving both metals nearly identical chemical behavior [2]. Taking additionally into account the very high reactivity of Zr and Hf with respect to O, N, C and H, it becomes clear why it is especially difficult to produce high-purity Hf-free Zr or Zr-free Hf. Both metals will be discussed together in this section.

In Zr for fuel cladding applications, an Hf content of less than 50 ppm is needed, compared with the Hf concentration in Zr ores, which is always of the order of 2–3%. A very complex industrial processing route for the separation of Zr and Hf is described in [14] and depicted in Fig. 4.2.

Starting from ore (zircon), the first step consists of a chlorination in the presence of carbon, according to the reaction

$$ZrO_2(+SiO_2 + HfO_2) + 2C + 2Cl_2 \rightarrow ZrCl_4 + HfCl_4 + SiCl_4 + 2CO \,. \tag{4.1}$$

This operation is carried out in a tube furnace on a bed fluidized by chlorine gas, at a temperature of about 1470 K. A selective condenser allows a first purification by fractionate condensation of the chlorides and collection of the mixed Zr + Hf chloride. The separation of Zr and Hf by distillation of the chlorides at atmospheric pressure is not practicable, due to the small difference in sublimation temperatures: 604 K for $ZrCl_4$ and 592 K for $HfCl_4$.

For the separation, two processes are currently in use:

1. By addition of water, the chlorides are transferred into the oxychlorides $ZrOCl_2 + HfOCl_2$. These are separated by liquid–liquid extraction with methyl isobutyl ketone (MIBK) in the presence of thiocyanate. The Hf passes into the organic phase, whereas the Zr remains in the aqueous phase. Both the Hf-containing and the Zr-containing phases are then treated with sulfuric acid and ammonia, and calcinated to form pure ZrO_2 or HfO_2. The oxides are chlorinated again to obtain pure water-free, $ZrCl_4$ and $HfCl_4$, respectively. In [15] it is pointed out that this process has some inherent environmental problems: the high aqueous solubility and low

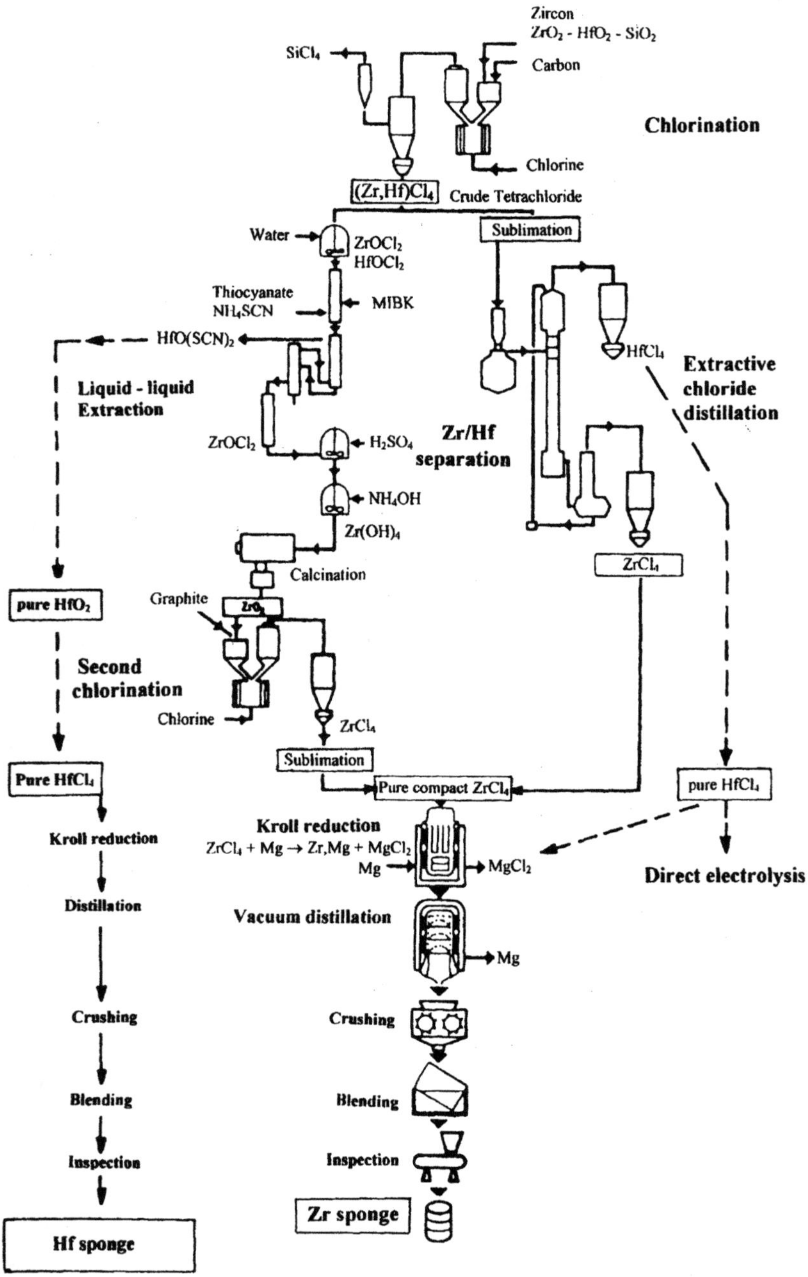

Fig. 4.2. A zirconium and hafnium sponge production flow-sheet [14]

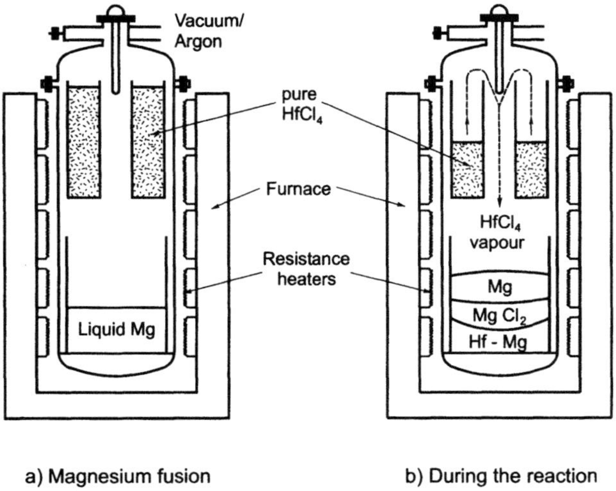

Fig. 4.3. The reduction of $HfCl_4$ to Hf by the Kroll process [14]

flashpoint of MIBK, and the formation of possible gaseous contaminants generated by the decomposition of the thyocyanic complexes in the acid medium (H_2S, HCN, mercaptans). It was shown that these problems can be avoided by using organophosphorous extractants instead of MIBK.

2. Extractive distillation of the chlorides in a bath of molten KCl–$AlCl_3$ at 623 K under atmospheric pressure [14–16].

The pure chlorides are then reduced by the classical Kroll process with high-purity Mg under a slight overpressure of pure argon. Figure 4.3 gives a schematic picture of the reaction vessel, which has two heating zones: the upper one for evaporating the chloride, and the lower one for melting the Mg acting as the reducing agent. The exothermic reaction is controlled by the sublimation rate of the chloride. At the bottom of the vessel, a pseudo-alloy of Hf + Mg or Zr + Mg is formed. After its removal from the Kroll vessel, the elimination of the magnesium is effected by vacuum distillation at 1173–1273 K. After crushing, the sponge Zr or Hf has to be stored under Ar, since it is very pyrophoric. Further purification of the sponge can be carried out by the van Arkel process (distillation in the presence of iodine, see Fig. 4.4). The iodine reacts at low temperature with the impure metal, yielding a volatile ZrI_4 or HfI_4 which is then decomposed to metallic Zr or Hf on a hot wire of the same metal heated to about 1570–1870 K by passing a current through it. The productivity is of the order of about 75 g of metal per hour and linear meter of filament. Another effective way to produce high-purity Hf is by electrorefining [14, 17–19]. This is carried out by direct electrolysis of $HfCl_4$ in a molten salt bath of NaCl–KCl–NaF at 1023–1123 K.

Typical values of the chemical composition of differently processed Hf metals are given in Table 4.5 [14]. On the one hand, they illustrate the progress

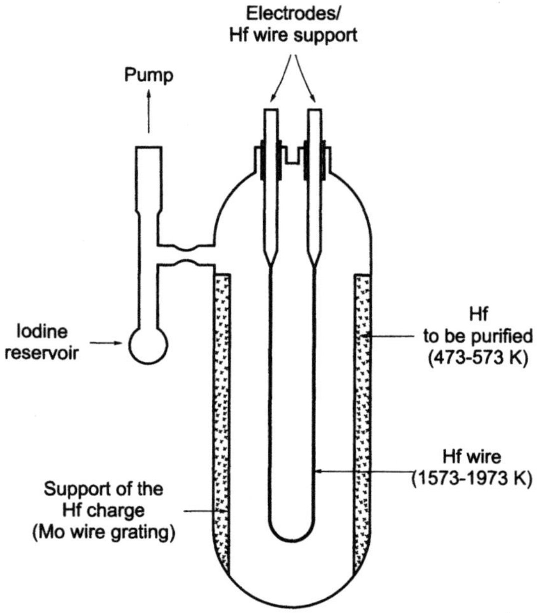

Fig. 4.4. The van Arkel refining process for the preparation of pure hafnium rods [14]

achieved in the preparation of pure Hf and, on the other hand, the tremendous and as yet unsolved problems in the preparation of, strictly speaking, ultra-high-purity Hf with impurity concentrations reduced down to the ppm level.

The removal of metallic and nonmetallic impurities from Zr with a nominal purity of 99.8% by EBFZM in high vacuum (total pressure about 2×10^{-6} Pa) has been investigated in [20]. Metallic impurities such as Fe and Cr, being more volatile than Zr, volatilized preferentially, and their contents were reduced to below a few mass ppm after three zone passes at a speed of $2\,\mathrm{mm/min^{-1}}$. Although no decrease of Hf (which has a slightly lower vapor pressure than Zr) by volatization was observed (the Hf concentration in the starting rod was 45–70 mass ppm), a redistribution of Hf caused by the zone melting segregation effect was found. The minimum content determined at the rod end was about 70% of the initial value. Similarly, a reduction of the O content below half of the initial value of 450–630 mass ppm was noticed near the final end of the zone-refined portion, whereas carbon (initial value 30–40 mass ppm) was enriched in this region.

The effective distribution coefficients of O and C at a zoning speed of $2\,\mathrm{mm/min^{-1}}$ were estimated to be 1.5–2.0 and 0.7–0.8, respectively. The content and distribution of N (about 20–35 mass ppm) remained unchanged. No reduction of the O content by ZrO evaporation and of the C content by CO evaporation was noticed [20]. The RRR value of the starting material, of 11,

Table 4.5. Typical values of the chemical composition of differently processed Hf metal (Hf and Zr in mass %, otherwise in mass ppm) [14]

Element	Hf sponge	Sponge + EBM	Sponge + van Arkel	Electrolytically refined Hf
Hf	>95%	>95.5%	>96%	>95.5%
Zr	<4.5%	<4.5%	<4.5% (<1.5%)	<4.5% (<1.5%)
O	1500 ppm	500 ppm	100 ppm	500 ppm
C	500	100	50	75
N	100	100	20	60
Al	200	50	50	75
B	–	5	5	5
Cl	500	50	50	250
Cr	200	20	20	100
Cu	100	25	25	50
Fe	750	300	250	250
Mg	600	20	20	50
Mo	20	10	10	–
Nb	100	50	50	50
Ni	50	25	25	50
P	–	5	5	5
Pb	–	10	10	10
Si	100	50	50	100
Sn	100	–	–	–
Ta	200	–	100	100
Ti	100	50	25	50
U	10	5	5	5
V	10	10	10	10
W	50	–	10	50

was increased to about 26 in the last-solidified zone after between three and five zone-melting passes.

In [21], Hf metal, produced either by calciothermic reduction of HfO_2 or fused salt electrolysis of $HfCl_4$, has been purified by the iodide decomposition process. The results obtained from the purification studies indicate that a metal that meets the AEC specifications can be obtained.

Table 4.6. Analyses of feed material and refined vanadium (in mass ppm) [22–24, 77]. n.d., Below detection limit of emission spectrography

Impurity element	Feed material	Refined product	
		First cycle	Second cycle
Al	30	n.d.	n.d.
C	410	20	10
Ca	15	n.d.	n.d.
Cs	6	6	n.d.
Cu	5	22	n.d.
Fe	320	n.d.	n.d.
H	20	10	10
Mg	8	2	n.d.
Mn	20	n.d.	n.d.
Mo	125	n.d.	n.d.
N	2416	10	10
Ni	20	130	10
O	1310	110	10
Si	498	21	15
Ti	24	n.d.	n.d.

4.3 Group Va Metals

4.3.1 Vanadium

The production of high-purity V by molten salt electrolysis in a LiCl–KCl–VCl_2 electrolyte at 993 K has been reported in [22–25]. Typical analyses of the Ca-reduced feed and the electrolytically deposited metal after one and two refining cycles, which yielded a V of 99.99% purity, are given in Table 4.6 [24]. The refined product had a porous dendritic structure still containing appreciable amounts of electrolyte, which could be removed by leaching with dilute acid and a following arc melting.

The iodide process described in [26] was primarily developed for the preparation of UHP-Nb (see detailed description in the next section). It was, however, also applied to the purification of V. The impurity concentrations determined in the purified V by chemical analysis and instrumental neutron activation analysis are given in Table 4.7.

4.3.2 Niobium

As mentioned above, in the preparation of UHP-Nb the control of the Ta content is a crucial problem, due to the chemical and physical similarity of

Table 4.7. The analysis of V purified by the iodide process (in mass ppm) [26]

Element	Chemical analysis	Instrumental neutron activation analysis
Ag	n.d.	<0.0004
Ce	n.d.	0.067
Co	<1	0.28
Cr	<10	1.9
Cu	1	n.d.
Fe	13	21
Hf	<3	0.32
Mo	150	340
Ta	<4	0.72
Sc	n.d.	<4
W	<5	n.d.
Zr	3	n.d.

both metals. The processes that have proved to be most suitable to remove Ta from Nb are liquid–liquid extraction and molten salt electrolysis.

The complex and relatively complicated purification process developed in Dresden made use of the earlier experiences with Mo (see Sect. 4.4.2) and was based on a combination of liquid–liquid extraction, chlorination and thermal decomposition of niobium pentachloride, followed by EBFZM and degassing treatments of the chemically pre-purified Nb [4, 7, 27–30]. The starting material used was a pre-refined Nb_2O_5 which contained Ta, Zr, Mo, W and Fe, each in the range of 100–700 at ppm (other substitutional impurities were < 10 at ppm each).

Liquid–liquid extraction with methyl isobutyl ketone from a weakly acid aqueous solution (2M HF/1M H_2SO_4) allowed an effective separation of Ta. In a second extraction step with methyl isobutyl ketone, this time from a strongly acid aqueous solution (6M HF/3M H_2SO_4), the niobium was separated from the other remaining elements. After back-extraction with water, precipitation with ammonia, filtration and heat-treatment at 1373 K, purified Nb_2O_5 was obtained. The yield was 70%.

This Nb_2O_5 was mixed with pure activated charcoal (in a stoichiometric amount with respect to CO formation) and chlorinated in a stream of chlorine gas at 973–1073 K to $NbCl_5$. The coal addition, combined with a coal-filled column situated behind the chlorination vessel, allowed a quantitative reaction of Nb_2O_5 to $NbCl_5$, thus avoiding the formation of $NbOCl_3$.

The reduction of the $NbCl_5$ to high-purity metallic Nb was carried out by the van Arkel process; that is, by thermal decomposition in vacuum (initial pressure < 0.7 Pa) at 1670–2670 K on a pure resistance-heated Nb wire of

0.5 mm diameter. In this way, rods of 5–6 mm diameter and 250 mm in length were produced. The compact rods had a density that approximated the theoretical value. One of the advantages of this method is that the niobium does not come into direct contact with any crucible material at high temperature. The growth rate was 0.5–1.0 mm h^{-1}, and the Nb yield amounted to 40–50%.

These rods could directly be used as the starting material for EBFZM, to produce high-purity single crystals with a predetermined orientation. In most

Table 4.8. The impurity element concentration after various steps of the preparation of high-purity Nb (in at ppm with respect to Nb) [4, 7, 30]

Element analyzed by:	Nb_2O_5 after extraction 1)	Polycrystalline Nb after thermal decomposition 2)	Nb single crystal after zone melting 2)
B		0.04	<0.008
Na		1.3	0.77
Mg	<9	<0.36	<0.36
Al		0.4	0.1
Si		1.5	0.3
S		0.34	0.27
Cl		0.85	0.23
K		2	0.5
Ca		0.5	0.15
Ti		0.3	0.05
V	<1.5	<0.04	0.03
Cr	<3	0.08	0.06
Mn	<0.9	0.05	<0.03
Fe		1.4	0.3
Co	<3	0.03	<0.03
Ni	<0.5	0.2	0.05
Cu		0.21	<0.02
Zn		0.08	<0.03
Ga		0.06	0.33
As		0.1	0.03
Zr		0.1	0.1
Mo	<0.2	0.21	0.5
Ag	<0.1	<0.27	<0.14
Ba		0.08	<0.03
Hf		0.17	0.6
Ta	<0.5	0.25 3)	0.8 3)
W	<1	0.33 3)	0.5 3)
Pb	<0.6	<0.09	<0.05

1) Analyzed by spectrophotometry or spectral analysis
2) Analyzed by spark source mass spectrometry
3) Analyzed by neutron activation analysis

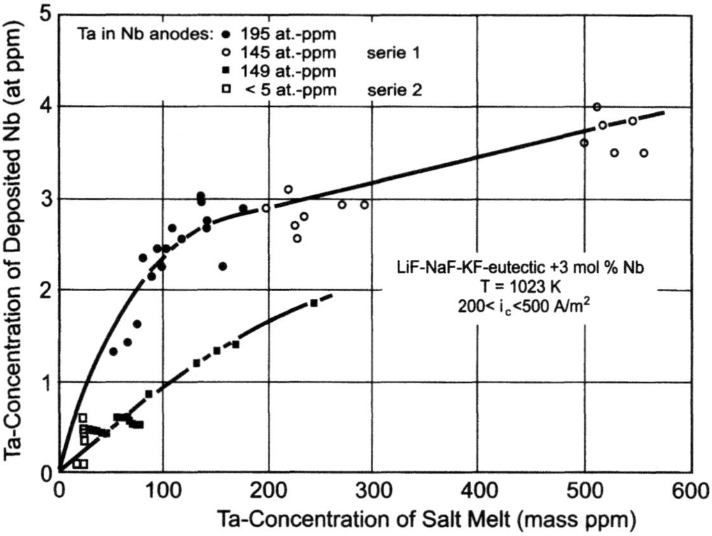

Fig. 4.5. The Ta concentration in Nb purified by molten salt electrolysis as a function of the Ta concentration in the Nb anodes and in the salt melt [5]

cases, one melting pass in a vacuum of 10^{-3} Pa at a speed of 3 mm/min^{-1} proved to be sufficient. The progress in purification over the various processing steps can be seen from the impurity concentrations given in Table 4.8.

A much simpler, but equally effective, purification technology for Nb is the molten salt electrolysis developed by K.K. Schulze in Stuttgart [5,31], on the basis of the earlier results of Senderoff and Mellors [32,33]. The purification is carried out by electrodeposition of Nb from LiF–KF–NaF melts containing 2.5 mol% K_2NbF_7 at about 1000 K. The reduction of the Nb^{5+} ions occurs in the sequence $Nb^{5+} \rightarrow Nb^{4+} \rightarrow Nb^{+} \rightarrow Nb$. The current efficiency is zero at the beginning of the electrolysis for a Nb valence of $\nu = 5$ and reaches 100% for $\nu = 4$ [5]. It has been stated that any oxidation of Nb ions caused by oxygen or water traces introduced into the bath will reduce the current efficiency of the electrolysis and degrade the purity of the Nb deposit.

Figure 4.5 shows the relation between the Ta concentration of the cathodically deposited Nb and the Ta concentration in the electrolyte. In the lower curve the experiments were optimized to exclude traces of oxygen. The Ta content of the deposit was shown to depend on the Ta and oxygen content of the electrolyte, but not on the Ta content of the anode. The lowest Ta concentration obtained in the deposit was 0.05 at ppm.

In Table 4.9 the impurity concentrations of the starting material, of the electrodeposited Nb and of the rods after EBFZM are compiled. The data also show that, besides that of Ta, a reduction of the W and Mo content is attained. The slight increase of Fe and Al is due to some contamination from structural materials [5]. In [6] it has been reported that a further increase in

Table 4.9. The impurity concentration in Nb[a] [5]. TWCA Nb, crude material; Nb-R-1, after electrolysis; Nb-R-2, after EBFZM

(TWCN Nb) $\xrightarrow{\text{Electrolysis}}$ Nb-R-1 $\xrightarrow{\text{EBFZM}}$ Nb-R-2			
Element	(TWCA Nb)[b]	Nb-R-1[c]	Nb-R-2[c]
Ta	83.0 ppm	0.05–3.00 ppm	0.05–3.00 ppm
W	3.8 ppm	1–30 ppb	1–50 ppb
Mo	1.2 ppm	96 ppb	96 ppb
Cr	3.0 ppb	6–28 ppb	17–44 ppt
Fe	26.0 ppb	1–3 ppm	3–11 ppb
Co	85.0 ppt	0.3–2.5 ppb	6.3–30 ppt
Cu	95.0 ppb	44 ppb	<20 ppb
V	5.0 ppb	0.4 ppb	0.3 ppb
Ti	0.3 ppm	<0.1 ppm	<0.1 ppm
Al	90.0 ppb	110 ppm	20 ppb

[a] In atomic fraction, [b] Averages from three determinations, [c] Content ranges of separately purified samples

purity can be obtained by a second electrolysis of the EBFZM-refined rods in an electrolyte of extremely low Ta and W content. Material produced by this technique had impurity concentrations of below 0.3 mass ppm Ta and 0.01 mass ppm W. Three EBFZM passes reduced the content of volatile metallic impurities down to the sub-ppm or even sub-ppb level. After EBFZM, decarburization and UHV degassing RRR values of 25 000 were attained [6].

Inspired by these results of Schulze [3, 32] also in Dresden fused salt electrolysis has been studied further in [29, 30, 34]. As the starting material, commercial sintered rods have been used. In order to begin the process with a metal that already has high purity, these rods were at first remelted in an electron-beam furnace repeatedly (at least three times). The main effect of this remelting was, besides the evaporation of other volatile metals, the reduction of the concentration of iron, which is difficult to remove by electrolysis. The eutectic LiF–NaF–KF electrolyte used contained 16 mass% of K_2NbF_7. The electrolysis was carried out at a temperature in the range of 973–1098 K. In situ purification of the electrolyte was made possible by a preliminary electrolysis, preferably carried out at a current density of $j \geq 200\,\mathrm{A\,m^{-2}}$. This treatment led to a pronounced reduction of the noble metal contamination introduced into the electrolyte with the chemicals. By means of an electrolysis with a reduced current density of $j \leq 100\,\mathrm{A\,m^{-2}}$, it was possible to enhance the concentration of the Nb^+ ions, which allowed the impurity concentration of the niobium to be decreased by at least one order of magnitude. The details of the experimental setup are described in [34].

Table 4.10. Impurity element concentrations after various steps of the preparation of ultra-high-purity Nb by electrolysis (in at ppm) [29,30,34]

Element	Anode	Cathode	Zone-refined single crystal
Al	70	<0.17	0.24
Ca	2.7	0.2	0.55
Co	1	0.3	0.03
Cr	<0.7	<0.15	0.2
Cu	2	1.2	0.2
Fe	1	3	0.5
K	1	1.4	0.37
Li	<0.05	0.06	0.03
Mg	0.6	0.5	0.36
Mo	4	<0.8	<2.2
Na	1	0.2	0.4
Ni	5.3	0.1	0.2
P	0.2	0.2	0.25
S	6	3.3	2.8
Si	80	0.3	0.6
Ta	72	<0.005	<0.005
Ti	0.1	<0.2	0.01
W	23	<0.005	<0.005
Zn	0.5	0.3	0.6
Zr	5	<0.4	0.1

The residual gas composition, and thus the properties of the electrolytic cell (i.e. the leakage rate and the inert gas quality), had an important effect on the impurity concentration of the separated niobium through its influence on the oxidation rate of the Nb^+ ions and the solubility behavior of impurities such as Ni, Mo and W. The impurity concentrations found in the cathodic niobium, deposited at 1073 K with $j = 400\,\mathrm{A\,m^{-2}}$, are presented in Table 4.10. The problematic elements Ta and W could not be detected by instrumental neutron activation analysis, and the electrolyte components were only present in concentrations of ≤ 1 at ppm. Table 4.10 also shows the chemical composition after EBFZM. It can be concluded that both purification methods (liquid–liquid extraction versus molten salt electrolysis) are able to yield single crystals of comparable high quality.

After the removal of interstitial impurities (see below) by careful decarburization in an oxygen atmosphere of 10^{-3} Pa at 2300 K and UHV degassing at $p < 10^{-7}$ Pa, 2600 K/30 h, the RRR of the crystals was > 10 000. The best crystal obtained had an extrapolated RRR of 90 000 which corresponds to a

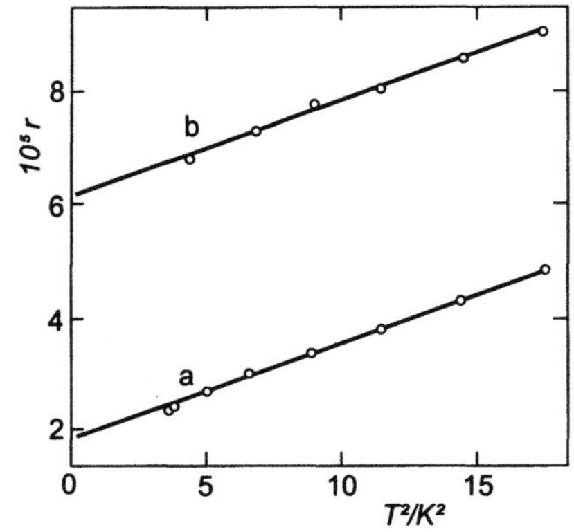

Fig. 4.6. Extrapolation of the reciprocal resistivity ratio of two UHP-Nb samples to $T \to 0\,K$ [31]. $r = \rho(T)/\rho$ (293 K) [30,35]

residual content of interstitial impurities of about 0.4 at ppm and of substitutional impurities of about 2.5 at ppm.

Since Nb is superconducting below 9.29 K, the RRR measurements of these samples have been carried out in a longitudinal magnetic field of $5.1 \times 10^5\,A\,m^{-1}$. This suppressed the superconducting state and led to saturation of the longitudinal magnetoresistance. The magnetoresistance was corrected for by factors of 1.24 and 1.9 for the <100> and <110> axis orientations, respectively. The resistance of the samples was measured as a function of temperature between 1.9 K and 4.2 K and extrapolated to $T \to 0\,K$ [29,30,35]. Two examples of this extrapolation are given in Fig. 4.6 [30].

Another process for the preparation of UHP-Nb, yielding about 6 mass ppm Ta and < 1 mass ppm of other metallic impurities (intended for application in monitoring the neutron fluence in nuclear reactors), has been described in [26]. It consists of the following steps:

1. Iodination of commercially available crude Nb pellets (diameter 1–5 mm) at 773–873 K. The NbI_5 is cooled to 473–523 K and condensed in a trap. In this step the NbI_5 is purified from iodides of trace impurities by the difference in their precipitation temperatures.
2. A heat treatment step (separation of Ta from Nb). The produced NbI_5 containing TaI_5 and a few other impurities is placed in a reactor for heat treatment under Ar at 623–673 K. The NbI_5 is reduced to NbI_3 or Nb_3I_8 during this step. TaI_5 that is not reduced and that has a higher volatility is transferred to another part of the reactor, where it is condensed and cooled. The Ta separation factor was determined to be 10–500.
3. Re-iodination of NbI_3 or Nb_3I_8 to NbI_5.
4. Thermal decomposition. The evaporation vessel ($P < 1.3 \times 10^{-4}$ Pa) is heated to 673–823 K. The vaporized NbI_5 is fed on to the surface of a Nb metal pellet, which is inductively heated to 1373–1673 K. The Nb metal pellet is rotated and raised slowly. Ar gas is used as a carrier gas. UHP

Nb (> 99.99%) deposits on the surface of the Nb pellet at a rate of a few tens of grams per hour.
5. Electron-beam melting.

The refining of crude Nb by molten salt electrolysis and electron-beam melting has also been discussed in [36]. The electrolysis was carried out in the common LiF–NaF–KF electrolyte containing 16.2 mass% K_2NbF_7 at 963 K under an Ar atmosphere. The cathodic current density was 480–700 A m^{-2}. The main remaining impurities in mass ppm were about 100–1000 Fe, 100 Ni and 1000–3000 O. After electron-beam melting (initial vacuum 1.3×10^{-3} Pa, final vacuum 6.7×10^{-3} Pa, sample mass 100–150 g, melting time 28.8–43.2 ks) the material still contained 15–20 mass ppm of metallic impurities and 80–85 mass ppm of interstitial elements. So the initial crude Nb of 99.8% could be refined to 99.99% purity.

4.3.3 Tantalum

The fused salt electrolysis of Ta has been described in [37]. By electrorefining in a melt of 20 mass% K_2TaF_7; 65 mass% NaCl; 15 mass% NaF at 1213 K, it has been found that elements more electropositive than Ta (W, Mo, Mn, Ni and Fe) do not dissolve from the anodic material, whereas impurities with a greater negative potential (Al, Zr, Ti, Cr and Si) pass into the electrolyte. But up to a certain level of concentration they are not reduced at the cathode. After electrorefining of sodium – thermal tantalum powder bricks in a melt of 80 mass% (NaCl–KCl–CsCl); 10 mass% K_2TaF_7; 10 mass% NaF, a refined tantalum with the following metallic impurity concentrations has been obtained (in at ppm): 15 Fe; 300 Nb; 5 Cu; 7 Mn; 40 Ti; 13 Al; 9 Mo; 6 Ni; 30 Si; 6 Co; 15 Mg; 7 V; 8 Sn; 5 W; 35 Cr; 45 Ca.

4.4 Group VIa Metals

4.4.1 Chromium

At CNRS-CECM in Vitry,France, research on chromium purification dates back to the 1960s and 1970s [11]. Aluminothermic Cr, electrolytic Cr and Cr purified by thermal dissociation of the iodide (van Arkel process) were used as starting materials. The best results have been obtained by electrolysis of an aqueous chromic anhydride solution, followed by heat treatment and zone refining. It has been found that on HF heating in a pure hydrogen atmosphere, nitrogen purification starts at 1220 K and carbon purification at about 1470 K, while oxygen removal is more difficult and only starts at about 1770 K, with much slower kinetics. A 20 h annealing at this temperature lowers the O content from 1800 to 10 mass ppm and the carbon content to 15 mass ppm. Zone refining in vacuum is hampered by the high vapor pressure of Cr at the melting temperature of about 931 Pa (7 torr). In order to avoid major distillation losses of Cr (4–5% per pass), zone melting is carried out in a copper crucible under a high-purity hydrogen atmosphere. Optimum results have been obtained with three passes at a melting rate of $4.5\,\text{cm}\,\text{h}^{-1}$. Although the distribution coefficients are not very favorable, Fe, Ni, Cu and N are enriched at the bottom of the bar, while Al, Sb and Na are eliminated by distillation. The main impurities of the purified Cr are (in mass ppm) 2 Fe, 0.2 Na, 0.15 Cu, 7 C, 10 N and 3.3 O. The resistivity ratio measured at liquid hydrogen temperature R(293 K)/R(20.4 K) was in the range of 220 to 420 [11].

At ENSM St. Etienne [38], pure Cr is prepared by electrolysis of commercial CrO_3 dissolved in sulfuric acid. The temperature and current density are chosen to obtain a minimum oxygen content in the metal. This, however, requires a low deposition rate, allowing a production rate of only 0.2 kg per week. The Cr cathodes are remelted in cold crucibles under an Ar/H_2 atmosphere. The purified Cr contains < 30 mass ppm of summed impurities (73 elements analyzed). In detail, these are as follows (in mass ppm):

- $C < 10$, $N < 5$, $O < 5$,
- with < 3: Al, Fe, S,
- with < 1: K, Mg, Ca, Mn, Sm, Cu, Ag, In, Si (all other elements < 0.2).

4.4.2 Molybdenum

Using commercial sintered Mo rods as a starting material, just by use of EBFZM, single crystals with a RRR of about 5000–10 000 could already be prepared in the early 1960s [4, 7, 39]. These crystals contained about 130 mass ppm tungsten and about 30 mass ppm of other metallic elements. The concentration of the interstitial impurities C, N and O was of the order of 1–10 mass ppm of each element (however, since C, N and O were bound in precipitate compounds, their influence on the RRR was insignificant). Taking into account the resistivity contribution of W in Mo, of $0.076\,\mu\Omega\,\text{cm/at\%}$ (or $0.145\,\mu\Omega\,\text{cm/mass\%}$, see Table 4.2) it is obvious that W dominated the residual resistivity of this material [4]. The tungsten contamination cannot be removed from the Mo by EBFZM, due to the distribution coefficient of W in Mo of 1.5, and due to the fact that the vapor pressure of W is the lowest of all metals; that is, lower than that of Mo. Therefore, it was clear that any further increase in purity of the Mo would require a chemical pre-purification of the starting material by chemical and/or physical methods, with special emphasis laid on the removal of W.

The best results have been obtained by liquid–liquid extraction of MoO_3 [4, 7, 40]. The whole preparation process is compiled in Table 4.11. After liquid–liquid extraction, the purified and dried MoO_3 is reduced with H_2 to Mo powder, which is then chlorinated to $MoCl_5$. The $MoCl_5$ is thermally decomposed on a hot Mo wire of high purity (van Arkel process) to polycrystalline Mo rods of a diameter of about 6 mm.

Table 4.11. Processing steps for the preparation of ultra-high-purity molybdenum [4, 7, 40]

Material	Processing step	Condition
Technical purity $MoO_3 \cdot H_2O$	Dissolution in 6M HCl	
	Extraction with tributyl phosphate/kerosene	Counterflow extraction with pulsating column, 2 l/h per phase
Wet MoO_3 with low W content	Back-extraction with ammonia, precipitation of $MoO_3 \cdot xH_2O$ with HCl	Yield 70%
	Conversion of pelletized wet MoO_3 into $MoCl_5$ in a mono-reactor process	Universal reactor
Dry MoO_3	Dehydration of $MoO_3 \cdot xH_2O$	Purging with N_2, 873–1123 K/3 h
Mo powder	Reduction of MoO_3 with H_2	Purging with N_2, 1073–1173 K/3 h
$MoCl_5$	Chlorination of Mo powder	
Polycrystalline Mo rod, 6 mm dia.	Thermal decomposition of $MoCl_5$ on high-purity Mo wire, 0.7 mm dia.	Open system, 1470–1670 K, evaporation rate 200–300 g $MoCl_5$/h, deposition rate 1 mm h^{-1}, yield 45%
High-purity Mo single crystal	Crucible-less electron-beam zone melting, three passes	Growth rate 3 mm min^{-1}, vacuum 10^{-3} Pa

Table 4.12. Typical impurity contents at various stages of the preparation of high-purity Mo (in at ppm with respect to Mo) [4,7,46]

Element	$MoO_3 \cdot H_2O$ starting material[a]	MoO_3 after extraction[a]	Mo single crystal[b]
W	60–150	<1	<1
Zr			<0.1
Si	1–60	7.2	<0.1
Co	2.8	<0.05	<0.05
Fe	3–150	<6.7	0.5
Mg	1.5	0.29	0.03
Cu	2.3	0.8	<0.06
Cr	<10	<10	
Ni	4	<2.5	
K			<0.05
Na			<0.1

[a] Analyzed by spectrophotometry or spectral analysis
[b] Analyzed by spark source mass spectrometry

In [41] this process has been further improved in such a way that the processing steps of drying of the MoO_3 powder, reduction to Mo, chlorination and evaporation of the $MoCl_5$ are performed in a single reactor without intermediate handling of the product, but only separated by intermediate flushing with inert gas.

From the polycrystalline Mo rods, high-purity single crystals with the desired orientation were prepared by EBFZM with three passes in a vacuum of 10^{-3} Pa at a zone speed of 3 mm min^{-1} and using oriented seeds.

The purification effects of some of the various processing steps are documented by the analyses given in Table 4.12. The essential effect of the chemical purification consists of the removal of tungsten. All other metallic elements will evaporate during zone refining.

In the previous chapters, which describe the preparation processes for UHP metals, no information was given concerning the structural perfection of the materials. Therefore, a few remarks will be added here for Mo, just to illustrate this aspect. The substructure of the as-grown Mo single crystals was characterized by means of etch pit investigations, transmission electron microscopy and Berg–Barrett X-ray topography. The dislocation density was of the order of 10^5–10^6 (sometimes up to 10^7) cm^{-2} with inhomogeneous distribution. The crystals had a subgrain structure with equiaxed subgrains of 500 μm mean diameter and a misorientation of 1–2 (in few cases up to 10) minutes. Sometimes small areas of subgrains with a diameter of about 50 μm have been found [40]. Single crystals with a much smaller dislocation density of $< 10^3$cm^{-2} could be obtained by means of a modified "strain-anneal" technique (critical plastic deformation followed by recrystallization annealing in a temperature gradient moving through the crystal).

The nitrogen content of the single crystals was about 0.2 mass ppm. The oxygen analysis yielded a value of about 0.3 mass ppm which was, however, mainly caused by surface-bound oxygen.

The zone-refined crystals had a RRR of > 250 000, which corresponds to a sum of solute impurities of about 3 at ppm. Even this value was still dominated by the remaining tungsten content, which was responsible for an RRR of about 550 000. The best RRR value obtained in a UHP-Mo single crystal was 1 000 000 (with size effect correction), which corresponds to a solute impurity content of less than 1 at ppm [42–45]. The RRR contribution of other structural lattice defects was thus very low. This is one of the best RRR values of any UHP metal obtained so far. In the "routine production" of Mo single crystals, RRR values > 100 000 could be realized [4, 7].

4.4.3 Tungsten

UHP tungsten single crystals can comparatively easily be prepared by EBFZM of commercial sintered rods. The melting point of tungsten, at 3695 K, being the highest of all metals, leads to evaporation of practically all impurity elements during electron-beam melting in high vacuum. Therefore, no chemical pre-purification is necessary. Already, in 1960–1961, RRR values of > 100 000 were obtained [47], and a few years later values of up to 300 000 could be achieved (RRR value with size effect correction) [48,49]. Single crystals of 1.2–6.0 mm in diameter and up to 100 mm in length were prepared by between one and three zone-melting passes. This proved to be sufficient to evaporate most of the impurities. A higher number of passes yielded no further improvement of the residual resistivity.

4.5 Group VIIa Metals

4.5.1 Rhenium

Re is one of the heaviest (density 21 g cm^{-3}) and one of the most strongly strain-hardening metals (580 HV after 30% cold working). It has one of the highest melting points (3459 K) and the third-highest modulus of elasticity of any metal. Furthermore, it forms no carbide phases and has no ductile–brittle transition down to room temperature, so that it is comparatively ductile [50]. Formerly used mainly as an alloying element to improve the strength, ductility and toughness of other refractory metals (mainly Mo and W), it is now increasingly recognized as an attractive structural material for high-temperature and energy systems applications such as space and missile propulsion systems, since it is virtually inert to thermal shock loading [51]. The two primary methods for fabricating metallic Re are by powder metallurgy (PM) and chemical vapor deposition (CVD).

The CVD process most often used to deposit Re is with chlorine gas [51]. The chlorine is fed into a chamber heated to 773 K which contains Re chips or pellets. The chlorine reacts with Re to form $ReCl_5$, which is volatile at this temperature. The substrate on which the Re is to be deposited is heated to about 1470 K. At this temperature, the $ReCl_5$ is decomposed: the Re metal precipitates on the substrate, and the Cl_2 gas evaporates and is exhausted from the reaction vessel. The deposits have 99.99–99.999% purity. With a suitable choice of processing conditions (temperatures, pressures and gas flow rates), impurities tend to be either left behind in the precursor or to bypass the substrate and flow out the exhaust rather than deposit [51].

The production of Re metal with a purity of more than 99.99% (based on metallic impurities; interstitial impurities and adsorbed gases are not taken into account) is also possible by the PM route using a sophisticated technology [51]. The starting material is ammonium perrhenate NH_4ReO_4 which is placed in a reactor to decompose the salt to rhenium heptoxide Re_2O_7. The heptoxide is captured in a recovery section and transformed with H_2O to perrhenic acid $HReO_4$. By passing the perrhenic acid through an ion-exchange column, alkali metals and other cation impurities are removed. The purified acid is then neutralized with ammonia to pure ammonium perrhenate. After grinding it to a fine powder, it is reduced in a tube furnace with hydrogen in a two-stage reduction process. In the first step, partial reduction and stripping away of the ammonia occurs, whereas in the second step (with other temperatures and hydrogen flow rates) the reduction to metallic Re is carried out [51]:

$$\text{First reduction:}\quad NH_4ReO_4 + H_2 \rightarrow Re_xO_y + NH_3 + H_2O\,, \tag{4.2}$$

$$\text{Second reduction:}\quad Re_xO_y + H_2 \rightarrow Re + H_2O\,. \tag{4.3}$$

The powder can then be compacted, sintered and zone-refined to yield a material of very high purity.

4.6 Interstitial Impurities

4.6.1 General Remarks

The group IVa–VIIa refractory metals react intensively with gases such as oxygen, nitrogen, hydrogen, water vapor, carbon monoxide and hydrocarbons. The details depend on the metal under consideration as well as on temperature, partial pressure and time. Since the main topic of this section is the presentation of methods for the preparation of UHP refractory metals, the discussion will concentrate on these problems, leaving aside details of controlled doping with interstitial impurities and of conditions leading to the pronounced formation of second phases – such as carbides, oxides and so on – on the metal surface.

The behavior of refractory metals in gaseous environments, especially the thermodynamic and kinetic aspects of metal–gas reactions that describe

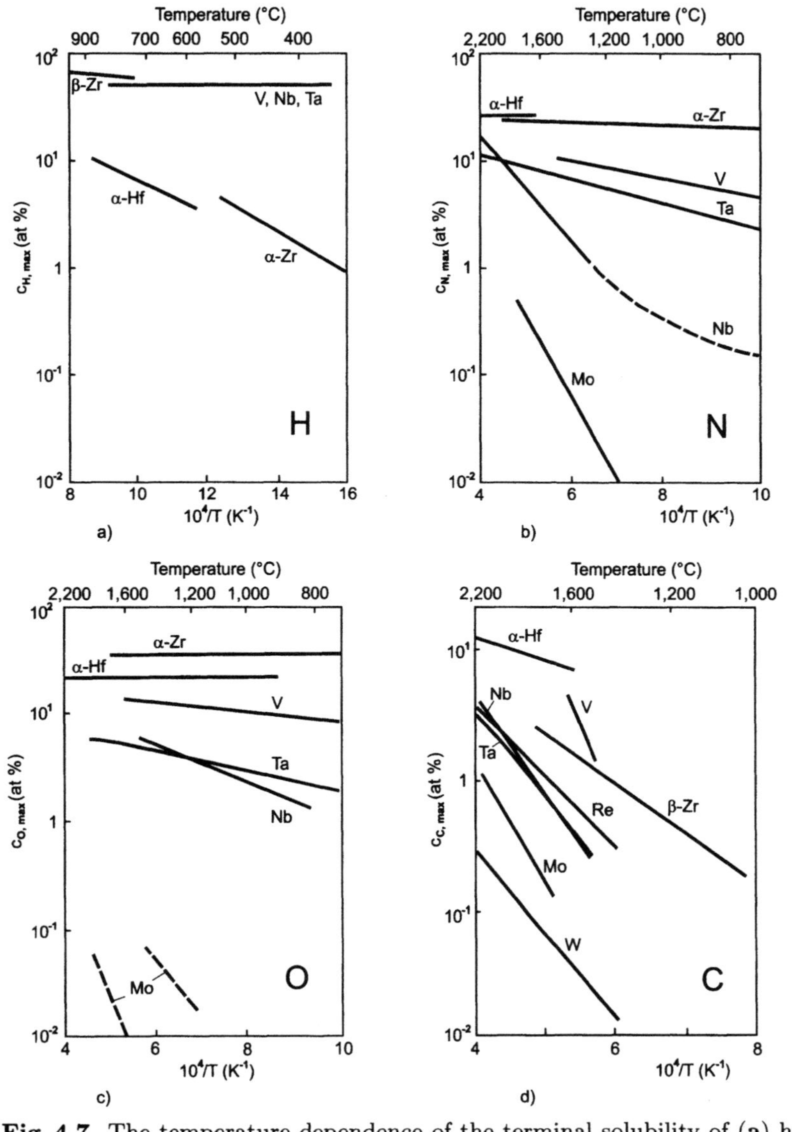

Fig. 4.7. The temperature dependence of the terminal solubility of (**a**) hydrogen, (**b**) nitrogen, (**c**) oxygen and (**d**) carbon in group IVa–VIIa metals [63]

absorption and desorption processes at high temperatures and low gas pressures, has been very thoroughly investigated by E. Fromm, E. Gebhardt, G. Hörz, H. Jehn, K. Schulze and co-workers at the Max-Planck Institute of Metals Research in Stuttgart, Germany [52–63]. One of the most recent and comprehensive descriptions primarily referred to here has been given by G. Hörz [63].

Table 4.13. Electrical resistivity contributions caused by interstitial impurities in refractory metals

Group	Metal	ρ (273 K) [$\mu\Omega$ cm]	ρ_{imp} [$\mu\Omega$ cm/at%]			
			C	N	O	H
IVa	Ti	40[a]	11.4[b]	8[c]	7[c]	3.2[b]
				8.9[b]	10.7[b]	
	Zr	42[a]		9.1[c]	8.9[c]	
	Hf	30[a]	14.0[c]	11.2[c]	9.6[c]	
Va	V	18.9[d]		4.9[d]	5.3[d]	1.12[e]
	Nb	13.3[d]	4.3[f]	5.2[f]	4.5[f]	0.8[f]
	Ta	12.1[d]	5.5[a]	5.2[d]	5.5[d]	0.92[d]
VIa	Cr	11.8[d]	18.2[g]	12[g]		
	Mo	4.88[h]	3[i]	2.4[c]	1[c]	
	W	4.84[j]	3.8[d]			
VIIa	Re	18[a]				

References: [a] [3]; [b] [67]; [c] [58]; [d] [68]; [e] [60]; [f] [5]; [g] [68]; [h] [65]; [i] [69]; [j] [12]

The terminal solid solubilities of H, N, O and C in refractory metals are shown in Figs. 4.7a–d [63]. Generally, the group IVa and Va metals exhibit large solubilities for interstitial impurities, and even at room temperature appreciable amounts of H, O and N can be retained in solid solution. The solid solubility of carbon in these metals (especially in group Va) is much more temperature-dependent, leading to low values at ambient temperature. In group VIa metals, the interstitial solubility is very low, so that C, N and O are normally present as carbides, nitrides or oxides in these metals. Very high quenching rates from high temperatures are necessary to retain up to a few hundred ppm of N and C in solid solution in Mo and W.

Since the measurement of the residual resistivity ratio is a rapid and frequently applied method to evaluate the overall level of purity of UHP metals, the specific contributions of interstitial impurities to the electrical resistivity of refractory metals are compiled in Table 4.13. Together with the values for substitutional impurities given in Table 4.2, they can help to define the RRR values necessary to achieve a desired degree of purity or to check globally the success of special steps or measures in the purification process. (However, it should be kept in mind that the resistivitiy contributions listed in Table 4.13 are for impurities in interstitial solid solution. If they are present as precipitates, as is mostly the case in Mo and W, their contribution to the electrical resistivity is insignificant.)

The different types of basic reaction mechanisms of gases with refractory metals are presented schematically in Figs. 4.8a–d [63]. As will be discussed below, in the reactions with nitrogen and hydrogen, true thermodynamic equilibria can be established, whereas the processes of pick-up and removal of oxygen and carbon in most cases follow different reaction paths, which at low pressure and high temperature can lead to steady states.

4.6.2 Nitrogen and Hydrogen

As is shown in Fig. 4.8a, d, the introduction of nitrogen and hydrogen from N_2- and H_2- containing atmospheres into refractory metals occurs via the following steps:

- adsorption of gas molecules from the gas phase at the metal surface,
- dissociation of the molecules and chemisorption of the N and H atoms,
- transfer of the gas atoms or ions through the metal surface,
- diffusion of the N and H atoms into the bulk forming an interstitial solid solution.

For degassing at high temperature, the same reaction steps are passed in the reverse direction:

- diffusion of N and H atoms from the bulk solid solution to the surface,
- recombination of N and H atoms at the surface into molecules,
- desorption of N_2 and H_2 molecules into the gas phase.

If the number of adsorbed and desorbed molecules per unit of time is equal, a thermodynamic equilibrium between the metal and the gas phase is reached.

The equilibrium concentration of interstitially dissolved nitrogen can be described by Sievert's law (which is valid for an ideal solid solution):

$$x_{\mathrm{N}} = \sqrt{p_{\mathrm{N}_2}} \exp\left(\frac{-\Delta S^{\circ}_{\mathrm{N}}}{\mathbf{R}}\right) \exp\left(\frac{-\Delta H^{\circ}_{\mathrm{N}}}{\mathbf{R}T}\right), \tag{4.4}$$

where x_{N} is the atomic fraction of nitrogen interstitially dissolved in the metal, p_{N_2} is the partial pressure of N_2 in the gas phase, T is the temperature of the metal sample, and $\Delta S^{\circ}_{\mathrm{N}}$ and $\Delta H^{\circ}_{\mathrm{N}}$ are the entropy and heat of solution of nitrogen in the metal.

Some available experimental data of the heat and entropy of solution of N and H in group Va and VIa bcc metals, as well as in α-Hf and Re (hcp), are given in Table 4.14 [63]. As can be seen from Table 4.14, the heat of solution for N and H in group IVa and Va metals is negative – that is, the dissolution is an exothermic process – and the dissolved concentration decreases with increasing temperature. Opposite to this, the dissolution of N in group VIa metals and in Re is an endothermic reaction, which causes an increasing concentration with increasing temperature.

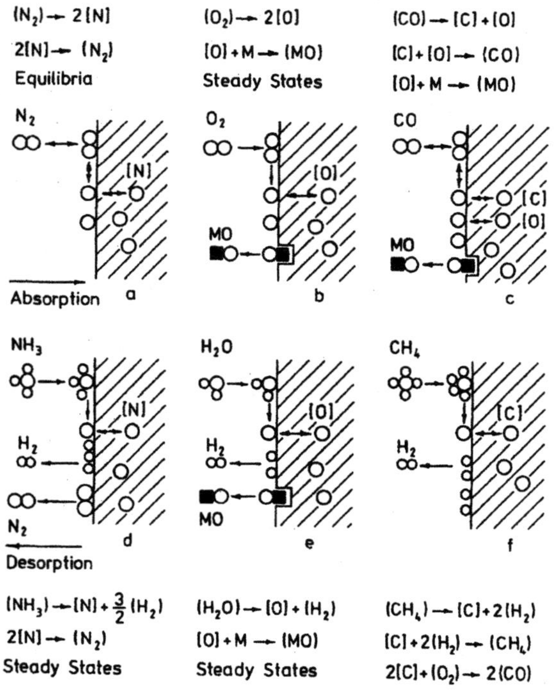

Fig. 4.8. Reaction mechanisms for basic refractory metal–gas reactions [58, 59, 63]

Table 4.14. The heat and entropy of solution of nitrogen and hydrogen in α-Hf, V, Nb, Ta, Mo, W and Re [63]

Group	Metal	Nitrogen		Hydrogen
		$\Delta H^\circ_\mathrm{N}$ (kJ mol^{-1})	$\Delta S^\circ_\mathrm{N}$ (K mol^{-1})	$\Delta H^\circ_\mathrm{H}$ (kJ mol^{-1})
IVa	α-Hf			−37.6
Va	V			−32.5
	Nb	−178	−70.3	−39.6
	Ta	−182	−60.2	−36.4
VIa	Mo	+92	−21.5	+52.2
	W	+195	−11.5	+100.4
VIIa	Re	+66.9		

In Fig. 4.9 [63] nitrogen concentration isotherms for Nb, Ta, W, Mo and Re are shown as a function of N_2 partial pressure (solid lines; the dot-dashed lines represent the terminal solid solubility).

At high temperatures, Nb and Ta can dissolve large amounts of N up to about 10 at% already at low N_2 pressures of 0.1–100 Pa. In group VIa the

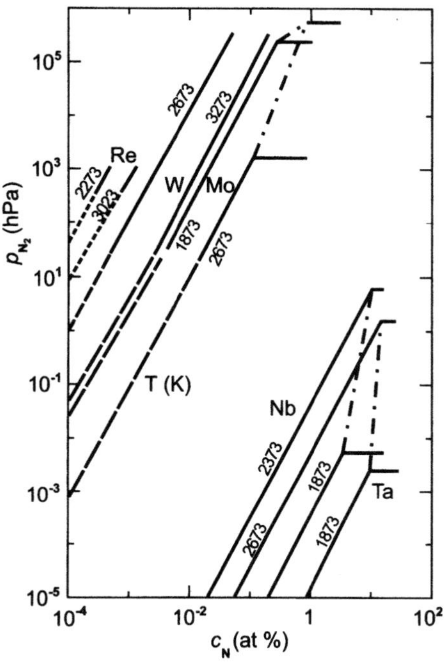

Fig. 4.9. Isotherms of the nitrogen concentration C_N as a function of the nitrogen pressure in systems of group Va, VIa and VIIa metals with nitrogen [63]

solubility is low even at high external N_2 pressure [for example, in W at 3270 K and 1 bar N_2, no more than about 140 at ppm (or 10 mass ppm) can be dissolved].

With respect to the purification of refractory metals from N by means of high-temperature annealing, it is clear that a diminution in the concentration dissolved in the metal can only occur if the equilibrium N_2 pressure calculated from (4.4) is higher than the N_2 partial pressure in the gas atmosphere surrounding the sample. The sign and rate of the concentration change of N in Nb and Ta is determined by the concurrent reactions of absorption and desorption described by the equation [7, 60]

$$\frac{\mathrm{d}c_N}{\mathrm{d}t} = k_1(T)p_{N_2} - k_2c_N^2(T)\,. \tag{4.5}$$

For Mo and W, the rate law is

$$\frac{\mathrm{d}c_N}{\mathrm{d}t} = \sqrt{p_{N_2}}k_3(T) - k_4c_N(T)\,. \tag{4.6}$$

The different laws for group Va and VIa metals are due to the fact that in Nb and Ta the introduction is controlled by the dissociative absorption of nitrogen, whereas in Mo and W it is controlled by the transfer of N atoms through the surface and their diffusion into the bulk metal. In the degassing reaction for Nb and Ta, the recombination of N atoms to N_2 molecules at the surface (bimolecular reaction) and their subsequent desorption is rate-controlling. In Mo and W, the diffusion of N atoms to the surface is the determining step [60].

Figure 4.10 [63] shows in the lower part the final concentrations of N and H which can be reached by annealing under a partial pressure of N_2 and H_2, respectively, of 10^{-4} Pa (i.e. under high-vacuum conditions). In group IVa and Va metals, due to the exothermic reaction, the annealing temperature should be as high as possible near the melting temperature, whereas the endothermic reaction in the group VIa and VIIa metals leads to lower concentrations at lower temperatures. However, taking the kinetics of the degassing reaction into account, also in the case of Mo and W, an intermediate temperature is recommended. As can be seen from Fig. 4.10, final concentrations of 1 at ppm and less can be obtained for H in refractory metals and for N in Mo, W and Re. The final concentrations of N in Nb and Ta are of the order of 20 at ppm.

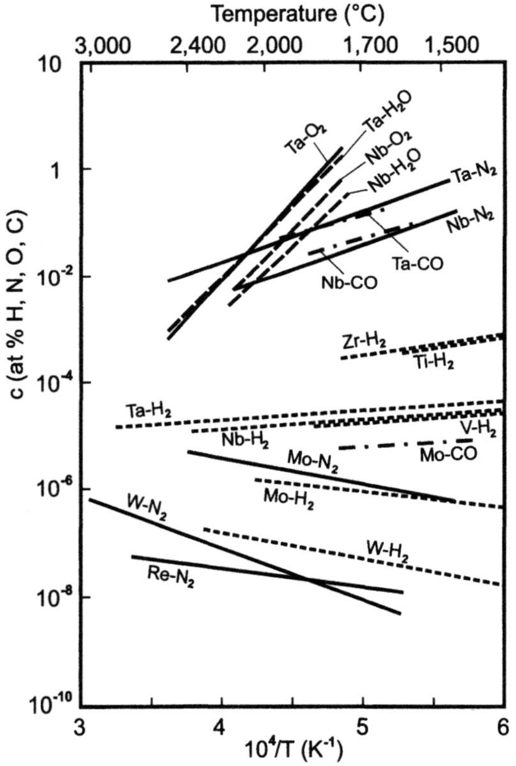

Fig. 4.10. The final concentrations of H, N, O and C in refractory metals after degassing at residual partial pressures of 10^{-4} Pa H_2, N_2, O_2, H_2O and CO, respectively [53, 55, 58, 62, 63]

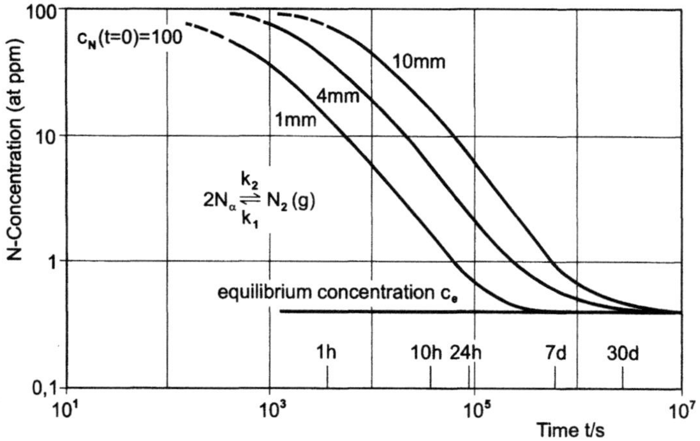

Fig. 4.11. The degassing of nitrogen from Nb cylinders at 2400 K; $p_{N_2,e} = 5 \times 10^{-6}$ Pa and $S_{N_2} = 10\lambda k_1(T)$ [5]

Thus, UHV conditions are necessary to reduce the N concentration to the sub-ppm level. Since the rate-controlling process of the degassing reaction is, in group Va metals, the recombination of N atoms to N_2 molecules on the metal surface, this may – for such low final N concentrations – require rather long and inconvenient annealing times, of the order of days or even weeks. Figure 4.11 shows the degassing kinetics of Nb cylinders 1–10 mm in diameter on annealing at 2400 K under a N_2 partial pressure of 5×10^{-9} Pa [5].

However, it should be mentioned that in spite of the results of the thermodynamic and kinetic investigations discussed above, it has been found by RRR measurements and special UHV extraction techniques that at least 0.1–0.5 at ppm of N, C and O remained in solid solution in the bulk of cylindrical rods of Nb [5].

In Mo and W, the diffusion of N to the surface is the rate-controlling step [57], as is also the case in other systems when the temperature is lower and the metal samples are thick. In these cases, the degassing time depends on r^2/D (where r = sample radius and D = diffusion coefficient of N).

In vanadium, due to its lower melting temperature and its high vapor pressure, it is not possible to reduce the N concentration by means of high- or ultra-high-vacuum annealing treatments in the solid state. Instead, a slow increase in the N content is observed due to V metal evaporation.

4.6.3 Oxygen

As illustrated in Fig. 4.8b, the introduction of oxygen into refractory metals from an O_2-containing atmosphere occurs in the same way as for N and H; that is, by the reaction steps absorption–dissociation–chemisorption–diffusion. Other possible sources of O are H_2O and CO in the environmental

Table 4.15. The parameters K_o and ΔQ of (4.7) describing the steady states in the systems Nb–O_2, Nb–H_2O, Ta–O_2 and Ta–H_2O ($n = 1$, x_o in at%, p in Pa) [63]

Parameter	System			
	Nb–O_2	Nb–H_2O	Ta–O_2	Ta–H_2O
$-\Delta Q$ (kJ/mol)	502	480	560	530
K_o (at% Pa)	9.1×10^{-10}	1.5×10^{-10}	1.35×10^{-10}	6.4×10^{-10}

atmosphere (see Fig. 4.8c, e). The removal of O, however, does not occur by recombination of O atoms to O_2 molecules and their desorption (this occurs only at a very low O_2 partial pressure), but by evaporation of metal oxides or sub-oxides.

Thus in metal–oxygen systems no true thermodynamic equilibrium is established. In the case in which the number of oxygen atoms incorporated per unit of time in the metal equals that evaporating as MeO_x, a steady state is established in which the oxygen concentration interstitially dissolved in the metal depends on the oxygen (or H_2O or CO) partial pressure in the gas phase, on the reaction probability at the surface and on the temperature. This steady state is connected with a continuous metal loss that may be orders of magnitude higher than that due to the vacuum evaporation of the pure metal. This type of reaction is most important for Nb and Ta at temperatures > 2000 K and O_2 and H_2O partial pressures < 10^{-2} Pa. It can be described by the equation

$$x_o = \left[K_o p_{O_2} \cdot \exp\left(\frac{-\Delta Q}{RT}\right)\right]^{1/n} , \qquad (4.7)$$

where $n = 1$ for evaporation of NbO and TaO (as is mostly prevailing), $n = 2$ for evaporation of NbO_2 and TaO_2, K_o is a constant, and $\Delta Q = Q_{ab} - Q_{des}$, the difference between the activation energies of absorption and desorption of oxygen.

The reaction parameters that describe the steady states for Ta–O and Nb–O are given in Table 4.15 [63].

As an example, the isotherms of the steady state oxygen concentration in Ta as function of the oxygen pressure are shown in Fig. 4.12 [63]. The upper dashed line characterizes the formation of a surface oxide at the solubility limit. It can be seen that in the steady state interstitially dissolved oxygen concentrations of up to about 7 at% may be reached. The minimum oxygen concentrations which can be extrapolated for Ta and Nb with annealing at a temperature near the melting point and an O_2 partial pressure of 10^{-6} Pa of the order of 0.1 at ppm or less.

In vanadium, the possibilities of removing oxygen by vacuum annealing are restricted due to the lower melting temperature and high vapor presure of V. In our own experiments, it was observed that an 1 h annealing just

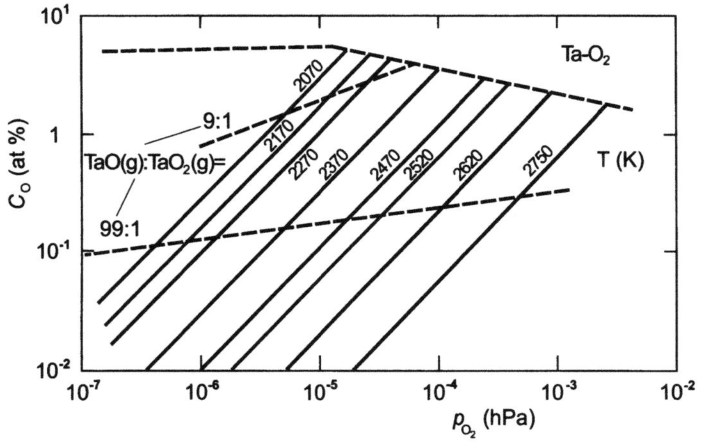

Fig. 4.12. Isotherms of the steady state concentration of oxygen in Ta as a function of the oxygen partial pressure and the temperature [63]

below the melting point of a 1 mm dia. wire in a vacuum of 2×10^{-4} Pa reduced the oxygen concentration from a starting value of 890 at ppm to about 220 at ppm, whereby the sample lost about 30% of its initial weight by evaporation. Similarly, in [70, 71] a reduction to 30% of the initial O concentration was found after an 80 min annealing at 1823 K in a vacuum of 1×10^{-4} Pa. At a temperature of 1823 K an optimal compromise was found between oxygen removal and metal evaporation. The oxygen content decreased exponentially according to

$$\ln \frac{c}{c_o} = -\frac{F}{m} \cdot t \cdot 2.55 \times 10^{13} \exp\left(\frac{6.88 \times 10^4}{T}\right), \tag{4.8}$$

where c_o is the initial O concentration, F is the surface area, m is the mass, t is the time in min, and T is the temperature.

In contrast to the group Va metals, the oxygen solubility in group VIa bcc metals is very low. However, also in the systems of these metals with O_2 and H_2O atmospheres, steady states are observed at temperatures above 1300 K. Here, several types of metal oxides are involved in the evaporation reaction, with their mixture being strongly dependent on temperature and pressure [63].

The effective removal of oxygen from group VIa metals is also possible by high-temperature annealing in a pure H_2 atmosphere. For example, with Mo rods of 5 mm in diameter, this could be accomplished by a 2 h anneal at 2800 K in a H_2 stream at 1 bar [72].

4.6.4 Carbon

In the group IVa metals Ti, Zr and Hf, the carbon solubility in the hcp α-phase is said to be low, very low and rather high, respectively, and it is low in all of the bcc β-modifications [61]. V, Nb, Ta and Re exhibit high terminal solubilities for C at high temperatures, amounting to 3.5–7 at% which, however, decrease rapidly with decreasing temperature. This is also the case for Mo and W, but at a level of about an order of magnitude lower, reaching a value of the order of 1 at ppm at 1273 K [61].

The pick-up of carbon by refractory metals occurs mainly through the adsorption and decomposition of hydrocarbons or CO on the metal surface. The reaction probability r for the irreversible decomposition of hydrocarbons (i.e. the ratio of the number of molecules decomposed to that impinging on the metal surface) decreases with increasing stability of the hydrocarbon. In the reaction with Nb, for example, r is near 1 for C_2H_2, of the order of 10^{-2} for C_3H_8 and about 10^{-4} for CH_4 [63].

The reverse process – that is, the removal of carbon by formation of CH_4 in a H_2 atmosphere – is possible with Mo and W (not in group IVa and Va metals), but requires temperatures of 1800–2300 K, a high partial pressure ratio of p_{H_2}/p_{CH_4} and a large amount of H_2 gas [63, 73].

In group Va metals that contain C and an excess of O in solid solution ($c_O \gg c_C$), decarburization can occur via the combination of C and O atoms at the surface and evaporation of CO and MeO_x occuring concurrently [63]. The ratio of CO to MeO_x formation depends on the oxygen coverage of the metal surface. Volatile oxide formation will prevail at low oxygen coverage. Decarburization by this process is, however, only possible down to a critical residual carbon concentration of the order of 10–40 at ppm [5]. The most effective decarburization can be achieved by annealing in a low-pressure oxygen gas stream, removing the carbon by CO formation. For example, this may be carried out for Nb at a pressure of 10^{-2}–10^{-3} Pa O_2 at 2300 K (the ratio of CO and O_2 partial pressures p_{CO}/p_{O_2} should be < 0.1 [27]). Even for thin samples, at least 4 h is required [5] to obtain C concentrations < 1 at ppm.

For decarburization of Nb through CO formation from dissolved C and O, the final C concentration can be estimated from the equation [24, 25, 74, 75]

$$x_{[C]} = \frac{p_{CO}}{x_{[O]}} \cdot \exp\left[\frac{\Delta G}{\mathbf{R}T}\right] = \frac{p_{CO}}{p_{O_2}} \cdot \exp\left[-\frac{5600}{\mathbf{R}T} - 11.2\right] , \tag{4.9}$$

and for degassing in an O_2 atmosphere by the equation

$$x_{[C]} = \frac{p_{CO}}{p_{O_2}^{\frac{1}{2}}} \cdot \exp\left[-\frac{12750}{\mathbf{R}T} - 5.30\right] \qquad p \text{ in bar, } T \text{ in K} . \tag{4.10}$$

Decarburization in a low-pressure O_2 atmosphere introduces oxygen into the metal, which must be removed afterwards by UHV degassing. This is more difficult in group VIa metals than in group Va metals. Therefore, for decarburization of group VIa metals, annealing in a $H_2 + H_2O$ atmosphere ("wet hydrogen") is recommended [73]. This allows a sufficiently high amount of oxygen to be brought to the metal surface and at the same time reduces the chemical potential of oxygen in the gas atmosphere, thus reducing the metal losses considerably by oxide evaporation. In 1.8 mm dia. Mo rods doped with 130 at ppm of radioactive ^{14}C, after a 4 h anneal at 2100 K in an atmosphere of 1.3×10^4 Pa H_2 + 700 Pa H_2O, no residual radioactivity could be detected (^{14}C content < 0.1 at ppm). The minimum carbon concentration that can be obtained by this treatment is primarily dependent on the purity of the H_2 gas with respect to carbon-containing compounds (hydrocarbons, CO) [71].

The desorption behavior of C and O from liquid Nb has been investigated in [76], using an electromagnetic levitation technique under various conditions of temperature, CO partial pressure and time. The pick-up of C and O from a CO atmosphere resulted in a higher C content against O due to the fact that some O re-evaporated as niobium oxide or sub-oxide. When excess carbon exists in the melt, only the CO evolution reaction is responsible for the decrease in the C and O concentrations. Vaporization of Nb oxides or sub-oxides is negligible in this case. The desorption process is second order, which implies that the overall reaction rate is controlled by the recombination of C and O on the liquid surface.

4.6.5 Purification by Gettering with a More Reactive Metal

Here, only external gettering will be discussed (i.e. gettering by an reactive element that is not alloyed to the metal to be purified but is outside of it).

In certain cases it is possible to reduce the interstitial impurity content of refractory metals by annealing them in a sealed capsule or in vacuum together with a more reactive metal, if the thermodynamic data (dissociation pressures, Gibbs free energies) are suitable [77].

In [78] the equilibrium concentration of O in V in contact with Ca, Mg and Ba at 1473 K has been studied. The calculated equilibrium values were 0.03, 0.7 and 7.8 at ppm for Ca, Mg and Ba, respectively. Experimentally, the O content was decreased from 225 at ppm to 8 at ppm with Ca or Mg and to 14 at ppm with Ba.

Vanadium wires were wrapped in Ti or Zr foils and heated in a sealed quartz capsule to 1273–1473 K [79]. From the height of the oxygen Snoek peak, a final oxygen content of 25 at ppm was deduced, whereas the N peak disappeared completely [77].

In [80] Nb samples ($2 \times 2.5 \times 100\,mm^3$, RRR = 170) were placed in a Ti cylinder 40 mm in diameter and heated to 1273–1573 K in a vacuum furnace at 4×10^{-4} Pa for up to 10 000 min. The purification occurred by evaporating a Ti layer on to the Nb specimen that was reacting with the interstitial impurities dissolved in the Nb. Higher RRR values could be obtained at lower annealing temperatures: however, this required a longer annealing time. The best RRR values obtained were about 1900.

A “cold removal” of hydrogen traces from niobium and tantalum, being especially suitable for thin specimens, was described in [81]. The samples were coated with palladium and subsequently exposed to air at about 420 K for several days. This enabled the complete removal of H by H_2O formation initiated by the catalytic action of the palladium.

4.7 Purification by Electrotransport

In electrotransport that occurs in metals at higher temperatures under the action of a high electric current passing through a rod or strip, the random diffusional motion of impurity atoms is biased according to the direction of the applied electric field. The force on an atom induced by the electric field consists of two components: the direct force is due to a net effective charge (either positive or negative with respect to the host lattice) of the moving atom, and the wind force is due to scattering of the current-carrying electrons at the impurity atom. The sum of both forces is proportional to the electric field and can be characterized by an effective valence Z^* of the impurity atom [70, 71]:

$$F = (Z_{\text{direct}} + Z_{\text{wind}}) \cdot e \cdot E = Z^* \cdot e \cdot E\,. \tag{4.11}$$

The electrotransport results in the enrichment of mobile impurities at one end of the sample. According to [70], the steady state distribution of a mobile solute follows the equation

$$c = k \cdot \exp\left(\frac{-UEx}{D}\right)\,, \tag{4.12}$$

where c is the concentration at distance x along the sample; k is a constant, being a function of the initial concentration c_o, the specimen length L and UE/D; D is the diffusivity; E is the electric field; U is the electrical mobility of the impurity atoms; and U/D is the purification ratio.

In V it has been found [82] that the interstitial impurities migrate in the direction opposite to that of the electron current. A vanadium of 99.93% purity (RRR = 138) could be refined to 99.99% purity (RRR > 1880) with residual contents in the refined portion of < 1 mass ppm O, < 0.5 mass ppm N and < 2 mass ppm C. In order to reduce the evaporation of V at high temperature, the treatment was carried out under a partial atmosphere of pure He [77, 82].

In Nb refined by electrotransport at 2173 K under a pressure of 1.3×10^{-8} Pa, the RRR increased from 35 to 7200. This effect was, however, at least partly due to vacuum degassing of O and N [83].

A combination of EBFZM with electrotransport was used in [84] for further purification and single crystal growth of 99.999% Nb sheet (4–5 mass ppm Ta). The current density was roughly 1300 A cm^{-2}, heating the samples to nearly 1870 K in a vacuum of $4 - -7 \times 10^{-5}$ Pa. The balance energy to melt the floating zone was supplied by the electron gun. It was shown that by this combined treatment the purity and quality of the Nb single crystals could be enhanced by almost a factor of two in RRR (best values of RRR up to 2600) over crystals that had only been zone-floated if the direct current was parallel to the zone melting direction. A reduction of RRR was observed if the directions were opposite. The electrotransport effect was supposed to be due to N atoms, which have a positive value of Z^* at all temperatures. A secondary effect of the direct current flowing through the sample can be due to the larger length of the molten zone, the higher temperature of the whole rod and the lower cooling rate, which may cause higher crystal perfection (lower dislocation density).

In principle, purification by electrotransport is also possible for substitutional impurities. However, the relatively low diffusion coefficients of most substitutional systems require extremely long times for purification to take place on a macroscopic scale, thus making this method impractical [85].

4.8 Deterioration of the Purity of UHP Refractory Metals During Preparation or Application

In closing this chapter on UHP refractory metals, a few examples illustrating some aspects of recontamination will be discussed briefly, taking Nb as an example:

- When preparing UHP-Nb via the van Arkel process described in Sect. 4.3.2, the thermodynamic equilibria of the gas–metal reactions (see above) are so unfavorable that during thermal decomposition of the niobium chloride nearly all of the C and N present in the starting material or in the reaction gas will be incorporated in the deposited niobium. Also, when preparing UHP-Nb by molten salt electrolysis, the content of the interstitial impurities in the deposited Nb (especially that of oxygen) is by far too high to obtain high RRR values, even when – in the construction of the electrolysis cell – the principles of high-vacuum or ultra-high-vacuum techniques are observed.
- If electron-beam zone-refining is carried out in high vacuum only, then the water vapor of the residual gas atmosphere contributes to an increased oxygen content in the molten rods. It has been observed that a partial pressure of 1×10^{-3} Pa H_2O yields an oxygen concentration of about 50 at ppm [27]. On the other hand, on EBFZM in ultra-high vacuum, the conditions are unsuitable for decarburization, and the time for degassing of nitrogen is mostly too short (Fig. 4.10). Thus subsequent decarburization and UHV degassing are necessary.
- During the long times required for nitrogen degassing in UHV, care must be taken to avoid a parallel increase in carbon content from the dissocia-

tion of CO and hydrocarbons from the pumping oil present in the residual gas atmosphere.

- The residual gas atmosphere in UHV vessels contains mainly H_2. Since, during and after cooling from UHV degassing, the metal surface is not protected by an oxide layer, after shut-off of the vacuum pumps quite large amounts of hydrogen may be dissolved in the metal, depending on the surface-to-volume ratio of the sample (the hydrogen solubility of Nb at 320 K and 10^{-6} Pa is about 3000 at ppm) [5]. This hydrogen can be removed by the Pd coating method, described in Sect. 4.6.5 [81].
- During storage in air, refractory metals are quickly covered by an oxide layer. It has been shown that at ambient temperature, 0.4 $\mu g\,cm^{-2}$ of O_2 and 0.05 $\mu g\,cm^{-2}$ of N_2 can be chemisorbed on the surface of clean Nb [5,86]. Large quantities of water and hydrocarbons may be adsorbed on to these oxide layers.
- During subsequent heating, the oxygen from these oxide surface layers can diffuse into the metal, even if the annealing is carried out in high vacuum and at a relatively low temperature. In our own investigations of the isochronal recovery ($\Delta T = 20$ K, $\Delta t = 10$ min) of deformed UHV-degassed Nb and Ta wires of 1 mm dia., it has been observed that already at $T_{ann} \geq 570$ K the residual resistivity starts to increase due to oxygen diffusing from the surface layer into the bulk (the diffusion distance of O in Ta and Nb at 573 K/10 min is about 0.2 μm). Thus any heating of UHP-Nb and Ta to higher temperatures will inevitably degrade the purity of the metal by oxygen diffusing from the surface into the bulk. With Ti, similar observations have been reported for temperatures beyond 773 K [67].

References

1. J.L. Wittemauer, K.N. Liddell: J. Metals **43**, 6 (1991)
2. D.R. Sadoway: J. Metals **43**, 15 (1991)
3. M.Winter: *The Periodic Table on the web,* http://www.shef.ac.uk/chemistry/webelements
4. J. Barthel, K.-H. Berthel, G. Ehrlich, C. Fischer, G. Günzler, P. Mühl, H. Oppermann, G. Sobe, G. Weise: In *10 Jahre Zentralinstitut für Festkoerperphysik und Werkstofforschung Dresden* (Wissenschaftliche Berichte des ZFW Dresden, 1979) p. 245
5. K.K. Schulze: J. Metals **33**, 33 (1981)
6. K.K. Schulze: Bull. Tantalum–Niobium Int. Study Center **49**, 2 (1987)
7. A. Köthe: In *Ultra high purity base metals*, Proc. UHPM' 94, ed. by K. Abiko, K. Hirokawa, S. Takaki (The Japan Institute of Metals, 1994), pp. 291–322
8. Y. Shindo, H. Miyazaki, Y. Yamakoshi: In *Ultra high purity base metals,* Proc. UHPM' 94, ed. by K. Abiko, K. Hirokawa, S. Takaki (The Japan Institute of Metals, 1994) pp. 413–416
9. H.W. Rosenberg, J.E. Green: In *Titanium '92 – science and technology*, ed. by I.H. Froes, I.L. Caplan (TMS, Warrendale 1993), Vol. 3, pp. 2371–2376

10. J. Bigot: Le Vide Suppl. **177**, 151 (1975)
11. J. Bigot: *In Proceedings of the sixth internal conference on ultra-high purity base metals, UHPM' 99*, May 31 – June 4, 1999, Sendai/Japan, ed. by K. Abiko, S. Takaki (The Japan Institute of Metals, Sendai 2000), pp. 1–6
12. Y. Yoshimura, Y. Shimizu, Z. Imamura: In *Titanium '92 – Science and Technology*, ed. by I.H. Froes, I.L. Caplan (TMS, Warrendale 1993), Vol. 3, pp. 2459–2464
13. C. Ouchi, H. Iizumi, S. Mitao: Mater. Sci. Eng. **A243**, 186 (1998)
14. R. Tricot: Mém. Etudes Scient. Rev. Métall. **88**, 747 (1991)
15. A.B.V. Dasilva, P.A. Distin: CIM Bull. **91**, no. 1018, 221–224 (1998)
16. L. Moulin, P. Thouvenin, P. Brun: *6th Conference on Zr in the nuclear industry,* Vancouver, June 28, 1982, ASTM STP 824, p. 37
17. P. Moinard: Thesis, Institut National Polytechnique de Grenoble, January 28, 1983
18. J.P. Garnier: Thesis, Institut National Polytechnique de Grenoble, September 20, 1983
19. A.P. Lamaze, D. Charquet: In *Refractory metals,* ed. by K. Nona et al. (The Minerals, Metals and Materials Society, 1990)
20. K. Mimura, M. Kasuga, M. Nanjo, M. Isshiki: J. Japan Inst. Metals **59**, 1124 (1995)
21. J.C. Sehra, R.H. Rakhasia, V.D. Shah: High Temp. Mater. Process. **16**, 123 (1997)
22. T.A. Sullivan: J. Metals **17**, 45 (1965)
23. K.P.V. Lei, T.A. Sullivan: J. Less-Common. Metals **14**, 145 (1968)
24. K.P.V. Lei, R.E. Campbell, T.A. Sullivan: J. Electrochem. Soc. **120**, 211 (1973)
25. E. Fromm, G. Spaeth: Z. Metallkd. **59**, 65 (1968)
26. K. Nishizawa, H. Sudo, T. Akiyama, M. Sugimori: In *Reactor dosimetry: methods, applications and standardization*, ed. by H. Farrar IV, ASTM STP 1001 (Lippincott, Philadelphia 1989) pp. 688–695
27. J. Barthel, K.-H. Berthel, K. Fischer, R. Göbel, G. Günzler, M. Jurisch, W. Neumann, J. Kunze, P. Mühl, R. Petri, G. Sobe, G. Weise, U. Wiesner: Fiz. Met. Metalloved. **35**, 921 (1973)
28. J. Barthel, K.-H. Berthel, C. Fischer, G. Günzler, J. Kunze, P. Mühl, H. Oppermann, R. Petri, G. Sobe, G. Weise: GDR patent No. 83 842, IPC C 22 b, 61/00, appl. May 4, 1970, publ. July 12, 1971
29. J.I. Mönch, K. Fischer, G. Enderlein, O. Großmann, M. Kubsch, C. Rodig, U. Stahlberg: In Proc. 6th Intern. Symposium *High Purity Materials in Science and Technology*, May 6–10, 1985, Dresden, (AdW der DDR, Zentralinstitut für Festkoerperphysik und Werkstofforschung Dresden, 1985); Poster Abstracts, p. 54
30. B. Fellmuth, D. Elefant, J.I. Mönch: Phys. Stat. Sol. (a) **100**, 597 (1987)
31. K. Schulze, E. Fromm, W. Grünwald, F. Haessner, S. Hofmann, R. Kirchheim, K.-D. Rasch, H. Schultz, M. Winterkorn: In *Reinststoffprobleme*, Proc. 4th Intern. Symposium *Reinststoffe in Wissenschaft und Technik*, Octber 14–17, 1975, in Dresden, ed. by M. Balarin (Akademie-Verlag Berlin, 1977), Vol. V, pp. 205–222
32. S. Senderoff, G.W. Mellors: J. Electrochem. Soc. **112**, 266 (1965)
33. S. Senderoff, G.W. Mellors: J. Electrochem. Soc. **113**, 66 (1966)
34. J.I. Mönch: Dr.-Ing. Dissertation, Akademie der Wissenschaften der DDR, 1984

35. K.H. Berthel: Dissertation, Technische Universität Dresden, 1976
36. K. Shimakage, T. Sato, S. Hirai, T. Toda, T. Takahasi, Y. Sakuta: J. Japan. Inst. Metals **59**, 947 (1995)
37. A. Kononow, S. Kuznetsov, E. Polyakov: J. Alloys Compounds **218**, 173 (1995)
38. J.L. LeCoze, R. Tardy, A. Kobylanski, M. Biscondi: In *Ultra high purity base metals*, Proc. UHPM '94, ed. by K. Abiko, K. Hirokawa, S. Takaki (The Japan Institute of Metals, 1994), pp. 371–389
39. H. Oppermann, G. Weise, J. Barthel, K.-H. Berthel: Phys. Stat. Sol. **22**, K151 (1967)
40. C. Fischer, P. Mühl, G. Günzler: GDR patent No. 48 589, IPC C 01 g, 39/00, appl. June 5, 1965, publ. May 20, 1970
41. H. Hammer, C. Fischer, H. Hauff, W. Gaertner, U. Hammer, R. Hentschel: GDR patent No. 140 668, IPC C 01 G 39/04, appl. Dec. 12, 1978, publ. August 08, 1982
42. K.-H. Berthel, D. Elefant: 4th Intern. Symp. *Reinststoffe in Wissenschaft und Technik*, Oct. 14–17, 1975, Dresden, volume of the extended abstracts of the original contributions, p. 279
43. D. Elefant: Thesis, Akademie der Wissenschaften der DDR/Zentralinstitut für Festkörperphysik und Werkstofforschung Dresden, 1971
44. K.-H. Berthel: Proc. 3rd Intern. Symp. *Reinststoffe in Wissenschaft und Technik*, May 4–8, 1970, Dresden, in *Reinststoffprobleme*, Vol. IV (Akademie-Verlag Berlin, 1972) p. 71
45. K.-H. Berthel, D. Elefant: Wissenschaftliche Berichte des ZFW Dresden, 1974, p. 2
46. P. Mühl: Proc. 4th Intern. Symp. *Reinststoffe in Wissenschaft und Technik*, Oct. 14–17, 1975 in Dresden, series *Reinstoffprobleme* (Akademie-Verlag, Berlin 1977) vol. V, p. 131
47. K.-H. Berthel, H.E. Longo, K. Schlaubitz: In Proc. Internat. Symp. *Reinststoffe in Wissenschaft und Technik*, Nov. 30 – Dec. 2, 1961, Dresden (Akademie-Verlag, Berlin 1963) p. 293
48. H. Oppermann, G. Günzler: GDR patent No. 48 588, IPC C 01 g, 39/00, appl. June 5, 1965, publ. May 5, 1971
49. K.-H. Berthel: Phys. Stat. Sol. **5**, 159 and 399 (1964); and K.-H. Berthel: In *Realstruktur und Eigenschaften von Reinststoffen*, Proc. 2nd Intern. Symp. *Reinststoffe in Wissenschaft und Technik*, ed. by E. Rexer, Sept. 28 – Oct. 2, 1965 in Dresden, part 3, series *Reinststoffprobleme*, Vol. III (Akademie-Verlag, Berlin 1967) p. 875
50. B.D. Bryskin: Adv. Mater. Process. **10**, 83–86 (1998)
51. A.J. Sherman, R.H. Tuffias, R.B. Kaplan: J. Metals **43**, 20 (1991)
52. E. Gebhardt, E. Fromm, D. Jacob: Z. Metallkd. **55**, 432 (1964)
53. E. Fromm: J. Less-Common Metals **14**, 113 (1968)
54. E. Fromm, H. Jehn: Metall. Trans. **3**, 1685 (1972)
55. G. Hörz: Z. Metallkd. **60**, 121 (1969)
56. E. Fromm: Vacuum **21**, 321 (1971)
57. G. Hörz: Metall. Trans. **3**, 3069 (1972)
58. E. Fromm, E. Gebhardt (ed.), *Gase und Kohlenstoff in Metallen* (Springer, Berlin Heidelberg New York 1976)
59. H.J. Grabke, G. Hörz: Ann. Rev. Mater. Sci. **7**, 155 (1977)
60. E. Fromm, G. Hörz: Intern. Metall. Rev. **25**, 269 (1980)

61. G. Hörz: J. Less-Common Metals **100**, 249 (1984)
62. H. Jehn, K. Schulze: in Proc. Symp. *Physical metallurgy and technology of molybdenum and its alloys*, ed. by K.H. Miska et al. (Ann Arbor, Michigan 1984) pp. 107–117
63. G. Hörz: In *Klaus Schulze symposium on processing and application of high purity refractory metals and alloys*, ed. by P. Kumar, H.A. Jehn, M. Uz (The Minerals, Metals & Materials Society, Warrendale, PA, 1994) pp. 127–151
64. H. Schultz: In *Ultra high purity base metals*, Proc. UHPM '94, ed. by K. Abiko, K. Hirokawa, S. Takaki (The Japan Institute of Metals, 1994) pp. 323–351
65. H. Oppermann, G. Weise, J. Barthel, K.-H. Berthel: Phys. Stat. Sol. **22**, K151 (1967)
66. K. H. Berthel: Phys. Stat. Sol. **5**, 399 (1964)
67. S. Komatsu, M. Ikeda, M. Yoshimura: In *Ultra high purity base metals*, Proc. UHPM '94, ed. by K. Abiko, K. Hirokawa, S. Takaki (The Japan Institute of Metals, 1994) pp. 452–457
68. J. Bigot, F. Faudot: In *Ultra high purity base metals*, Proc. UHPM '94, ed. by K. Abiko, K. Hirokawa, S. Takaki (The Japan Institute of Metals, 1994) pp. 478–481
69. A. Köthe, P. Burck: in Proc. Intern. Conf. *Defects in refractory metals*, ed. by R. DeBatist, J. Nihoul, L. Stals, Mol/Belgium, S.C.K./C.E.N. Mol, 1972, p. 117
70. D.T. Peterson, in *Electro- and thermo-transport in metals and alloys*, ed. by R.E. Hummel, H.B. Huntigton, (TMS-AIME, Warrendale, PA, 1977) p. 54
71. J.P. Dekker, A. Lodder: J. Phys. Condens. Mater. **10**, 6687 (1998)
72. D. Müller, Thesis, Akademie der Wissenschaften der DDR/Zentralinstitut für Festkörperphysik und Werkstofforschung Dresden, 1972
73. J. Kunze, W. Reichelt: J. Less-Common Metals **20**, 327 (1970)
74. E. Fromm, H. Jehn: Z. Metallkd. **58**, 61 (1967)
75. J. Elliot, M. Gleiser: *Thermochemistry for steelmaking* (Addison-Wesley, Reading, MA, 1960)
76. Hyun Gyoon Park, R. Abbaschian: Metall. Mater. Trans. B **29B**, 1309 (1998)
77. K.K. Schulze, O.N. Carlson, M. Uz: In *Klaus Schulze symposium on processing and application of high purity refractory metals and alloys,* ed. by P. Kumar, H.A. Jehn, M. Uz (The Minerals, Metals & Materials Society, Warrendale, PA, 1994), pp. 229–244
78. D.T. Peterson, R.G. Clark, W.A. Stensland: J. Less-Common Metals **30**, 169 (1973)
79. O. Yoshinari, T. Konno, K. Suma, M. Koiwa: J. Less-Common Metals **12A**, 1127 (1981)
80. H. Safa, D. Moffat, B. Bonin, F. Koechlin: J. Alloys Compounds **232**, 281 (1996)
81. J.A. Rodriguez, R. Kirchheim: Scr. Metall. **17**, 159 (1983)
82. O.N. Carlson, F.A. Schmidt, D.G. Alexander: Metall. Trans. **3**, 1249 (1972)
83. F.A. Schmidt, O.N. Carlson: J. Less-Common Metals **26**, 247 (1972)
84. E.U. Malang, C.S. Oglesby, T. Siegrist, E. Bucher: Physica B **204**, 363 (1995)
85. T.W. Ellis, F.A. Schmidt, K.A. Gscheidner Jr.: In *Ultra high purity base metals*, Proc. UHPM '94, ed. by K. Abiko, K. Hirokawa, S. Takaki (The Japan Institute of Metals, 1994) pp. 76–85
86. E. Fromm, O. Mayer: Surf. Sci. **74**, 259 (1978)

5 Purification of the Rare Earth Metals

D. Fort

5.1 Introduction

The rare earth elements have a history of intriguing scientists with their rich variety of physical properties. Early attempts to investigate these properties were hampered, however, by a lack of suitably pure samples, for it soon came to be realized that the intrinsic properties of these notoriously reactive metals could only be measured if their typical impurity levels were significantly reduced. Thus, for the past few decades, purification and property measurement have gone hand in hand, a process that is still not complete, as is regularly demonstrated by the discovery of new phenomena as higher-purity samples are produced. This chapter looks at the sources and nature of impurities in rare earth elements and discusses those properties of the rare earths that are relevant during their purification. Several purification methods are described, including vacuum melting, zone refining, vapor techniques and electrotransport. Since the aim of preparing very-high-purity metal is normally to perform some sort of experiment, rather than purification for its own sake, consideration is also given to sample preparation and crystal growth, for these operations generally take place alongside the purification and can sometimes be combined with it. The final section comprises an element-by-element review of the published literature.

5.2 Background Information

5.2.1 Impurities in Rare Earth Metals

Rare earth ores typically contain compounds of several rare earth elements, a situation that for over a century complicated preparation of the individual metals. This problem was overcome during the 1940s and 1950s when, using liquid–liquid extraction, ion exchange or a combination of the two techniques, it became possible to separate rare earth salts from each other to a high state of purity [1]. Thus oxides of the rare earths of remarkably good purity are available today. To prepare a rare earth metal, its oxide is usually converted to a fluoride, which is then metallothermically reduced to the metal using Ca, although direct reduction of the oxide is favored for those metals with high vapor pressures, in which case La is the normal reductant [2]. Since the rare earth oxides are available in higher purity than the metals, it follows that either impurities are introduced during the reduction process and/or all of the oxygen is not removed in the preparation of the metal. Gschneidner [3] lists the major impurities in the oxides as other rare earths and metals such as Ca, Si, Fe, Cu, Al and Na, while contaminants that can be introduced into the final metals during the reduction stage(s) are H and O from water in the HF used in the fluoride conversion, F itself, Ca (if used) and pick-up of crucible materials. Also, if the reductant itself is not pure, any impurities within it are likely to be transferred to the rare earth metal. Further contamination

Table 5.1. Start metal analyses for rare earth metals used in the author's laboratory. The metals were either purchased from commercial manufacturers (C) or supplied by the Ames Laboratory, Iowa State University (A). Impurities denoted "na" were not analysed for. The absence of an analysis value indicates that the impurity was at a very low level. All values are in ppm atomic

Impurity	Host						
	La	Pr	Nd	Nd	Gd	Gd	Tb
H	275	na	na	571	1560	472–708	445–730
C	na	1000	na	84	320	1000	na
N	30	500	203	21	11	na	60–350
O	295	4700	12 700	360	550	5000	3800
F		30	28			1–3	
Al		50	56		4	10–300	10–25
Si		15	<6	<6	2	15–30	3–7
S		15	<5			<7	
Cl	3		2	6		1–6	
Ca		6	2			<3	
Sc			1	4.3			210–550
Cr				3		<2	
Fe	10	10	120	11	2	100	10–100
Co			365		10	1	
Ni	2	75	34			1–3	
Cu	10	15	8	3	1	15	20
Y			10		2	3–30	55–60
Zr			82				
La			17		9	1	10–20
Ce			1	3.6		1	2–6
Pr			17			3–5	6
Tb			6	<2			
Ta	3.4		6	88		3	
W					5		
Source	(A)	(C)	(C)	(A)	(A)	(C)	(C)

can arise from the consolidation of the metals and from the preparation of samples, for all of the rare earth metals have a high affinity for impurities such as O, N, H and C, which can be readily introduced by the presence of water vapor, hydrocarbons or air leaks into the processing equipment.

Before giving examples of analyses from commonly available start metals, it should be noted that analyses in this article are normally quoted in the form of atomic % (at%) or parts per million atomic (ppm.at), rather than weight % (wt%) or parts per million by weight (ppm.wt). For heavy metals such as the lanthanides, there can be a significant difference between at% and wt%, especially as the impurities in these metals are usually of fairly low mass, so care must always be taken to ascertain whether impurity levels in a metal refer to weight or atomic values. Table 5.1 shows analyses for a number of rare earth metals from different sources. These analyses illustrate the fact that preparation techniques have advanced to the stage whereby the most abundant contaminants in currently available rare earths are not other metallic species but impurities such as O, H, N and C, which typically account for over 95% of the total impurity content found in these metals.

To fully understand rare earth purification, it is helpful to know how impurities are contained within the metals. Gases such as H, O and N are soluble in all of the rare earths, being present in the form of atoms (rather than molecules) that occupy interstitial sites within the metal lattice [4]. It is, of course, the relatively small size of these atoms that allows them to occupy interstitial sites, a feature also shared by C, another common impurity in rare earths. Metallic impurities, on the other hand, generally occupy substitutional sites in the rare earth elements. Interstitials usually have higher diffusion coefficients in rare earths than larger atoms, while H has an exceptionally high diffusivity, a phenomenon explained by assuming that it moves between interstices as a proton. With all but the best commercially available rare earths, the levels of some of the "gaseous" impurities (particularly oxygen) may exceed the solubility limit, meaning that the metals also contain precipitates (e.g. oxides).

5.2.2 The Basic Properties of the Rare Earths

The choice of a purification technique, or techniques, for a rare earth metal will depend upon the species of impurities to be removed and the properties of the rare earth element itself. The properties of relevance to purification are vapor pressure, melting point, crystal structures, transformation temperatures and density. It is often not appreciated just how much the vapor pressures of the rare earth elements differ, so Table 5.2 lists the vapor pressure of each rare earth at its melting point, together with the approximate time that would be required for 1% of the sample to evaporate at the melting temperature under both vacuum conditions and 0.5 bar argon. These evaporation rates assume a rod-shaped sample of the approximate dimensions of a pencil (a shape used in both zone refining and electrotransport purification).

Table 5.2. Basic properties of the rare earth metals showing melting points, vapor pressures at the melting point, the approximate time to lose 1% of a pencil-shaped sample by volatilization at the melting point in vacuum and under 0.5 bar argon, and density

	Melt. pt[a] (°C)	V press at m.p. (mbar)	Time to lose 1% at m.p. in vac.	Time to lose 1% at m.p. in 0.5 bar Ar	Density[a] ($g\,cm^{-3}$)
La	918	8×10^{-10}	23 years	20 000 years	6.145
Ce	798	2×10^{-11}	700 years	600 000 years	6.770
Pr	931	8×10^{-7}	9 days	20 years	6.773
Nd	1021	4×10^{-5}	4 hours	150 days	7.007
Sm	1074	5	0.1 sec	30 sec	7.520
Eu	822	1	0.3 sec	1.5 min	5.243
Gd	1313	1×10^{-4}	1.2 hours	35 days	7.900
Tb	1365	1×10^{-3}	15 min	5 days	8.229
Dy	1412	6×10^{-1}	1.3 sec	6 min	8.550
Ho	1474	6×10^{-1}	1.5 sec	6 min	8.795
Er	1529	3×10^{-1}	2.8 sec	12 min	9.066
Tm	1545	2×10^{2}	0.004 sec	1 sec	9.321
Yb	819	1×10^{1}	0.02 sec	7 sec	6.965
Lu	1663	1×10^{-2}	1.5 min	10 hours	9.840
Sc	1541	1×10^{-1}	5 sec	20 min	2.989
Y	1522	2×10^{-3}	4 min	1.5 days	4.469

[a] Data taken from Beaudry and Gschneidner [2]

The evaporation rates in vacuum are calculated using the Langmuir equation [5], while those in argon are taken from the author's own experimental data, and should only be considered as a very approximate guide.

Figure 5.1 shows the melting points, crystal structures and transformation temperatures of the rare earth elements; the crystal structures have an effect on some techniques, particularly electrotransport, while the densities, listed in Table 5.2 are of significance in zone refining. Although all of the rare earths will degrade if heated to high temperatures in anything but very clean conditions, the reactivities of the elements in bulk form at room temperature vary significantly, a factor to be considered when handling these metals. While the heavy rare earths are virtually inert at room temperature, Eu can only be handled for a few seconds in air, while La and Ce will begin to degrade after a few minutes.

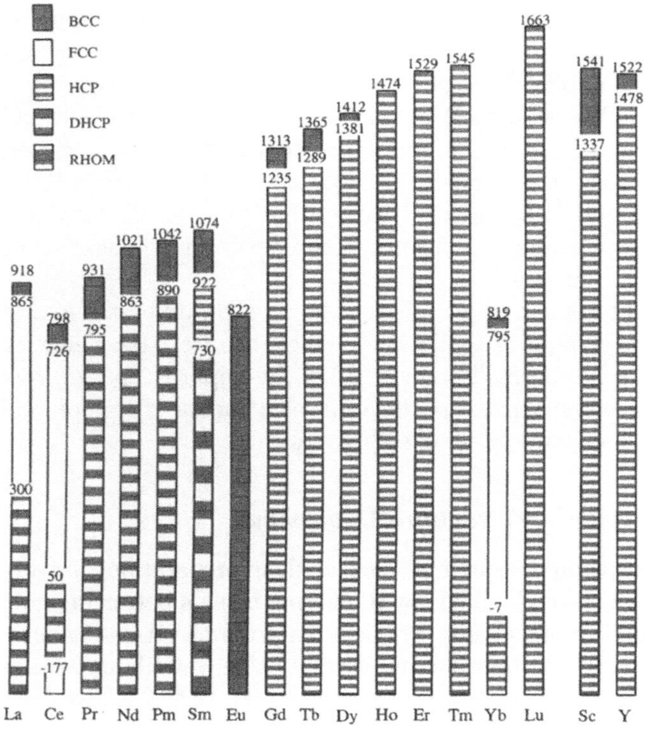

Fig. 5.1. Crystal structures, transformation temperatures (the intermediate numbers in each column) and melting points (the figure at the top of each column) of the rare earth elements. All temperatures are in °C

5.3 Purification Techniques for Rare Earth Metals

The first step to be taken when aiming to prepare high-purity rare earth metal is to decide what to take as a starting point. The choices are (1) to start with high-purity oxide and reduce this to metal, (2) to start with low-purity commercially available metal or (3) to start with good commercially available metal. Although the process of forming the metal from the oxide is a crucial step in the preparation of high-purity rare earths, few laboratories have the facilities to perform this operation better than the best commercial suppliers, so the vast majority of researchers will begin their purification process with the metal itself. Using poor-quality metal as a starting point and aiming to improve its quality is, however, a false economy, for it is invariably far more efficient, in terms of both time and money, to expend effort making good metal better rather than simply trying to make poor metal less poor. For the purposes of this chapter, therefore, it is assumed that the starting point for purification will be relatively good-quality metal.

Because all rare earths at elevated temperatures will combine with active gases in the purification equipment, any purification must be performed under conditions in which these potential contaminants are absent, or at very low levels. Effectively, this implies processing under ultra-high-vacuum (UHV) conditions (i.e. 10^{-8} mbar or better) or an inert gas of equivalent cleanliness. Attempts to refine rare earths in anything other than these conditions will inevitably lead to the metal being contaminated rather than purified, a fact that is well illustrated by the observation that prior to the introduction of UHV technology, several purification techniques were thought to be unsuitable for rare earth metals.

A general rule to be kept in mind when purifying the rare earths is that no single purification technique is applicable to all of the rare earth elements, while no one technique will normally remove all of the impurity species from any one rare earth. Comprehensive purification of a metal, therefore, will often involve using two or more refining techniques.

5.3.1 Vacuum Melting and Vacuum Degassing

Many rare earths are vacuum melted as part of their preparation, but improvements in purity can often be made by re-melting using a better vacuum, and possibly at a higher temperature. Publications from the Ames Laboratory [2, 6] report that residual fluoride and certain other volatile impurities can be removed by vacuum melting of rare earths at 500–1000°C above their melting points, where vapor pressures permit. To reach such temperatures, a hot crucible will probably be necessary; the favored crucible material is tungsten, since this is less soluble in liquid rare earths than tantalum [7]. The tungsten contamination of the metal may be sufficient, however, to require its subsequent removal by zone refining, or possibly by sublimation or distillation. Crucible heating can be achieved using either resistive or induction heating, although the latter is probably easier to adapt to UHV conditions, since there are no heating elements to degas within the vacuum system.

Electron-beam float melting (EBFM) is also a viable method of refining by vacuum melting. The technique is essentially containerless, so pick-up of crucible contaminants is avoided, but the metal cannot be heated greatly above its melting point. Mimura and Isshiki [8] report the removal by vaporization of Ca and Mg from La and Nd during EBFM in which the temperature within the molten zone was estimated to be about 100°C above the melting point. The induction heating of rare earth metals in a cold boat (also described later, in the zone refining section) can also be a means of vacuum melting of rare earths, but, like float zoning, it is difficult to achieve temperatures significantly above the melting point.

The complete removal of most gaseous impurities from a rare earth metal is, unfortunately, not simply a case of melting or heating the metal in a good vacuum and allowing gaseous impurities to diffuse to the surface and desorb. To facilitate the release of the major dissolved gases requires not only the movement of the gas atoms to the surface of the metal, but also their combination to molecules and escape from the surface. The behavior of gaseous impurities at the surface of a rare earth metal is complex, since it includes adsorption, desorption and chemical reactions at the surface (the effects of these processes have been considered in some detail by Ionov et al. [9]). As a general rule, however, it can be said that, with the exception of H, dissolved gaseous impurities in rare earths are not extensively liberated when the metals are heated, or even melted, under vacuum conditions [4]. Isshiki [10], however, has noted some reductions in O and C contents in Nd during float-zone melting under vacuum, a situation attributed to the segregation of C and O to the surface, leading to the formation and release of gaseous molecules such as CO.

Although not a vacuum melting technique, it seems appropriate to mention here the proposal by Isshiki [10] that hydrogen plasma arc melting could be used to purify some of the rare earths that have low vapor pressures at their melting points, since this technique successfully removes both metallic and gaseous impurities from some refractory metals, such as Zr [11].

5.3.2 Zone Refining

Zone refining is a good example of a purification technique thought to be unsuited to the rare earths until the advent of UHV technology and cold boats. The principle behind zone refining is the phenomenon whereby the solute concentration in a freezing solid differs from that in the liquid from which it forms [12]. In zone refining, a molten zone of typically 5–10% of the length of a rod-shaped sample is moved repeatedly along the sample, always in the same direction. Impurities that lower the melting point of the host metal will tend to be transported in the direction of zone travel toward the "end" of the sample, while impurities that raise the melting point are transported in the opposite direction toward the "beginning". Thus impurities are not removed from the sample by zone refining, only redistributed within it. An ultimate distribution is eventually approached for any impurity, after which further zone passes will not result in further redistribution. After zoning, it is often possible to select a part of the sample that is of enhanced overall purity compared with the start metal, or in which the levels of a specific impurity are significantly reduced.

The actual efficiency of zone refining depends upon the distribution coefficient of any particular impurity within the host, the number of zone passes, the zone speed and the sample length to zone length ratio. The equilibrium distribution coefficient k_0 for an impurity in a solvent is defined by the ratio C_s/C_l where C_s is the composition of the first solid to freeze, under equilibrium conditions, from a liquid of composition C_l (Fig. 5.2). When the impurity raises the melting point of the host, $k_0 > 1$; when the melting point is lowered, $k_0 < 1$. Zone refining is least effective when the impurity

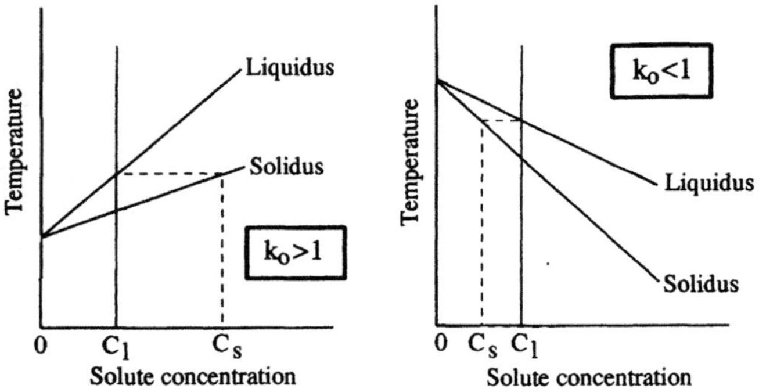

Fig. 5.2. The derivation of k_0 from an equilibrium phase diagram. If the melting point of the host is raised by the addition of solute (*left*), the first solid to form upon cooling a liquid of composition C_l will have a higher solute concentration, C_s, and k_0 (i.e. C_s/C_l) will be greater than unity. During zone refining, such solute will move in the direction opposite to the zone travel. If the solute lowers the melting point (*right*), k_0 will be less than unity and solute will move in the direction of zone travel

has a k_0 value close to unity. While it is possible to derive k_0 from a host-impurity phase diagram, in practice the solidification process during zone refining is rarely ideal; for example, the rate of zone travel may be too fast to allow complete mixing in the liquid zone. Thus the actual zoning process is described by an effective distribution coefficient, k_{eff}, which must be determined experimentally. For any given distribution coefficient, it should be possible to compute the impurity profile along the sample after passing a certain number of molten zones [12]. Complications can arise, however, if the level of an impurity in the host exceeds the solubility limit and precipitation occurs. This can happen during refining as impurities concentrate at one end of the sample; with anything but very good quality rare earth start metals, precipitation, particularly of oxides, may occur as zoning proceeds.

5.3.2.1 Zone Refining of Rare Earth Metals. When zone refining the rare earths, it is essential that the environment within the refining system is sufficiently clean to prevent degradation of the sample during its supposed purification. Early zone refining of rare earths using high vacuum, rather than UHV-rated apparatus, successfully moved many metallic impurities, but any potential for purification with respect to interstitial impurities was negated by contamination [13, 14]. During zone refining, any one part the sample spends a significant amount of time at or close to its melting point, so zone refining in UHV is only possible with the lower-vapor-pressure rare earths. Operation in an inert gas environment extends the range of the method to the medium-vapor-pressure elements, but it is imperative that very pure gas is used.

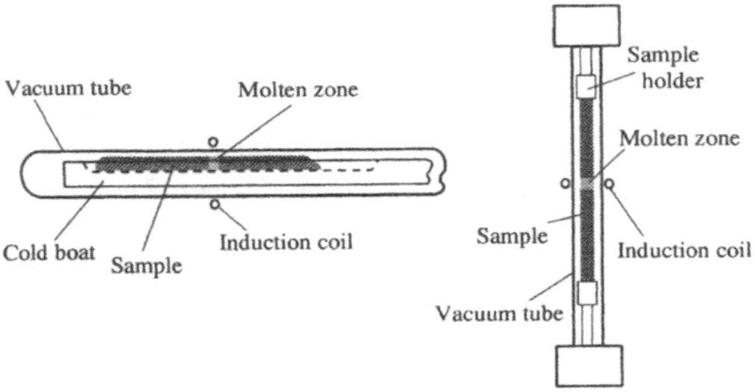

Fig. 5.3. Horizontal zone refining with the sample in a cold boat (*left*) and vertical float zoning (*right*). In both cases illustrated, induction heating is used to create the molten zone, while the zone is passed along the sample by moving the sample relative to the induction coil

During zone refining the sample can either be horizontal, in which case it has to be held in some sort of container, or vertical, in which case it need only be held at its ends, the molten zone being kept in place primarily by surface tension (Fig. 5.3). The reactivity of rare earth metals effectively precludes the use of hot containers during horizontal zone refining, because all of the typical crucible materials (e.g. W, Ta, Mo, alumina and BN) will react with, and so contaminate, the sample to some degree [2]. The usual alternative is to use water-cooled copper or silver cold boats, an innovation that dates back to the early 1960s, since when a number of different designs have appeared [16]. Usually, induction heating is used to create the molten zone, in which case either a vacuum or an inert gas environment can be used, although it is also possible to use arc heating, in which case an inert gas environment is necessary to maintain the arc. The zone refining of rare earth metals in a cold boat can be complicated, however, by temperature gradients between the top and bottom of the sample, which are introduced by the proximity of the boat. Such gradients can result in impurity redistribution not only from end to end of the bar but also from top to bottom. Fort et al. [16] have noted that, with rare earths, the stability of the molten zone can also vary with the purity of the sample. (With an unstable zone, the molten metal has a tendency to flop from side to side, and in so doing touches the cold boat and momentarily freezes.) Perversely, low-purity samples of La and Ce were found to have a higher apparent surface tension, attributed to a "skin" of impurity precipitates, which greatly improved the stability of the molten zone. In vertical zoning of rare earths, several heating methods are possible, including induction heating, electron-beam bombardment and laser heating. Zone widths are generally narrower in vertical zoning than horizontal zoning, which increases the efficiency of the technique, although control is significantly more difficult. This is particularly the case with higher-density metals, and probably explains the predominance of horizontal zone refining for many of the rare earths.

There is a general scarcity of information for k_{eff} values for impurities in rare earths, so refining these metals tends to be a trial and error process, although certain trends can be anticipated. Evaluation of zone-refining results from Gd, Nd [15], La, Ce, Gd [16] and Tb [17] indicates that zone refining is efficient at redistributing many typical metallic impurities, but its usefulness with regard to interstitial impurities such as O, N and C is limited. Jones et al. [18] reported that a total of 20–30 passes at rates between $2\,\text{cm}\,\text{h}^{-1}$ and $12\,\text{cm}\,\text{h}^{-1}$ for a rare earth sample 10 cm long and a zone width of 1 cm could give reductions of metallic impurity levels by factors of several hundred, although it was rare for interstitial levels to be reduced by a factor of more than two or three. Overall, therefore, the main value of zone refining for the rare earths is probably its ability to refine relatively rapidly with respect to many metallic impurities.

5.3.3 Vapor Techniques

The essence of vapor techniques is that the metal to be purified is heated in a container under vacuum conditions and vapor from it is collected by condensation on to a cooler surface. Impurities with lower vapor pressures than the host metal will tend to remain in the container, leaving the condensed metal of enhanced purity; thus vapor techniques are most effective with the higher-vapor-pressure metals. If the vapor pressure of the host metal is sufficiently high that it does not have to be melted before a reasonable volatilization rate is achieved, the technique is known as sublimation, whereas if the metal has to be melted before volatilization rates are acceptably high, the method is known as distillation. Either sublimation or distillation can be used to purify all of the rare earth metals with the exception of La, Ce, Pr and Nd, which have the lowest vapor pressures.

Several of the rare earth metals are sublimed or distilled by manufacturers during their preparation, so to make it worthwhile repeating this step for further purification, the techniques must be performed under a better vacuum and probably at a slower rate. Manufacturers tend to sublime or distil rapidly to save time, but reducing the rate of volatilization will improve results by lessening the possibility of impurities being carried over in the metal vapor. On a laboratory scale, it is not unreasonable to aim to purify by sublimation or distillation using a vacuum four to five orders of magnitude better, and at rates up to two orders of magnitude slower, than those used by the many suppliers. Further improvements can be obtained by collecting the condensate on a relatively hot surface rather than a cold surface. Collection on a cold surface, such as a water-cooled copper plate, is favored by some suppliers, as it makes the condensate easier to retrieve with little loss, but collecting on a hot surface yiel ds metal that is better consolidated and less prone to subsequent contamination upon exposure to air. However, condensate collected on a hot surface has to be removed mechanically from the surface, usually by machining off the condenser material (typically Ta or W foil) from the purified metal, a process that inevitably leads to the loss of some of the purified metal.

The approximate rate of volatilization for the host metal can be calculated using vapor pressure charts and the Langmuir equation [5], suitably modified (in the case of sublimation) to take account of the fact that molecular distillation (i.e. the assumption that vaporized metal has a high probability of reaching the condensing surface without colliding with another molecule) is not actually achieved [20]. It is also possible to calculate the degree of separation attainable between host and impurity atoms during vaporization purification for the ideal case where the impurities are other simple metals dissolved in and completely mixed within the sample [21]. However, with the rare earths, the major impurities are not simple metals, but nonmetals, possibly together with oxides or sub-oxides whose volatilization characteristics are less predictable.

5.3.3.1 Sublimation. During purification by sublimation, impure metal is heated in an evacuated container to a temperature below its melting point, and vapor from it is collected by condensation on to a cooler surface. Gschneidner [3] reports that the more volatile rare earths (Sm, Eu, Dy, Ho, Er, Tm, Yb and Sc) can be significantly purified with respect to O, N and C because the vapor pressures of the respective oxides, nitrides and carbides are quite low, meaning that they tend to remain in the residue. The removal of certain metallic impurities, particularly Ta or W, can also be expected. At its most basic, sublimation can be performed using a can made of thin Ta sheet in which the impure metal is heated, with a loose-fitting Ta sheet lid upon which the condensate is collected (Fig. 5.4). Since the can does not have to be liquid tight, it can be formed by spot welding, or even by simply folding the Ta sheet. It is possible to heat the can using either induction or resistance heating, the aim being to achieve the correct temperature differential between the heater and condenser sections. The heater temperature needs to be set to give a relatively low sublimation rate (e.g. of the order of $1\,\mathrm{g\,h^{-1}}$ for a laboratory-scale setup), while the condenser temperature needs to be high enough to collect the vapor in a well-consolidated form, but low enough to allow condensation. Sublimation purification can be enhanced by adopting a multi-stage approach, whereby the condensate from an initial run is used as the start metal for a second run. Alternatively, a condenser with a temperature gradient can be employed to give what may be termed fractional sublimation. As always with the rare earths, an ultra-high-vacuum environment within the sublimation apparatus is the key to success. Ionov et al. [22] have demonstrated the benefits of employing UHV conditions during sublima-

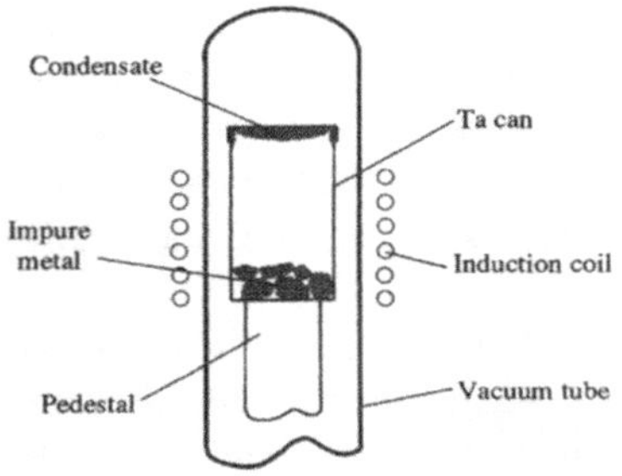

Fig. 5.4. A simple sublimation can for refining high-vapor-pressure metals. Vapor from impure metal is condensed on the lid of the can, higher-vapor-pressure contaminants tending to remain in the residue. The position of the induction heating coil determines the relative temperatures of the charge and condensate

tion purification of Sm, whereby start metal with a residual resistance ratio (RRR) of 5 yielded condensate with a RRR of 23 after sublimation under a vacuum of 5×10^{-6} mbar; whereas when this metal was further sublimed under a vacuum of 5×10^{-8} mbar, the RRR went up to 107. For successful purification, the heater and condenser need to be made from high-purity materials (Ta is usually favored) and to be thoroughly degassed at high temperatures in vacuum prior to use. For an induction-heated arrangement, the can is likely to be contained within a silica vacuum tube, with the induction coil on the outside, an arrangement that allows quick pumpdown and degassing times. Resistance heating of the can is also possible, but may involve longer pumpdown times, as the heater has to be degassed.

A drawback to sublimation, noted by Murphy et al. [23], who studied purification of Sm, is that purification with respect to metals with higher vapor pressures than the host, such as Ca and Mg, is limited, since these contaminants are trapped in the sample and released gradually as sublimation proceeds, rather than subliming before the host metal, which would allow their removal. Likewise, Ionov et al. [22] noted, from extensive experimentation on the sublimation of Eu, Sm, Tm and Yb, the formation of a web-like coating on the surface of the residual part of the metal in the crucible. This coating was reported to consist of metallic impurities with low vapor pressures, sub-oxides, oxides and other compounds formed between the host metal and gaseous impurities. The authors speculated that the accumulation of such impurities on the surface of subliming metal may produce additional solid state reactions on the surface, which could lead to the formation of compounds that could co-sublime with the host metal and so contaminate the condensate. Such contaminants could also become entrapped in the vapor, and carried on to the condensate, especially at higher sublimation rates. Rytus and Nikiforova [24] and Ionov et al. [22] investigated ways of counteracting these effects, by subliming through a porous W filter pressed from high-purity

50 micron W powder and held in the middle of the crucible. Comparison with sublimation without a W filter for Sm gave appreciably better results, the improvement being attributed to the minimization of the processes of entrainment, entrapment and co-sublimation. A further refining effect during sublimation purification of rare earth metals comes from degassing of H from the host metal, which occurs as the charge is heated, while Ionov et al. [22] also report meaningful degassing of hydrocarbons, CO, CO_2 and possibly N during rare earth sublimation.

5.3.3.2 Distillation. It is not possible to sublime the medium-vapor-pressure rare earths Gd, Tb and Y, but they can be purified by distillation (i.e. condensing vapor from a molten charge). Distillation adheres more closely to the theoretical model for separation by vaporization, as contaminants continue to be well mixed within the molten charge during a run, rather than congregating at the surface, as happens during sublimation. However, for those rare earths amenable to distillation, the process takes place at higher temperatures than sublimation, and Gschneidner [3] states that no purification with respect to oxygen can be expected, since the RO species is volatile at the temperatures necessary to distil these metals. The removal of Ta and W can, however, be achieved, although zone refining may be considered as an alternative for this purpose. The vapor pressure of Lu makes it a borderline case between sublimation and distillation, but in either case the high processing temperature means that little purification with respect to O can be expected.

Experimentally, distillation is similar to sublimation, except that the container has to be more robust, since it must be liquid-tight and also relatively thick-walled to allow for some erosion by the melt. W crucibles are preferable to Ta for this purpose, while Gschneidner and Beaudry [2] also recommend the use of a W lining in the condenser if this is to be made of Ta. During the initial stages of a distillation run on rare earths, a considerable degree of purification of the melt by vacuum degassing of impurities such as Ca, fluorides and H occurs. It is important that the condenser is protected from the evolving vapor during this period using, for example, a shutter mechanism. An alternative to a shutter is the arrangement described by Gschneidner [25], whereby a small hole (about 1 mm) is drilled in the top of the condenser, which allows volatile impurities to escape during the initial stages of a run. When the temperature is sufficiently high for the rare earth to vaporize at an appreciable rate, the hole becomes closed off by a build-up of condensate. Gschneidner further states that a W filter may enhance purification by tying up F, Fe and Ti impurities so that they do not co-distil with the host metal.

5.3.3.3 Liquid Metal-Through Distillation (LMTD). As noted above, the disadvantages of sublimation are that higher-vapor-pressure impurities such as Ca and Mg are released gradually as sublimation proceeds (rather than at first, as in distillation), while lower-vapor-pressure impurities complicate the purification process by building up at the surface of the subliming metal. Murphy et al. [23] suggested improving the efficiency of sublimation purification of Sm by dissolving it in a solvent or flux metal that has a very low vapor pressure and a relatively low melting point, so effectively distilling Sm at a temperature below its melting point. With this arrangement, the higher-vapor-pressure impurities would not be locked into the metal being purified, but would be free in the liquid, where they could diffuse quickly to the surface and evaporate at the start of the run. Experiments by Murphy et al. confirmed that this was indeed the case for Ca and Mg impurities in Sm dissolved in a Ce–10 wt% Fe flux. Unfortunately, the equipment used had a relatively poor vacuum capability (around 1×10^{-6} mbar), while their fluxes contained approximately 2000 ppm.wt oxygen, so purification with respect to gaseous impurities could not be expected.

More recently, the idea of improving upon conventional sublimation of rare earths by dissolving them in another metal has been pursued by Ionov et al. [22], who named the technique liquid metal-through distillation (LMTD). Their research involved distilling Sm, Eu, Tm and Yb through a liquid metal (typically La or Ce), with the aim of improving upon simple sublimation by minimizing the segregation processes on the surface of the metal and reducing clustering. La and Ce were considered to have ideal melting-point and vapor-pressure characteristics to act as the liquid metal, as they would react with residual gases within the system and possibly reduce some compounds (e.g. sub-oxides) of the metal being sublimed. In a comparison with conventional sublimation for Sm purification, Ionov et al. [22] noted significantly better results for LMTD as judged by RRR values. They also found that the degrees of purification for Fe, Mn and Cu impurities were, respectively, three, ten and two times higher using the LMTD technique.

5.3.4 Solid State Electrotransport (SSE)

SSE has become one of the most successful techniques for refining all but the highest-vapor-pressure rare earth metals, a success achieved because SSE is particularly adept at purification with respect to interstitial impurities, while it is also quite efficient with regard to some of the more important substitutional contaminants. Because SSE is generally less familiar than other purification methods, it will be described in some detail here. SSE is the phenomenon whereby both host and impurity atoms in a metal migrate under the influence of an electrical field. In the rare earths, the movement of host atoms is slow compared with the movement of many impurity atoms, particularly those on interstitial sites, resulting in an effective redistribution of impurity atoms within the host. In practice, SSE simply involves passing a high direct electrical current along a rod or strip of metal held in a suitable container. Those impurities amenable to electrotransport will migrate toward one end of the sample, leaving the remainder of enhanced purity (Fig. 5.5).

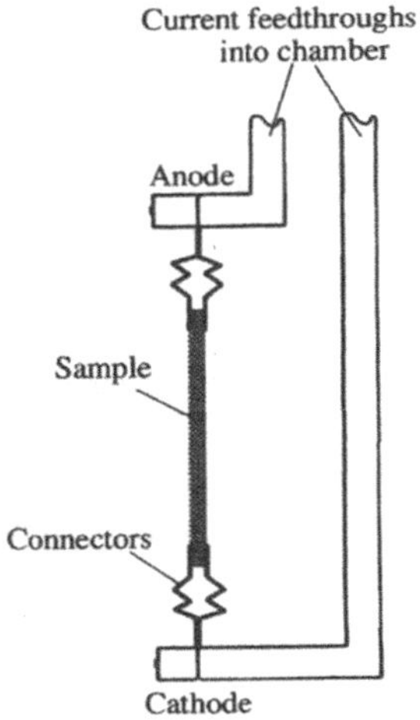

Fig. 5.5. Solid state electrotransport. A high direct current is passed along a sample held under vacuum or inert gas. The current both heats the sample and induces the electromigration of certain impurities. In the rare earths, the major interstitial impurities migrate toward the anode end of the sample, leaving the cathode of enhanced purity

Thus, like zone refining, impurities are not actually removed from the sample as a result of electrotransport, but only redistributed within it.

5.3.4.1 The Theory of SSE. While the first explanations of SSE were based upon the electrostatic forces between ions in the metal and the applied electric field, it was subsequently realized that the momentum transfer that occurs when current carriers collide with ions is an important factor. The force exerted on the ions by the current carriers is commonly called the "electron drag force" or "electron wind" and theories to explain its nature have been reviewed by Verhoeven [26], Pratt and Sellors [27], Jordan [28] and Sorbello [29]. The actual rate and direction of migration of an ion during SSE is determined by the balance of the drag and electrostatic forces on the ion. Thus electrotransport forces vary in magnitude and direction according to the species of host and impurity atoms, meaning that in any host it is possible for different impurities to migrate at different velocities in different directions. The effective valence (Z^*) of an ion during SSE represents a combination of the electrostatic and drag effects and comprises either the difference between, or the sum of, the two component forces depending upon their relative directions. A negative Z^* indicates transport in the cathode to anode direction. Schmidt and Carlson [30] have proposed a simple rule to predict the sign of Z^* for interstitial impurities in metals by noting that the sign of Z^* for such impurities in metals whose atomic radius is greater than Ti is negative, while Z^* is positive if the atomic radius is smaller than Ti.

The mobility of an ion during SSE (U) is defined its velocity per unit electric field (E) and is normally quoted in units of $cm^2\,V^{-1}\,s^{-1}$. Always opposing the effects of electrotransport, however, are the forces of diffusion, which tend to even out any impurity gradients resulting from electromigration. Thus the ratio U/D, where D is the diffusion coefficient, determines the maximum degree of redistribution attainable during SSE, a high value of U/D signifying greater potential redistribution. During SSE a steady state is eventually approached when the forward movement of impurity atoms due to electrotransport is matched by the backward movement due to diffusion. Peterson [31] has calculated the impurity concentration at any point along a sample at steady state as:

$C_x = C_0(U/D)El \exp(-UEx/D)$, where x is the distance along a sample of length l with an initial impurity concentration C_0.

Peterson [33] has also proposed guidelines to predict the time required for an impurity species to approach steady state. These can be summarized as follows:

if $UE/D > 2$, then the time required for the purest 75% of the sample to be within 1% of the steady state value is approximately $1.25l/UE$

if $UE/D \ll 2$, then the time required for concentration to be within 5% of the steady state value is approximately $0.4l^2/D$.

The above equations can be used as a guide when deciding the values of the experimental parameters during SSE (i.e. sample dimensions, run length and temperature), although it should be emphasized that theory and practice do not always agree as closely as would be desired, so theoretical predictions should be tempered with some regard to practical realities.

The above equations show that the longer the sample is, the greater is the impurity distribution theoretically obtainable, while a high value of E is also advantageous. The current that can be passed during SSE, and hence the E value, is limited by the amount of Joule heating that the sample can withstand without melting, although processing under an inert gas (rather than vacuum) or having a sample with a high surface-to-volume ratio will allow higher currents by increasing heat losses from the sample. Typical E values, as a function of temperature, for a variety of rare earth samples during SSE have been published by Fort [34]. For the principal impurities in rare earths, U and D values increase with increasing temperature, and consistently show rapid increases on passing into the high-temperature bcc phase (where one exists), although the crucial U/D ratio is often not strongly temperature dependent. However, the higher U and D values at higher temperature will significantly reduce the time required to approach steady state. Even at high temperatures, however, runs of the order of hundreds of hours can be calculated as being required for many of the common impurities in rare earths to approach steady state for laboratory-scale SSE samples.

When typical U and D values for an impurity such as oxygen in a rare earth are used in the Peterson equations, impurity levels of a fraction of a ppm at the cathode end of a sample are predicted after a few hundred hours of processing, given a starting concentration of, say, 1000 ppm. In practice, however, it is rare for values significantly below 30 ppm.at for oxygen to be reported. Reasons suggested for this discrepancy include samples being contaminated from the electrodes or the environment during SSE [32,35], uneven temperatures along samples giving cold spots from which impurities slowly bleed into pure regions [35] and that the mobility of some impurity ions may be reduced by trapping at vacancies within the host lattice [34]. The effects of contamination from the environment have been illustrated by Schmidt et al. [36], who reported a progressive increase in RRR values at the cathode ends of SSE refined thorium (an element similar to the rare earths) samples from below 200–2000 as the processing vacuum was improved from 10^{-5} mbar to 10^{-11} mbar. While there is no direct evidence to confirm the notion that impurities may be trapped by lattice vacancies, the equilibrium vacancy concentration in a typical metal can be calculated as being approximately 10 ppm.at at SSE temperatures [34] (although Blackburn [37] has postulated that one effect of electrotransport is to increase vacancy concentrations), meaning that the lower practical limit found for interstitial impurity contents in SSE-refined rare earths would appear to be of the same order of magnitude as the number of vacancies at processing temperatures. This similarity gives some credence to the idea that vacancies may trap interstitials and inhibit their migration.

While the electromigration of interstitial impurities in rare earth metals has been highlighted above, significant transport of some important substitutional contaminants can also occur. With such impurities there appears to be a correlation between migration velocity and the atomic size of the solute, with smaller atoms such as Fe, Ni and Co exhibiting fast transport in a number of rare earth hosts [38–40]. The fast migration of these impurities is fortuitous, for Fe, in particular, is a common contaminant in many rare earths. As regards migration directions for substitutional impurities in rare earths during SSE, Carlson and Schmidt [41] have noted that while the Group I metals Cu and Ag typically exhibit positive Z^* values, the Z^* values of the first transition group metals Fe, Co, Ni and Ti are negative.

5.3.4.2 Experimental Aspects of SSE Refining of Rare Earths. The cross-sectional shape of an SSE sample is normally determined by the dimensions of the required specimen. Typical sample shapes for rare earths are rods 8–20 cm in length and 4–10 mm in diameter; such dimensions imply the use of currents in the 50–400 A range. During SSE, electrical connectors, capable of carrying high currents, must be attached to the sample at its ends. Luckily, rare earth elements are ductile, so samples are not crushed by using clamp-type connectors, while it is also relatively easy to drill and tap a sample so that connectors can be screwed into it. However, connectors present a potential source of contamination during SSE, since the connector material, or impurities within it, may diffuse into the sample. Such contamination can be lessened by water-cooling the connectors so that they operate at relatively low temperatures, but this means that there will be a temperature gradient within the sample, for its ends will be at a lower temperature than its middle. Temperature gradients of this type inhibit purification, as impurities will only migrate slowly into or out of the cooler ends of the sample. Such problems can be reduced by carrying out a two-stage processing whereby the cooler, less-refined cathode end is cut off and discarded after an initial period of processing, a new connection is made and the remainder of the sample is processed further. It is usually preferred, however, to use "hot" connectors that run close to the sample temperature, so maintaining a fairly even temperature along the length of the sample and eliminating any slow bleed of impurities from cooler regions. With hot connectors, the electromigration direction of the major impurities in the connector material should be opposite to the migration direction in the sample, so reducing the possibility of impurities migrating from connector to sample at the pure end. Nikoforova and Volkov [42], who studied the relative merits of using Ta, Mo and Zr connectors during SSE of Y, concluded that connector-to-sample contamination should be minimized for the rare earths if Ta is used as the cathode connector material with Zr for the anode.

While SSE processing at the highest temperature that can be reached without melting the sample theoretically gives advantages in both the degree of purification attainable, and the time required to reach it, this is not an option for most rare earth metals because of high volatilization rates at such temperatures. Excessive volatilization can lead to the sample thinning (i.e. necking) and eventually melting. The choice of an optimum temperature/time program for a SSE run becomes progressively more difficult as the vapor pressure of the host element increases. With the lowest-vapor-pressure rare earths (La, Ce and Pr), long runs in UHV at temperatures close to the melting point are feasible and run lengths can approach the hundreds of hours required by theoretical considerations. The next batch of elements in ascending order of vapor pressure (Nd, Gd, Tb, Y and Lu) can be processed under UHV conditions for reasonably lengthy periods sufficient for meaningful purification, although it may be necessary to reduce the processing temperatures slightly below the ideal to counteract volatilization. Alternatively, they can be processed at optimum temperatures and long times in an inert gas environment without weight loss problems. For Sc, Er, Ho, Dy and Eu, processing under an inert gas is essential to allow meaningful SSE, although even with an inert gas at moderate pressures (i.e. that which can be maintained in a conventional UHV-type system), some compromise on time and/or temperature will have to be made. The remaining rare earths, Sm, Yb and Tm, all have vapor pressures that are too high for conventional SSE purification.

Since SSE refining of rare earths is a lengthy process, it is vital that the environment within the processing chamber is ultra-clean. Whether the environment is vacuum or inert gas, the chamber will invariably be a vacuum system capable of operating in the 10^{-10} mbar range or better. SSE refining in an inert gas introduces several practical difficulties, the most obvious one being contamination of the sample from the environment, so it is generally only used if volatilization losses cannot be kept within bounds during UHV refining by using a slightly lower processing temperature or reduced run length. Two methods of refining under inert gas are possible, the first being to use a continuous flow of gas, and the second to seal the system once it has been filled with gas and to operate under static gas. The former method has the advantage of continually purging any contaminants degassing from the system walls, but it relies upon having a source of extremely pure gas in some quantities. In the latter method, gas of extreme purity is not so important, for it is only admitted once and impurities in can be gettered prior to the SSE run commencing, but degassing products are not purged so they can build up within the system. Fort et al. [43, 44], who studied the degree of contamination of Dy and Er samples during SSE under argon, concluded that the static gas option was preferable, and that contamination could be minimized by thoroughly degassing the system by baking under UHV conditions and by degassing the sample in vacuum to moderate temperatures prior to processing under inert gas. A final period in vacuum was also found to be beneficial to degas any dissolved H. They did, however, find that C contamination was a particularly difficult problem to eradicate (this was believed to come from methane degassing from the system walls during the run), meaning that overall purification with respect to C was minimal. Another phenomenon peculiar to processing higher-vapor-pressure metals in inert gas is whiskering. Whiskers are clusters of dendrite-like growths that form on the surface SSE samples, and presumably result when vapor from the sample re-condenses upon the sample as it is "reflected" back from the gas. Whiskers tend to form in clusters and can significantly lower the temperature of the parts of the sample so affected.

Over the years, several ideas have been proposed to improve SSE purification. Marchant et al. [45] found that pulsed currents could be used to enhance SSE, although the technique does not seem to have been widely applied, despite claims of success when applied to La and Ce. Studies by Ionov et al. [9] of the surface of Y during SSE using Auger electron spectroscopy indicated the segregation of O and C impurities to the surface at temperatures below 623°C. This led to the proposal that SSE of rare earths could be enhanced by periodically cooling the sample and argon-ion etching the surface. Preliminary experiments confirmed this to be the case. Although the technique has not been widely applied to the rare earths, it is also possible to combine zone refining with SSE [46, 47]. Here a molten zone, produced by an electron beam or an induction coil, is passed along a vertically held sample during SSE processing. Care must be taken to ensure that the zone direction and SSE polarity are such that they move major impurities in the same direction.

5.4 Retaining Purity During Sample Preparation and Crystal Growth

More often than not, the purification of a metal is just one step in the route to preparing a sample for a particular experiment, a procedure that may also involve melting the metal into a particular shape or crystal growth. Retaining purity during these operations is clearly just as important as the purification itself, but growing a single crystal using as-purified metal is a potentially risky process, for it is all too easy to re-contaminate the metal. Fortunately, it is often possible to combine purification with crystal growth, or to grow the crystal first and then to refine it. If neither of these routes can be followed, crystal growth must be carried out under the same ultra-clean conditions as the initial purification. There is, of course, a strong link between purification and crystal growth, in that it is far easier to grow large crystals in pure metals, so purification is often a prerequisite to preparing a crystal. Melting as-purified metals – for example, to cast a particular shape – must also be carried out under clean conditions – for example, in a cold boat – while melting of the highest-vapor-pressure elements may necessitate arc melting under high inert gas pressures [48].

5.4.1 Combining Purification and Crystal Growth

The sublimation, distillation, zone refining and SSE processes all have the potential for simultaneous purification and crystal growth, while SSE has the further advantage that it can be used to purify an as-grown crystal. Crystal growth from the vapor, which is a well-established technique for many of the higher-vapor-pressure rare earth elements [49], mirrors sublimation or distillation purification in that vapor from a heated charge of metal is condensed upon a cooler surface. Clearly the vapor has to be collected in a more controlled manner for crystal growth than is required for purification alone, but it should be possible to combine the two operations. Many of the rare earth elements are amenable to crystal growth using either horizontal zoning or vertical float zoning, possibly in conjunction with zone refining. Float zoning is usually preferred, since seeding for a particular orientation is straightforward, although large crystals have been grown using cold boats. Crystal growth by zoning is applicable to all but the highest-vapor-pressure rare earths, although the crystallographic phase changes near to room temperature will create problems with La, Ce and possibly Yb. Conversely, the bcc to hcp phase changes at high temperature exhibited by several heavy rare earths do not prevent successful crystal growth or seeding. Like zone refining, float zoning for crystal growth only became suited for rare earths when UHV technology became available, earlier work in less clean environments [50] meeting with limited success due to oxidation problems. UHV-rated float-zoning equipment suitable for the rare earths using either electron-beam [51] or induction heating [52] has been described in the literature. Two routes are possible for the preparation of high-purity crystals of rare earth metals using SSE. The first is to SSE refine a rod-shaped single crystal previously grown, for example, by float zoning. The second is to use a temperature/time programme for the SSE processing that encourages the growth of large grains during SSE. Both options have been extensively described by Fort [34].

5.4.2 Crystal Growth After Purification

Apart from vapor growth, which is limited to high-vapor-pressure rare earths, the techniques for combining crystal growth with purification are restricted to rod-shaped samples. It is not always convenient to have such a shape, so crystal growth may have to be carried out on previously purified metal. Several rare earths would appear to be candidates for crystal growth from the melt using the familiar Bridgman or Czochralski techniques [53], but although the Bridgman method is easily uprated to UHV specifications, it has not found favor for the growth of rare earth crystals because of the difficulty in finding a container material that will not react with or dissolve into the melt. Hot crucible Czochralski growth also suffers from crucible compatibility problems, so for the rare earths it is preferable to employ cold crucibles with induction heating. UHV-rated Czochralski systems have been described by Brown et al. [54] and Fort [55], the latter using a cold crucible that is a development of the containers earlier reported by Sterling and Warren [56], Hukin [57] and McEwen and Touborg [58]. Another variant of the Czochralski method that is suitable for rare earths is the tri-arc technique [59], which uses arc heating with the charge supported on a cold copper hearth. Details of a tri-arc crystal puller, built to UHV specifications, have been published recently [60].

Solid state crystal growths based upon the strain/anneal technique are applicable to the majority of rare earths. Such growth involves taking a polycrystalline ingot and heat treating it in such a way that some of the grains grow at the expense of others, to leave – if not a single crystal – at least large grains. The advantages of this technique are that crystal growth can be attempted if only a small amount of high-purity start metal is available, that crystal quality is generally better than for melt growth and that volatilization losses are reduced for higher-vapor-pressure elements. McEwen and Touborg [58] found that the controlling parameters for growing crystals of Sm and Gd to Tm by this method were the purity of the metal and the annealing temperature, while Fort [61] proposed a recipe for hcp rare earths that involved straining ingots by fast cooling from the melt, followed by annealing at a temperature equivalent to 85% of the absolute melting temperature or 95% of any hcp–bcc transformation temperature, whichever was the lower. Good results have also been reported for the dhcp metals Pr and Nd using a program that cycled the sample temperature about the dhcp–bcc transition temperature [2, 62].

5.5 Purification Routes for Rare Earth Metals

As has been demonstrated, more than one purification technique may be needed to remove all of the desired impurities from a rare earth metal, so to prepare a sample it may be necessary to plan a route involving perhaps two or three purification methods, and possibly a crystal growth technique. The route chosen is likely to be determined principally by the vapor pressure characteristics of the rare earth in question, so Fig. 5.6 illustrates the applicability of the various purification and crystal growth techniques to the rare earth elements in order of ascending vapor pressure for the host metal.

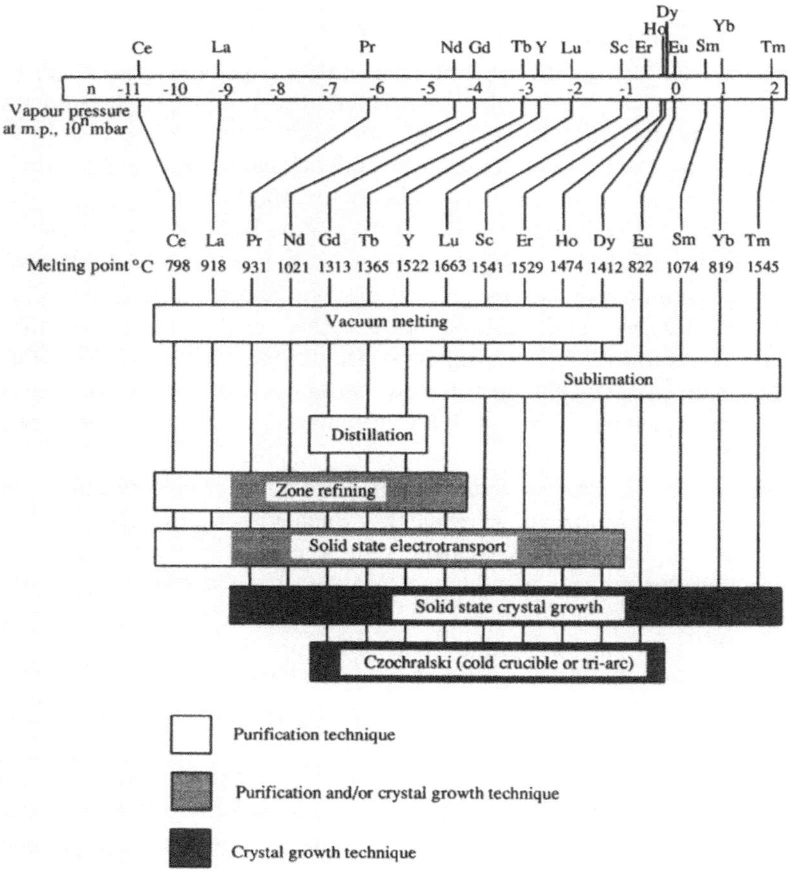

Fig. 5.6. Purification and crystal growth routes for the rare earth metals. The metals are shown in order of ascending vapor pressure at their respective melting points. Depending upon the final requirement, it may not be necessary to employ all of the techniques shown on a particular route

Depending upon the final requirement, it may not be necessary to employ every purification technique on the indicated route, while, clearly, only one crystal growth method will be required.

5.6 Individual Elements

La has a low vapor pressure at its melting point, making it unsuited to purification by vaporization methods, while the crystal structure transitions at relatively low temperatures complicate crystal growth. The electrotransport parameters of C, N and O have been measured in the bcc and fcc phases of La by Schmidt et al. [38]; all three solutes exhibited negative effective valences with significantly higher mobilities and diffusivities in the bcc phase. Fe and Cr were also found to have negative effective valences and high mobilities in bcc-La. Pan et al. [63] prepared metal with a RRR of 260 using SSE, the major impurities in the refined metal being H (276 ppm.at), F (180 ppm.at), C (92 ppm.at) and O (78 ppm.at). Pure single-phase dhcp La was prepared by cooling the SSE sample very quickly to room temperature, heating to 400°C for 2 h, cooling to 100°C and finally heating the sample to 265°C for 10 days. Mimura and Isshiki [8] refined La using EBFM, and found that C moved in the zone direction while S moved in the opposite direction. No significant redistribution of O or N was found after five passes, although the levels of Ca and Mg were reduced along the whole sample as a result of volatilization. Zone refining of La in a cold boat [64] revealed distinct differences in the behavior of the metal according to its purity. While relatively low-purity La with a "skin" of precipitates could be zoned controllably, higher-purity metal, with a lower surface tension, tended to flop about in the boat leading to an inconsistent zone width. Despite the unstable zone, some impurity redistribution was found after 25 zone passes with O moving toward the "beginning" of the bar while Ta, Cu, Fe, Ni and Co moved quite markedly in the opposite direction.

Ce has an exceptionally low vapor pressure at its melting, precluding the use of volatilization methods for purification. Fort et al. [16, 65] zone refined Ce under UHV conditions in a cold boat using induction heating, and found that O, N and H all moved in the direction opposite to the zone travel, with k_{eff} values of 1.2, 1.15 and 1.1–1.4 respectively. Early SSE work on Ce [66–68] showed that Fe migrated very efficiently toward the anode at 600°C. Co, Ni and Mn also moved toward the anode, but less effectively than Fe, while Cu and Ag moved in the opposite direction. Most early SSE work on Ce concentrated on metallic impurities (probably because systems were not sufficiently clean to prevent "gaseous" contamination) although Marchant et al. [45] did note the movement of O toward the anode when a pulsed current was used. One of the main problems when working with Ce is to prepare single-phase samples, a procedure complicated by the phase changes either side of room temperature (at normal pressures). Koskimaki et al. [19] report that close to 100% beta-Ce (i.e. the dhcp form) can be prepared by repeated steps of thermal cycling between room temperature and liquid helium temperature, followed by annealing for long periods at 75°C. Single-phase alpha-Ce was also prepared using a pressure-cooling–depressurizing cycle, which involved annealing the sample at 400°C for 3 days, compressing at 9000–10 000 bar, cooling to −196°C and releasing the pressure.

Pr. The vapor pressure of Pr is low, limiting its purification to nonvapor techniques. Muirhead and Jones [69] refined what they termed commercially available Pr using SSE to give material better than 99.95 at% with a RRR of 240. Refining was carried out under UHV for a total of 1050 h in three stages. Cold connectors were employed with the hottest center section of the sample running at about 830°C (i.e. in the bcc phase). The O content was reduced from 4700 ppm.at in the start metal to a best value of around 100 ppm.at after SSE. The comparative figures for N were 500 and 5 ppm.at respectively. Fort [34] managed to combine SSE purification of Pr with simultaneous solid state crystal growth by cycling the processing temperature about the dhcp–bcc transformation temperature. A typical run comprised 500–600 h in the bcc phase at 810–820°C, before cycling from three to five times about the transformation at 795°C followed by 200 h at 730°C and slow cooling to room temperature. Using this recipe, Pr with a RRR of 400 was prepared from Ames Laboratory start metal, or 250 from lower-purity commercial metal, with grains up to 2 cm^3. McMasters et al. [70] grew large crystals using a strain anneal method on metal that had been previously zone refined, noting that crystal growth did not occur in metal that had been simply arc melted rather than zoned.

Nd has a moderate vapor pressure, probably just a little too low for distillation purification. Mattocks et al. [71] refined Nd to a purity of 99.98 at% and a RRR of 120 using SSE. Using start metal from the Ames Laboratory, they reduced the O content from 410 to 130 ppm.at at the cathode end, while the N content was reduced from 560 to 10 ppm.at. The total processing time, over two stages, was 789 h at temperatures up to 888°C under UHV. Similar experiments using metal from Rare Earth Products Ltd. saw the O and N contents reduced from 12 700 and 203 ppm.at to 550 and 130 ppm.at, respectively, at the cathode end. The levels of Co, Fe and Ni were also substantially redistributed, all three impurities migrating toward the anode. From a comprehensive study of SSE processing in Nd, Fort et al. [15] concluded that the low RRR values yielded by refined Nd were not due to high overall impurity levels, but were a result of scattering from low-temperature magnetic transitions. They also managed to prepare a sample of Nd with an overall purity of 99.97 at% from 99.88 at% start metal, using a combination of zone refining and SSE. Fort [34] has reported simultaneous purification and crystal growth during SSE processing for Nd, using a transformation cycling method similar to that described for Pr. He found, however, that while high-purity grains up to 2.5 cm^3 could be prepared, crystal quality was often affected by the presence of subgrain boundaries. Float zoning in vacuum by Mimura and Isshiki [8] reduced the level of O in a Nd sample from 1300 ppm.wt to around 1000 ppm.wt along the whole sample after five zone passes. N was similarly reduced from 140 to 70 ppm.wt., while C was found to move against the zone direction, the level at the end section of the sample being reduced to just over half of the initial value of 180–200 ppm.wt after 10 passes. Cu, Si and Fe were observed to move in the zone direction, while both Ca and Mg were very significantly reduced over the whole sample as a result of volatilization.

Sm. Ionov et al. [22, 72] have performed a considerable amount of research into sublimation purification of Sm, whose high vapor pressure makes it ideally suited to such refining. With a simple sublimation cell under UHV, they improved the RRR from 3 to 107, using heater and condenser temperatures of 900°C and 427°C, respectively. By subliming the same material through a W filter (using a sublimation temperature of 650°C and a condenser at 427°C), a RRR of 300 was achieved. A similar RRR was recorded after LMTD using La flux at 950°C, again using a condenser temperature of 427°C. Beaudry and Gschneidner [2] suggest a sublimation temperature of 800°C, with a condenser at 500°C, to give a rate of 3 g h^{-1} for laboratory-scale apparatus. Sm is an interesting element, in that it can be prepared at room temperature in bulk form in two different crystal structures. At room temperature, Sm normally adopts a rhombohedral structure with an ABABCBCACA stacking sequence. Pure Sm transforms to the hcp structure (i.e. ABAB stacking) at 731°C [73], and a further phase change to bcc is generally accepted to occur at 922°C, although resistivity studies by Hiemstra et al. [74] found no evidence for this transformation. Shi and Fort [75] found, however, that by suitable heat treatment it is possible to prepare Sm in the dhcp form (i.e. ABAC stacking) at room temperature and pressure. The heat treatment involved annealing in the hcp phase, quick cooling to around 400°C and reheating to 700°C, followed by cooling to room temperature. This structure variant was found to form in metals of different purities, including material that had been purified by sublimation under UHV conditions. Both Daane et al. [76] and Schieber [49] have grown crystals of Sm by vapor deposition, a technique that has the advantage that the deposition temperature can be set below the rhombohedral to hcp transformation temperature. It may be possible to combine such crystal growth with sublimation purification. Sm crystals have also been grown by McEwen and Touborg [58], using a solid state recrystallization method.

Eu. One of the major problems with Eu is its extreme reactivity, which precludes it being handled in air for more than a few seconds. Thus, while it is possible to purify Eu and to grow crystals, the preparation of specimens can necessitate great ingenuity. Busch et al. [77] tackled the problem of handling Eu by using an argon-filled glove box gettered with a liquid Na–K alloy. They also reported very good results for the purification of Eu by distillation under UHV conditions, using a vacuum system that made use of the gettering action of Eu to increase the pumping speed. The distillation columns were made from Mo foil and heated by an induction coil; distillation took place at 1000°C and the condensate was collected either upon Mo baffles or on the water-cooled quartz vacuum tube. The best material produced was analyzed to have a total impurity content of 170 ppm.at., with O at 5 ppm.at and H (the major contaminant) at 140 ppm.at. Beaudry and Gschneidner [2], however, have suggested that their mass spectrographic results for H and O could be in error by a factor of 10, since the spectrometer was not checked against standard methods. Eu with a RRR of 225 has been prepared by Ionov et al. [22] from start metal with a RRR of 5, using sublimation through a W filter under UHV. The sublimation and condenser temperatures were 600°C and 427°C, respectively. Metal with a RRR of 275 was achieved from similar start metal using the LMTD process, with La as the solvent at a temperature of 950°C and condensation at 427°C. Beaudry and Gschneidner [2] indicate that a sublimation temperature of 700°C, with a condenser at 400°C, will give a rate of $3\,g\,h^{-1}$ for laboratory-scale apparatus. Crystal of Eu can be grown from the vapor, Schieber [49] reporting crystals 2–3 mm long, so for high-purity crystals, sublimation and crystal growth by vapor deposition could possibly be combined. Alternatively, McEwen and Touborg [58] have grown an Eu crystal 20 mm long and 8 mm in diameter by the Czochralski technique from a cold crucible under an Ar atmosphere.

Gd. The moderate vapor pressure of Gd makes it amenable to purification by a range of techniques. Beaudry and Gschneidner [2] list a vacuum melting temperature of 1800°C as being suitable for Gd, while a distillation temperature of 1725°C (with a condenser at 900°C) is recommended. Fort and co-workers [15, 16, 65] have published the results from several zone-refining experiments, all using a cold boat, induction heating and an argon environment. Fe, W and Al were all found to move markedly with k_{eff} values in the 0.5–0.7 region. C also moved in the zone direction, opposite to the other interstitials, O and N. By SSE processing after zone refining, metal with an overall purity of 99.94 at% and an O content of 29 ppm.at was achieved [15]. Peterson and Schmidt [78] measured the electrotransport velocities of O, N and C in both the hcp and bcc phases, finding significantly faster rates in the cubic phase, while all three impurities migrated toward the anode at all temperatures tested. A sample with an initial RRR of 40 was improved to 340 after 150 h of SSE processing at 1245°C in a He environment. Hasegawa et al. [79] SSE refined commercial-grade Gd for 100 h at 0.8 Tm, lowering the O content at the cathode end to about 1500 ppm.wt from an initial level of

3600 ppm.wt. Jordan and Jones [80] managed to reduce the O content at the cathode end of a Gd SSE sample from around 5000 ppm.at to 106 ppm.at and found that the levels of several other impurities, including C, Al, Si, Fe, Ni and Cu, were reduced by factors of 5–10. Jordan et al. [81] investigated the quality of crystals grown during SSE in an initially polycrystalline rod of Gd, with a starting purity of 99.3 at%, reporting the growth of grains in excess of 20 mm in length after a total of 600 h processing at 1100°C in vacuum. The crystals were measured by X-ray Laue methods to have a mosaic spread of about 1 minute of arc, and an overall purity better than 99.9 at% with a RRR of 250. In further research, single-crystal rods of Gd, previously grown by EBFZ, were SSE refined to give crystals that were claimed to be superior to those produced by other methods [82]. Gd is also amenable to crystal growth by solid state and float-zoning methods [52, 58, 61, 83].

Tb can be refined by a variety of methods; Beaudry and Gschneidner [2] suggest a vacuum melting temperature of 1750°C, and a distillation temperature of 1575°C (with a condenser at 800°C). Hukin et al. [84] zone-refined a Tb sample, reducing the O content at the "end end" to 210 ppm.wt from a start level of 740 ppm.wt, which equates to a k_{eff} of 1.2. Jordan et al. [85] prepared metal of better than 99.9 at% purity from start metal with a purity of about 99.5 at%, using SSE under UHV conditions. The processing involved heating the sample, which was actually a single-crystal rod grown by EBFZ, to 1050°C for a total of 350 h over two stages. The O content was reduced from 3800 ppm.at in the start metal to 260 ppm.at, although H and N were not markedly reduced, the final contents being 350 and 180 ppm.at respectively. Fort [34] prepared high-quality crystals of Tb with a RRR of 400 by combining strain anneal crystal growth with SSE purification. Hasegawa et al. [86] report the reduction of the O content in a Tb sample from 1400 ppm.wt to about 700 ppm.wt (at the cathode end) by SSE refining for 100 h at 0.8 Tm, while Whittaker [50] grew crystals of Tb during SSE in metal too impure for crystal growth by other means. Tb is also amenable to crystal growth by solid state methods and zoning [52, 58, 61, 83].

Dy has a fairly high vapor pressure, but can be vacuum melted to a limited degree – Beaudry and Gschneidner [2] recommend a temperature of 1440°C for 45 min – while for sublimation they suggest a temperature of 1175°C with a condenser at 700°C. Hasegawa et al. [86] used a He environment to SSE refine Dy for 80 h at 1120°C, reducing the O content at the cathode end from an initial 4000 ppm.wt to below 300 ppm.wt. N and H were also reduced, but C did not migrate appreciably. Fort et al. [34, 43] SSE refined Dy using start metal that was initially sublimed. The best results came from a three-stage SSE run, the first stage being a degassing stage of 350 h under UHV with a maximum sample temperature of 730°C, and the second stage being under 1.5 bar Ar for 225 h with a maximum temperature of 1115°C, with a final degassing stage under UHV. After sublimation, the metal had a RRR of 85 and an overall purity of 99.93 at%, these values rising to 124 and 99.96 at%, respectively, at the cathode end after SSE. The thermodynamic behavior of this high-purity Dy near to the ferromagnetic-to-antiferromagnetic transition was found to exhibit both superheating and supercooling, effects not observed in less pure metal [87]. Dy is amenable to crystal growth by solid state methods [51, 61, 83], while RFFZ was found to be possible using 0.7 bar Ar [52], the best crystals coming from re-sublimed start metal.

Ho. Beaudry and Gschneidner [2] indicate that Ho can be purified to some degree by vacuum melting at 1480°C for 45 min, or sublimed at 1220°C, with a condenser at 725°C. Fort [34] has reported the preparation of Ho with a RRR value of 90 after SSE processing for 200 h at 1100°C under Ar. Two grains, each 2 cm in length and 6 mm diameter, were encouraged to grow in the initially polycrystalline sample by straining it prior to SSE. Good-quality crystals of Ho have been reported [52] using the RFFZ technique with an environment of 0.7 bar Ar, while the metal is also amenable to crystal growth by solid state methods [58, 61, 83].

Er. Beaudry and Gschneidner [2] list a vacuum-melting temperature for Er of 1540°C for 30 min, and a sublimation temperature of 1300°C with the condenser at 825°C. Maezawa et al. [88] SSE refined Er for 70 h at 1050°C, the initial RRR of 7.5 being raised to between 31 (anode end) and 46 (cathode end); this overall increase in the RRR suggests that vacuum degassing of this relatively low-purity metal gave greater benefit than electrotransport. Fort et al. [44] SSE refined previously sublimed Er to a total purity of 99.97 at% (99.996 wt%). The SSE was carried out mainly under 0.7 bar Ar, together with some vacuum degassing stages. The total refining time was about 2350 h, of which 1020 h were under vacuum; the maximum temperature reached in the Ar stages was 1210°C. Extensive analysis of the sample showed that O, H, C, N and F migrated in the cathode-to-anode direction. Of the metallic impurities, which were generally at low levels, Fe, Ni and Pb migrated toward the cathode, while Cu moved in the opposite direction. The O content was reduced from 638 ppm.at prior to SSE to 168 ppm.at at the cathode end, while H and N were below the limits of detection. F was reduced from 79 to 26 ppm.at, but C showed a marginal increase, a feature attributed either to the use of hydrocarbon oils used in the swaging operation during rod preparation or to methane degassing from the system walls during the SSE stage under Ar. This high-purity Er was the second metal (after Dy) in which both superheating and supercooling were observed, effects that have only been found in ultrapure SSE-purified lanthanide metals [89]. Er is amenable to crystal growth by solid state and zoning methods [52, 58, 61, 83].

Tm. Every aspect of purification and sample preparation work with Tm is dominated by its exceptionally high vapor pressure, which effectively limits its purification to vaporization methods. Ionov et al. [22] increased the RRR of Tm from 10 to 330 by sublimation under UHV conditions. A boiler temperature of 850°C was used, with the condenser at 427°C, although Beaudry and Gschneidner [2] recommend a slightly higher sublimation temperature of 950°C. Schieber [49] has grown Tm crystals from the vapor, a process that could presumably be incorporated into the sublimation purification process. Tm is also amenable to crystal growth by solid state methods [58, 61, 83].

Yb. Despite its low melting point, Yb has a high vapor pressure, so most published purification research has involved sublimation; Beaudry and Gschneidner [2] give a sublimation temperature of 625°C, with the condenser at approximately 350°C. Ionov et al. [22] report the preparation of Yb with a RRR of 48 by sublimation through a W filter under UHV conditions. The boiler and condenser temperatures were 480 and 427°C, respectively. Sample preparation with Yb, particularly for low-temperature work, is complicated by the martensitic fcc–hcp transformation, which starts on cooling at about −13°C [90]. The transition can be reversed by heating and is completed by 92 ± 15°C. Kayser [90] reports that under favorable conditions, specimens can be converted to ∼ 90% alpha-Yb. Schieber [49] has grown a Yb crystal 5 mm long by vapor deposition, while small crystals of both hcp and fcc-Yb have been grown by Winarski [91] using vapor condensation under argon. However, the fcc crystals were found to transform slowly to the hcp structure when kept at room temperature.

Lu has the highest melting point of any rare earth, but has a moderate vapor pressure, which means that it can be purified by vacuum melting at 1800°C [2]. Peterson and Schmidt [92] have measured the electrotransport velocities of C, N and O in Lu at 1330, 1450 and 1600°C; all three impurities were found to migrate toward the anode. SSE purification was performed at 1150°C for 168 h in UHV, the RRR increasing from 21 to 151 at the cathode end. Similar Lu was SSE refined under helium at 1450°C for 184 h, but the RRR values were found to be approximately the same as those obtained in the UHV run, despite the higher current density. This was attributed to contamination from residual impurities in the inert gas and from impurities degassing from the inner surface of the sample chamber. Sublimation is reported to be possible with Lu, Beaudry and Gschneidner [2] quoting a temperature of 1645°C with a condenser at 850°C. Lu is a menable to crystal growth by zoning and solid state methods [52, 58, 61, 83].

Sc. The vapor pressure of Sc is sufficiently high to allow purification by sublimation at 1425°C (with collection at 900°C), while improvements can also be made by vacuum melting at 1550°C for 10 min [2]. Schmidt and Carlson [93] studied the electrotransport behavior of C, N and O in both the hcp and bcc phases of Sc, all three solutes migrating toward the anode, while mobilities in the bcc phase were measured to be an order of magnitude higher than those at the upper end of the hcp phase. Fe and Ni were also observed to move toward the anode. SSE experiments saw Sc purified to a RRR of 520 at the cathode end after 1250 h processing at 1220°C under 130 torr He. No excessive loss of Sc was encountered under these conditions, or from a run at 1360°C for 100 h. The authors believed that higher RRR values could be achieved if no contamination of the specimen was introduced from the residual gaseous impurities on the inner surface of the sample chamber and from the He gas. Axtell et al. [94] report that Fe is a fast-diffusing element in Sc, with a mobility during SSE comparable with that of interstitial solutes. Both zoning and solid state crystal growth methods are possible with Sc [52,58,61,83].

Y. Having a moderate vapor pressure, Y can be vacuum melted at 1850°C and distilled at 1725°C, with the condenser at 900°C [2]. Carlson et al. [35,95] determined the U, D and Z^* values during SSE for C, O, N and H in Y at various temperatures. All four interstitials were found to move toward the anode. A sample with an initial RRR of 12 was SSE refined at 1175°C for a total of 280 h under a relatively poor vacuum of 2×10^{-7} mbar, to give a RRR of 46 near the cathode end. The O content was reduced from 780 to 90 ppm.wt. Volkov et al. [96] studied the behavior of a number of impurities in Y during SSE, and found that O, Si, Al, Fe, Ta and V migrated toward the anode, while Cu moved in the opposite direction. By SSE refining metal with a starting RRR of 26 for a total of 780 h in three stages at temperatures up to 1240°C under UHV conditions, they produced metal with a RRR of 570. Nikoforova and Volkov [42] achieved an RRR of 1200, from start metal with an RRR of 30, by SSE refining at 1170°C for 900 h under UHV. Crystal growth by float zoning under Ar was reported by Fort [52] to give reasonable results for high-purity samples, while a good-quality crystal with an RRR value of over 800 was prepared by SSE refining a crystal that had been previously grown using RFFZ [34]. Y is also amenable to crystal growth by solid state methods [58,61,83].

References

1. F.H. Spedding *The rare earths*, ed. by F.H. Spedding, A.H. Daane (John Wiley: In, New York, 1961), p. 3
2. B.J. Beaudry, K.A. Gschneidner, Jr.: In *Handbook on the Physics and Chemistry of Rare Earths*, ed. by K.A. Gschneidner, Jr., L. Eyring, Vol. 1 (North Holland, Amsterdam 1978), p. 173
3. K.A. Gschneidner, Jr.: In *Science and Technology of Rare earth Materials*, ed. by E.C. Subbarao and Wallace(Academic Press, 1980), p. 25

4. J.D. Fast: *Gases in metals* (Macmillan Press, London 1976)
5. I. Langmuir: Phys. Rev. **2**, 329 (1913)
6. C.E. Habermann, A.H. Daane, P.E. Palmer: Trans. Metall. Soc. AIME **233**, 1038 (1965)
7. D.H. Dennison, M.J. Tschetter, K.A. Gschneidner, Jr.: J. Less-Common Metals **11**, 423 (1966)
8. K. Mimura, M. Isshiki: In Proceedings of the 1st International Conference on Processing Materials for Properties, ed. by H. Hansen, T. Oki (The Minerals, Metals and Materials Society, 1993) p. 897
9. A.M. Ionov, V.T. Volkov, T.V. Nikiforova: J. Alloys Compounds **223**, 91 (1995)
10. M. Isshiki: Vacuum **47**, 885 (1996)
11. K. Mimura, Sang-Woon Lee, M. Isshiki: J. Alloys Compounds **221**, 267 (1995)
12. W.G. Pfann: Zone melting, 2nd edn. (Robert E. Krieger, New York 1966)
13. C.L. Huffine, J.M. Williams: In *The rare earths*, ed. by F.H. Spedding, A.H. Daane, (John Wiley, New York 1961) p. 145
14. G. Revel, J-C. Rouchand, J-L. Pastol: *Proceedings ofthe 11th Rare Earth Research Conference*, ed. J.M. Haschke and H.A. Eick (US Atomic Energy Commission Technical Information Ceter, 1974) p. 602
15. D. Fort, B.J. Beaudry, K.A. Gschneidner, Jr.: J. Less-Common Metals **134**, 27 (1987)
16. D. Fort, D.W. Jones, B.J. Beaudry, K.A. Gschneidner, Jr.: J. Less-Common Metals **81**, 273 (1981)
17. D.W. Jones, D. Fort, D.A. Hukin: in *The rare earths in modern science and Technology*, ed. G.J. McCarthy, J.J. Rhyne (Plenum Press, New York 1978) p. 309
18. D.W. Jones, J.S. Abell, D. Fort, J.K. Hulbert: J. Magnetism Magnetic Mater. **29**, 20 (1982)
19. D.C. Koskimaki, K.A. Gschneidner, Jr., N.T. Panousis: J. Crystal Growth **22**, 225 (1974)
20. G. Burrows: J. Appl. Chem. **7**, 375 (1957)
21. A.J. Martin: Vacuum **7–8** 346 (1959)
22. A.M. Ionov, T.V. Nikiforova, N.N. Rytus: Vacuum **47**, 879 (1996)
23. J.E. Murphy, E. Morrice, M.M. Wong: U.S. Bureau of Mines, report no. 7466 (1970)
24. N.N. Rytus, T.V. Nikiforova: Russian Metall. **6**, 74 (1980)
25. K.A. Gschneidner, Jr.: In *Transactions of Vacuum Metallurgy Conference,* ed. L.M. Bianchi (American Vacuum Society, 1966) p. 99
26. J.D. Verhoeven: J. Metals **18**, 26 (1966)
27. J.N. Pratt, R.G.R. Sellors: *Electrotransport in metals and alloys* (Trans Tech SA, Rienen, Switzerland 1973)
28. R.G. Jordan: Contemp. Phys. **15**, 375 (1974)
29. R.S. Sorbello: In *Electro- and thermo-transport in metals and alloys*, ed. by R.E. Hummel, H.B. Huntington (American Institute of Mining, Metallurgical and Petroleum Engineers., New York 1977) p. 2
30. F.A. Schmidt, O.N. Carlson: Metall.Trans. **7A**, 217 (1976)
31. D.T. Peterson: In *Atomic transport in solids and liquids*, ed. by A. Lodding, T. Lagerwall (Verlag der Zeitschrift für Naturforschung, Tubingen, Germany 1971) p. 104
32. D.T. Peterson, F.A. Schmidt, J.D. Verhoeven: Trans. Metall. Soc. AIME **236**, 1311 (1966)

33. D.T. Peterson: In *Electro- and thermo-transport in metals and alloys*, ed. by R.E. Hummel, H.B. Huntington (American Institute of Mining and Petroleum Engineers, New York 1977) p. 54
34. D. Fort: J. Less-Common Metals **134**, 45 (1987)
35. O.N. Carlson, F.A. Schmidt, D.T. Peterson: J. Less-Common Metals **39**, 277 (1975)
36. F.A. Schmidt, R.A. Outlaw, B.K. Lunde: J. Electrochem. Soc. **126**, 1811 (1979)
37. D.A. Blackburn: In *Electro- and thermo-transport in metals and alloys*, ed. by R.E. Hummel, H.B. Huntington (American Institute of Mining and Petroleum Engineers, New York 1977) p. 20
38. F.A. Schmidt, G.A. Martsching, O.N. Carlson: J. Less-Common Metals **68**, 75 (1979)
39. F.A. Schmidt, R.J. Conzemius, O.N. Carlson: J. Less-Common Metals **59**, 53 (1978)
40. W.N. Weins, O.N. Carlson: J. Less-Common Metals **66**, 99 (1979)
41. O.N. Carlson, F.A. Schmidt: J. Less-Common Metals **53**, 73 (1977)
42. T.V. Nikoforova, V.T. Volkov: J. Less-Common Metals **115**, 91 (1986)
43. D. Fort, V.K. Pecharsky, K.A. Gschneidner, Jr.: J. Alloys and Compounds **236**, 190 (1995)
44. D. Fort, B.J. Beaudry, K.A. Gschneidner, Jr.: J. Less-Common Metals **166**, 173 (1990)
45. J.D. Marchant, E.S. Shedd, T.A. Henrie, M.M. Wong: U.S. Bureau of Mines Report of Investigators 7480 (1971)
46. C.A. King, S.V. Brown: Rev. Sci. Instum. **63**, 3185 (1992)
47. E.U. Malang, C.S. Oglesby, T. Siegrist, E. Bucher: Physica B **204**, 363 (1995)
48. A.R. Bradshaw, D. Fort: Rev. Sci. Instrum. **63**, 5459 (1992)
49. M. Schieber: In *Crystal growth*, ed. by H.S. Perser (Pergamon Press, New York 1967) p. 271
50. M. Whittaker: J. Crystal Growth **3/4**, 317 (1968)
51. K.J. Bachmann: J. Crystal Growth **18**, 13 (1973)
52. D. Fort: J. Crystal Growth **94**, 85 (1989)
53. R.A. Laudise: *Growth of single crystals* (Prentice-Hall, Englewood Cliffs, New Jersey 1970)
54. S.A. Brown, B.K. Howard, S.V. Brown, R.S. Julian: Rev. Sci. Instrum. **61**, 2427 (1990)
55. D. Fort: Rev. Sci. Instrum. **64**, 3209 (1993)
56. H.F. Sterling, R.W. Warren: Nature **192**, 745 (1961)
57. D.A. Hukin: Report 24/71, Clarendon Laboratory, University of Oxford (1971)
58. K.A. McEwen, P. Touberg J. Phys. F, Metal Phys. **3**, 1903 (1973)
59. T.B. Reed and E.R. Pollard, J. Crystal Growth **2**, 243 (1968)
60. D. Fort: Rev. Sci. Instrum. **68**, 3504 (1997)
61. D. Fort: J. Alloys Compounds **177**, 31 (1991)
62. J.J. Tonnies, K.A. Gschneidner, Jr.: J. Crystal Growth **10**, 1 (1971)
63. P.H. Pan, D.K. Finnemore, A.J. Bevolo, H.R. Shanks, B.J. Beaudry, F.A. Schmidt, G.C. Danielson: Phys. Rev. B **21**, 2809 (1980)
64. D.W. Jones, D. Fort, D.A. Hukin: In *The rare earths in modern science and technology*, ed. by G.J. McCarthy, J.J. Rhyne (Plenum Press, New York 1978) p. 309

65. D. Fort, B.J. Beaudry, D.W. Jones, K.A. Gschneidner, Jr.: In ed. by G.J. McCarthy, J.J. Rhyne, H.B. Silber *The rare earths in modern science and technology*, **2**, (Plenum Press, New York 1980) p. 33
66. R.H. Moore, F.M. Smith, J.R. Morrey: Trans. AIME **233**, 1259 (1965)
67. J.D. Marchant, E.S. Shedd, T.A. Henrie: US Bureau of Mines, Report 6894 (1967)
68. W.N. Cathey, J.E. Murphy, J.R. Woodyard, Sr.: Metall. Trans. **4**, 1463 (1973)
69. C.M. Muirhead, D.W. Jones: J. Less-Common Metals **50**, 73 (1976)
70. O.D. McMasters, G.E. Holland, K.A. Gschneidner, Jr.: J. Crystal Growth **43**, 577 (1978)
71. P.G. Mattocks, C.M. Muirhead, D.W. Jones, B.J. Beaudry, K.A. Gschneidner, Jr.: J. Less-Common Metals **53**, 253 (1977)
72. A.M. Ionov, N.N. Rytus, T.V. Nikiforova: Russian Metall. **1**, 85 (1979)
73. B.J. Beaudry, K.A. Gschneidner, Jr.: In *The rare earths in modern science and technology*, ed. by G.J. McCarthy, J.J. Rhyne (Plenum Press, New York 1978) p. 303
74. C. Hiemstra, P. Keegstra, W.T. Masselink, J.B. Van Zytveld: J. Phys. F **14**, 1867 (1984)
75. N.L. Shi, D. Fort: J. Less-Common Metals **113**, L21 (1985)
76. A.H. Daane, D.H. Dennison, F.H. Spedding: J. Amer. Chem. Soc. **75**, 2272 (1953)
77. G. Busch, E. Kaldis, J. Muheim, R. Bischof: J. Less-Common Metals **24**, 453 (1971)
78. D.T. Peterson, F.A. Schmidt: J. Less-Common Metals **29**, 321 (1972)
79. R. Hasegawa, K. Kamihira, K. Goto: Rare Earths, [14], March 1989, abstracts from the March 1989 meeting of the Rare Earth Society of Japan
80. R.G. Jordan, D.W. Jones: J. Less-Common Metals **31**, 125 (1973)
81. R.G. Jordan, D.W. Jones, P.G. Mattocks: J. Less-Common Metals **34**, 25 (1974)
82. R.G. Jordan, D.W. Jones, M.G. Hall: J. Crystal Growth **24/25**, 568 (1974)
83. H.E. Nigh: J. Appl. Phys. **34**, 3323 (1963)
84. D.A. Hukin, R.C.C. Ward, D.K. Morris, K. Davies: In *The rare earths in modern science and technology*, Vol. 2, ed. by G.J. McCarthy, J.J. Rhyne, H.B. Silber (Plenum Press, New York 1980) p. 25
85. R.G. Jordan, D.W. Jones, V.J. Hems: J. Less-Common Metals **42**, 101 (1975)
86. R. Hasegawa, K. Kamihira, K. Goto: Rare Earths [16], May 1990, Abstracts from the May 1990 meeting of the Rare Earth Society of Japan
87. V.K. Pecharsky, K.A. Gschneidner, Jr., D. Fort: Scr. Mater. **35**, 843 (1996)
88. K. Maezawa, Y. Saito, S. Wakabayashi: Jpn. J. Appl. Phys. **24**, 28 (1985)
89. K.A. Gschneidner, Jr., V.K. Pecharsky, D. Fort: Phys. Rev. Lett **78**, 4281 (1997)
90. F.X. Kayser: Phys. Rev. Lett **25**, 662 (1970)
91. A. Winarski: J. Crystal Growth **57**, 443 (1982)
92. D.T. Peterson, F.A. Schmidt: J. Less-Common Metals **18**, 111 (1969)
93. F.A. Schmidt, O.N. Carlson: J. Less-Common Metals **50**, 237 (1976)
94. S.C. Axtell, C.I. Okafor, R.J. Conzemius, O.N. Carlson: J. Less-Common Metals **115**, 269 (1986)
95. O.N. Carlson, F.A. Schmidt, D.T. Peterson: J. Less-Common Metals **10**, 1 (1966)
96. V.T. Volkov, T.V. Nikiforova, A.M. Ionov, A.N. Pustovit, G.G. Sikharulidse: J. Less-Common Metals **79**, 199 (1981)

Part II

New Methods for Purification

6 Hydrogen Plasma Arc Melting

K. Mimura and M. Isshiki

6.1 Introduction

The thermal plasma generator, generally known as the plasma torch, is already in practical use in various fields of the metals industry, such as welding, cutting and metal spraying. In addition, a number of studies have been made for the utilization of thermal plasmas in smelting and melting since the early stages of its development. This is indicated by numerous related publications and by the introduction of several processes into practice [1–6]. The temperature of a thermal plasma generated by a plasma torch ranges from 5000 to 20 000 K at pressures around 0.1 MPa. Plasmas can be considered to be the result of converting electric energy into a gas-flow that contains high-density thermal energy. Since various gases, such as H_2, O_2 and N_2, as well as inert gases like Ar, are used as the plasma-generating gas, the thermal plasma is useful not only as an excellent heat source but also as an efficient source of reagents, by means of dissociated an d activated species that are formed at high temperature [2, 3].

The arc-discharged thermal plasma is used commonly for melting, this being represented by the term "plasma arc melting". This technique is well known as a practical and useful melting method for various metals and alloys. In most cases, Ar is used as a plasma-generating gas, which implies that a significant refining effect cannot be expected on Ar plasma arc melting. It has been found, however, that the addition of H_2 to Ar as the plasma gas enhances the removal of not only nonmetallic impurities such as O and N but also metallic impurities from several molten metals. Then, H_2+Ar plasma arc melting, the so-called "hydrogen plasma arc melting (HPAM)", has been recognized as a useful melting and refining method, which is expected to become available for the production of pure metals.

It is well known that electron-beam melting (EBM) under high vacuum is also useful for the melting and refining of metals. The main purifying effects are deoxidation by evaporation of volatile oxides and removal of metallic impurities with higher vapor pressures [7, 8]. However, high-vacuum melting processes involve the evaporation loss of the molten metal and the technical difficulty of maintaining a high vacuum state for a long period of time. On the other hand, HPAM is operated at around atmospheric pressure: therefore,

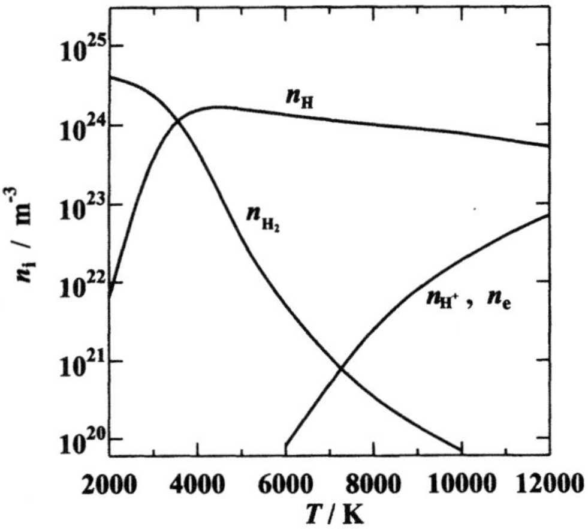

Fig. 6.1. The equilibrium composition of species in hydrogen gas at high temperature under atmospheric pressure

its operation is easier and the refining effect is expected to progress without considerable evaporation loss of the melted metal.

Figure 6.1 shows the equilibrium composition of the main species that constitute hydrogen gas at high temperature under atmospheric pressure [3, 9]. The dissociation of H_2 according to (6.1) is up to above 95% at 5000 K. Then, for HPAM, it is considered that dissociated and activated hydrogen atoms formed within the high-temperature plasma arc (> 5000 K) play an important role in the refining procedure by blowing H_2+Ar plasma arc on to the molten metal:

$$H_2 = H + H\,. \tag{6.1}$$

This chapter reviews the removal of nonmetallic impurities (such as O and N) and also metallic impurities from melted metals (mainly refractory metals) by HPAM.

6.2 The Removal of Nonmetallic Impurities by Hydrogen Plasma Arc Melting

When producing high-purity metals, the elimination of nonmetallic impurities such as O, N and C to a low concentration level is an important subject, as well as the elimination of metallic impurities. It is no easy task to remove nonmetallic impurities from pure metals without contamination. In the case of deoxidation of Fe, for example, deoxidation of solid and liquid Fe under an H_2 gas flow is not practical, because a long treatment time is needed to reduce the oxygen concentration to ppm levels, and because this method

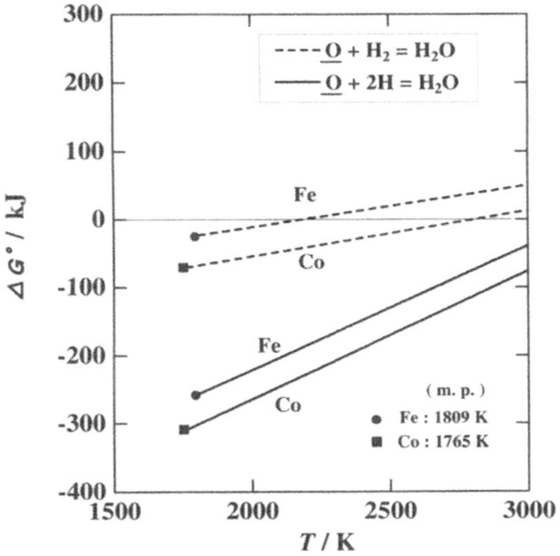

Fig. 6.2. Free energy changes of deoxidation of liquid iron and cobalt with H_2 and H as a function of temperature

cannot produce enough quantitatively [10–12]. On the other hand, EBM is useful for deoxidation of refractory metals such as Nb and Ta, due to their volatile oxides. However, the oxides of Fe and Co are not volatile compared with the metal, so that deoxidation of Fe and Co by EBM is difficult [7,8,13].

By HPAM, on the other hand, deoxidation of molten metals is expected to be promoted through the activated hydrogen atom. Figure 6.2 shows free energy changes as a function of the temperature for deoxidation of liquid Fe and Co by H_2 and H, according to (6.2) [14–16]. It is evident that the deoxidizing power of atomic H is considerably higher than that of molecular H_2. Then, HPAM is expected to deoxidize Fe and Co rapidly, and the oxygen concentration must be reduced to a very low concentration level. In the same way, denitrogenization and decarburization of the melted melts during HPAM will be also expected to proceed effectively, as described later.

$$\underline{O}(\text{in liq. metal}) + 2H = H_2O\,. \tag{6.2}$$

The removal of nonmetallic impurities by HPAM, deoxidation of Fe, Co, Cu, V, Mo and Ta [17–23], denitrogenization of Fe, Co, Mo, Ta and Fe–Cr alloys [17, 24, 25], and decarburization of Fe and Fe–Cr alloys [17, 25] have already been reported. This section deals with the main typical results of removing nonmetallic impurities such as O and N from molten Fe, Co and Ta by HPAM.

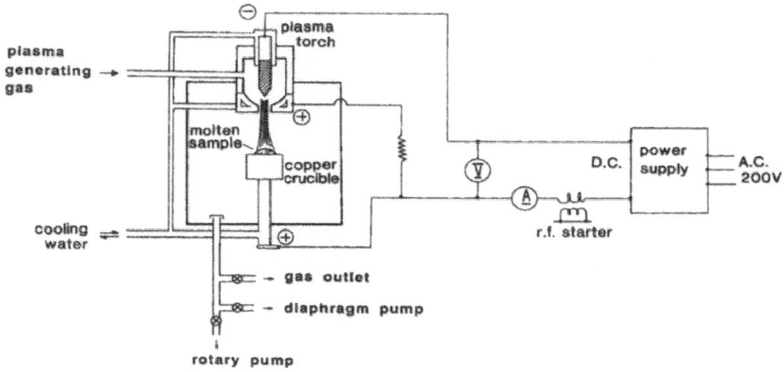

Fig. 6.3. A schematic drawing of a laboratory-scale plasma arc furnace

6.2.1 The Removal of Oxygen, Nitrogen and Carbon from Iron

Mimura et al. [17] have investigated deoxidation, denitrogenization and decarburization of Fe by Ar and H_2(0.5–20%)+Ar plasma arc melting of electrolytic Fe ($>$ 99.9%; O = 157, N = 22, C = 5 ppm) as the specimen. Experiments were carried out using a laboratory-scale plasma arc furnace, as shown in Fig. 6.3, equipped with a transferred-arc type plasma torch with a maximum power of 20 kW. The Fe specimen of about 40 g was set on the water-cooled copper crucible of 45 mm diameter and 4 mm depth and melted at around 2200 K under atmospheric pressure. A gas mixture of pure Ar and H_2 was used as the plasma gas, with a flow rate of 5 l/min. During the process of plasma arc melting, no evolution of gas bubbles from the Fe melt was observed.

6.2.1.1 Deoxidation of Fe. Figure 6.4 shows the changes in oxygen concentration of Fe melted by Ar and 0.5%H_2+Ar plasma arcs. The decrease of the oxygen concentration was practically negligible when only Ar was used, but remarkable deoxidation was observed by the addition of a small amount of hydrogen, such as 0.5%H_2, to the plasma gas. Deoxidation progresses by the following two steps, as indicated in Fig. 6.4: the fast first step followed by the slow second step – both of which obey a first order rate law.

In order to elucidate the influence of the addition of hydrogen to the plasma gas on deoxidation efficiency, Fe specimens were melted in 1%H_2 to 20%H_2+Ar plasma arcs. As shown in Fig. 6.5, the deoxidation rate increased extremely with the hydrogen content. The oxygen concentration was reduced to less than 10 ppm in 0.18 ks, and was then expected to reach below 1 ppm by using 20%H_2+Ar plasma gas.

The apparent rate constant of deoxidation obtained from the slopes at the first step was found to be proportional to roughly the square root of

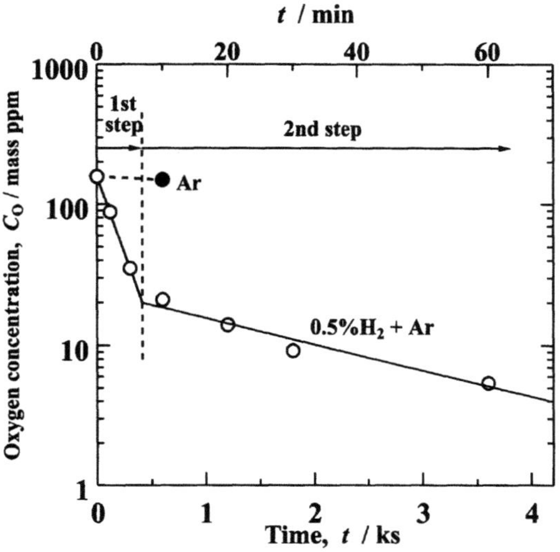

Fig. 6.4. Changes in the oxygen concentration of liquid iron during Ar and 0.5%H_2+Ar plasma arc melting

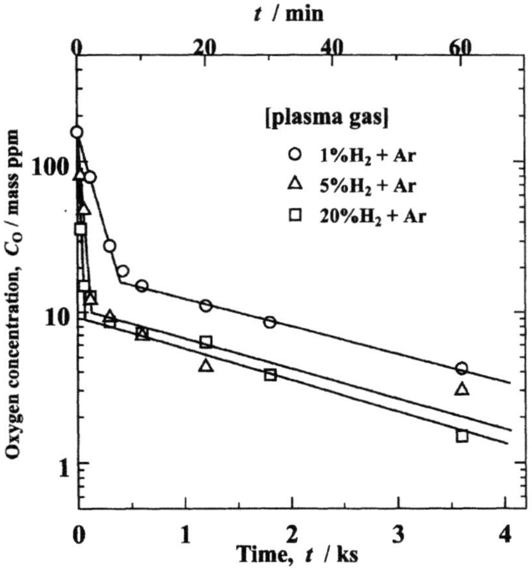

Fig. 6.5. Changes in the oxygen concentration of liquid iron during H_2 (1–20%)+Ar plasma arc melting

the hydrogen content in the plasma gas. This means that deoxidation rate at the first step can be represented by (6.3). This result corresponds to the fact that P_H is proportional to the square root of P_{H_2}, as shown in (6.4), and it suggests that the reaction by H as (6.2) plays a main part in efficient deoxidation during HPAM:

$$-\frac{d[\%O]}{dt} = k'(P_{H_2})^{1/2}[\%O]\,, \tag{6.3}$$

$$1/2H_2 = H, \quad P_H = K(P_{H_2})^{1/2}\,. \tag{6.4}$$

It has been concluded from the above-mentioned results that the chemical reaction on the surface of the liquid Fe is rate-determining in the first step of deoxidation, while oxygen transfer within the boundary layer in the liquid side is rate-determining for the second step.

6.2.1.2 Denitrogenization of Fe. Figure 6.6 shows the changes in nitrogen concentration of Fe melted with Ar, 1%H_2+Ar and 5%H_2+Ar plasma arcs. A remarkable decrease of the nitrogen concentration was observed by HPAM, even at a low hydrogen content in the plasma gas, but it was practically not seen using only Ar plasma. The nitrogen concentration was reduced to a few mass ppm at 1%H_2 and further lowered to less than 1 ppm at 5%H_2, indicating a great increase in the denitrogenizing power of HPAM with the hydrogen content. To consider the denitrogenization of HPAM thermodynamically, equilibrium nitrogen concentrations in liquid Fe are calculated for

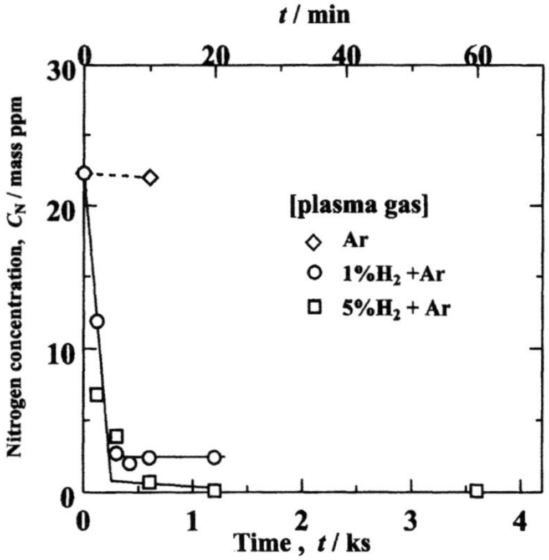

Fig. 6.6. Changes in the nitrogen concentration of liquid iron during Ar and H_2(1–5%)+Ar plasma arc melting

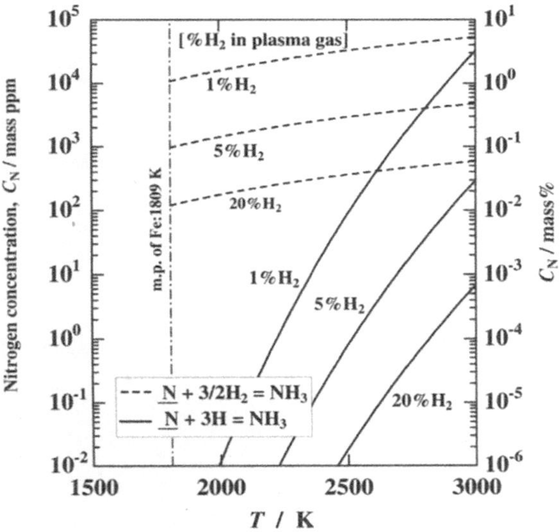

Fig. 6.7. The equilibrium nitrogen concentration of liquid iron calculated for $\underline{N} + 3/2H_2 = NH_3$ and $\underline{N} + 3H = NH_3$ at $p_{NH_3} = 0.1\,Pa$

the reactions with H_2 and H, as represented by (6.5) and (6.6), and indicated in Fig. 6.7 [16,26,27]. In the temperature range during the HPAM treatment, NH_3 is said to be predominant as the reaction product [25]:

$$\underline{N}(\text{in liq. Fe}) + 2/3H_2 = NH_3, \quad (6.5)$$

$$\underline{N}(\text{in liq. Fe}) + 3H = NH_3, \quad (6.6)$$

$$\underline{N}(\text{in liq. Fe}) = 1/2N_2. \quad (6.7)$$

Figure 6.7 suggests that denitrogenization of liquid Fe cannot be enhanced by (6.5). In fact, Narita et al. [28] have reported that, when an Ar and H_2(5–10%) gas mixture was blown on the surface of the Fe melt during induction heating, no significant enhancement was observed in the reduction of nitrogen concentration, besides a denitrogenization progress by (6.7). This confirms that the high denitrogenizing power of activated hydrogen atoms results in the strong denitrogenization effect of HPAM. The nitrogen concentration of liquid Fe could be reduced quickly by (6.6) during HPAM to lower levels than the equilibrium value set by (6.7).

6.2.1.3 Decarburization of Fe. Figure 6.8 demonstrates that even Ar plasma arc melting has an observed effect on decarburizing liquid Fe. In this case, the reaction, represented by (6.8), resulting in the formation of CO must have progressed because of the high initial oxygen concentration. The carbon concentration was further reduced to below 1 ppm by $1\%H_2$+Ar

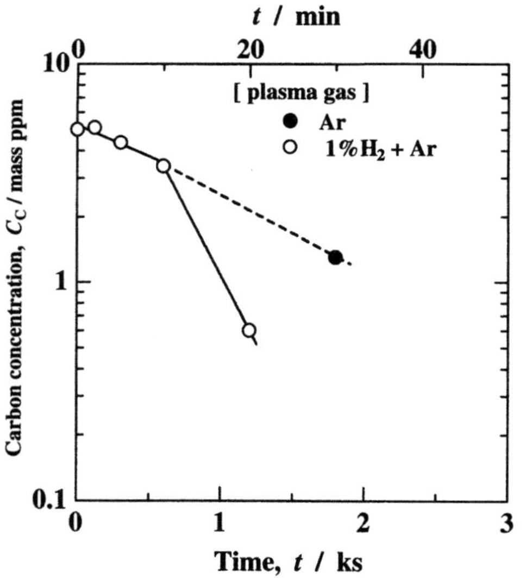

Fig. 6.8. Changes in the carbon concentration of liquid iron during Ar and 1%H_2+Ar plasma arc melting

plasma arc melting. Decarburization must be accelerated by the participation of H, according to (6.9), as has been reported by Kaneko et al. [25] and Matsuo et al. [18]:

$$\underline{C}(\text{in liq. Fe}) + \underline{O}(\text{in liq. Fe}) = CO, \tag{6.8}$$

$$\underline{C}(\text{in liq. Fe}) + 4H = CH_4. \tag{6.9}$$

Although mass loss due to evaporation of Fe during the HPAM treatment increased with melting time and with the hydrogen content of the plasma gas, it did not amount to much. As a result, the mass loss of the melted Fe was only 0.4% even after 3.6 ks of 20%H_2+Ar plasma arc melting. Consequently, nonmetallic impurities such as O, N and C can be removed from liquid Fe efficiently without excessive evaporation loss of Fe.

6.2.2 The Removal of Oxygen and Nitrogen from Cobalt

Deoxidation and denitrogenization of electrolytic Co (> 99.9%; O = 290, N = 26 ppm) have been examined by applying Ar and H_2 (0.5–20%)+Ar plasma arc melting [17]. Experiments were carried out in the same manner as the above-mentioned case of Fe.

6.2.2.1 Deoxidation of Co. Figure 6.9 shows the changes in oxygen concentration of liquid Co during Ar and 0.5%H_2 to 20%H_2+Ar plasma arc

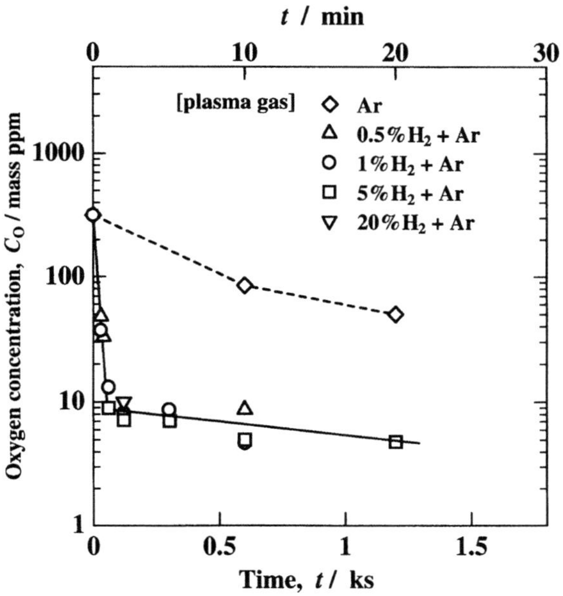

Fig. 6.9. Changes in the oxygen concentration of liquid cobalt during Ar and H_2 (0.5–20%)+Ar plasma arc melting

melting. A slight reduction of the oxygen concentration was observed even by Ar plasma arc melting. This result is probably caused by the evaporation of Co oxide such as CoO, because its vapor pressure is a little higher than that of liquid Co, unlike in the case of Fe. Moreover, by the addition of a small amount of H_2 to the plasma gas, the oxygen concentration decreased rapidly to nearly 10 ppm in 60 s and to 5 ppm in 1.2 ks. Deoxidation proceeded in two steps, in the same way as for Fe; however, its rate at the first step was about ten times faster than that of Fe. As also expected from Fig. 6.2, it is noted that HPAM is more effective for deoxidation of Co than that of Fe.

6.2.2.2 Denitrogenization of Co. The changes in nitrogen concentration of liquid Co during Ar and H_2(0.5–5%)+Ar plasma arc melting are shown in Fig. 6.10. Unlike the case of Fe, effective reduction of the nitrogen concentration was observed even by Ar plasma arc melting. This result must arise from a denitrogenization reaction expressed in (6.10), because the equilibrium nitrogen concentration of liquid Co is about one order of magnitude lower than that of liquid Fe under the same conditions [29]. By the addition of 0.5%H_2 or more to the plasma gas, the denitrogenization rate became markedly fast in the early stage, and the nitrogen concentration was reduced to a few ppm in 0.12 ks, and to less than 1 ppm in 0.3 ks. The excellent denitrogenization

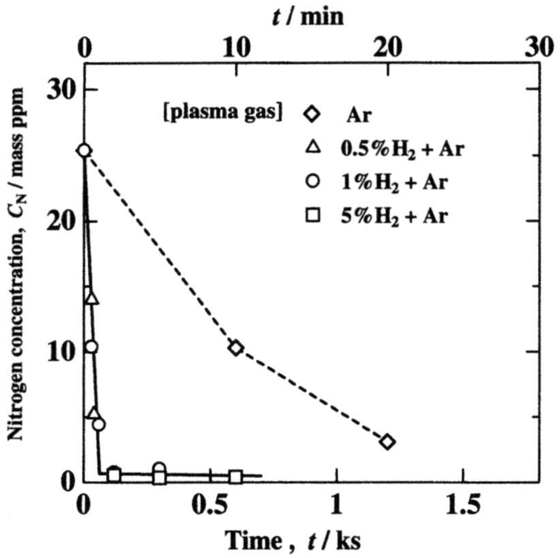

Fig. 6.10. Changes in the nitrogen concentration of liquid cobalt during Ar and H_2 (0.5–5.0%)+Ar plasma arc melting

effect of HPAM can be explained thermodynamically by (6.11):

$$\underline{N}(\text{in liq. Co}) = 1/2N_2, \tag{6.10}$$

$$\underline{N}(\text{in liq. Co}) + 3H = NH_3. \tag{6.11}$$

The evaporation loss of Co during HPAM is as low as about one third of that in the case of Fe. It will therefore be concluded that HPAM is very efficient and practical for the removal of nonmetallic impurities, such as O and N, from the Co melt.

6.2.3 The Removal of Oxygen and Nitrogen from Refractory Metals

The removal of oxygen from a crude Ta (O = 0.4–1.2%, C < 0.02%, Vickers hardness, H_v = 300–600), prepared by carbothermic reduction of Ta_2O_5, has been examined using 10%H_2 and 20%H_2+Ar plasma arc melting [22]. As a result of the excellent deoxidation by HPAM, ductile Ta (O + C < 100 ppm, H_v = 95–100) could be obtained from the crude starting material.

Based on this good result, Mimura et al. [23] have attempted the recovery of pure Ta metal from Ta-capacitor scrap by HPAM. Scrap from used Ta capacitors has become a significant resource of Ta, because Ta metal is used mainly for Ta capacitors. Since it consists of sintered Ta powder, Ta_2O_5 film as a charge trapping and a MnO_2 coating as a solid electrolyte, the scrap can be considered as a crude Ta metal which contains more than 90% Ta

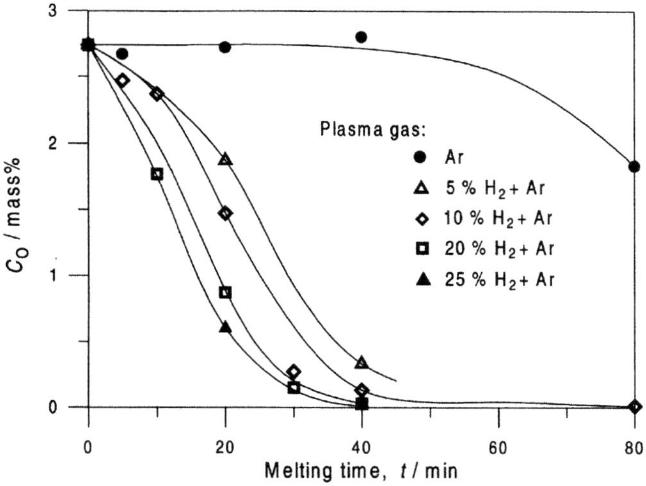

Fig. 6.11. Changes in the oxygen concentration of Ta-capacitor scrap during Ar and H_2 (5–25%)+Ar plasma arc melting

and a small amount of O, Mn and so on. Therefore, pure Ta metal would be recovered directly from the scrap if the O and Mn could be removed easily.

About 15 g of Ta-capacitor scrap (Ta = 93.7%, O = 2.7%, Mn = 3.4%) was melted by Ar and H_2(5–25%)+Ar plasma arc using a laboratory-scale furnace (Fig. 6.3). The melting temperature was set to be about 3500 K. The oxygen concentration was reduced from above 2.5% to less than 100 ppm during HPAM, as shown in Fig. 6.11, while deoxidation was scarcely observed by the Ar plasma arc. Also, the Mn concentration could be reduced quickly to less than 20 ppm by the effect of evaporation. Furthermore, other minor impurities such as N, Fe and Ni in the scrap were also removed to low levels. Consequently, ductile pure Ta metal (Ta $>$ 99.9%, O $<$ 100 ppm, Mn $<$ 20 ppm) was recovered directly even from the scrap. Figure 6.12 shows the Ta-capacitor scrap and the sample melted by HPAM. It gained the shape of a button and a clear metallic luster as the melting time was increased. Thus, the progress of melt refining could also be confirmed by this change in appearance.

In practice, EBM has been used for the refining step of the scrap after acid leaching to remove MnO_2. However, there is a problem about Ta loss by evaporation of Ta oxide such as TaO and Ta itself during melting in high vacuum. On the contrary, above 90% recovery of Ta can be attained by the above-discussed HPAM treatment. It is evident, therefore, that HPAM is superior to EBM with regard to diminishing losses by evaporation of the melted metal.

Good results of deoxidation by HPAM have also been reported for V and Mo [20,21]. Those deoxidation rates were also found to increase in proportion to about the square root of the hydrogen content (%H_2) in the plasma gas.

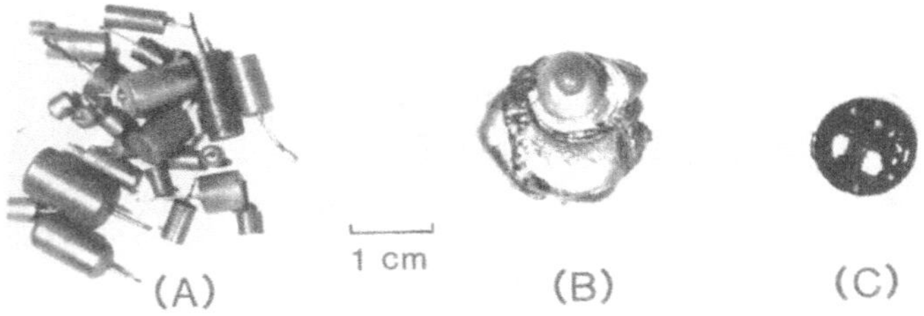

Fig. 6.12. Ta-capacitor scrap (**A**), and the material obtained by 10%H_2+Ar plasma arc melting for 5 min (**B**) and 80 min (**C**)

Moreover, the removal of other nonmetallic impurities, such as N and S, in addition to O by HPAM from liquid Mo and Ta has also been observed [23,24].

6.3 The Removal of Metallic Impurity Elements by Hydrogen Plasma Arc Melting

The rapid removal of Fe and Cr from molten Nb–Fe, Nb–Cr, Mo–Fe and Mo–Cr binary alloys by HPAM treatment was first pointed out about 15 years ago [30]. Since then, it has became evident that more volatile elements (impurities) than the molten base metal have a tendency to evaporate preferentially and be liberated from the melt during HPAM. This result means that, despite melting under atmospheric pressure, HPAM has a similar refining effect to that of EBM under high vacuum. Furthermore, the refining effect of HPAM has been found to be significantly enhanced under reduced pressures and higher hydrogen contents.

As a typical result, Fig. 6.13 shows the changes in the concentration of the alloying element X in melts of Nb – 1 mass%X ($X =$ Fe, Ti, V, Zr, Mo) binary alloys by 40%H_2+Ar plasma arc melting at a reduced pressure of 6.6 kPa. To maintain a constant reduced pressure in the furnace chamber, a diaphragm pump, installed on the exhaust system, was used, as shown in Fig. 6.3. The removal rates of the alloying elements from the melted alloys were found to increase in the following order: Mo $<$ Zr $<$ V $<$ Ti $<$ Fe, and this order corresponds approximately to that of the corresponding ideal vapor pressures in the pure state. It is clear that Zr and Mo, having low vapor pressures similar to that of Nb, cannot be removed, but Fe, due to a significant higher vapor pressure than that of Nb, can be removed by preferential vaporization during HPAM. Based on these results, HPAM is expected to be suitable for purifying refractory metals, because the vapor pressures of most impurities are considerably higher than those of the melted refractory metals at the melting temperature.

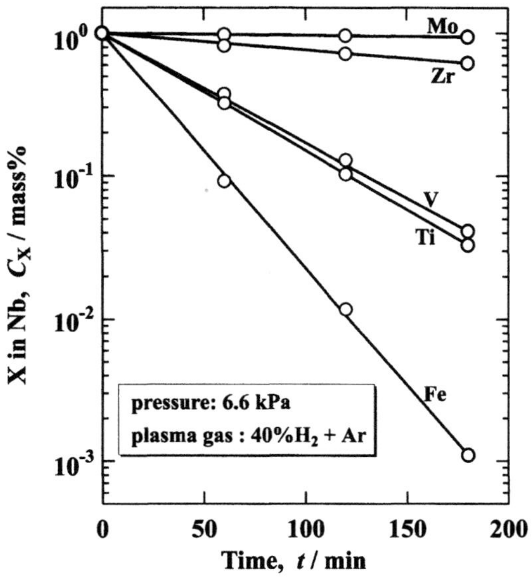

Fig. 6.13. Changes in the alloying element concentrations of Nb – 1 mass% X (X = Fe, Ti, V, Zr, Mo) binary alloys during hydrogen plasma arc melting at reduced pressure

This section deals with the purification of refractory metals by the reduction of trace impurities, the removal of alloying elements from refractory metal based alloys and the recovery of pure refractory metals from refractory metal based alloy scrap, by applying the HPAM treatment. The mechanism of removing metallic impurity elements during HPAM is also discussed.

6.3.1 The Purification of Commercially Pure Refractory Metals

Commercially pure Cr (99.4%), Ti (99.9%), V (99.7%), Zr (99.6%), Nb (99.8%), Ta (99.9%) and Fe (99.9%) were melted in a 30%H_2+Ar plasma arc under the pressures of 5.3 kPa and 101.3 kPa using the furnace as shown in Fig. 6.3 [31, 32]. The values in parentheses indicate the nominal purity. Each specimen (7 cm^3 in volume) was melted at a temperature set about 200–400 K higher than the melting point by controlling the power of the plasma arc. During the total 3.6 ks (1 h) melting time, the specimen was turned upside down once to allow equal duration in both positions. Impurities of the specimens before and after the HPAM treatment were determined by glow discharge mass spectrometric (GDMS) analysis.

The removal degree RD (%) of each impurity is discussed using the definition

$$RD(\%) = 100(C_\mathrm{i} - C_\mathrm{f})/C_\mathrm{i}\,, \tag{6.12}$$

Table 6.1. The removal degree of metallic impurities from iron and refractory metals by 30%H_2+Ar plasma arc melting for 60 min at 5.3 kPa

Base	Removal degree, *RD*%				
metal	0–1%	1–25%	25–50%	50–75%	75–100%
Fe	Ti, Co, Ni Zr, Mo	Si	Cr	Al	Mn, Cu
Cr	Al, Ti, V, Fe Co, Ni, Zr Nb, Mo, Y	Si, P, Ti	S, Cu	Mn	Pb
Ti	Si, V, Y, Zr Nb, Mo	–	Al, Fe, Ni	Cr, Co, Pb	Mn
V	Ti, Y, Zr Nb, Mo	Si	Fe, Co, Ni Pb	Al, Cr	Mn, Cu
Zr	Mo	–	V	Si, Ti, Co Ni	Al, Cr, Mn Fe, Cu, Pb
Nb	Zr	Mo	–	Ti	Al, Si, Cr, Fe Co, Ni, Cu
Ta[a]	Nb	Mo	–	–	Al, Si, Ti, Zr Cr, Mn, Fe Co, Ni, Cu, V

[a] Ta was melted at 101.3 kPa.

where C_i and C_f are the initial and final concentrations, respectively. The *RD* (%) is a useful indicator of the semi-quantitative tendency for the removal of each impurity from the molten metal, although C_i varies from below 1 ppm to nearly 100 ppm. Practical impurities (C_i > 1 mass ppm) are classified according to the typical *RD* (%) range for each melted metal by HPAM in Table 6.1. In the case of Fe and Cr as the melted metal, a few metallic impurities such as Cu and Mn can show high *RD* (%) values. On the other hand, the *RD* (%) of almost all impurities, except for Mo and Zr, becomes higher than 75% for the cases of Nb and Ta. Therefore, further purification of commercially pure Nb and Ta will be achieved by HPAM treatment.

Recently, ultra-pure refractory metals such as Ta, Mo and W have been required as electrode materials for ULSI components. For this purpose, some impurities, such as Na, Fe, U and Th, which degrade the ULSI performance, must be eliminated down to ppb levels. Elanski et al. [33] have investigated the preparation of ultra-pure Ta through further purification of commercially pure Ta (> 99.9%) by 30%H_2+Ar plasma arc melting under atmospheric pressure. Instead of the thoriated W commonly used as a cathode inside the plasma torch, Ta metal was used in the experiment, to avoid contamination of the Ta melt with W and Th. Table 6.2 shows the impurity concentrations

Table 6.2. Concentrations of impurities in Ta (in mass ppm) before and after 30%H_2+Ar plasma arc melting for 60 min at 101.3 kPa

Impurity	Before melting	After melting
Si	0.75	0.064
P	0.01	0.001
Ti	0.018	<0.001
Cr	0.08	0.001
Fe	0.34	0.001
Ni	0.075	0.001
Cu	1.4	0.004
Ga	0.14	0.004
Y	2.8	<0.001
Nb	21.2	21.4
Mo	0.92	0.81
W	12.1	12.9
Th	0.71	<0.001

of Ta (about 30 g in mass) before and after HPAM treatment. Most of the impurities, except for Nb, Mo and W, are removed to a level of a few ppb, and their final concentrations are almost lower than those achieved by EBM. In addition, since the concentration of U in the Ta specimen was below the detection limit of GDMS, the possibility of the removal of U from Ta during HPAM has been studied by thermodynamic calculation. As a result, it is suggested that the elimination of U from Ta by HPAM will proceed efficiently, as well as that of Th shown in Table 6.2, because the vapor pressures of U and Th are a few orders higher than that of Ta at the temperature of the Ta melt during HPAM. Therefore, HPAM is expected to be efficient for the ultra-purification of Ta, and probably of other refractory metals, by a simpler process than EBM.

6.3.2 The Removal of Alloying Elements from Refractory Metal Based Alloys

Alloying elements with high vapor pressures vaporize preferentially and are supposed to be removed from the melt of refractory metal based alloys during HPAM, as seen in Fig. 6.13. So, reduction of some alloying elements from Zr-based binary alloys, such as Zr – 1 mass%Fe and Zr – 1 mass%Sn, during Ar and 10%H_2 to 50%H_2+Ar plasma arc melting under atmospheric and reduced pressures has been examined [34], to clarify the influence of hydrogen in the plasma gas and the pressure during melting. Figure 6.14 shows the changes in the Fe concentration of the Zr–Fe binary alloy melt. The removal rate of Fe was slightly raised even by Ar plasma arc melting as the pressure

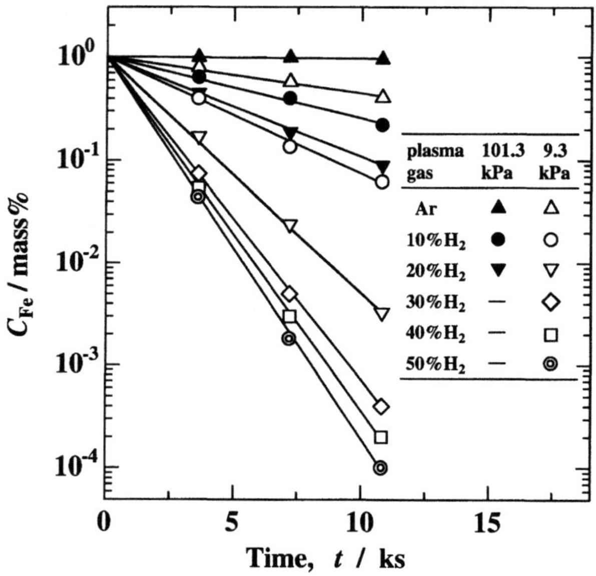

Fig. 6.14. Changes in the Fe concentration of Zr-1 mass%Fe binary alloy during Ar and H_2 (10–50%)+Ar plasma arc melting at 101.3 kPa and 9.3 kPa

was decreased, and it was remarkably enhanced by increasing the hydrogen content in the plasma gas. Under these conditions, the Fe concentration decreased rapidly from 1% to below 10 ppm. Since the Fe concentration on a logarithmic scale decreases linearly with the time, the removal of Fe obeys a first order rate law. From the relationship between the apparent rate constant obtained from the slopes of the straight lines in Fig. 6.14 and the hydrogen content in the plasma gas, the removal rate of Fe was found to increase proportionally to about the three-fourths power of the hydrogen content in the plasma gas at 101 kPa and 9.3 kPa.

Similar good results have been obtained for the removal of Al from a crude V metal containing small amounts of Al and oxygen by HPAM [35]. The crude V metal had been produced commercially by the aluminothermic reduction method. EBM has been used in practice for refining, but there is a problem of significant V vaporization loss during high-vacuum melting. Figure 6.15 shows the changes in the Al concentration in the melt of a crude V metal (Al = 0.64%, O = 1.55% initially) during Ar and H_2(10–50%)+Ar plasma arc melting under atmospheric and reduced pressures. It was also confirmed that increasing the percentage of H_2 in the plasma gas and decreasing the pressure during melting remarkably raised the removal degree of Al. The removal of Al obeys a first order rate law, indicated by the linear decrease of the Al concentration on a log scale with time. From the relation between the slopes of the straight lines in Fig. 6.15 and the hydrogen content in the plasma gas,

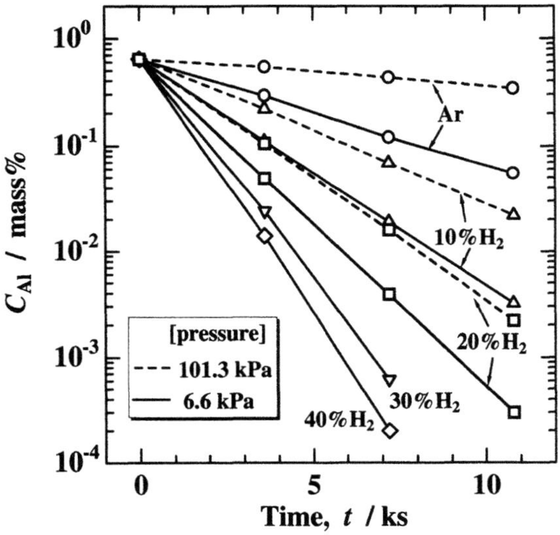

Fig. 6.15. Changes in the Al concentration of crude V during Ar and H_2 (10–40%)+Ar plasma arc melting at 101.3 kPa and 6.6 kPa

the removal rate of Al was also confirmed to increase proportionally to about the two-thirds power of the percentage of H_2 in the plasma gas at 101 kPa and 9.3 kPa. At the same time, the oxygen concentration in the melt during the HPAM treatment was reduced to less than 0.15%. The method of HPAM was, therefore, capable of producing pure V metal with above 99.5% purity from the crude V metal.

Using this feature, HPAM will also be suitable for recovering the pure refractory metal from a scrap of refractory metal based alloy. As an example, to recover pure Zr metal from a zircaloy scrap, removal of the alloying elements such as Sn, Fe, Ni and Cr has been examined by the use of HPAM, yielding good results [34]. Zircaloy scrap is a valuable source of Zr, since the metal has been mainly consumed to produce this alloy. The scrap has been generated in large quantities during the fabrication of zircaloy tubes and rods for nuclear reactors [36]. Figure 6.16 shows the changes in the Sn, Fe, Cr and Ni concentrations of the melt of a zircaloy scrap sample (about 13 g in mass; Sn = 1.42%, Fe = 0.17%, Ni = 0.06%, Cr = 0.10% as the initial composition) during 50%H_2+Ar plasma arc melting at reduced pressure. The melting temperature was about 200–300 K higher than the melting point of Zr. No significant reduction of Sn, Fe and Ni was observed when Ar plasma was used. However, their removal rates were raised by increasing the hydrogen content of the plasma gas and by decreasing the pressure. After this, pure Zr metal (>99.9%; Sn < 100 ppm, Fe < 10 ppm, Ni < 10 ppm, Cr < 10 ppm) could be recovered directly from the zircaloy scrap, and the mass loss of Zr was as low as below 3% in this case.

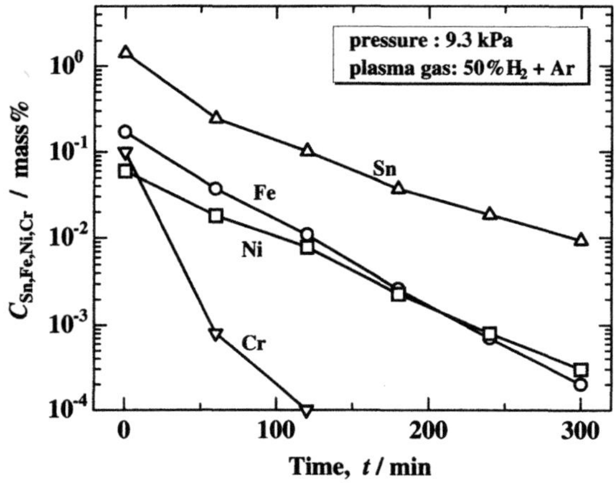

Fig. 6.16. Changes in the Sn, Fe, Ni and Cr concentrations of zircaloy scrap during 50%H_2+Ar plasma arc melting at reduced pressure

As a further example, rapid removal of Fe from the melt of Nb–Fe and Mo–Fe binary alloys containing 10–30% Fe has been achieved by HPAM [30]. By analogy with the other cases, the removal rate of Fe increased proportionally to about the one-third power of the percentage of H_2 in the plasma gas.

In addition to the refractory metal based alloys, the removal of Cu and Sn from molten steel containing small amounts of Cu (0.4%) and Sn (0.06%) has been investigated using HPAM by Matsuo [37]. The maximum removal degrees of Cu and Sn were about 90% and 60%, respectively. It should be noted, therefore, that HPAM will be capable of removing metallic impurities from various metals such as Fe, besides refractory metals, if the impurities are sufficiently more volatile than the melted metal.

6.3.3 The Mechanism for Removing Metallic Impurity Elements from the Melt by HPAM

In spite of not using a high vacuum, HPAM has been confirmed to be very efficient for removing metallic impurity elements which are more volatile than the base metal. A similar effect has not been observed during Ar plasma arc melting [30, 34] and during induction-melting under an H_2 gas flow [37]. In the above-mentioned results on HPAM, the removal rates are found to increase proportionally to the hydrogen content in the plasma gas. From these results, dissociated and activated hydrogen atoms formed within the high-temperature plasma arc must play an important role in this unique refining process, as well as in the removal of nonmetallic impurities such as O and N. For the refining mechanism during HPAM, the following effects that are due to activated hydrogen atoms have been considered:

(a) The elimination of an oxide thin layer and surface-active agents on the melt surface by the hydrogen atoms.
(b) An increase in the temperature of the melt caused by the high thermal conductivity of the hydrogen plasma (including the heat of recombination of H, supersaturated in the melt, to H_2 at the molten metal surface) [37–39].
(c) Enhancement of the transfer of metal vapor within the gaseous boundary layer by a dynamic interaction between the hydrogen atoms and the metal vapor [34].
(d) Enhancement of the mass transfer in the melt by the effect of supersaturated hydrogen atoms in the melt.

According to effect (a), it is supposed that vaporization of the volatile elements can progress more easily when the melt surface is not covered with a thin oxide layer and surface-active elements. However, this effect cannot explain why the removal of the metallic impurity elements is enhanced remarkably, and why its rate increases proportionally with the hydrogen content of the plasma gas.

Furthermore, if the removal rates could be enhanced by increasing the temperature of the melt due to effect (b), a similar enhancement should be observed even in the case of Ar plasma arc melting with an increase in the power of the plasma arc. To confirm this prediction, the Ar plasma arc melting of Nb–10 mass%Fe and Zr–1 mass%Fe binary alloys has been carried out by raising the power [30, 34]. As a result, the removal of Fe was raised slightly by increasing the melting temperature, but the removal rate of Fe during Ar plasma arc melting – even at higher temperatures – was much lower than that during HPAM. So, it is difficult to interpret reasonably the enhancement of the removal rate during HPAM solely by an increase in temperature of the melt. Therefore, activated hydrogen atoms should be considered to have not only thermal but also chemical significance in the refining mechanism during HPAM.

The vaporization rates of metals melted in a stream of inert gas are generally about four orders of magnitude lower than those in high vacuum [40]. This is because the transfer of metal vapor within the gaseous boundary layer (from the melt surface to the gas phase) is hindered under atmospheric pressure. Nevertheless, the removal rates of metallic elements (impurities) with higher vapor pressures than that of the melt can be raised remarkably even under atmospheric pressure during HPAM. This fact suggests a certain driving force, chemically induced by the activated hydrogen atoms, as in effect (c). The most probable mechanism is that the counter-diffusion of hydrogen atoms will trap the vaporized metal within the gaseous boundary layer and transfer it to the gas phase, as expressed by (6.13). In the expression, $(M_x - H_y)$ indicates the metal vapor trapped by the hydrogen atoms, which is supposed to form clusters. In other words, the activated hydrogen atoms are considered to act as a carrier medium of the metal vapor, M(vap):

$$x\mathrm{M(vap)} + y\mathrm{H(from\ plasma\ arc)} \longrightarrow (\mathrm{M}_x - \mathrm{H}_y) \longrightarrow x\mathrm{M(g)} + y/2\mathrm{H}_2\,. \qquad (6.13)$$

This mechanism can be regarded as being similar to that proposed by Turkdogan et al. [41] for the enhancement of vaporization of metals melted in a stream of $Ar+O_2$ mixture gas under atmospheric pressure. The enhanced vaporization has been concluded to be caused by a vaporization–oxidation process in which the counter-diffusion of oxygen and metal vapor interacts to form a metal oxide mist within the gaseous boundary layer. As the typical result, the vaporization rates were found to increase proportionally to the partial pressure of oxygen in the gas stream. This tendency is analogous to the result obtained during HPAM. In the case of HPAM, therefore, it is presumed that the activated hydrogen atoms interact with the metal vapor within the gaseous boundary layer and promote the vaporization process.

In addition, mass transfer in the melt may also be enhanced by supersaturated hydrogen atoms, as expressed by effect (d). In order to make the still obscure mechanism clearer, therefore, further HPAM examination for various binary alloys with different vapor pressures and with different hydrogen solubilities of each component will be needed. Spectral analyses of the plasma gas–molten metal interface during HPAM is also required.

6.4 Summary

The removal of nonmetallic impurities such as O and N and metallic impurity elements from melted metals by H_2+Ar plasma arc melting (HPAM) has been reviewed. The results obtained indicate that dissociated and activated hydrogen atoms formed within the high-temperature plasma arc take a main part in the unique refining progress during HPAM. It has been made clear that HPAM is a very useful melting and refining method for the production of pure metals. However, those good results discussed above have been obtained by the use of a laboratory-scale plasma arc furnace, so the refining effect of HPAM must be proved by further examination on a practical scale.

By the way, in order to develop a new production technique for solar-grade Si (SOG-Si), for example, the removal of B from the melted metallurgical grade Si (MG-Si) has been examined under weak oxidizing plasma conditions by using $Ar+O_2$ and $Ar+H_2O$ as a plasma gas [42]. As a result, B in MG-Si can be removed effectively as a gaseous boron oxide. In this case, dissociated and activated oxygen atoms formed within the plasma arc are considered to play an important role, like the hydrogen atoms during HPAM. By utilizing the activated species formed within a high-temperature plasma arc, another enhanced purification effect will be expected.

In addition, a more advanced purification process may be developed by a combination of HPAM and other purification methods, such as a zone melting technique using the hydrogen plasma arc as a heat source [43].

References

1. J. Feinman (ed.): *Plasma technology in metallurgical processing* (The Iron and Steel Society, Inc., Warrendale, Pa 1987)
2. K. Mimura, M. Nanjo: High Temp. Mater. and Process. **7**(1), 1 (1987)
3. V. Dembovsky: *Plasma metallurgy*, Materials Science Monographs, Vol. 23 (Elsevier Science, Prague 1985)
4. S.M.L. Hamblyn: Mater. Sci. Eng. **9**(3), 151 (1977)
5. K. Upadhya, J.J. More, K.J. Reid: J. Metals **36**, 46 (1984)
6. N.N. Rykalin: Pure Appl. Chem. **48**, 179 (1976)
7. J. Kruger: In *Vacuum metallurgy*, ed. by O. Winkler, R. Bakish (Elsevier, Amsterdam 1971) p. 268
8. L. Brewer, G.M. Rosenblatt: Trans. Metall. Soc. AIME. **224**, 1268 (1962)
9. H.A. McGee, Jr, G. Heller: In *Electric propulsion development*, ed. by E. Stuhlinger, G.C. Marshall (Academic Press, New York 1963) p. 443
10. M. Isshiki, K.Igaki: Trans. JIM. **18**, 412 (1977)
11. K. Ito, K.Sano: J. Japan Inst. Metals **29**, 626 (1965)
12. H. Wada, K.Sakamoto: J. Japan Inst. Metals **41**, 981 (1977)
13. S.R. Seagle, R.L. Martin, O. Bertea: J. Metals **14**, 812 (1962)
14. H. Sakao, K.Sano: Trans. JIM **1**, 38 (1960)
15. E.S. Tankins, N.A. Gocken, G.R. Belton: Trans. Metall. Soc. AIME. **230**, 820 (1964)
16. M.W. Chase Jr. et al.: *JANAF thermochemical tables*, 3rd edn. (American Chemical Society and American Institute of Physic for National Bureau of Standards 1985) p. 1211
17. K. Mimura. K. Saito, M. Isshiki: J. Japan Inst. Metals **63**, 1181 (1999)
18. T. Matsuo, S. Masuda, R. Ikeda: J. Iron Steel Inst. Japan **71**, S135 (1985)
19. O. Barin: Plasma Chem. Plasma Process. **13**, 211 (1993)
20. K. Mimura, W.L. Gu, K. Taniuchi: J. Min. Mater. Proc. Inst. Japan **98**, 1153 (1982)
21. T. Noda, T. Kainuma, M. Okada: J. Japan Inst. Metals **48**, 604 (1984)
22. K. Mimura, M. Nanjo: Mater. Trans. JIM. **31**, 293 (1990)
23. K. Mimura, T. Takahashi, M. Nanjo: J. Min. Mater. Proc. Inst. Japan **106**, 187 (1990)
24. K. Mimura, M. Nanjo: J. Min. Mater. Proc. Inst. Japan **108**, 111 (1992)
25. K. Kaneko, N. Sano, Y. Matsushita: J. Iron Steel Inst. Japan **62**, 43 (1976)
26. H.D. Kunze, E. Schurmann, N.A. Parlee: Metall. Trans. **1**, 281 (1970)
27. O. Kubaschewski, C.B. Alcock: *Metallurgical thermochemistry*, 5th ed., (Pergamon Press, Oxford 1979) p. 380
28. K. Narita, S. Koyama, T. Makino, M. Okamura: J. Iron Steel Inst. Japan **57**, 2207 (1971)
29. R.G. Blossey, R.D. Pehlke: Trans. Metall. Soc. AIME. **236**, 28 (1966)
30. K. Mimura, M. Nanjo: In *Proceedings of the 8th International Symposium on Plasma Chemistry, Tokyo*, Japan, ed. by K. Akashi, A. Kinbara (International Union of Pure and Applied Chemistry, 1987) p. 1886
31. D. Elanski: Refining of metals by Ar–H_2 plasma arc melting, Dissertation, Tohoku University, Sendai, Japan (1995)
32. D. Elanski, K. Mimura, M.Isshiki: Vacuum **47**, 849 (1996)
33. D. Elanski, K. Mimura, T. Ito, M. Isshiki: Mater. Letts **30**, 1 (1997)

34. K. Mimura, S.W. Lee, M.Isshiki: J. Alloys Compounds **221**, 267 (1995)
35. K. Mimura, S.W. Lee, M.Isshiki: In *Proceedings of the 1st International Conference on Processing Materials for Properties*, Hawaii, ed. by H. Henein, T. Oki (The Minerals, Metals and Materials Society (TMS), Warrendale, PA 1993) p. 853
36. P. Pande: In *Proceedins of the Conference on Electron Beam Melting and Refining*, Reno, NV (Bakish Materials Co., Englewood, NJ 1984) p. 113
37. T. Matsuo: J. Iron Steel Inst. Japan **75**, 82 (1989)
38. A.M. Fudolig, H. Nogami, J. Yagi, K. Mimura, M. Isshiki: ISIJ International **37**, 623 (1977)
39. S. Ohno, M. Uda: J. Japan Inst. Metals **48**, 640 (1984)
40. E. Fromm: Metall. Trans. A **9**, 135 (1978)
41. E.T. Turkdogan, P. Grieveson, L.S. Darken: J. Phys. Chem. **67**, 1647 (1963)
42. K. Suzuki, K. Sakaguchi, T. Nakagiri, N. Sano: J. Japan Inst. Metals **54**, 161 (1990)
43. K. Mimura, M. Kishida, M.Isshiki: In *Proceedings of the 2nd International Conference on Processing Materials for Properties*, San Francisco, CA, ed. by C. Yamauchi, B. Mishra (The Minerals, Metals and Materials Society(TMS), Warrendale, PA 2000) p. 1059

7 Ion-Beam Deposition

K. Miyake

7.1 Introduction

Ion-beam deposition (IBD) is a direct method for forming high-purity solid films on a substrate by irradiating a low-energy, mass-separated ion beam (ions of solid materials) in an ultra-high-vacuum environment [1–4]. By this method, we can easily obtain isotopically enriched high-purity metal or semiconductor films. Furthermore, we can control the film properties by selecting the ion irradiation energy. These features are completely different from other conventional purification methods, such as floating-zone purification or arc melting in ultra-high vacuum. Currently, three terms have been used to denote this technique; *ion-beam deposition* [1,2,5], *direct ion-beam deposition* [3] and *primary ion-beam deposition* [4].

Historically speaking, the IBD method is based partly on the electromagnetic isotope separation technique, which was developed for uranium isotope separation during World War II under the Manhattan Project [6–8]. In the electromagnetic isotope separation method, the ions to be mass-separated are extracted from a high-intensity metal ion source at high energy (about 30–50 keV) and selected by using an electromagnet. Mass-separated metal ions such as $^{235}U^+$ are implanted into a target, which is usually made of carbon or copper. After accumulating the mass-separated isotope ions (for example, $^{235}U^+$) in the target, the target undergoes a chemical procedure to produce the mass-separated isotope element as oxide or oxyhydride powders.

In contrast to the electromagnetic isotope separation technique, in the IBD method the mass-separated ions are finally decelerated by an electric field to very low energy to be deposited on a substrate – typically below 100 eV. Therefore, the bombarding ions penetrate only a few monolayers below the substrate surface. This deceleration procedure enables us to deposit any material directly on various substrates at a low substrate temperature (usually room temperature) with high adherent strength. Of course, no post-chemical treatment is necessary.

In this chapter, the general design concepts of the IBD system will be discussed from the viewpoint of low-energy ion-beam transport and film formation mechanisms. Then, the recent application to high-purity Fe film formation will be presented. Finally, a summary of the IBD method and future trends will be described.

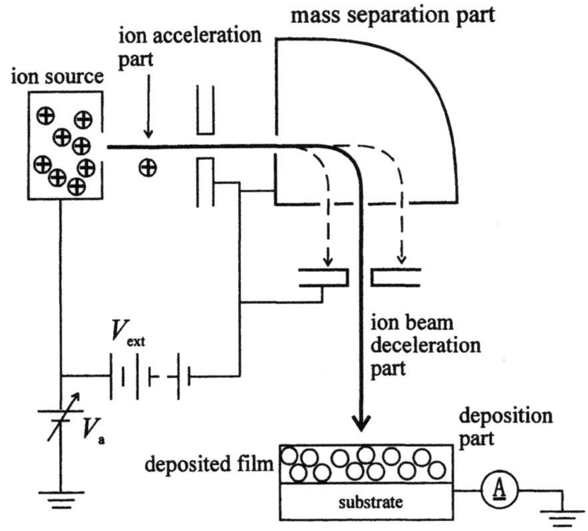

Fig. 7.1. A schematic of an ion-beam deposition system

7.2 The Basic Principle of the Ion-Beam Deposition Method

As shown in Fig. 7.1, the IBD system consists of five major ion-beam components: (i) the ion source, (ii) the ion-extraction part, (iii) the mass-separation part, (iv) the ion-deceleration part, and (v) the deposition part. In the deposition part, a substrate is located in a deposition chamber and connected to a current meter in order to monitor a deposited ion quantity. In addition, vacuum pumping and power supply systems are also needed.

A typical electrical potential layout of these ion-beam components is shown in Fig. 7.2. The ion source, in which ions are produced, has a potential V_a, an ion-acceleration voltage (usually 10–100 V) that determines the final deposition energy. The ion-extraction and mass-separation parts, including the major ion-beam transport lines, are held at a negative high potential to the ion source, V_{ext} (the ion extraction voltage, usually −20 to −40 kV), when positive ions are used. When negative ions are used, the extraction voltage has to be reversed to positive. The deposition chamber and the substrate are held at ground potential. Then the ions extracted from the ion source have a kinetic energy of eV_{ext} and are mass-separated when passing through the mass-separating part. One mass-selected ion beam is decelerated from a high kinetic energy (eV_{ext}) down to a low kinetic energy (eV_a) just before the substrate by a bipotential lens in the deposition chamber. This potential layout is completely different from that of the isotope mass separators [6–11] and ion implantation machines [8], in which an ion source is held at a high

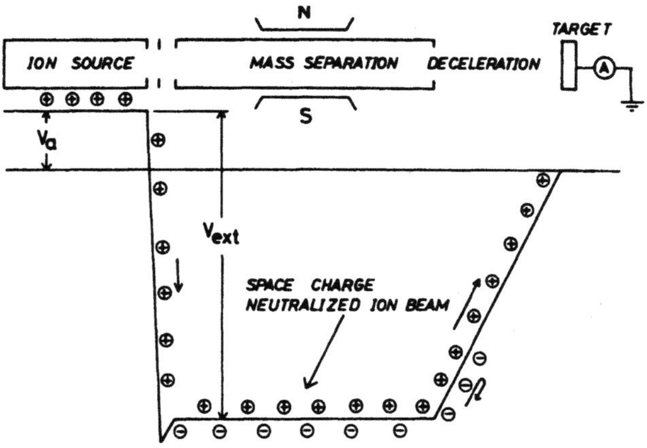

Fig. 7.2. The potential layout of ion-beam deposition

voltage to the grounded target and the mass-separated, high-energy ions are implanted deep into the target.

From the viewpoint of thin film formation, the IBD method has several distinguishing features that cannot be achieved in conventional physical vapor deposition methods, such as sputter deposition or ion plating [12]. The IBD system can give us a variety of ions: atomic (e.g. Ar^+, Fe^+), molecular (e.g. $SiCl^+$, $SiCl_2^+$), gas or solid ions, positive or negative ions, and so on. We can choose ion species across a wide range, from metals to semiconductors. High-purity compound film formation is made possible by combining ion irradiation with other reactive gas sources.

In the IBD method, independent control of the deposition parameters, such as the ion species, the ion flux (J_i = 1–500 μA cm^{-2}), ion-irradiation energy (eV_a = 10–1000 eV), the ion dose, the substrate temperature, the vacuum conditions and so on, is possible. In conventional deposition methods such as sputter deposition or ion plating, these parameters are related in such a complicated way that it is very difficult to change one parameter from the other. However, in the IBD method the ion current density at the substrate can be easily changed by adjusting the ion-source output current while keeping the ion energy constant. These deposition parameters significantly affect the film properties – mechanical, electrical, optical, electrochemical, and so on.

It is well known that sputter-deposited metal films in an argon plasma inevitably contain argon atoms as impurities (up to a few %). However, no such argon atom inclusion problem exists in IBD films, because of the absence of argon gas. Since the ion kinetic energy used in the IBD method is so low (usually 10–100 eV), the substrate temperature rise due to ion bombardment is negligible. Therefore, one can directly deposit various films at a low substrate temperature – even at room temperature. Furthermore, no

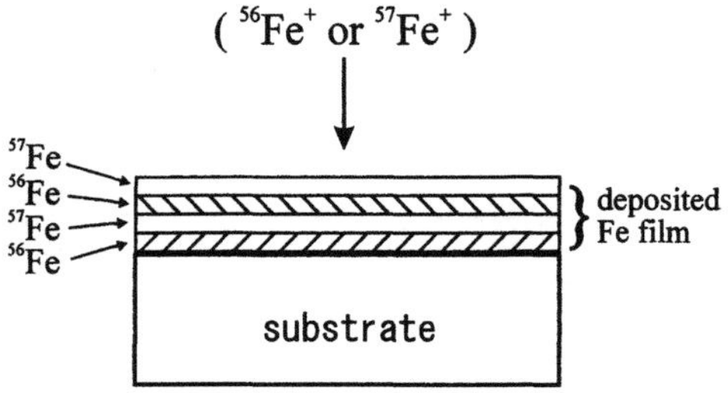

Fig. 7.3. The multilayer formation of different isotope elements

post-treatments, such as wet chemical purification processes, are necessary to obtain high-purity films.

If one irradiates ions of different mass (for example, $^{56}Fe^+$ and $^{57}Fe^+$ ions) sequentially on a substrate, as shown in Fig. 7.3, one can easily obtain a multilayered Fe film of different isotope elements (^{56}Fe, ^{57}Fe).

Due to these favorable features, the IBD method has been successfully used not only to obtain high-purity materials but also as tool for a fundamental research on thin film formation [4].

7.3 The Design Concepts of Ion-Beam Deposition Systems

In order to design an IBD system, one has to know what kinds of ion species are necessary and what ion energy and current ranges are required. Of course, the ion species depend on the composition of the films required. When we think of film formation processes, it should be noted here that not only deposition but also etching reactions are occurring simultaneously during the film growth, as shown in Fig. 7.4 [3, 4]. Ion bombardment induces collision cascades in the surface layer, which result in the sputtering of substrate atoms and the creation of damage in the surface layer. Depending on the ion energy, deposition, sputtering and implantation processes become dominant. Surface-atom migration may be enhanced due to an increased energy input in the surface layer. This enhanced atom migration is believed to enable metastable phase formation (for example, diamond-like carbon film formation) at a low substrate temperature. If chemically reactive elements such as hydrogen or halogen elements are included in the incident ion, they may contribute to surface etching instead of deposition. Therefore, one has to consider these complicated surface processes when one selects the ion species.

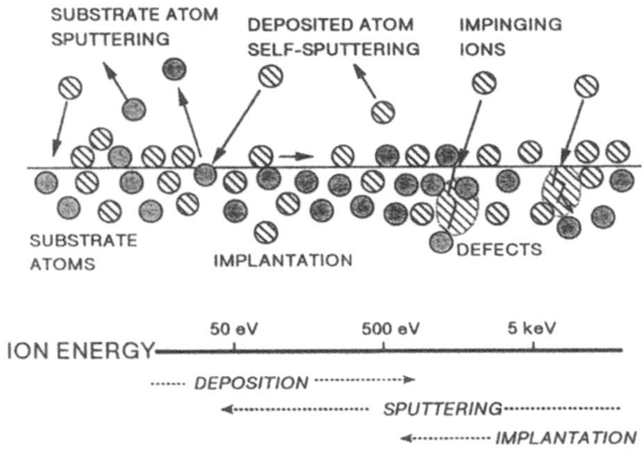

Fig. 7.4. The fundamental surface processes that occur in ion-beam deposition [3]

7.3.1 The Ion Energy Range

The ion energy range used in the IBD method is determined from both theoretical and experimental requirements. An upper energy limit (E_{max}) is determined by a theoretical condition that the deposited film should not be sputtered away by the incoming ions. That is, for film build-up, the self-sputtering yield should not exceed unity [3,4]. According to Sigmund's theory on physical sputtering [13], the sputtering yield, $S(E)$, is given as a function of energy by the following formula for ion energies smaller than 1 keV:

$$S(E) = 3\alpha(M_1 \cdot M_2)E/\pi^2(M_1 + M_2)^2 V_0 \,, \tag{7.1}$$

where M_1 is the mass of the incident ion, M_2 is the mass of the target atom, α is a factor depending on the mass ratio M_1/M_2, E is the energy of the incident ion, and V_0 is the height of the surface potential of the target.

In the case of $^{74}Ge^+$ IBD ($V_0 = 7.63$ eV), 400 eV gives an upper threshold energy for deposition. Actually Yagi et al. [1] found that $^{74}Ge^+$ ions deposited on Ge substrates below 500 eV. Above 500 eV, sputter etching of the Ge substrate became dominant instead of deposition. Therefore, the upper energy limit depends on the ion species (i.e. ion mass). Some of the upper energy limits obtained experimentally are summarized in Table 7.1 [12, 14]. The large energy variation is due to experimental error.

From an application point of view, the upper energy limits may also be determined by the degree of damage generated in the deposited films. This restriction seems to depend on the application itself. So one has to consider this issue case by case.

A lower energy limit (E_{min}) is usually determined by the experimental evidence that the ion intensity decreases drastically in the low-energy region (in particular, below 100 eV). If the ion-beam intensity is not sufficient, the

Table 7.1. The upper energy limits for ion-beam deposition [3]

Ion species	Upper energy limit (keV)
Fe^+	1.5–2.0
Co^+	1.0–1.5
Ni^+	0.8–1.0
Cu^+	0.3–0.4
Zn^+	0.3–0.4
Sn^+	0.45–0.5
Si^+	~ 0.7
Ge^+	~ 0.5
C^+	~ 1.2

purity of the deposited films becomes poorer, due to contamination from the ambient vacuum background. In addition, accurate measurement of the incident ion energy calls for great care, because the energy spread of the ion source becomes comparable with the incident ion kinetic energy. It should be noted that the incident ion beam sometimes irradiates the substrate obliquely, or some of the incident ions are reflected before reaching the substrate. These phenomena will be demonstrated later, in the deceleration technique.

In the author's system for pure Fe film formation [15–17], the acceleration voltage (V_{a}, which determines the deposition energy) was set to be controllable from $E_{\mathrm{min}} = 10\,\mathrm{eV}$ to $E_{\mathrm{max}} = 3000\,\mathrm{eV}$. For Fe film deposition, energies below 100 eV are successfully used in order to avoid heavy damage formation.

7.3.2 Ion Current Density and Vacuum Requirements

In the IBD system, an ion source is usually operated under considerable gas pressure to maintain a stable discharge. Therefore, it is necessary to consider the deposition atmosphere carefully. In order to realize high purity and high-quality film formation, the incident ion flux should be much larger than the residual gas molecule flux at the growing film surface [2–4]. Otherwise, impurities coming from the residual gas background would be found in the deposited films. If we denote the incident ion flux and the residual gas molecule flux at the surface by Γ_i (ions/cm$^2\cdot$s) and Γ_n (molecules/cm$^2\cdot$s), the above requirement can be expressed as follows:

$$S_i \cdot \Gamma_i \gg S_n \cdot \Gamma_n . \tag{7.2}$$

Here, S_i and S_n denote the sticking probabilities of the deposited ions and the residual gas atoms, respectively. This relationship represents the worst-case criterion, where the depositing ions are assumed not to sputter the adsorbed residual gas atoms and all residual gas atoms are thought to act as impurities.

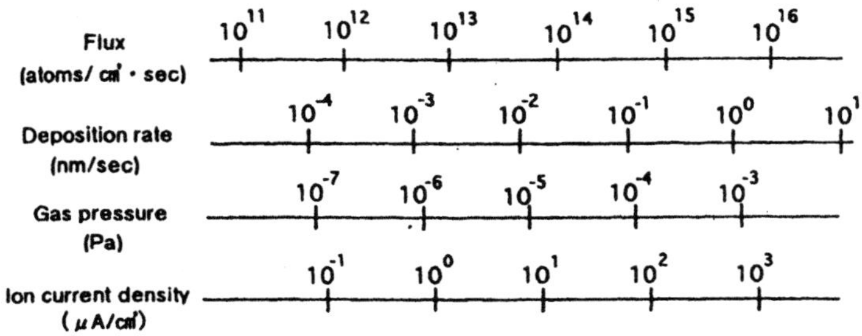

Fig. 7.5. Relationships of flux (ion and residual gas), deposition rate, gas pressure, and ion beam flux [4]

We can calculate the incident ion flux and the residual gas molecule flux from the following relationships:

$$\Gamma_i \text{ (ions/cm}^2 \cdot \text{s)} = 6.3 \times 10^{12} \ J_i \tag{7.3}$$

and

$$\Gamma_n \text{ (molecules/cm}^2 \cdot \text{s)} = 5.3 \times 10^{20} p \,, \tag{7.4}$$

using an ion current density, J_i (μA cm^{-2}) and a background vacuum pressure, p (Pa). An useful diagram to calculate these parameters is given in Fig. 7.5.

If we assume both sticking probabilities, S_i and S_n, to be unity, (7.2) can be expressed as follows:

$$J_i \ (\mu\text{A cm}^{-2}) \gg 8.4 \times 10^7 p \text{ (Pa)} \,. \tag{7.5}$$

In the author's IBD system for pure Fe film formation [15], the operating vacuum pressure, p, was chosen to be 10^{-6} Pa and the ion current density, J_i, was about 100 μA cm^{-2}. The base vacuum pressure was 10^{-8} Pa. Then (7.4) was fulfilled and the ion flux density, Γ_i, became two orders of magnitude larger than the residual gas molecule flux density, Γ_n.

7.3.3 Deposition Time and Dose Calculation

The deposition time, T (s), and the dose, D (ions/cm^2), required for the build-up of a film of thickness d (nm), can be calculated from the following formula [12]:

$$T\text{(s)} = 3.2 \cdot d \text{ (nm)}/J_i(\mu\text{A cm}^{-2}) \tag{7.6}$$

and

$$D \text{ (ions/cm}^2) = 2.0 \times 10^{14} \ d\text{(nm)} \,. \tag{7.7}$$

Here, one monolayer was assumed to have a density of 10^{15} atoms/cm^2 with a thickness of 0.5 nm.

7.3.4 The Vacuum Pumping System

In the IBD system a steady gas flow (usually from 0.1 to a few sccm) is introduced into the ion source to generate intense ion beams (1–10 mA). Therefore, closed-type pumping components such as ion pumps, Ti-sublimation pumps and cryopumps are not suitable. Instead, open-type pumping components, such as turbo molecular pumps or oil diffusion pumps compatible with ultra-high vacuum (UHV), are more favorable. Of course, a combination of these pumps is acceptable.

In order to realize a UHV condition during deposition, efficient differential pumping in the ion-beam line is a must in IBD systems. The vacuum pressure in the beam drift tube should be at least less than 10^{-3} Pa. Otherwise, the ion-beam transport efficiency becomes significantly small due to charge exchanges with residual gas molecules. For this purpose, several orifices are commonly installed in beam drift tubes. Furthermore, in order to prevent charge-exchanged high-energy neutral particles from hitting a substrate, the mass-separated ion beam is often deflected before the substrate [18–20].

In most IBD systems, the ion beam drift tubes are held at a high negative potential (typically −10 to −40 kV). Ceramic or epoxy insulators are used to isolate the beam drift tube electrically from the pumping stages, which are held at ground level. In this case one should be careful to prevent breakdowns due to electrons emitted from vacuum gauges. Great care is also needed to avoid high-voltage surge attack from high-voltage power supplies.

7.3.5 The Production of a High-Intensity, Low-Energy Metal Ion Beam

As described in the previous section, the production of low-energy and high-intensity metal ion beams is the most important issue in IBD. In choosing an ion source, low gas consumption is preferable in order to realize UHV conditions during deposition. A small energy spread is also a crucial issue to get a higher ion intensity in a low-energy region.

For compact IBD systems, an arc discharge-type ion source has been used by several groups. Figure 7.6 shows a linear-type IBD system developed by Amano et al. [21]. They used a small-arc discharge-type ion source developed by Menzinger and Wahlin [22]. However, the ion current output is very small (a few μA). The energy spread is reported to be as low as 0.1 eV when the arc power is 50 W. However, it might become larger if one uses the ion source for a larger current application due to plasma instabilities in the ion source arc plasma. Since an $E \times B$ type Wien filter is used for mass separation in this system, the maximum mass resolution is at most 10–20.

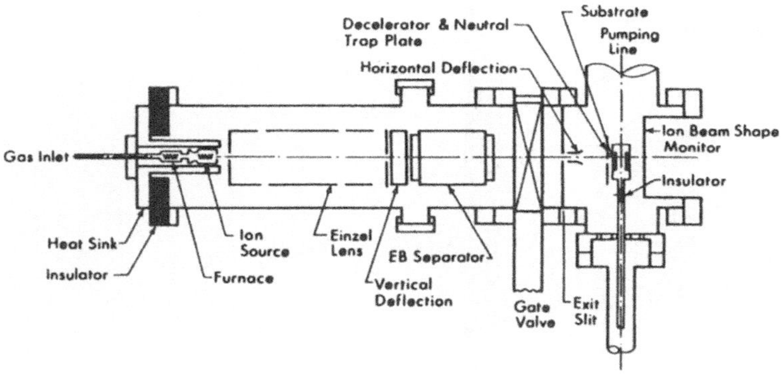

Fig. 7.6. Linear type ion beam deposition system using $E \times B$ Wien filter [21]

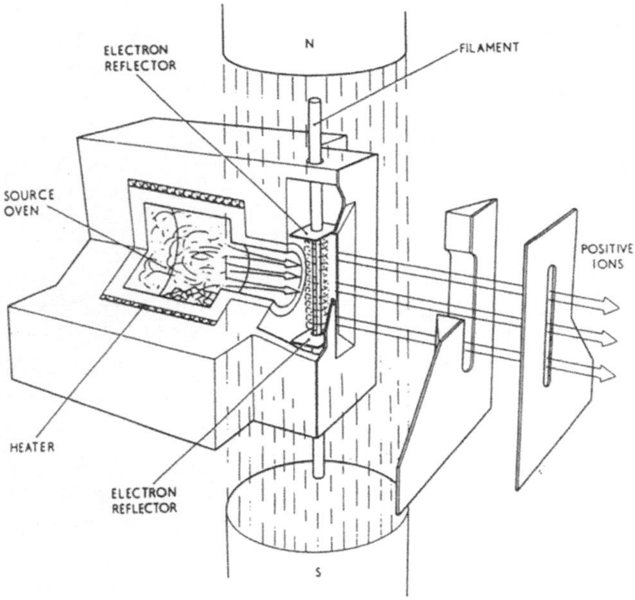

Fig. 7.7. A schematic of the Freeman-type ion source [7–9]

For larger ion-beam current systems, the Freeman-type ion source [7–10, 18, 19] or a microwave ion source [15–17] have been widely used.

Figure 7.7 shows a schematic illustration of the Freeman-type ion source. A rod filament (usually tungsten, 2 mm in dia.) is located in a cylindrical arc

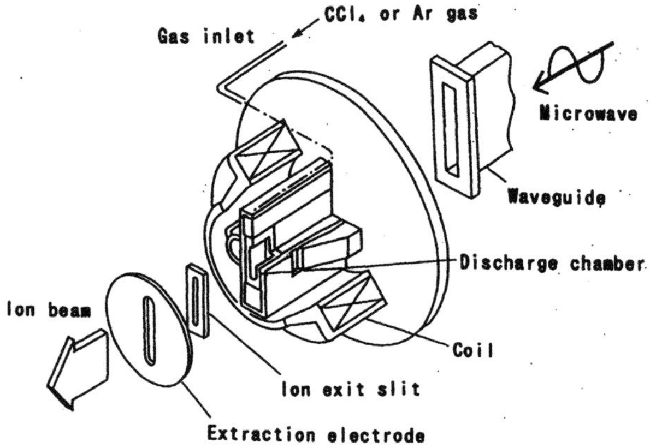

Fig. 7.8. A schematic of the microwave ion source [15]

chamber (carbon or tantalum, about 25 mm in dia.) and a weak magnetic field is applied parallel to the direction of the filament. In order to produce metal ions, metal chloride gas (for example, $SiCl_4$ for Si^+ ion production) is introduced into the arc chamber. Alternatively CCl_4 gas is introduced into the arc chamber through a heated source oven, in which metal oxide powders are placed. In this configuration, a stable dc arc discharge is obtained, because the location of the arc discharge is essentially fixed by the filament in the most effective position just behind the extractor slit. Details of the Freeman-type ion source are presented in [7,8].

Figure 7.8 shows a schematic illustration of a microwave ion source. A microwave (2.45-GHz) discharge is struck in a boron nitride box in which ion source material gas is introduced. An electron cyclotron resonance magnetic field (875 gauss or more) is applied in the discharge region to increase the plasma density.

Both ion sources extract rectangular-shaped mA metal ion beams through slit extraction electrodes. A typical gas load to these ion sources is about 1 sccm or less. The energy spreads of these ion sources were reported as being small as a few eV. The discharge characteristics of Freeman-type ion sources are strongly dependent on a magnetic field generated by a filament current [7,8,12]. A useful list of ion source materials was reported by Chivers and Freeman [4,23].

7.3.6 Ion Beam Transport and Deceleration

In order to extract a high-intensity ion beam, an accel–decel-type potential layout is commonly employed, as shown in Figs. 7.2 and 7.9. An ion beam is extracted initially at a high extraction voltage, V_{ext}, and then mass-separated.

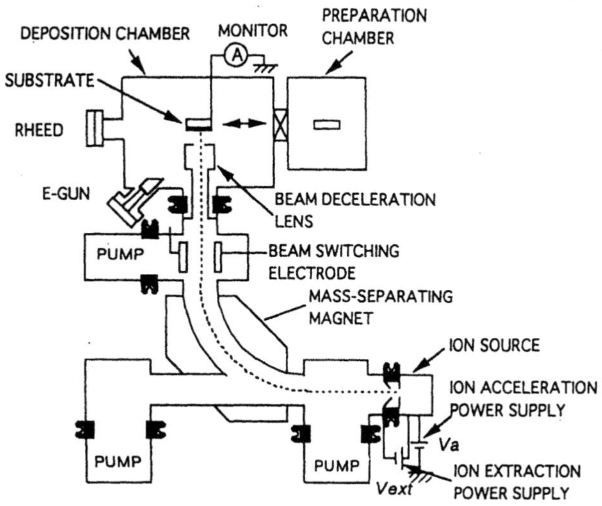

Fig. 7.9. A schematic of an ion-beam deposition system for high-purity Fe film formation

The mass-separated ion beam, which has a kinetic energy of eV_{ext}, is finally decelerated to the deposition energy (eV_a) just in front of the substrate. The potential difference, V_a, between the ion source and the substrate is called the acceleration voltage. This potential layout is advantageous in safely performing in situ observation of the deposited film. However, some workers apply a high positive voltage (about +20 to +40 kV) to both the ion source and the substrate, in order to hold the mass-separation part at ground potential [10, 21]. In this case, the ion-acceleration voltage is supplied by a floating dc power supply, such as batteries.

In most cases a shutter is located in front of the substrate and is used to check the ion mass spectrum of the extracted ion beam before deposition. We can monitor the ion dose during deposition by measuring the ion current flowing to the substrate.

In designing an ion-beam transport system for IBD, it should be noted that space-charge neutralization of the ion beam by electrons is necessary for the transport of a high-intensity ion beam. According to a theoretical calculation, a non neutralized cylindrical Ar^+ ion beam (beam radius 1 cm, parallel kinetic energy 40 keV, ion current 10 mA) diverges to a radius of 30 cm after passing through a distance of 100 cm along the beam axis [12]. This is the worst-case calculation.

However, in actual cases, the beam divergence is not as large as the one calculated. The space charge of the extracted ion beam is automatically neutralized after a few microseconds with secondary electrons, which are generated by collisions between the extracted ion beam and the vacuum walls. A radial potential difference (about a few tens of eV) induced by the ion beam itself also acts on the secondary electrons to accelerate and multiply neutralizing electrons through collisions with residual gas molecules.

Instead, the use of an electrostatic focusing lens or deflection electrodes degrades the space-charge neutralization. This is because an electrostatic field induced by those electrostatic optics reduces the number of space-charge neutralizing electrons, and this induces a divergence of the ion beams. In particular, in high-current IBD machines (typically mA-class), electromagnetic focusing and deflection components are preferable.

For mass separation, two types of method have been commonly used. One is an $E \times B$ type mass filter (Fig. 7.6) and the other is a sector-type electromagnet (Fig. 7.9), which was used by the author for high-purity Fe film formation. The former type is used for low-ion-current IBD systems. This method has an advantage of making the machine compact and linear. However, the mass resolution is not as high as with the sector type. The latter type is used for larger-ion-current IBD machines. A high mass resolution (over 100) is easily obtained by using an electromagnet with a deflection angles of 60° or 90°. Therefore, sector-type mass separation is suitable for isotopically enriched film formation. Mass-separation using electrostatic quadrupoles is not employed because of the space-charge neutralization problem.

The ion-beam deceleration technique is one of key issues in IBD. Since the mass-separated metal ion beam is space-charge neutralized by electrons, it becomes necessary to suppress an acceleration of space-charge neutralizing electrons when the ion beam is decelerated. A deceleration electric field for

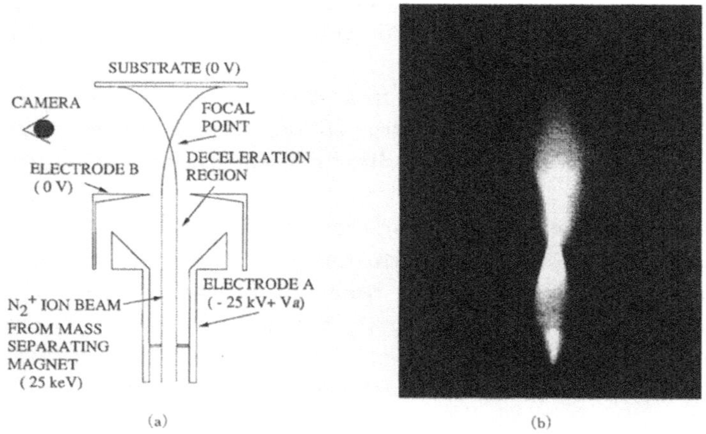

Fig. 7.10. (**a**) A schematic of the deceleration lens in the IBD system (Fig. 7.9). (**b**) Optical emission from N_2^+ beams decelerated from 25 keV to 500 eV

the ion beam acts at the same time as an acceleration electric field applied to the space-charge neutralizing electrons. One efficient method to suppress the electron acceleration is to apply a weak magnetic field (about 0.03 Tesla) perpendicular to the ion-beam axis [3,12,24]. In this case, the electrons follow magnetron motion and are excluded from the ion-beam region. The ion-beam trajectories, however, are not influenced so much, because of their larger mass-to-charge ratio.

We can directly observe the shapes of the ion-beam trajectories by exploiting optical emission in the visible wavelength region. A typical example of an N_2^+ ion-beam trajectory decelerated from 25 keV to 500 eV in an IBD system is shown in Fig. 7.10 [4,24]. The strong focusing effect of the deceleration electric field is shown. Ion-beam divergence due to positive space-charge repulsion is not as large, even in a background pressure of 10^{-6} Pa.

In IBD it is of great concern to assess how much ion current can be obtained when the ion energy is reduced below 100 eV. Direct observation of the decelerated ion beam indicates that the decrease in the ion current at the substrate in the low-energy region is primarily due to the ion-beam crossover effect [4,24]. Ion-beam divergence due to a positive ion space charge is not as serious. Instead, an energy spread of the mass-separated ion beam causes the ion current decrease in the low-energy region. It is due to the fact that the ions not having enough energy to reach the substrate are repelled before reaching the substrate, these ions are re-accelerated backward to the beam drift tube: they and hit the deceleration electrodes somewhere and produce metal impurities due to sputtering [4]. Furthermore, a lot of secondary electrons are produced, and these electrons are accelerated to the substrate. Sometimes sample deterioration takes place (such as substrate heating, or substrate melting). Therefore, it is very important to choose an ion source that has a small energy spread.

7.3.7 Sample Manipulation and Film Deposition

In IBD it is necessary to hold a substrate electrically isolated from the deposition chamber, in order to monitor the depositing ion current. This requirement is entirely different from the case in MBE. In order to obtain uniformly deposited films, uniform substrate heating and some kind of beam scanning such as substrate rotation are needed. In situ observation tools, such as RHEED, XPS/UPS, AES, STM and SIMS, are frequently installed in the deposition chamber to monitor how films are deposited.

7.4 High-Purity and High-Corrosion-Resistance Fe Film Formation

IBD has been attracting great interest as a method for obtaining high-purity metal films. A typical example is the formation of high-corrosion-resistance

Fe films. It is generally believed that iron is so easily corroded that alloy designing or surface treatment is essential in order to suppress corrosion problems, such as chromium oxide formation in the case of stainless steel case. However, recent metallurgical studies have revealed that high-purity iron metal is not corrodible at all. Instead, the corrosion property should be attributed to small amounts of impurity elements, such as sulfur, manganese and silicon, contained in normal-grade (usually below 99.5%) iron metals. Nevertheless, corrosion is a crucial problem in all industries. In particular, in micro-electronics products such as magnetic devices, a corrosion-free iron metal based magnetic film is really needed. Therefore, the formation of high-purity iron metal is thought to be a very interesting and important subject.

The IBD system used in this study is, in principle, composed of five major parts, as shown in Fig. 7.9; a microwave ion source, ion-extraction electrodes, a sector-type mass-separating magnet (deflection angle 90°, deflection radius 40 cm, pole gap 40 mm), an electrostatic ion deceleration lens and an ultra-high-vacuum deposition chamber. A 2 inch Si(100) wafer was used as a substrate and was located in the middle of the deposition chamber with the surface downward. Six turbomolecular pumps were used to evacuate the whole system. The base pressure of the deposition chamber was 10^{-8} Pa and the operating pressure was 10^{-6} Pa. The voltage V_a, which determines the deposition energy, was chosen to be controllable from 10 eV to 3000 eV. For Fe film formation, deposition energies of 50 eV and 100 eV were used in order to avoid heavy damage formation.

An ion current density up to $100\,\mu A\,cm^{-2}$ and a vacuum pressure of 10^{-6} Pa were obtained. The base pressure was in the 10^{-8} Pa range. If we assume that the sticking probability of incoming Fe^+ ions is unity, an ion current density of $100\,\mu A\,cm^{-2}$ will give a deposition rate of $0.2\,nm\,s^{-1}$.

The Fe^+ ions were produced in a microwave ion source (see Fig. 7.8) using a conventional CCl_4 method. Hematite Fe_2O_3 powder (about 500 mg) was placed in a discharge box made of boron nitride and then CCl_4 gas was introduced into the box. An intense CCl_4 plasma was generated in the box by applying 2.54 GHz microwave power of 300 W and an axial magnetic field of 0.1 Tesla. The CCl_4 plasma reacted with hematite Fe_2O_3 to produce $FeCl_x$ ($x = 1, 2, 3$) molecules, which were eventually dissociated into $FeCl_2^+$, $FeCl^+$ and Fe^+ ions by electron impact.

All ions produced in the ion source were extracted at an energy of $eV_{ext} =$ 20 keV and mass-separated by the magnet. The magnetic flux needed to deflect $^{56}Fe^+$ ion was 0.3 Tesla.

The mass-separated iron isotope ions, in this case $^{56}Fe^+$, with a kinetic energy of 20 keV, were decelerated just in front of the substrate from 20 keV to 50 eV or 100 eV, and the ion current flowing to the substrate was monitored with a current meter. A mechanical shutter was located between the substrate and the ion deceleration lens to check the ion mass spectrum of the extracted ion beam before deposition. A typical mass spectrum of the extracted ion

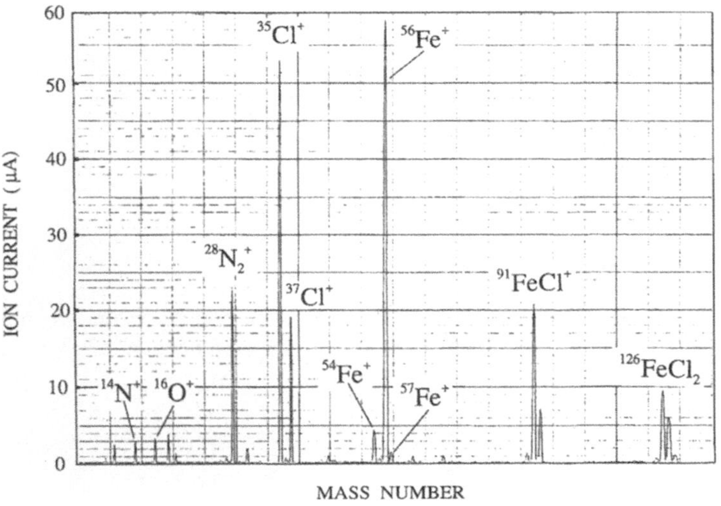

Fig. 7.11. The ion mass spectrum of the ion-beam-generated plasma chemical reaction in a microwave ion source. The ion energy is 50 eV [16]

beams from the $CCl_4 + Fe_2O_3$ discharge is shown in Fig. 7.11. The ion energy was 50 eV. A large $^{56}Fe^+$ ion peak of 62 μA was obtained. The $^{54}Fe^+$ and $^{56}Fe^+$ intensities appeared with their natural abundances: $^{54}Fe^+ : {}^{56}Fe^+ : {}^{57}Fe^+ = 5\% : 92\% : 2\%$. The $FeCl_2^+$ and $FeCl^+$ ion peaks, as well as a Cl^+ ion peak, were obtained from chemical reactions of Fe_2O_3 in the CCl_4 reactive gas plasma. The B^+ and $^{28}N_2^+$ ion peaks are considered to be due to thermal dissociation of the boron nitride discharge box.

The mass-separated $^{56}Fe^+$ ion beam of 50 μA ion current was irradiated on Si(100) substrates (n-type, 0.001Ω· cm) at room temperature. The ion energy was chosen to be 50 eV or 100 eV. The deposition area was about 10 mm in diameter and the deposition time was varied from 30 min to 60 min. The deposition rate was about 0.1–0.3 nm s^{-1}.

The IBD-Fe films deposited using the mass-separated $^{56}Fe^+$ ion beam for 30 min at ion energies of 50 eV and 100 eV were 300 nm in thickness. Both Fe films showed a shiny and metallic color, and no appreciable change in color was observed for the air-exposed films even after 6 months. SEM (Hitachi S-900) plane observations indicated that both Fe surfaces were so smooth that no appreciable surface roughness could be observed within the resolution. Cross-sectional SEM micrographs of the air-cleaved IBD-Fe films showed extremely fine and homogeneous morphology. No columnar structure was observed [16]. This property is completely different from those often observed with sputter deposition. In the latter case, columnar film growth is more common, and might be partly due to argon gas incorporation in the

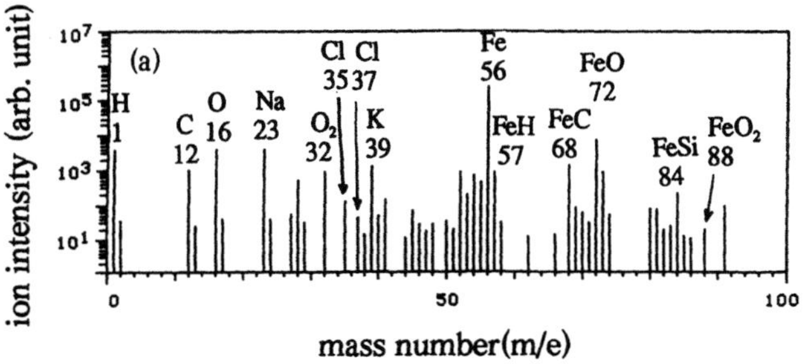

Fig. 7.12. The SIMS mass spectrum of the $^{56}Fe^{+}$ ion-beam-deposited film on an Si substrate. A 10 keV oxygen ion beam was used as a primary ion beam and sputtered secondary positive ions were detected

films. In general, sputter-deposited films contain a small percentage of argon (sputtering gas) atoms as impurities.

Figure 7.12 shows the SIMS mass spectrum for the IBD-^{56}Fe films. In this case, the oxygen ion beam was used as the primary ion beam. The figures indicate that only an ^{56}Fe ion peak was observed in the IBD-Fe films. ^{54}Fe and ^{57}Fe ion peaks were not observed within the detection limit. Therefore, it was concluded that the IBD-^{56}Fe films were isotopically enriched high-purity Fe films with a single isotope of ^{56}Fe [25].

The electrochemical behavior of the IBD-^{56}Fe films deposited at 50 eV and 100 eV was measured by a polarization method in 0.001 mol NaCl solution at room temperature. For comparison, several high-purity iron samples were measured in the same way: 99.5% purity Fe sheet, 99.995% purity Fe sheet and FZ(floating zone)-purified Fe sheet, whose purity is more than 99.995% with an RRR_H (residual resistivity ratio) of 6000 and 8000. The IBD-Fe film deposited at 100 eV gave a −100 mV corrosion potential, which was the same as that of the FZ-Fe sheet of $RRR_H = 6000$. It is interesting to note that the IBD-Fe film deposited at 50 eV showed a much lower corrosion current density (below $10^{-5}\,A\,m^{-2}$). Corrosion current densities for these iron specimens, derived from the intersection of extrapolated lines for the cathode and anode curves in the Tafel regions, are summarized in Fig. 7.13: $10^{-2}\,A\,m^{-2}$ for 99.5% Fe; $10^{-4}\,A\,m^{-2}$ for 99.995% Fe, FZ-Fe ($RRR_H = 6000$) and IBD-Fe (100 eV); and below $10^{-5}\,A\,m^{-2}$ for FZ-Fe ($RRR_H = 8000$) and IBD-Fe (50 eV). The corrosion current density for SS 304 is also shown as a reference.

It is clearly shown that as the purity became higher, the corrosion current density became lower. In other words, the corrosion resistance was much improved for higher-purity Fe specimens. Here we can find an interesting fact that the IBD-Fe film (50 eV) had a two orders off magnitude higher corrosion resistance than the IBD-Fe film (100 eV). A comparison of the 100 eV IBD-Fe film with the FZ-Fe ($RRR_H = 8000$) specimen indicates that the purification level for the latter seems more improved than that for the IBD-Fe films.

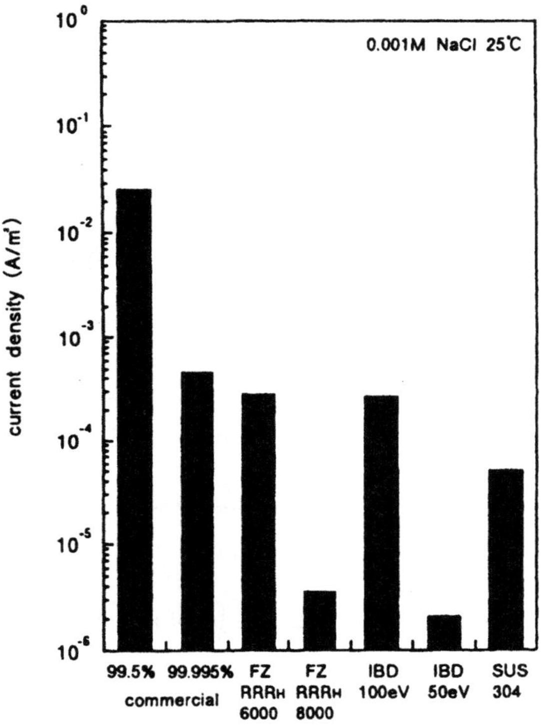

Fig. 7.13. A comparison of the corrosion current densities of various iron metals [16]

This difference was explained by different microcrystalline structures of the IBD-Fe films. Cross-sectional TEM observations for the Fe films deposited at 100 eV (a) and 50 eV(b) showed a clear difference in crystalline structure, as shown in Fig. 7.14. The IBD-Fe film [(b), 50 eV] was composed of three layers. An Si native oxide layer existed between the Fe film and the Si(100) substrate. The top Fe layer (III) was considered to be an oxidized Fe layer, presumably Fe_2O_3, having a continuous single crystalline-like image parallel to the surface. The intermediate Fe layer (II) clearly showed a bcc polycrystalline Fe structure from the TED (transmission electron diffraction) pattern, which is in good agreement with the XRD (X-ray diffraction) results. No columnar grain structure was observed in this region. Instead, rectangular shaped sub-grains were observed. The lateral lattice image in the rectangular-shaped sub-grain regions is continuously connected to the one in the outer region at their boundaries. Therefore, it is expected that surface atoms were very mobile laterally on the surface during film growth. The third Fe layer (I),

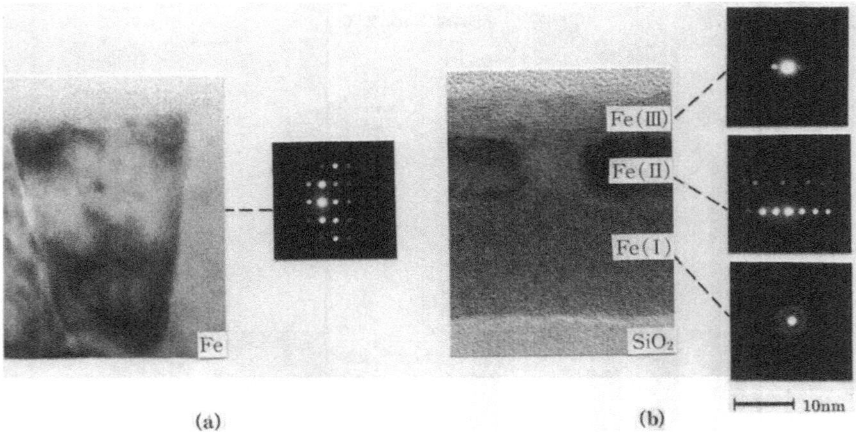

Fig. 7.14. Cross-sectional TEM micrographs of the ion-beam-deposited Fe films at 100 eV (**a**) and 50 eV (**b**) [25]

the interface layer, was found from the TED results to be of an amorphous structure or of very fine polycrystalline structure.

The IBD-Fe film (100 eV), however, showed a columnar structure on the Si native oxide layer, as shown in Fig. 7.14a. The average grain size was about 40 nm. These columnar grains were so closely connected to each other that it was difficult to define the grain boundary dimension precisely. The 50 eV IBD-Fe film (b) having the highest corrosion resistance had a stable surface oxide layer, while the 100 eV IBD-Fe film (b) did not show such a clear oxide layer. Only an irregularly shaped thin oxide layer was observed in the latter film.

From the viewpoint of ion–solid interaction, 100 eV seems large enough to cause radiation damage as well as sputtering due to collision cascades On the other hand, 50 eV might be in a very critical energy region that causes such effects. The absence of a columnar structure and the presence of layer-by-layer type film growth in the case of 50 eV ion bombardment suggest that surface atom migration was largely enhanced even if the substrate temperature was not very high; that is, room temperature. With the help of such enhanced surface atom migration, a closely packed Fe(110) surface would have become possible. And stable iron oxide layer formed from the Fe(110) surface, which shows the highest work function. These mechanism is thought to be a rreason of the high corrosion resistance of the IBD-Fe films.

The fact that 50 eV ion-beam deposition created a closely packed Fe(110) surface indicates the possibility of controling the surface electronic structure as well as the impurity level by the IBD method. From this example, it is found that the IBD method is useful not only for synthesizing high-purity materials but also functional new materials.

7.5 Summary

The methods and recent applications of the ion-beam deposition (IBD) method were described in detail. The general design concepts of IBD systems were discussed from the viewpoint of ion–solid interactions. An IBD system employed for high-purity and high-corrosion-resistance Fe film formation was presented. The most distinguished advantage of the IBD method over conventional purification methods is the capability of forming isotopically enriched, high-purity films directly on various substrates at a low substrate temperature. Furthermore, the formation of an extremely smooth surface and control of the crystalline structures by the ion energy are also quite attractive for an investigation of various new functional materials and devices.

Acknowledgments

The author wishes to express his gratitude to his collaborator, Kenya Ohashi of Hitachi Research Laboratory. He thanks Professors Minoru Isshiki and Kenzo Igaki of Tohoku University for their kind provision of high-purity FZ Fe specimens.

References

1. K. Yagi, S. Tamura, T. Tokuyama: Jpn. J. Appl. Phys. **16**, 245–251 (1977)
2. T. Tokuyama, K. Yagi, K. Miyake, M. Tamura, N. Natsuaki, S. Tachi: Nucl. Instrum. Methods **182/183**, Part I, 241–250 (1981)
3. K. Miyake, T. Tokuyama: In *Direct ion beam deposition in ion beam assised film growth*, ed. by T. Itoh (Elsevier, Amsterdam, 1989) Chap. 8, pp. 289–318
4. K. Miyake: In *Handbook of thin film process technology*, ed. by D.A. Glocker, S.I. Shah (Institute of Physics Publishing, Bristol 1997) A3.4.1
5. J.M.E. Harper: Ion beam deposition. In *Thin film processes*, ed. by J.L. Vossen and W. Kern (Academic Press, New York 1978) pp. 175–206
6. A. Guthrie, R.K. Wakerling (eds.) (1949): (a) Electromagnetic separation of isotopes in commercial quantities. N.N.E.S. 1-4 T.I.D.5217: (b) The characteristics of electrical discharges in magnetic fields. N.N.E.S. 1-5: (c) Sources and collectors for use in calutrons N.N.E.S. 1-6, T.I.D.5218: (d) Problems of physics in the ion source. T.I.D.5219
7. J.H. Freeman: Radiation Effects **100**, 161 (1986)
8. G.D. Dearnaley, J.H. Freeman, R.S. Nelson, J. Stephens: *Ion implantation* (North-Holland, Amsterdam 1973) pp. 456–474
9. J.H. Freeman, W. Temple, D. Beanland, G.A. Gard: AERE Report 8287 (1976)
10. J.H. Freeman, W. Temple, D. Beanland, G.A. Gard: Nucl. Instrum. Methods **135** 1–11 (1976)

11. J.H. Freeman, D.J. Chivers, G.A. Gard, W. Temple: Nucl. Instrum. Methods **145**, 473–505 (1977)
12. K. Miyake: Ph.D. thesis, Kyoto University, (1981)
13. P. Sigmund: Phys. Rev. **184**, 383 (1969)
14. A. Fontell, E. Arminen: Can. J. Phys. **47**, 2405–2414 (1969)
15. T. Matsuo, K. Miyake: J. Vacuum. Sci. Technol. **A13**, 2138–2141 (1995)
16. K. Miyake, K. Ohashi, H. Takahashi, T. Minemura: Surf. Coating Technol. **65**, 208–213 (1994)
17. H. Takahashi, K. Ohashi, K. Miyake, T. Minemura: Advanced Materials '93, II/Information Storage Materials, edited by M. Abe et al., Trans. Mater Res. Soc. Japan **15B** (Elsevier, Amsterdam 1994) pp. 1165–1168
18. K. Yagi, K. Miyake, T. Tokuyama: Proc. Int. Conf. Low Energy Ion Beams, Salford, 1977, Conference Series No. 38 (Institute of Physics, Bristol 1978) Chap. 3, pp. 136–141
19. K. Miyake, T. Tokuyama: Thin Solid Films **92**, 123–129 (1982)
20. O. Tsukakoshi, S. Shimizu, S. Ogata, N. Sasaki, H. Yamakawa: Nucl. Instrum. Methods Phys. Res. **B55**, 355–358 (1991)
21. J. Amano, P. Brice, R.P.W. Lawson: J. Vacuum Sci. Technol. **13**, 591–595 (1976)
22. M. Menzinger, L. Wahlin: Rev. Sci. Instrum. **40**, 102 (1969)
23. D.J. Chivers: Rev. Sci. Instrum. **63**, 2501–2506 (1992)
24. K. Miyake, K. Yagi, T. Tokuyama: Nucl. Instrum. Methods **198**, 535–538 (1982)
25. K. Miyake, K. Ohashi, M. Komuro: Mater. Res. Soc. Symp. Proc. **279**, 787–793 (1993)

8 Purification and Isotope Separation by Laser Techniques

S. Sato

8.1 Introduction

Lasers have been utilized in various fields of study and have become one of the key components in our daily life since their discovery in 1960 by Maiman. In fact, we can find at least one or two laser diodes in our home or office, which are used in compact discs for music and computers to pick up information by the focused beam. The use of high-power continuous or pulsed lasers enables precise marking of letters and/or figures on a plate. Keratotherapy by means of lasers is applied for the improvement of eyesight.

The laser might be also an attractive tool for purification processes, by applying its excellent characteristics, which are unprecedented amongst conventional light sources. In particular, the spectral width of laser light is so narrow as to distinguish one atom from a different kind of atom or isotope. If a specific atom is excited by means of selective laser excitation, the atom can be spatially separated and collected by a proper method, thus providing a novel purification and isotope separation processes.

In this chapter, we will discuss first the salient features of laser light and then review newly developing processes that use laser light for purification and isotope separation, although its development has only just begun. We believe that these novel techniques might be able to achieve a superior degree of purification and isotope separation, because the laser process avoids contamination. Moreover, a higher separation coefficient in isotope separation may be possible in only a single-step procedure, due to the outstanding selectivity of the specific species.

8.1.1 Laser Light

When a material is excited in such a way as to provide more atoms or molecules at a higher energy level than at a lower level, the material will be able to amplify radiation at the frequency corresponding to the energy level difference by a stimulated emission process. The acronym "laser" derives its name from this process: "**L**ight **A**mplification by **S**timulated **E**mission of **R**adiation".

The light wave obtained by the above process has a high degree of coherence and is able to clearly interfere, providing a high-contrast interference

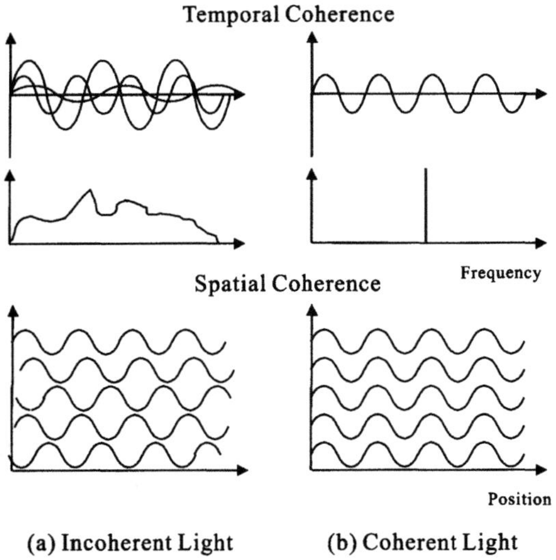

Fig. 8.1. A comparison of the characteristics of incoherent and coherent light. In the upper part, the temporal behavior of the electric amplitude of the light and its spectra are shown. Phase relations concerning spatial coherence are shown in the lower part

pattern. This feature of laser light is impossible to achieve with conventional light sources. As seen in Fig. 8.1, laser beams are different from conventional light sources in being both spatially coherent and temporally or spectrally coherent. The output beams from lasers can be highly monochromatic and can consist almost entirely of a single frequency. In other words, the output signal from a laser is a nearly pure, constant amplitude, highly stable, single-frequency sine wave. The directivity of the laser beam is excellent, and a laser beam can propagate for a long distance with a very small diffraction spread. By contrast, incoherent light from a thermal light source emits a generally broadband continuum of light at many different wavelengths. There are light sources, such as discharge mercury lamps, that emit only few spectral lines, but the spectral widths of the light are still limited by the linewidths of the atomic transitions. Moreover, the emission of the light occurs randomly, essentially in all directions.

Atomic transitions in the vapor phase typically have an atomic linewidth of the order of a few GHz, whereas the linewidths of single-frequency laser beams can range from a few tens of MHz down to only a few Hz in a very highly stabilized system. Therefore, selective excitation of specific atoms in a frequency domain is possible even if there are isotopes and other kinds of atoms that are regarded as impurities.

8.2 Absorption and Emission Processes of Light

In general, an atom in a vapor phase has intrinsic and discrete energy levels. If a photon encounters an atom whose energy levels differ by just the energy of the photon, the atom resonantly absorbs the photon and is excited to the upper state (excited state). If we suppose that the atom is a two-level atom, the atom in the excited state should decay to the previous state (ground state) after a specific time (lifetime) through radiative or nonradiative transition. These spectral and temporal characteristics can be used for characterization and identification of the atom. Elements such as alkali metals and alkaline earth metals have generally stronger transition probabilities that show dominant absorption and emission phenomena. They are used for both basic studies and practical applications.

Since absorption and emission processes play an important role in purification and isotope separation processes that use the laser techniques to be reviewed in this chapter, an overview from the point of view of the theoretical basis will be helpful in order to understand the processes.

The interaction between the electric field of light and an electron is mainly related to the absorption and emission of the light by atoms. The magnitude of the interaction can be expressed by the electric dipole moment:

$$\mu = \sum_i q_i \boldsymbol{r}_i \,, \tag{8.1}$$

where q_i and $\boldsymbol{r}_i$ are the quantity of the electric charge and the position vector of the ith electric charge, respectively. The transition probability is determined by the matrix element of the electric dipole moment $\boldsymbol{\mu}_{ul}$ between the states u and l. The optical transition is called *allowed* for large $\boldsymbol{\mu}_{ul}$. On the contrary, when $\boldsymbol{\mu}_{ul}$ is small or zero, the transition is *forbidden*. Thus the matrix element should take a finite value in order to induce the transition between the two energy levels of the atom.

Suppose that an atom has two energy levels and that an electromagnetic wave (light) of frequency ω is incident on the atom, as shown in Fig. 8.2. Here, E_{u} and E_{l} are the energies of the upper and lower levels, respectively. The probability that the atom will be found in the upper level is expressed by $|b_{\mathrm{u}}(t)|^2$. By a rotating wave approximation, we can deduce that

$$|b_{\mathrm{u}}(t)|^2 = \frac{X^2 \sin(\Delta t/2)}{\Delta^2} \,, \tag{8.2}$$

where Δ is $\omega - \omega_0$, ω_0 is the transition frequency of the atom, $X = \boldsymbol{\mu}_{ul} \cdot \boldsymbol{E}_0/\hbar$ and $\boldsymbol{E}_0$ is the amplitude of the electric field of the light. Figure 8.3 shows the result of the calculation for (8.2). As seen in the figure, the probability decreases sharply with detuning Δ. This means that the transition occurs dominantly when the energy of the incident photon $\hbar\omega$ is equal to the energy difference of the energy levels $E_{\mathrm{u}} - E_{\mathrm{l}}$. The transition probability w_{lu} from

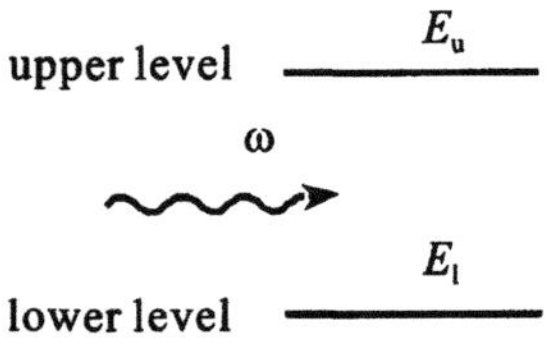

Fig. 8.2. An interaction between a two-level atom and resonant light. E_u and E_l are the energies of the upper and lower levels, respectively

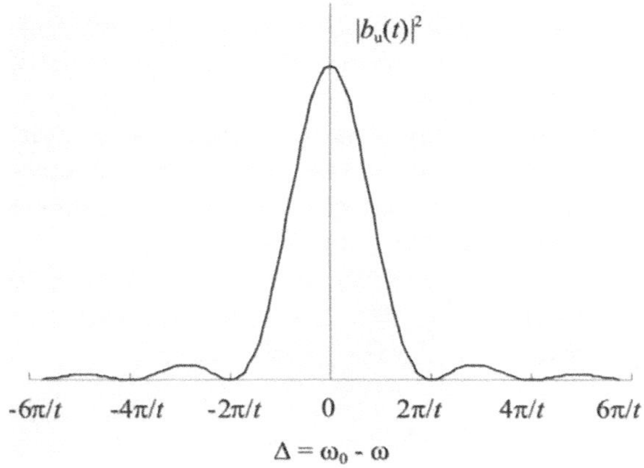

Fig. 8.3. A theoretical calculation of the probability of being excited to the upper state as a function of Δ

the upper to the lower state is derived as

$$w_{lu} = \frac{d|b_u(t)|^2}{dt} = \frac{\pi|\boldsymbol{\mu}_{ul}|^2 U(\omega_0)}{3\varepsilon_0\hbar^2}, \tag{8.3}$$

where $U(\omega)d\omega$ is the average energy density between ω and $\omega + d\omega$, and ε_0 is the dielectric constant of the vacuum. As seen in (8.3), the transition probability is proportional to $U(\omega)$ and the square of $\boldsymbol{\mu}_{ul}$. The same result as (8.3) is obtained for the transition from the lower to the upper states. This transition process is called stimulated emission. In other words, the absorption process and the stimulated emission process are essentially the same, except that the process is inverse.

In addition to the stimulated emission process, Einstein introduced another emission process (spontaneous emission) in order to explain Planck's law of radiation. The stimulated emission process occurs at the probability proportional to the resonant light intensity, at the position where the atom exists. The light emitted by this process has completely the same frequency, phase and polarization as that of the incident light. On the other hand, the spontaneous emission process cannot be controlled externally, because the process occurs regardless of the incident light. The light emitted by this process has no relationship with the incident light. The constants of the transition probability for the spontaneous process and the stimulated process (absorption and stimulated emission) were defined by Einstein as coefficients A and B:

$$A = \frac{{\omega_0}^3 |\boldsymbol{\mu}_{\mathrm{ul}}|^2}{3\pi\varepsilon_0 \hbar c^3}\,, \tag{8.4}$$

$$B = \frac{\pi |\boldsymbol{\mu}_{\mathrm{ul}}|^2}{3\varepsilon_0 \hbar}\,. \tag{8.5}$$

Since Einstein's coefficient A is proportional to the cube of ω_0, as seen in (8.4), spontaneous emission occurs with a larger probability in the higher-frequency region, such as visible and ultraviolet light, in contrast to the lower-frequency region, such as microwaves and radio waves. Stimulated emission plays a key role in the laser oscillation that generates so-called coherent light as a result of amplification action by the stimulated process, generating completely the same light wave as the incident wave.

In general, the energy level of an atom is spreading with a finite width. This width (the spectral width or linewidth) is required from the uncertainty principle between energy and time. An atom in the excited or upper energy level should decay to the lower level after a finite time and thus the energy should be spreading. The time at which the probability $|b_{\mathrm{u}}(t)|^2$ decreases to $1/\mathrm{e}$ is called the lifetime τ. The lifetime due to spontaneous emission should generate the linewidth called the natural width, that provides the Lorentzian lineshape. Collisions can also broaden the linewidth, due to random phase fluctuation providing the Lorentzian lineshape (collision broadening or pressure broadening). The atom in the vapor phase is moving in space with a certain velocity according to the Maxwell–Boltzmann distribution. Consequently, each atom absorbs light of different frequency, determined by the Doppler shift. Thus, the linewidth due to the Doppler effect is called Doppler broadening and provides the Gaussian lineshape.

8.3 Laser Manipulation of Atoms

Laser manipulation of atoms is one of the most attractive fields of study in the past decade. On the basis of atom–light interaction, manipulations of atoms by light, such as velocity control (acceleration, deceleration), reflection, focusing and trapping, have been demonstrated. The cooling technique of the atoms at temperatures down to in the order of nano-Kelvins are being used for precise measurement of optical constants, the improvement of a precise atomic clock, the development of an atomic wave interferometer, the observation of Bose–Einstein condensation and the so-called atom laser.

These atom optics are expected to be a promising technique in both research and industry in the very near future.

Although application of the laser manipulation of atoms in materials science has just started, it might be very useful to review the basic concept and related techniques that can be used for novel processing, analysis and purification techniques for the materials. In this subsection, we will summarize the basic theory of radiation force, laser manipulation techniques and their potential use in materials processing.

8.3.1 Radiation Force

Let us consider the radiation force acting on a two-level atom in a travelling light wave that has a nonuniform field distribution. Suppose that the laser beam propagates in the positive direction of the z-axis and that the electric field $\boldsymbol{E}$ of the laser beam is expressed as

$$\boldsymbol{E} = \tfrac{1}{2}\boldsymbol{E}_0(\boldsymbol{r})\exp(\mathrm{i}\omega t) + c.c.\,, \tag{8.6}$$

where $\boldsymbol{E}_0$ is the amplitude vector of the electric field of the laser beam, $\boldsymbol{r}$ is the position vector in cylindrical coordinates, ω is the angular frequency of the laser beam, t is the time and $c.c.$ refeers to complex conjugation. The radiation force can be expressed as [1]

$$\boldsymbol{F} = \boldsymbol{F}_{\mathbf{scat}} + \boldsymbol{F}_{\mathbf{grad}}\,, \tag{8.7}$$

$$\boldsymbol{F}_{\mathbf{scat}} = \frac{\hbar \boldsymbol{k}\gamma G(\boldsymbol{r})}{1 + G(\boldsymbol{r}) + (\Delta - \boldsymbol{k}\cdot\boldsymbol{v}_z)^2/\gamma^2}\,, \tag{8.8}$$

$$\boldsymbol{F}_{\mathbf{grad}} = -\frac{\hbar}{2}\nabla G(\boldsymbol{r})\frac{\Delta - \boldsymbol{k}\cdot\boldsymbol{v}_z}{1 + G(\boldsymbol{r}) + (\Delta - \boldsymbol{k}\cdot\boldsymbol{v}_z)^2/\gamma^2}\,, \tag{8.9}$$

$$G(\boldsymbol{r}) = \frac{1}{2(\mu\cdot\boldsymbol{E}_0(\boldsymbol{r})/\hbar\gamma)^2}\,, \tag{8.10}$$

where $\boldsymbol{k}$ is the wave vector of the laser light, $G(\boldsymbol{r})$ is the saturation parameter, $2\gamma = 1/\tau$ is the natural width, $\boldsymbol{v}_z$ is the z component of the atomic velocity, $\Delta = \omega - \omega_0$ is the detuning of the laser frequency ω from the atomic transition frequency ω_0, and $\boldsymbol{\mu}$ is the matrix element of the dipole moment of the transition. As seen in (8.7), the radiation force is divided into two components, $\boldsymbol{F}_{\mathbf{scat}}$ and $\boldsymbol{F}_{\mathbf{grad}}$, called the scattering force and the gradient force, respectively (Fig. 8.4). $\boldsymbol{F}_{\mathrm{scat}}$ directs the z direction and represents the force induced by the absorption that results from the momentum exchange between the atom and the light. On the contrary, $\boldsymbol{F}_{\mathrm{grad}}$ directs the direction of the gradient of the laser intensity and represents the force generated from the interaction between the electric field of the light and the dipole moment

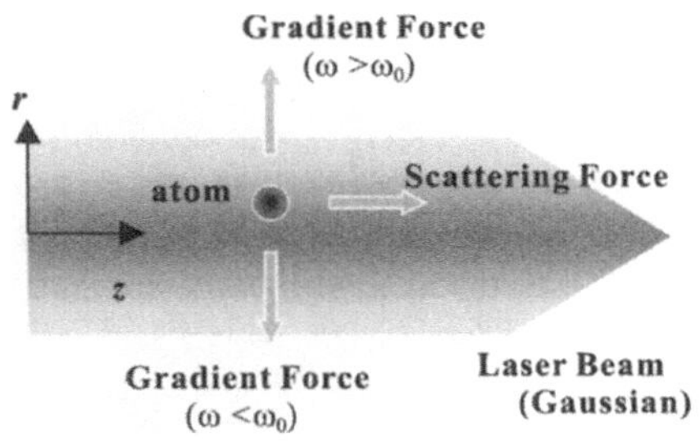

Fig. 8.4. The forces acting on an atom that is nearly resonant with the incident laser beam. Here a fundamental Gaussian (TEM_{00} mode) beam in the positive z direction is supposed. Note that the direction of the scattering force is the same as that of the beam, but the gradient force is perpendicular to the beam because the laser beam is supposed to be well collimated. If the laser beam is converging or diverging, the direction of the gradient force will change

of the atom, induced by the original electric field. The scattering force is associated with spontaneous emission, and the gradient force with a coherent redistribution of the incident electric field due to stimulated emission. Since the gradient force varies linearly with the gradient of the laser intensity, as seen in (8.9), a plane wave does not generate a gradient force but a scattering force. When the laser is, by contrast, a fundamental Gaussian (TEM_{00} mode) beam that has an intensity gradient in the $\boldsymbol{r}$ direction, in addition to the scattering force, a gradient force occurs, as shown in Fig. 8.4. The force direction depends on a sign of the detuning of the laser frequency.

When an atom absorbs a photon and is excited to the upper state, the velocity change of $\Delta \boldsymbol{v} = \hbar \boldsymbol{k}/m$ occurs to satisfy the conservation law of momentum, where m is the atomic mass. If the excited atom emits a photon by stimulated emission, the net velocity change due to a single absorption-and-stimulated emission cycle is zero, because the photon is emitted to the same mode and the velocity change is $-\Delta \boldsymbol{v}$. On the contrary, if the emission process is spontaneous, the net velocity change is still $\Delta \boldsymbol{v}$ because the direction of the photon emission is random, and the average change is zero after repeating many spontaneous absorption-and-emission cycles. Thus, the scattering force is the product of the momentum change $\hbar \boldsymbol{k}$ and the number of spontaneous emission events per unit time ($= \gamma$) as shown in (8.8).

When the laser intensity is strong enough ($G \gg 1$), the scattering force is

$$\boldsymbol{F}_{\mathbf{scat}} = \frac{\hbar \boldsymbol{k}}{2\tau} . \tag{8.11}$$

This equation indicates that the force saturates at a higher laser intensity, because the possibility that the atom exists in the excited state is 1/2 and the time needed to emit the energy is τ on average.

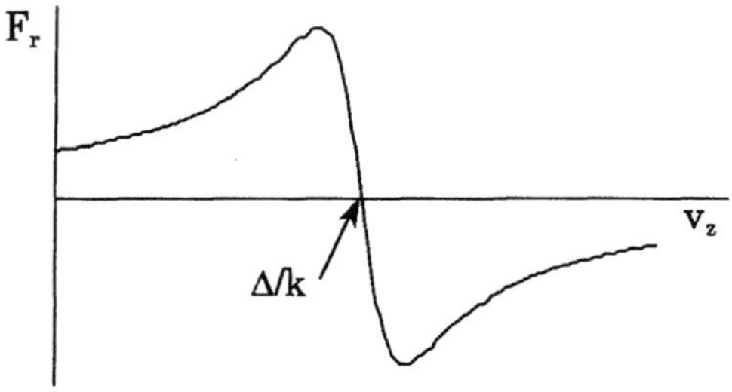

Fig. 8.5. A example of a calculated gradient force as a function of atomic velocity

The gradient force for a Gaussian beam, which is a typical laser beam, is expressed as

$$\boldsymbol{F}_{\mathbf{grad}} = \frac{\hbar r}{q^2} G(\boldsymbol{r}) \frac{\Delta - \boldsymbol{k} \cdot \boldsymbol{v}_z}{1 + G(\boldsymbol{r}) + (\Delta - \boldsymbol{k} \cdot \boldsymbol{v}_z)^2 / \gamma^2} . \quad (8.12)$$

Here q is the cross-sectional radius of the laser beam. As seen in Fig. 8.5, the gradient force shows a velocity dependence and the maximum value $\boldsymbol{F}_{\mathbf{grad,max}}$ is obtained at a velocity of $\boldsymbol{k} \cdot v_z = \Delta \pm \gamma [1 + G(\boldsymbol{r})]^{1/2}$:

$$\boldsymbol{F}_{\mathbf{grad,max}} = \frac{\hbar \gamma \boldsymbol{r} G(\boldsymbol{r})}{2q^2 (1 + G(\boldsymbol{r}))^{1/2}} . \quad (8.13)$$

It is found that the gradient force directs the laser beam axis when the laser is tuned below the atomic transition frequency ($\omega - \omega_0 < 0$). The maximum force is proportional to the square root of $G(\boldsymbol{r})$ for $G(\boldsymbol{r}) \gg 1$.

8.3.2 Deceleration of an Atom

For a sodium atom, $\tau = 16$ ns, $\lambda = 589$ nm and then $\boldsymbol{F}_{\text{scat}}$ is calculated to be as low as 3.5×10^{-20} N from (8.11). This value looks extremely small but is sufficient to manipulate atomic motion, because the magnitude of the force is larger than that of gravity, as shown below. Suppose that the initial velocity of a sodium atom, $\boldsymbol{v}_0$, is $900\,\mathrm{m\,s^{-1}}$. Since the sodium atom is decelerated by $\Delta \boldsymbol{v}_z = 0.03\,\mathrm{m\,s^{-1}}$ per absorption of one photon, the number of photons N needed to decelerate the velocity to zero is $N = \boldsymbol{v}_0 / \Delta \boldsymbol{v}_z = 3 \times 10^4$. When the laser intensity is strong enough to satisfy (8.11), the acceleration $a = \boldsymbol{F}_{\text{scat}}/m = 3.5 \times 10^{-20}/3.8 \times 10^{-26} = 9.2 \times 10^5\,\mathrm{m\,s^{-2}}$. It should be noted that this value is 10^5 times higher than that of gravity. Therefore, the sodium atom stops within approximately 1 ms after propagating nearly 0.44 m. However, the atom can be resonant with the laser beam only within the homogeneous linewidth of the atom.

Deceleration of the atom results in simultaneous annihilation of the atoms that absorb the photon within the linewidth. Since the homogeneous linewidth and the Doppler width are generally of the order of MHz and GHz, respectively, the deceleration of the atom is limited within a narrow region of the velocity distribution. A counter-propagating laser beam can thus only burn a hole in the distribution. Two well known methods, chirp slowing and Zeeman slowing [2], are used to solve this problem. In the former, the laser frequency is varied with time so as to compensate for the Doppler shift. In the latter, the atomic transition frequency is modified in space by an inhomogeneous magnetic field, which shifts the energy levels by the Zeeman effect.

The chirp method is rather easy to implement, especially with semiconductor lasers, which realize a fast frequency sweep by changing the injection current. Although the linewidth of the semiconductor laser output is of the order of tens of MHz at free running, even with precise temperature and current control, an external cavity configuration using a proper frequency selective element such as a grating and an etalon enables us to make the linewidth narrower than the natural linewidth. The velocity distribution of the atomic beam by the chirp method is pulsed in nature and complicated. The Zeeman method provides a continuous and steady flow of atoms, in which all of the atoms reach their final velocity at the same position at the end of the magnet. Careful design of the magnet is required in this method to maintain resonance. The intensity control of the magnetic field at the end of the magnet is generally difficult and the consequent velocity distribution is rather broader than that given by the chirped method.

The velocity distribution of atoms expected by the chirped deceleration method is depicted in Fig. 8.6. The initial distribution is the Maxwell–

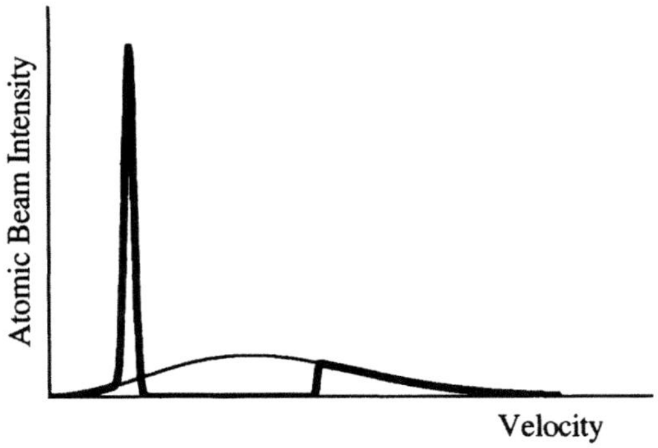

Fig. 8.6. The variation of the velocity distribution of an atomic beam induced by a chirped deceleration method. The bold line and thin line show the distributions after and before the deceleration, respectively

Boltzmann distribution, which is denoted by a thin line. By decreasing the laser frequency, atoms are decelerated to the lower-velocity side, generating a sharp and narrow peak at the end of the chirping of the laser frequency. It should be noted that the temperature of the atoms included in the velocity peak is very low in comparison with the initial temperature.

8.3.3 Deflection of an Atomic Beam

Since the divergence of the atomic beam is small in the direction perpendicular to the laser beam compared to the initial velocity distribution in the direction of the laser beam, the atomic beam is expected to be deflected by absorbing resonant laser light propagating perpendicular to the atomic beam. The first demonstration of light pressure was the small deflection of a sodium atomic beam in 1933. Due to the discovery of the laser, deflection with a larger angle became observable.

A simple experiment has been performed for sodium atoms [3]. A laser beam perpendicular to the atomic beam transfers the additional perpendicular momentum to the atoms. A fraction of the atomic beam is thus deflected. However, there is a limit in the deflection angle because the atom will be out of resonance when the Doppler shift of the atom is beyond the laser linewidth. This is the same problem that occurs in the deceleration of the atoms by the counter-propagating laser beam. The use of a laser beam with a

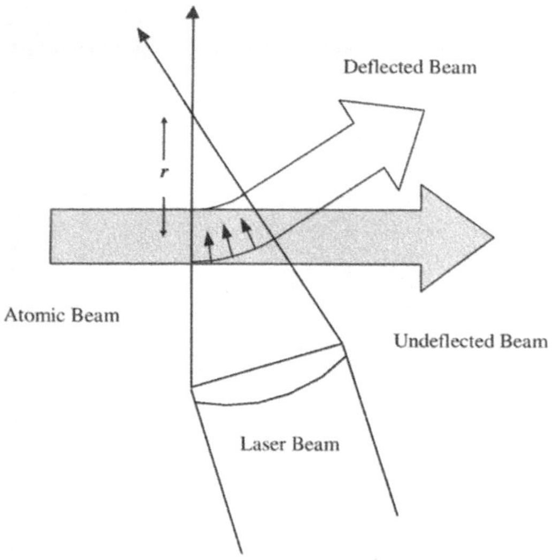

Fig. 8.7. The deflection of an atomic beam by a scattering force [3]. The deflecting atom always propagates perpendicularly to the laser beam, so that the atom maintains resonance with the laser beam of fixed frequency

broader linewidth may solve this problem, but a more powerful laser beam is necessary. Another solution is provided by a simple geometrical consideration of the laser beam configuration. If the wavefront of the deflection laser beam is convex so that the atom being deflected is always perpendicular to the laser beam, the resonance will be maintained and a larger angle of deflection will be produced, as shown in Fig. 8.7. In this case the spontaneous force $\boldsymbol{F}_{\text{spont}}$ of the laser beam satisfies the following equation:

$$\boldsymbol{F}_{\text{spont}} = \frac{m\boldsymbol{v}^2}{\boldsymbol{r}}, \tag{8.14}$$

where $\boldsymbol{r}$ is the distance between the focus of the laser beam and the atom and $\boldsymbol{v}$ is the velocity of the atom. A deflection angle up to 30° was demonstrated by this method. This deflection is expected to separate isotopes and/or impurities spatially and to be used for isotope separation and purification, because different kinds of isotopes and impurities will never absorb the laser light and be deflected. The efficiency of the separation and/or purification will depend on the angle and efficiency of the deflection. If the atoms are decelerated before deflection and its velocity distribution is sufficiently narrow, the deflection will be more efficient.

8.3.4 Reflection and Focusing of the Atomic Beam

An atomic beam can be reflected and focused by light. The evanescent field is well known as a phenomenon in which the light leaks at total reflection. Since this light field varies sharply from the initial value to zero within a very short distance, that is nearly equal to the wavelength of the light, the evanescent field provides an intense gradient force due to a large gradient of the light intensity, as shown in Fig. 8.8. Since the direction of the gradient force depends on the detuning of the laser frequency, as seen in (8.9), an atom approaching the evanescent field can be reflected near the surface where the laser beam is totally reflected if the laser frequency is detuned above the atomic transition frequency [4]. It should be mentioned here that the atom is attracted to the surface if the laser frequency is detuned below. This selective reflection has been demonstrated for hyperfine splitting sublevels of ground-state sodium atoms (F = 1,2). It was found that the reflectivity for atoms in the F = 1 sublevel is at least 100 times higher than for the F = 2 sublevel. This feature thus means that the separation of a specific atom and/or isotope may be possible with a separation coefficient of larger than 100 in a single process.

Let us consider an atomic beam crossing perpendicularly to two counter-propagating laser beams whose intensities are equal. If the laser beams are detuned toward the lower-frequency side from the atomic transition frequency, and an atom has a velocity component parallel to the laser beam, the atom absorbs more efficiently those photons which come from the direction opposite to that of the velocity vector of the atom. As a consequence, the atom

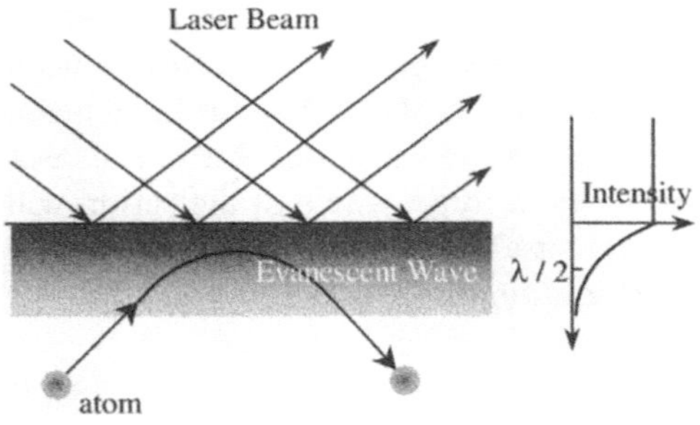

Fig. 8.8. The reflection of an atom by a gradient force [4]. A large gradient of the light field is generated by total reflection of the laser beam. Since detuning of the laser frequency is positive, a repulsive force drives the atom away from the interface

tends not to expand but to be collimated, maintaining the beam size over a finite distance. An axicon reflector can provide such a two-dimensional counter-propagating beam configuration, that acts as a convex lens in optics.

Another configuration using the scattering force is possible by tuning just to the atomic transition frequency. Two sets of counter-propagating laser beams that are diverging symmetrically to the atomic beam axis produce a repulsive force that acts on the atom moving away from the atomic beam axis. In this case, the atomic beam is collimated or focused, resulting in an increase of the atomic beam density.

Focusing of the atomic beam is also demonstrated by means of the gradient force. In this experiment, the axes of both an atomic beam and a Gaussian laser beam are taken to be coincident. Since the laser frequency is tuned below the atomic transition frequency, the gradient force acts on the atom as an attractive force, pulling the atom toward the beam axis, as indicated by (8.12). An atomic beam diameter of 26 μm was achieved by this method. The resolution is limited by momentum diffusion.

To improve the resolution, an atomic objective is proposed, as shown in Fig. 8.9. The configuration for this lens is similar to the previous one used for the focusing, except that the TEM_{01}^* (doughnut) mode laser beam detuned largely from the atomic transition frequency is tightly focused. The intensity of the TEM_{01}^* mode laser beam is zero on the beam axis, like a doughnut, and it has a sharp intensity gradient around the axis. The tight focusing provides a narrow interacting region, resulting in a thin lens. The TEM_{01}^* mode and the large detuning of the laser frequency allow us to regard the laser field as an ideal objective. Calculations show that the focal spot of the atomic beam may be of the order of a few angstroms, assuming that the dimension of the lens is equal to a few wavelengths and the laser power is about 1 W.

As seen in the collimating and focus of the atomic beam, the atoms can interact with only the specific laser light whose frequency is just the atomic

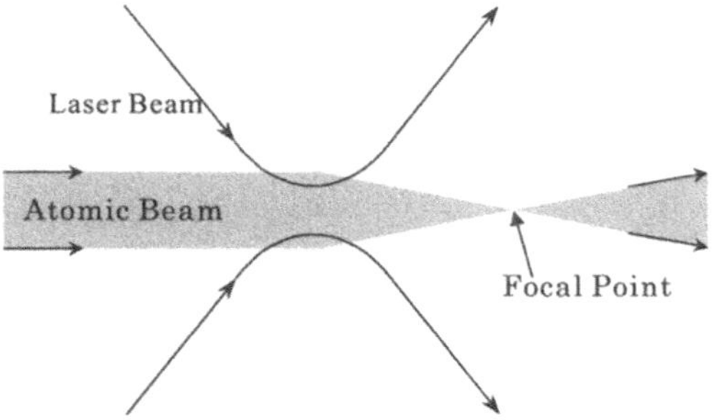

Fig. 8.9. The objective of an atomic beam, using a TEM^*_{01} mode laser beam [4]. The repulsive force near the laser beam focus is predicted to act on the atomic beam as an ideal lens in optics, providing an atomic beam focus as small as a few angstroms

transition frequency, or very close to it. Since isotopes and impurities should have different transition frequencies, these species are not manipulated by the laser field and eliminated from the atomic beam if a pinhole or an orifice is set at the focal point of the atomic beam. In other words, a pinhole may act as a spatial filter against the isotope and the impurity, using the analogy of the elimination of the incoherent fraction from a laser beam in optics. If a small focus of the atomic beam such as a few angstroms is realized, not only an ultra-high-density atomic beam but also a separation coefficient of nearly four figures in magnitude may be achieved.

8.4 Laser Isotope Separation

Although isotope compounds have quite similar physical and chemical characteristics, they often show different behavior in nuclear reactions. In consequence, separation of isotopes and enrichment are strongly required in the nuclear power industry. Uranium is the typical element of which isotope separation is important in practice. In this case, the enrichment of ^{235}U from 0.7% natural concentration up to 3% is necessary for its use as a fuel in commercial nuclear power plants. Deuterium can be used in nuclear reactors when the concentration is increased from 150 ppm in natural water to 99.8%.

8.4.1 Atomic Vapor Laser Isotope Separation (AVLIS) of Uranium

In AVLIS, uranium metal is vaporized at about 3000 K, as shown in Fig. 8.10. A laser beam at the wavelength of 590 nm is irradiated to the vaporized atoms, inducing selective excitation of ^{235}U. Since the isotope shift of ^{235}U is 0.18 cm^{-1} (5.4 GHz), as seen in Fig. 8.11, a dye laser pumped by a copper

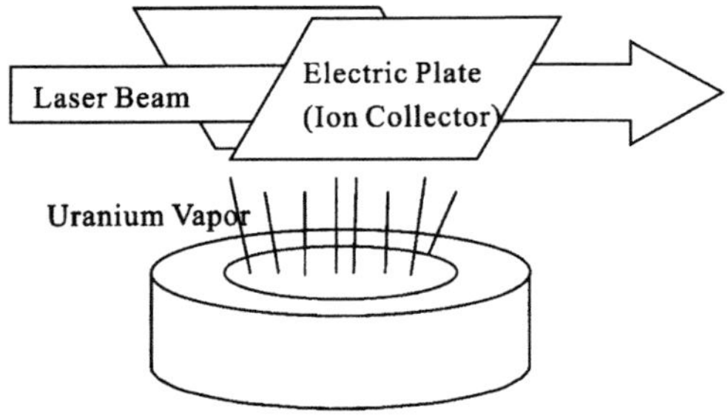

Fig. 8.10. A schematic diagram of an atomic vapor isotope separation (AVLIS) system for ^{235}U

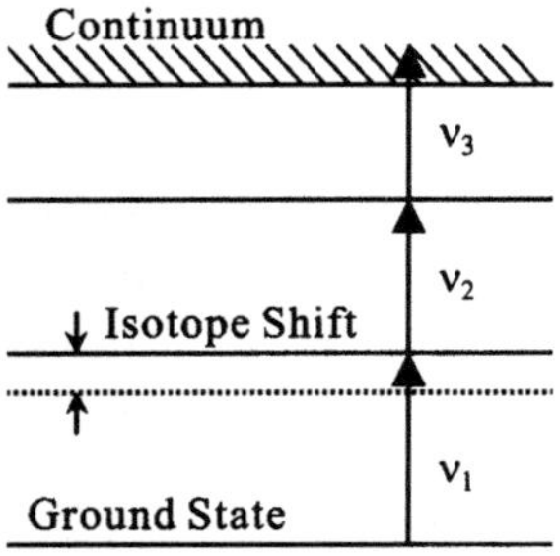

Fig. 8.11. A scheme of selective excitation in the atomic vapor isotope separation (AVLIS) of ^{235}U. Solid lines denote the energy levels of ^{235}U. The dotted line denotes the energy level of ^{238}U. This energy difference is called an isotope shift and is utilized for selective excitation. $\boldsymbol{v}_{1,2,3}$ denote the laser frequencies

vapor laser is used as the excitation light source in order to obtain sufficient laser power, linewidth and wavelength.

The selectively excited atoms are further excited by additional laser beams up to continuum level, resulting in ionization of ^{235}U that is easily collected by electric plates. The first excited level is chosen from the point of view of a larger absorption cross-section and isotope shift among nearly 900 energy levels of the uranium atom.

8.4.2 Molecular Laser Isotope Separation (MLIS) of Uranium

In MLIS, an infrared laser beam is irradiated to decompose a uranium compound by selective excitation of the molecular vibration level. Although many uranium compounds are available, UF_6 is often used. There are two types of excitation scheme that use lasers of different frequency as a second decomposition light source. One method, which uses an ultraviolet laser, has a problem in that the separation coefficient tends to be small due to absorption of the ultraviolet light by atoms that are not selectively excited by the first laser beam. This problem is much reduced in the other method, that uses an infrared laser for both selective excitation and decomposition by a multi-photon process.

The isotope shift of UF_6 is 0.6 cm^{-1} and is larger than in AVLIS, making the preparation of the laser source easier. The spectral width of the molecular absorption spectrum is, however, much wider than the isotope shift. In consequence, cooling of the molecular temperature down to 100 K is required by using adiabatic expansion to reduce the spectral width, in order to clearly separate the absorption spectra of the isotopes. A laser wavelength of 16 μm is thought to be the most effective for UF_6, due to its large absorption cross-section. A stimulated Raman scattering pumped by a CO_2 laser at a wavelength of 10 μm is used as an excitation source.

8.4.3 Other Elements

Isotope separation for more than 20 elements has been reported so far. In most of them, the multi-photon process is used to excite and decompose the isotope compound as seen in the MLIS. The selection of a proper material as a working compound is a key point in improving the performance of the isotope separation. Since the CO_2 laser is an efficient and powerful laser in the infrared region, strong absorption around 10 μm is required for the compound. CF_3CDCl_2 and CF_3D are chosen as promising compounds for deuterium separation. Compounds including C and F, such as CTF_3, are used for tritium separation.

Other isotopes of B, S, Os, Se, Si and Mo are studied by the multi-photon process with an infrared laser. Alkali metals and rare metals are studied by AVLIS. In general, the larger the atomic mass, the narrower is the isotope shift. Therefore, isotope separation for heavy elements becomes difficult.

8.5 Light-Induced Drift

Light-induced drift (LID) can occur when atoms in a buffer gas are velocity-selectively excited. The excited atoms have a flux in the direction of the selected velocity [5]. On the other hand, since the excitation makes a hole in the ground-state velocity distribution, the ground-state atoms have a flux in the opposite direction. Under such a condition that the collision cross-section of the excited state and the ground state with buffer gas atoms are different, a net drift of atoms occurs. The direction of the drift depends on the difference between the collision cross-sections of the excited and ground states.

In 1975, a strange phenomenon was reported in an atom-light system, but the mechanism was not investigated in detail. In 1979, Gel'mukhanov and Shalagin predicted that a combination of radiation and collision gives rise to a new mechanical effect. This is so-called light-induced drift. After this prediction, much work has been carried out both theoretically and experimentally. At the early stage of the study, the precise measurement of

LID was not easy, mainly because the effect is hidden by collisions between the cell wall and the atoms. A technique of coating the capillary wall with paraffin brought about an experimental breakthrough. By this technique, the first direct quantitative measurement of the drift velocity in an Na–noble gas mixture was carried out.

8.5.1 The Principle

As described earlier, each element has an intrinsic transition energy. This feature is utilized to identify the species of atom, molecule and ion by emission and absorption spectrum measurement. In other words, when atoms consist of the same species, they should have the same resonant frequency for light emission and absorption. Therefore, those atoms will be excited if a laser frequency is tuned to the resonant frequency of the atom. However, the velocity of the atoms in the vapor phase at finite temperature is distributed according to the Maxwell–Boltzmann distribution. The velocity difference between a stationary atom and a moving atom results in a Doppler shift. Then, a laser frequency, ν, that is resonant with a moving atom of the velocity, $\boldsymbol{v}_x$, is expressed by

$$\nu = (1 + \boldsymbol{v}_x/c)\nu_0 , \tag{8.15}$$

where ν_0 is the atomic transition frequency and c is the velocity of light. We suppose here that a laser beam is entered in the direction of the x-axis. The laser beam whose frequency corresponds to the Doppler shift with a narrower spectral width than the Doppler width can excite a fraction of the atoms with a specific velocity. Let us consider a situation where the atoms are in a buffer gas. The atoms will collide with the buffer gas atoms. In general, the collision cross-section of an atom differs between the excited and the ground states. Therefore, the atoms in one state will collide more frequently and suffer stronger dumping of the motion than those in the other state. When only a fraction of the atoms with a specific velocity in a certain direction is excited, the velocity distribution of the atoms both of the excited state and the ground state will be different from the initial ones, and will not be symmetrical with respect to zero velocity. Then the different collision cross-section in the gas mixture will cause an imbalance of the dumping of the atomic motion. This imbalance will produce a net drift of the atoms. This phenomenon is the so-called light-induced drift. Figure 8.12 shows the velocity distributions of the atoms in the gas mixture in the excited state and in the ground state. When the laser is off, the initial distribution is the symmetrical Maxwell–Boltzmann distribution, as indicated by a dotted line in the figure. Suppose that the laser is detuned to the higher frequency. In this case, the population of atoms whose velocity corresponds to the Doppler shift decreases and the distribution of the ground state changes, as indicated by the solid line. In contrast, the population of the excited state increases, as shown in the upper

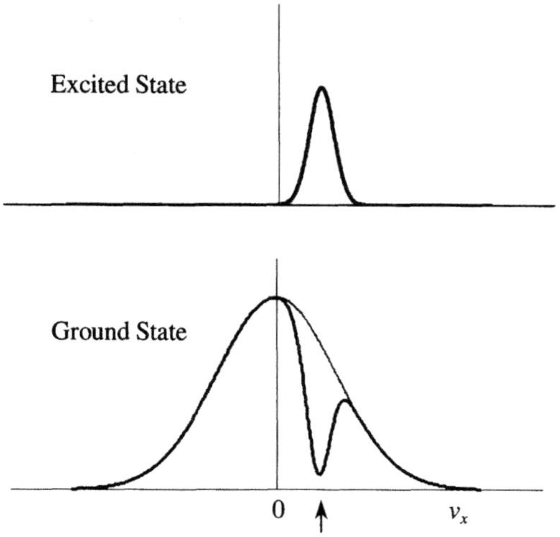

Fig. 8.12. The velocity distribution of atomic vapor. The thin solid line in the lower part of the figure shows the velocity distribution in thermal equilibrium. In contrast, the bold line shows, the distribution when the atomic vapor is velocity-selectively excited. Note that there is a dip around the atomic velocity to which the laser frequency corresponds, as indicated by an arrow. Here the laser linewidth is supposed to be narrower than the Doppler width

part of the figure. If the collision cross-section of the excited state is larger than that of the ground state, the atoms in the positive x direction will suffer stronger dumping than those in the negative x direction.

Then the atoms in the gas mixture will drift to the left-hand side (the positive x direction) as a whole. By detuning the laser to the lower frequency, the direction of the drift will be the positive x direction. Thus, the direction of the drift can be changed by the laser frequency tuning. The drift velocity, $\boldsymbol{v}_{\mathrm{dr}}$, for a two-level system is expressed as

$$\boldsymbol{v}_{\mathrm{dr}} = -\boldsymbol{v}_x \eta \frac{\sigma_{\mathrm{e}} - \sigma_{\mathrm{g}}}{\sigma_{\mathrm{g}}} . \tag{8.16}$$

Here η is the fraction of the absorbing atoms that has been excited but has not yet suffered a thermal collision, and σ_{e}, and σ_{g} are the collision cross-sections of the excited and the ground states, respectively. The factor $(\sigma_{\mathrm{e}} - \sigma_{\mathrm{g}})/\sigma_{\mathrm{g}}$ is the relative change of the collision cross-section due to excitation. As seen in this equation, the drift velocity depends on the difference $\sigma_{\mathrm{e}} - \sigma_{\mathrm{g}}$. Using the equation, we can roughly estimate the drift velocity for an element that can be regarded as a two-level system. For example, alkali metals are included in this category. η might be as large as 50% under sufficient excitation. One can also assume that the factor $(\sigma_{\mathrm{e}} - \sigma_{\mathrm{g}})/\sigma_{\mathrm{g}}$ is nearly 50%. Since the typical velocity of the rubidium atom is approximately $270\,\mathrm{m\,s^{-1}}$ at 373 K, one might estimate the drift velocity to be nearly $70\,\mathrm{m\,s^{-1}}$.

As is known from the above discussion, the atom flux can be pushed or pulled, depending on the laser detuning. This feature can be used for isotope separation when the laser frequency is properly tuned, because each isotope has a different absorption frequency. Here, we will concentrate our attention on some applications of LID to isotope separation as a novel processing technique in materials science.

Since the physical and chemical characteristics of isotopes resemble each other, isotope separation is generally difficult in comparison with separation with different kinds of atoms. However, slight differences in the resonant frequency of the isotopes can be used for separation, as seen in atomic manipulation. As a simple case, a configuration with only a single laser beam is capable of performing the separation. In the case of a two-beam configuration, for example, efficient isotope separation is expected when one laser is tuned to cause LID for one isotope in a certain direction and the other laser is tuned for the other isotope in the opposite direction.

Atutov et al. [6] demonstrated the isotope separation of Li by using a two-beam configuration. Figure 8.13 shows a block diagram of the experimental setup. Figure 8.14 shows the absorption spectrum of Li and the laser frequencies used for the LID experiment. Laser 1 is tuned to the higher-frequency side of the D_1 line of ^{6}Li. Laser 2 is tuned to the lower side of the ^{7}Li D_2 line. These two laser beams enter from one side of the drift cell, which contains Xe buffer gas and Li atoms. Since the collision cross-section of the excited state is larger than that of the ground state in the case of a Li–Xe gas mixture, laser beam 1 pulls the ^{6}Li atom and laser beam 2 pushes the ^{7}Li, resulting in the spatial separation of Li isotopes. The powers of lasers 1 and 2 are 160 and 240 mW, respectively. The Xe pressure in the cell is 40 Torr. The concentration of the Li isotope is estimated from the luminescence intensity at each resonance frequency as follows. First, one of the laser beams is suddenly blocked by a shutter while the LID is occurring. The other laser beam can be absorbed and the consequent luminescence is detected by a photo-detector near the cell. The position dependence of the luminescence

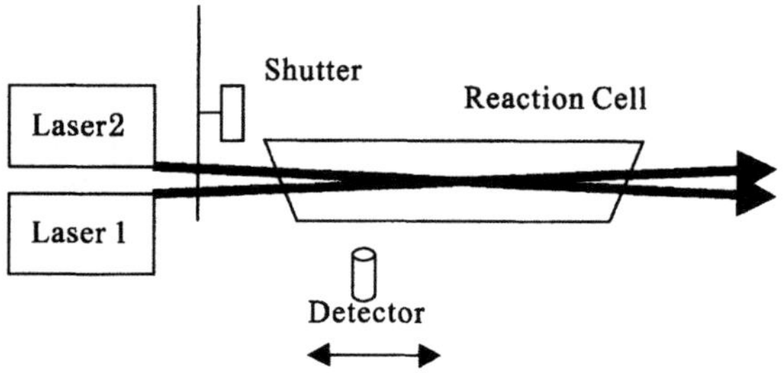

Fig. 8.13. A schematic diagram of the experimental setup for the isotope separation of Li by light-induced drift. Two laser beams with different frequencies are passing through the reaction cell. The emissions from the Li vapor atoms are detected and recorded by moving the detector horizontally

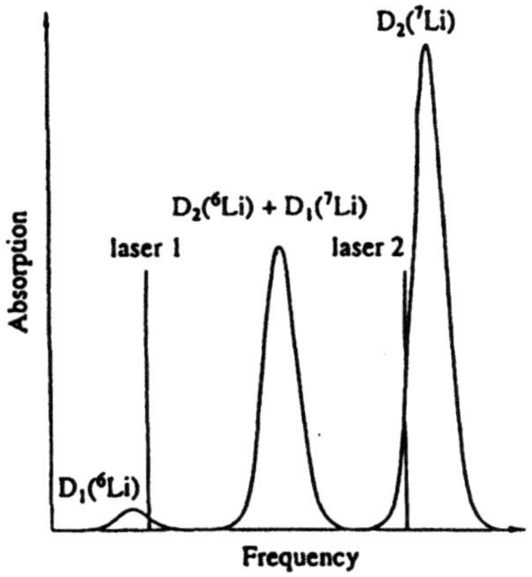

Fig. 8.14. The absorption spectrum of lithium atomic vapor [6]. Both isotope ^{6}Li and isotope ^{7}Li have two lines, D_1 and D_2. Since the D_2 line of ^{6}Li and the D_1 line of ^{7}Li are almost the same, there are three peaks in the spectrum. The solid lines also show the laser frequencies of lasers 1 and 2

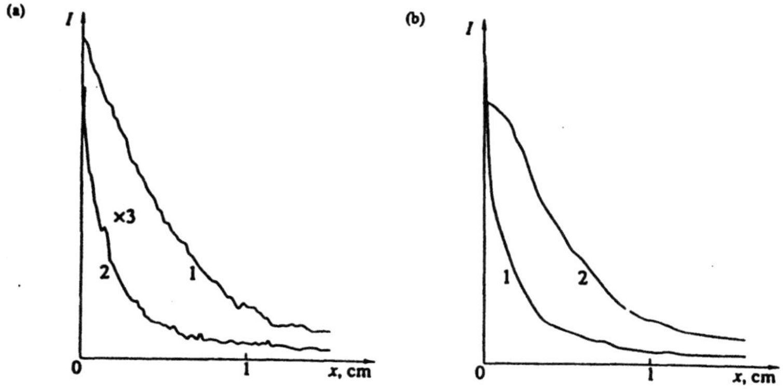

Fig. 8.15. Examples of the spatial distributions of lithium isotopes in the reaction cell in the absence of LID (**a**) and derived by LID (**b**) [6]. In the figure, the curves denoted by 1 and 2 correspond to the results for ^{7}Li and ^{6}Li, respectively

intensity is obtained by moving the detector position. Second, the other laser beam is blocked and the same procedure is performed for another laser beam within a short period. Comparison with the luminescence intensities for the non-LID condition will reveal the concentration variation of the isotopes. Figure 8.15a,b show a distribution of the luminescence from ^{6}Li and ^{7}Li in non-LID and LID conditions, respectively. It is seen that the concentration of ^{6}Li is predominantly increased by LID. For example, the concentration ratio of ^{6}Li to ^{7}Li increases from 10% to 80% at $x = 1\,\mathrm{cm}$, demonstrating a large isotope separation effect compared with conventional methods.

Isotope separation by LID is also performed for rubidium (Rb). It is shown that the hyperfine structure of the Rb atom plays an important role in LID. In general, atoms are excited by absorbing a photon, and then return to the ground state by emitting a photon. The atom sometimes collides with a buffer gas atom during this cycle. It is necessary for LID that this process is repeated over and over again. However, the situation is more complicated for the Rb atom. Since the Rb atom has a hyperfine structure in both the excited and the ground state, transition to the other ground state can occur. Then the atom will be out of the absorption–emission cycle, because the laser frequency is not tuned to the transition.

Figure 8.16 shows the energy levels of the Rb isotopes ^{85}Rb and ^{87}Rb. The values inserted on the arrows show the transition probabilities. Suppose that the laser frequency is tuned to the transition from $^2S_{1/2}$ (F = 3) to $^2P_{3/2}$ (F' = 4) of ^{85}Rb. If the atom is excited to $^2P_{3/2}$ (F' = 4), then the atom will absolutely decay to $^2S_{1/2}$ (F = 3) and the absorption–emission cycle will be a closed one. It happens, however, that the transition to $^2P_{3/2}$ (F' = 3) occurs because each energy level is broadened and overlapped in part. Consequently, decay to $^2S_{1/2}$ (F = 2) can surely occur with a small but finite probability. Since the laser frequency is far from the transition from $^2S_{1/2}$ (F = 2) to $^2P_{3/2}$ (F' = 1,2,3), the atom is no longer excited and remains in off-resonance. Therefore, the population of the atom in $^2P_{3/2}$ (F = 3) gradually decreases and the LID becomes weak.

To overcome this problem, Streater et al. [7] applied a back-pumping method, and demonstrated the suppression of off-resonance atoms by using a unique semiconductor laser that has a sideband whose frequency differs

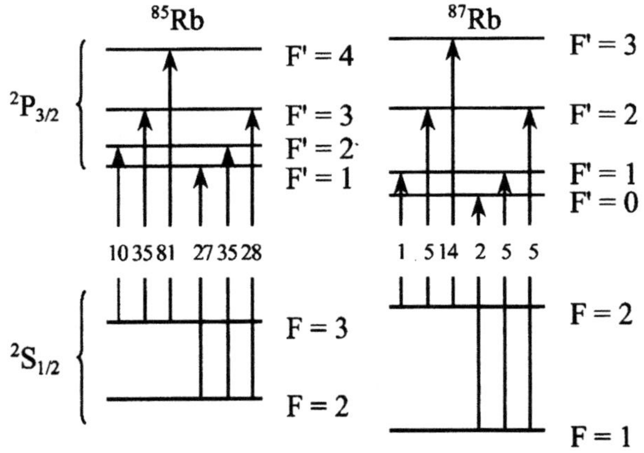

Fig. 8.16. The energy levels of rubidium isotopes, ^{85}Rb and ^{87}Rb. Both the fine structure and the hyperfine structure are shown. The numerical value on each arrow indicates its transition probability

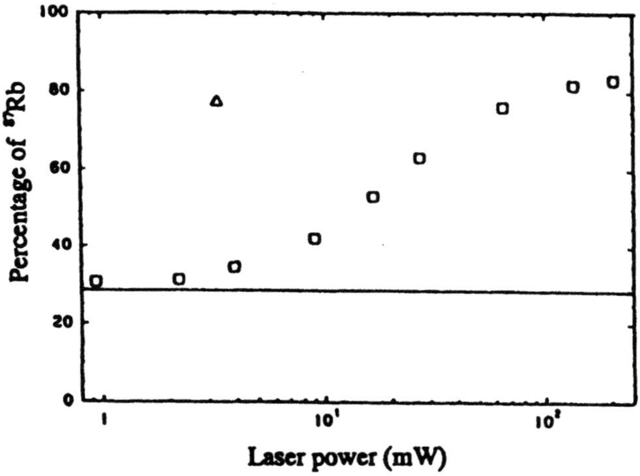

Fig. 8.17. The percentage of measured ^{87}Rb as a function of the laser power [7]. The open squares are results for the single-frequency dye laser. The triangle is the result for the strong relaxation sideband semiconductor laser

from that of the main line by 3.2 GHz. Since the hyperfine splitting of ^{85}Rb is 3.05 GHz, this sideband semiconductor laser is suitable for pumping the atoms from F = 2 to F = 3 of the $^2S_{1/2}$ ground state by exciting the atom to $^2P_{3/2}$ once. Figure 8.17 shows the experimentally observed percentage of ^{87}Rb as a function of the laser power. The square plots show the results for a single-frequency dye laser. It is seen that the percentage of ^{87}Rb increases from a natural value (27.8%) to nearly 80% with an increasing laser power. The laser frequency is tuned to the transition from $^2S_{1/2}$ (F = 3) to $^2P_{3/2}$ (F' = 4). In contrast to the value obtained at the dye laser power of 100 mW, almost the same value is obtained by the sideband semiconductor laser excitation with only 3.3 mW. It is considered that the off-resonance atoms accumulated in the F = 2 state come back to the F = 3 state by back-pumping, and consequently the efficiency of the LID is increased.

This experiment was carried out using a capillary cell and a single special semiconductor laser. Since it is generally hard to tune the sideband frequency just to the hyperfine splitting, the two-beam configuration is more convenient for back-pumping. In this case, one laser is used for the pumping of the absorption–emission cycle and the other laser for the back-pumping. Sato et al. [8] observed a strong LID phenomenon, which can be easily recognized by an IR scope, in a large-diameter separation cell. Figure 8.18 shows the luminescence images from the Rb atoms in the separation cell. The laser beams are coming from the left-hand side of the cell and passing through apertures of diaphragms that intersect the separation cell into three sub-cells. The luminescence image is seen like a narrow line inside the cell. The powers of the Ti : Al_2O_3 laser and the semiconductor laser were 65 and 12 mW,

(a)

(b)

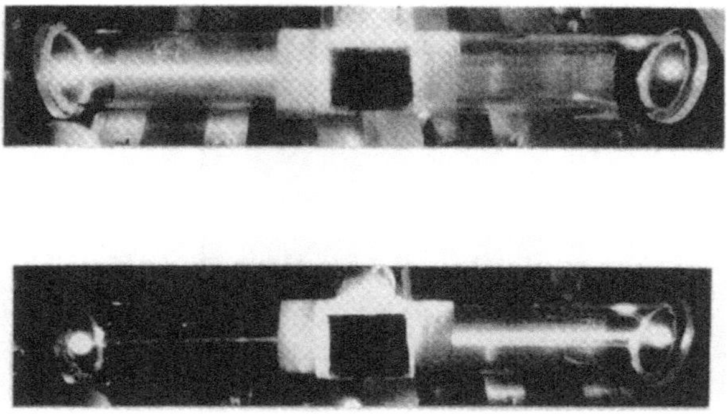

Fig. 8.18. Luminescence images from rubidium atomic vapor in the separation cell during LID using a Ti : Al_2O_3 and a semiconductor laser [8]. The laser beams are incident from the left-hand side of the cell. The luminescence image looks line-shaped, which reflects the shape of the laser beam. The ^{85}Rb atoms are pulled to the left-hand side in (**a**) and pushed to the right-hand side in (**b**). Note the dominant intensity change between (**a**) and (**b**), which indicates strong LID effect

respectively. In Fig. 8.18a, it is clearly seen that the luminescence is strong in the left sub-cell and weak on the right. In this case, the ^{85}Rb is being pulled to the left-hand side, and this phenomenon is called "pulling". On the other hand, by changing the laser frequency, the luminescence becomes strong in the right sub-cell, as shown in Fig. 8.18b. This is called "pushing". The ratio of the luminescence intensities between pulling and the pushing is nearly 30. When the semiconductor laser is not used, the ratio decreases to two. This observation indicates the effectiveness of back-pumping for LID. In this study, a stronger LID is predicted for a greater power of the semiconductor laser. From the fact that the absorption peak of ^{87}Rb was not detected, an ^{85}Rb concentration of at least 94% was estimated by considering the signal-to-noise ratio of the absorption measurement.

As one of the applications of LID, thin-film deposition is reported [9]. The atoms are transferred to a substrate by the LID and deposited on it inside the cell. One of the remarkable features of this thin-film-making technique is the controllability of the film deposition by varying the laser parameters. The position and the thickness can be controlled by the laser position and the laser power and/or the irradiation time, respectively.

Figure 8.19 shows the thickness of the potassium (K) film as a function of the horizontal position on the substrate, where the shape of the film looks like a dot. The film thickness is estimated from the transmitted light intensity through the film. The laser beam was approximately 1 mm in diameter. It was found that the dimensions of the deposited film correspond to those of the laser beam, although a spread of the thickness at the foot of the potassium dot is seen, due to horizontal diffusion of the potassium atoms. It was also found

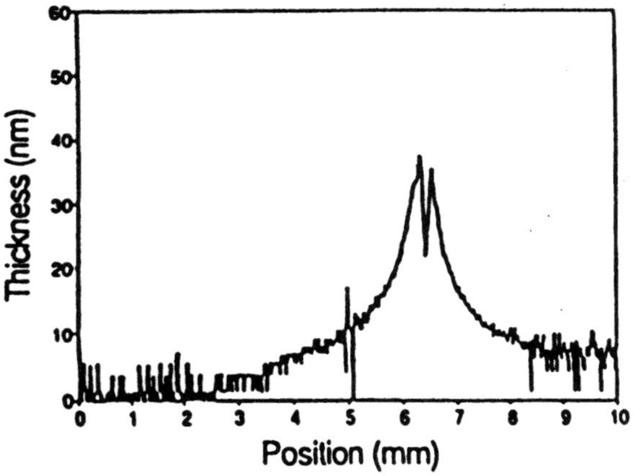

Fig. 8.19. The thickness profile of potassium film generated by LID as a function of the horizontal position on the substrate [9]. The peak at the center of the curve corresponds to a potassium dot. It is thought that excess heating at the laser intensity peak produces the dip in the center of the peak

that the film thickness increases with increasing laser irradiation time. It is thought that the dip seen at the center of the curve is caused by evaporation due to absorption of the laser light at the center of the beam, where the laser beam is strongest. This experiment demonstrates the possibility of collection of an isotope separated by LID.

References

1. V.G. Minogin, V.S. Letkhov: *Laser light pressure on atoms* (Gordon and Breach, New York 1987)
2. W.D. Phillips: In *Laser manipulation of atoms and ions*, ed. by E. Arimond, W. D. Phillips, F. Strumia (Elsevier, Amsterdam 1992)
3. W. Ertmer: In *Laser manipulation of atoms and ions*, ed. by E. Arimond, W.D. Phillips, F. Strumia (Elsevier, Amsterdam 1992)
4. V.I. Balykin: In *Laser manipulation of atoms and ions*, ed. by E. Arimond, W.D. Phillips, F. Strumia (Elsevier, Amsterdam 1992)
5. H.G.C. Werij, J.P. Woerdman: Phys. Rep. **169**, 145 (1988)
6. S.N. Atutov, P.V. Kolinko, A.M. Shalagin: Laser Phys. **3**, 855 (1993)
7. A.D. Streater, J. Mooibroek, J.P. Woerdman: Appl. Phys. Lett. **52**, 602 (1988)
8. S. Sato, M. Saito: Mater. Trans. JIM **37**, 1789 (1996)
9. J.A. Kleinfeld, A.D. Streater: Appl. Phys. Letts **62**, 3077 (1993)

Part III

Characterization

9 Electrical Properties of Ultra-High-Purity Metals

E. Hashimoto and Y. Ueda

9.1 Introduction

The electrical properties of metals are very sensitive to the presence of impurities, even in ppm amounts or less. It is therefore no exaggeration to say that the intrinsic properties of truly pure metals are still not known. In an ultra-high-purity metal with a residual resistance ratio RRR ($\equiv R_{300\,\mathrm{K}}/R_{4.2\,\mathrm{K}}$) of more than 100 000, the mean free path of conduction electrons reaches several millimeters at low temperatures. At the same time, the relaxation time is also enlarged. Therefore, unknown properties may possibly appear, especially in low-temperature electronic transport.

Aluminum is a typical example of a "simple metal" which can be described by a nearly-free-electron model, and is often chosen to test our understanding of the basic scattering processes of electrons and phonons in a metallic environment. However, our recent experiments in the region of ultra-high purities have revealed the following: anisotropy of the bulk residual resistivity with respect to the current direction; nonlinear current–voltage characteristics, which depend on the specimen shape; and a negative bend resistance due to ballistic electron transport and related phenomena. These phenomena are noteworthy in relation to a basic principle of the theory of electronic transport in metals. For example, it tells us that there should be no anisotropy of resistivity of a cubic metal such as aluminum.

In this chapter, we intend to give a detailed account of the "new phenomena" in electronic transport in ultra-high-purity aluminum. First, however, a brief explanation of the general features of electrical resistivity is given, since it is a major tool for the characterization of ultra-high-purity metals.

9.2 The Temperature Dependence of the Electrical Resistivity

The electrical resistivity of metals depends strongly on temperature. In highly pure metals, the resistivity at low temperatures is decreased by several orders of magnitude from that at room temperature. In an ideal metal, where the whole system has perfect periodicity, the electrical resistivity will be zero. In a real metal, however, the periodicity is violated by crystal imperfections,

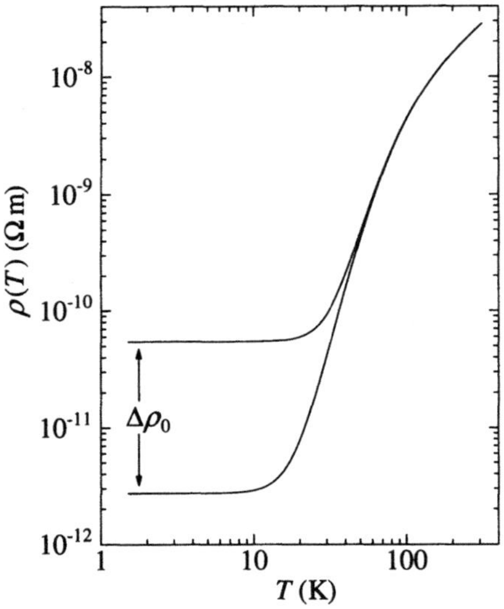

Fig. 9.1. Typical variations of the total electrical resistivity, $\rho(T)$, in pure metals with a temperature T

such as impurities, dislocations, and so on. The resistivity then becomes finite even at 0 K, due to the scattering of electrons at the imperfections. This term is the *residual resistivity*, ρ_0. When the temperature T is raised, the thermal vibrations of the ions affect the periodicity, giving rise to finite resistivity as the phonon part of the resistivity. This phonon part, $\rho_\mathrm{p}(T)$, is intrinsic to the metal at a temperature T. The features of the total resistivity, $\rho(T) = \rho_0 + \rho_\mathrm{p}(T)$, are schematically shown in Fig. 9.1.

An increase in the residual resistivity, $\Delta\rho_0$, at the low-temperature limit in Fig. 9.1 is an incremental function of the impurity concentration. For isolated impurities of each species, $\Delta\rho_0$ is proportional to its concentration. Therefore, the residual resistivity ρ_0 is a good index of the purity of pure metals. As ρ_0 is negligible compared with the phonon part of the resistivity at room temperature, a ratio of the room-temperature resistivity to ρ_0 is also suitable for an indication of purity.

The room-temperature resistivity is usually defined at 300 K, 20°C or 0°C, and denoted, for example, as $\rho_{300\,\mathrm{K}}$, $\rho_{20^\circ\mathrm{C}}$ or ρ_{0° C, respectively. The resistivity obtained at 4.2 K, $\rho_{4.2\,\mathrm{K}}$, is practically used for ρ_0, since the phonon part of $\rho_{4.2\,\mathrm{K}}$ is comparable with or less than ρ_0. In many cases, the ratio $\rho_{300\,\mathrm{K}}/\rho_{4.2\,\mathrm{K}}$, or more practically $R_{300\,\mathrm{K}}/R_{4.2\,\mathrm{K}}$, is called the *residual resistance ratio*, and is widely used as an index of the purity level of a pure metal by the abbreviation RRR. Here, $R_{300\,\mathrm{K}}$ and $R_{4.2\,\mathrm{K}}$ are the electrical resistance of the metal at 300 K and 4.2 K, respectively. If the thermal contraction of the specimen is negligible between room temperature and low temperatures, the ratio $\rho_{300\,\mathrm{K}}/\rho_{4.2\,\mathrm{K}}$ is equal to $R_{300\,\mathrm{K}}/R_{4.2\,\mathrm{K}}$. Hereafter, the ratio $R_{300\,\mathrm{K}}/R_{4.2\,\mathrm{K}}$ is employed as RRR.

The phonon part of the resistivity, $\rho_{\mathrm{p}}(T)$, is well approximated by the Bloch–Grüneisen law for many metals [1]:

$$\rho_{\mathrm{p}}(T) = C\left(\frac{T}{\Theta}\right)^5 \int_0^{\Theta/T} \frac{4z^5}{(\mathrm{e}^z - 1)(1 - \mathrm{e}^{-z})}\,\mathrm{d}z\,, \tag{9.1}$$

where C is a constant that is independent of T, and Θ is the Debye temperature.

At high temperatures, where $\Theta/T \ll 1$, the Bloch–Grüneisen law gives the resistivity $\rho_{\mathrm{p}}(T) \sim T$. At low temperatures, where $\Theta/T \gg 1$, the asymptotic form $\rho_{\mathrm{p}}(T) \sim T^5$ is strongly dependent on temperature. In this model, only the contribution of the normal processes is incorporated for electron–phonon scattering. For the investigation of $\rho_{\mathrm{p}}(T)$, however, other scattering mechanisms such as the Umklapp processes and electron–electron interaction must be considered. These mechanisms, which are intrinsic to the matter itself, are expected to become important for highly pure metals, where the electron–impurity scattering probability is very small. Impurities in metals also modify the intrinsic behavior of $\rho_{\mathrm{p}}(T)$, leading to deviations from Matthiessen's rule [2].

Especially at low temperatures, where $\rho_{\mathrm{p}}(T)$ is extremely small, highly pure metals are necessary to satisfy the condition that $\rho_0 \leqslant \rho_{\mathrm{p}}(T)$. In aluminum, the value of $\rho_{\mathrm{p}}(T)$ is nearly 0.1 and 0.7 pΩ m at 4.2 and 10 K, respectively. Therefore, pure aluminum with, for example, RRR $\geq$ 50 000 with respect to impurities, is desired for the experiment in that temperature range.

Using a single-crystal specimen, AL-1, prepared from ultra-high-purity aluminum with an RRR of 100 000 [3], we measured the temperature dependence of the resistivity, $\rho(T)$, between 4.2 K and room temperature. The results are shown in Fig. 9.2 and the characteristics of the specimen are listed in Table 9.1. The notations $\{\ \}$ and $\langle\ \rangle$ indicate the crystallographic orientation of the surface and the axis of the specimen, respectively. In Fig. 9.2, the results are also shown for another single-crystal specimen, AL-2, prepared from less pure aluminum with an RRR of 30 000.

We are now interested in the behavior of the $\rho_{\mathrm{p}}(T)$ of high-purity aluminum. Therefore, we must subtract ρ_0 at 0 K from $\rho(T)$. To obtain ρ_0, a polynomial expression,

$$\rho(T) = \rho_0 + AT^2 + BT^5\,, \tag{9.2}$$

was fitted to the data of specimen AL-1 for $T < 15$ K by the least squares method. The second term, AT^2, was added in anticipation of an electron–electron scattering contribution proportional to T^2. The third term is the contribution predicted from the Bloch–Grüneisen law for low temperatures. The extrapolated value of ρ_0 is listed in Table 9.1. It is noted here that ρ_0 includes a contribution from the scattering of electrons at the specimen surface, as well as the impurity scattering contribution in bulk. This increase

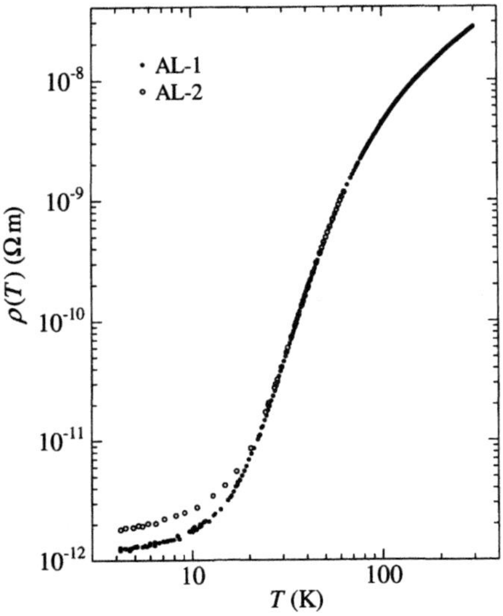

Fig. 9.2. The temperature dependence of the resistivity, $\rho(T)$, of aluminum single-crystal specimens AL-1 and AL-2

Table 9.1. The characteristics of the aluminum single-crystal specimens used for the measurement of the temperature dependence of the resistivity

Sp.	Size (mm)	Orientation	$\rho_{4.2\,\mathrm{K}}$ (pΩ m)	ρ_0 (pΩ m)	ρ_b (pΩ m)[a]
AL-1	0.44 × 1.7 × 16.5	$\{110\}\langle 110\rangle$	1.24	1.13	0.31
AL-2	0.55 × 1.4 × 23.0	$\{110\}\langle 110\rangle$	1.81	1.72	0.87

[a] Size-corrected value of $\rho_{4.2\,\mathrm{K}}$: bulk residual resistivity at 4.2 K

in resistivity, the *size effect*, will be discussed later. For the less pure specimen, AL-2, one more term proportional to T^3 was added to the polynomial expression for $T < 20\,\mathrm{K}$, in order to describe the small contribution of the deviations from Matthiessen's rule.

The phonon part of the resistivity $\rho_\mathrm{p}(T) = \rho(T) - \rho_0$ for specimen AL-1 is shown in Fig. 9.3. The main features of $\rho_\mathrm{p}(T)$ in specimen AL-1 are the clear T^2 behavior for $T < 10\,\mathrm{K}$, and the T^5 behavior for $20\,\mathrm{K} < T < 40\,\mathrm{K}$. At higher temperatures, $\rho_\mathrm{p}(T)$ approaches a T^1 dependence. The Bloch–Grüneisen law is also shown in the figure, as the solid curve. This curve was obtained by fitting the Bloch–Grüneisen law to the data in specimen AL-1 for $20\,\mathrm{K} < T < 300\,\mathrm{K}$, according to the least squares method. The curve is constrained to go through the fixed point $\rho_{300\,\mathrm{K}} = 27.33\,\mathrm{n\Omega\,m}$ at $T = 300\,\mathrm{K}$ [4]. The parameters Θ and C thus obtained are 421.3 K and 170.9 nΩ m, respectively.

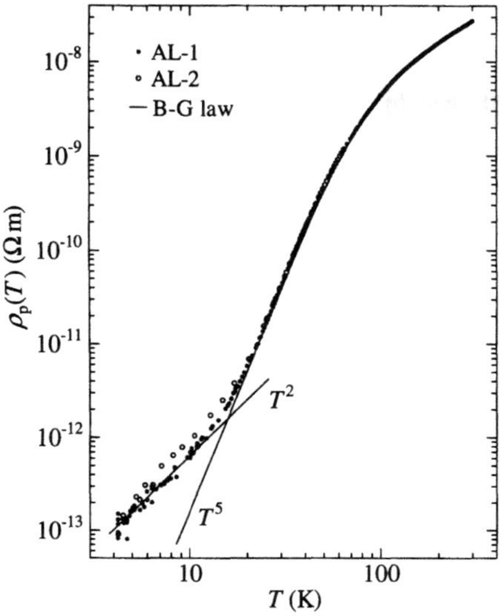

Fig. 9.3. The temperature dependence of the phonon part, $\rho_\mathrm{p}(T)$, of specimens AL-1 and AL-2. The solid curve shows the Bloch–Grüneisen law with $\Theta = 421.3\,\mathrm{K}$ and $C = 170.9\,\mathrm{n\Omega\,m}$

The deviation from the Bloch–Grüneisen law is notable for $T < 20\,\mathrm{K}$. For $T < 10\,\mathrm{K}$, the coefficient A of the T^2 term was determined as $6.0\,\mathrm{f\Omega\,m\,K^{-2}}$. This T^2 term is attributed to electron–electron scattering [5]. Most investigations of the T^2 term have been performed at low temperatures less than 4.2 K, in contrast to the present report. The value of A for specimen AL-1 is somewhat larger than that for specimen AL-2, $4.9\,\mathrm{f\Omega\,m\,K^{-2}}$, which shows the sample dependence of A. In the temperature range $50\,\mathrm{K} < T < 80\,\mathrm{K}$, the experimental results of $\rho_\mathrm{p}(T)$ in specimen AL-1 are larger than the prediction of the Bloch–Grüneisen law. In this temperature range, therefore, the important contribution of the Umklapp processes is suggested for the temperature dependence of $\rho_\mathrm{p}(T)$.

The Debye temperature Θ is also intrinsic to metals and is affected by impurities. The values of Θ obtained in aluminum by several methods are listed in Table 9.2. There is no extreme difference between the values of Θ in the table. However, the present data shows a tendency to increase with purity. This suggests a change in the phonon spectrum of aluminum with purity.

The temperature dependence of the resistivity is a powerful method for the characterization of pure metals and the study of electron–phonon interactions. Especially at low temperatures, where electron–phonon scattering events are reduced, the effects of impurities on electronic properties are not negligible even in so-called high-purity metal with an RRR of 30 000, as shown in Fig. 9.3. We need truly pure metal to know its intrinsic properties.

Table 9.2. The Debye temperature Θ of aluminum obtained by several methods

Θ (K)	Method	RRR	Reference
383	Entropy at 298 K		[6]
403	Elastic constant		[6]
430	Low-temperature phonon		[6]
394	Specific heat		[7]
395	Specific heat		[8]
428	Specific heat		[9]
400	Electrical resistance		[10]
415	Electrical resistance	4 100	[11]
408	Electrical resistance	30 000	Present
421	Electrical resistance	100 000	Present

9.3 Specimen Preparation and Resistivity Measurements

In the following two sections, we are concerned with "anisotropic phenomena", recently identified in ultra-high-purity aluminum. The topics to be presented are: the anisotropy of the dc size effect (Sect. 9.4); and that of the bulk residual resistivity with respect to the current direction (Sect. 9.5). Before proceeding to the main subject, we briefly describe the experimental procedure in order to make clear that the anisotropy is due to the intrinsic origin.

9.3.1 Specimen Preparation

Specimens were prepared by spark erosion from three single-crystal rods of high-purity aluminum, grown by zone melting. The residual resistance ratios, RRR, of the rods were about 10 000, 50 000 and 100 000 in the bulk.

Single-crystal plates, 0.2–2.0 mm thick, were first cut parallel to each of three crystallographic planes $\{100\}$, $\{111\}$ and $\{110\}$ from nearby parts of the rod. Then they were cut into rectangular strips, approximately 3 mm by 20 mm, to serve as the specimens. After cutting, the specimens were chemically etched with aqua regia to remove all spark damage, cleaned in distilled water and rinsed with acetone.

The crystallographic orientation of the specimens was determined by the transmission Laue method to an accuracy of $\pm 1°$.

Two pairs of electrodes, 0.3 mm thick wires made of zone-refined aluminum, were spot-welded on to each specimen for the dc four-probe measurement. The potential probes were placed sufficiently far inside to avoid end effects. The specimens thus prepared were lightly etched with aqua regia, annealed at 300°C for 3 h in air and furnace-cooled to room temperature.

Specimens other than the rectangular-shaped ones were also prepared in the same way.

The details of the specimens will be given whenever necessary.

9.3.2 Resistivity Measurements

The resistance measurement at 4.2 K was performed using a superconducting chopper amplifier with a voltage sensitivity of ±2 pV [12]. The specimen was shielded with a superconducting Pb can, to avoid the influence of external fields such as the Earth's field. A dc current through the specimen was set at 0.5–1.0 A after confirming the linear behavior of the I–V characteristics, as shown in Fig. 9.4. This means that the influence of Joule heat, as well as that of magnetoresistance due to the self-magnetic field of the measuring current I, is negligible in the measurement, since these quantities increase in proportion to I^2.

The resistivity of the specimen at 4.2 K was determined from the relation $\rho_{4.2\,\mathrm{K}} = (\rho_{300\,\mathrm{K}}/R_{300\,\mathrm{K}})R_{4.2\,\mathrm{K}}$, where $\rho_{300\,\mathrm{K}} = 27.33\,\mathrm{n\Omega\,m}$ for high-purity aluminum [4]. The resistance at 300 K was measured using a dc comparator potentiometer with a sensitivity of ±0.5 nV (Guildline, Model-9930).

The thickness d of the specimen was determined from the following relation: $d = (\ell_\mathrm{p}/w)(\rho_{300\,\mathrm{K}}/R_{300\,\mathrm{K}})$. The width w of the specimen and the distance ℓ_p between the potential probes were measured using a travelling microscope with a resolution of 0.01 mm.

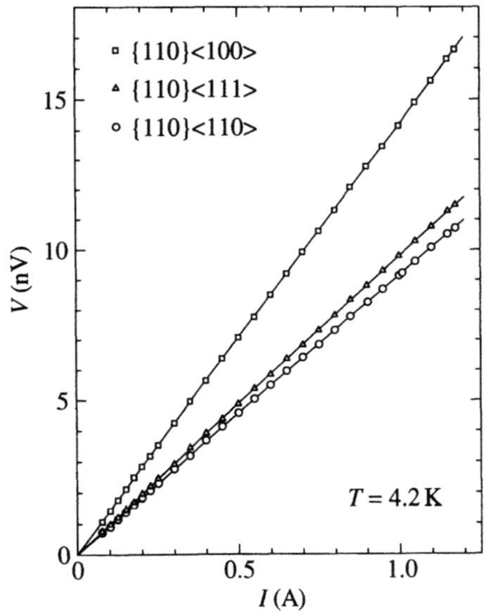

Fig. 9.4. The linear I–V characteristics in rectangular single-crystal strips of ultra-high-purity aluminum (RRR ≃ 50 000)

9.4 The Anisotropy of the Size Effect in dc Electrical Resistivity

Size effects in electrical resistivity are of importance when the bulk mean free path ℓ_b of conduction electrons is comparable with or larger than the characteristic dimension of the specimen; that is, the surface scattering of electrons increases the apparent resistivity, depending on the specimen size. This situation is encountered when a bulk property is to be measured on a pure metal; for example, the measurement of the contribution of T^2 electron-electron scattering to the bulk residual resistivity ρ_b, or the measurement of the specific resistivity of defects, such as impurities and dislocations.

According to the Fuchs–Sondheimer theory (F–S theory) [13], developed assuming a spherical Fermi surface and an isotropic relaxation time, the ratio of the resistivity of a metallic film of thickness d to that of the bulk metal, ρ/ρ_b, is given as follows:

$$\frac{\rho}{\rho_b} = \left[1 - \frac{3}{2\kappa}(1-p)\int_1^\infty \left(\frac{1}{t^3} - \frac{1}{t^5}\right)\frac{1-e^{-\kappa t}}{1-pe^{-\kappa t}}\,dt\right]^{-1}, \tag{9.3}$$

where $\kappa = d/\ell_b$, and p is the specularity parameter, which corresponds to the relative number of electrons specularly reflected from the specimen surface. The limiting form for thick films ($\kappa \gg 1$),

$$\frac{\rho}{\rho_b} = 1 + \frac{3}{8\kappa}(1-p), \tag{9.4}$$

is so convenient that it has often been used to evaluate the product $\rho_b\ell_b$ from size effect data. In this theory, there is clearly no orientation dependence.

In high-purity metals, however, the anisotropy of the size effect is pronounced [14]. This anisotropy is closely related to the following two effects: one is the effect due to the anisotropy of the Fermi surface, and the other is the effect due to the anisotropy of the relaxation time of the conduction electrons. However, the origin of the anisotropic size effect has still not been clarified, because of the lack of systematic studies on well-defined single crystals. In this section, we are concerned with the anisotropic size effect in ultra-high-purity aluminum single crystals [15].

9.4.1 Dependence on the Surface Orientation

Figure 9.5 shows $\rho_{4.2\,K}$ for the $\{100\}\langle 110\rangle$, $\{111\}\langle 110\rangle$ and $\{110\}\langle 110\rangle$ specimens, plotted as a function of the inverse of the reduced thickness, $P/2A$, where P is the perimeter for a cross-section and A the cross-sectional area of a specimen. The abbreviated notation $\{100\}\langle 110\rangle$, for instance, indicates that the surface orientation of the specimen is $\{100\}$ and the axis orientation is $\langle 110\rangle$. Since the axis orientation (the direction of current flow) is common, the bulk resistivities ρ_b of these specimens essentially agree with each other.

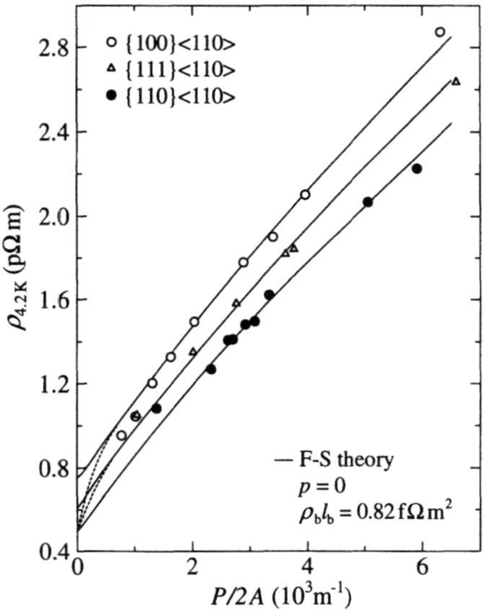

Fig. 9.5. Experimental values of $\rho_{4.2\,\mathrm{K}}$ for ultra-high-purity aluminum single crystals (RRR $\simeq$ 50 000), plotted as a function of the inverse of the reduced thickness, $P/2A$

Therefore, as is obvious from the figure, the size effect increases in the following order of the surface orientations: {110}, {111}, {100}. The anisotropy of ρ_b with respect to the current direction in high-purity aluminum will be described later.

The solid curves in the figure are drawn according to (9.3), with $p = 0$, using $\rho_\mathrm{b}\ell_\mathrm{b} = 0.82\,\mathrm{f\Omega\,m^2}$. As the specimens have been etched with aqua regia and annealed in air, the specimen surface is rather rough. Therefore, the surface scattering of conduction electrons is reasonably assumed to be diffuse; that is, we assume that $p = 0$. Sambles and co-workers [16] have amply shown that the theory given by Soffer [17] works better than the F–S theory. When the surface roughness is large, however, the two theories agree closely with each other over a wide range of d/ℓ_b, including the values for the specimens in Fig. 9.5.

The curves thus obtained fit the experimental data closely, except those on the thick specimens. Deviations from the F–S theory for the thick specimens suggest a contribution of the size effect due to the side surface. In fact, the side surface of the {110}⟨110⟩ specimen has the {100} orientation, yielding a size effect larger than the {110} main surface, as expected from the anisotropic size effect above.

As already mentioned, the resistivities ρ_b of the specimens in Fig. 9.5 are essentially the same. However, the ρ_b value adjusted to give a good fit increases with the contribution of the size effect. Thus, if the contribution from the side surface is negligible, a rapid increase in $\rho_{4.2\,\mathrm{K}}$ with decreasing thickness must take place at first in the very thick {100}⟨110⟩ and {111}⟨110⟩ specimens, as illustrated by dotted curves in Fig. 9.5. In this condition, it is very difficult to measure the resistivity of these specimens accurately. Nevertheless, for the size effect due to the {100} and {111} surfaces, there is substantial disagreement with the F–S theory, with $\rho_b \ell_b = 0.82\,\mathrm{f\Omega\,m^2}$, for the very thick specimens.

In our recent experiments on single-crystal specimens of dilute Al–Ag and Al–Si alloys [18], any orientation dependence of the size effect was less than the experimental error: cthe size effect in both dilute alloys was well described by (9.3), with $p = 0$, using $\rho_b \ell_b = 0.82\,\mathrm{f\Omega\,m^2}$. This implies that the anisotropy of the relaxation time due to the impurities is not the intrinsic origin of the anisotropic size effect, since the solute Ag in aluminum decreases the relaxation time near the Brillouin zone boundaries, while Si has the reverse effect.

A probable explanation for the disappearance of the anisotropy, caused by the addition of the impurities Ag and Si, is that the scattering of conduction electrons from impurities isotropizes the deviation of the electron distribution from the equilibrium in the steady state. In such a situation, the F–S theory is sure to be a good approximation, since the Fermi surface of aluminum is nearly a free-electron sphere, even though it extends beyond the first Brillouin zone. It is therefore reasonable to expect that the effect of the anisotropy of the distribution function leads to the anisotropic size effect.

The structure of the Fermi surface of aluminum is highly anisotropic near the zone boundaries, as shown in Fig. 9.6. As the electron velocity is proportional to the gradient of the energy in $\boldsymbol{k}$-space, the velocity distribution of the electron approaching the specimen surface will in general vary depending on the surface orientation. This is probably an important effect, which may cause the anisotropic size effect as shown in Fig. 9.5, since the surface scattering affects all of the electrons, especially those moving in the direction of the surface normal.

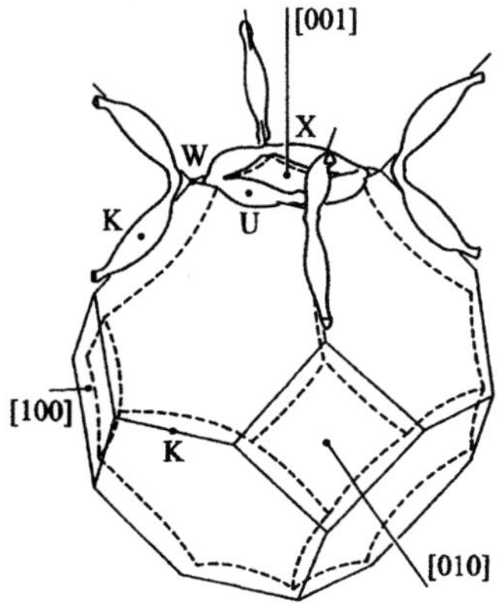

Fig. 9.6. The 4-OPW Fermi surface of aluminum in the reduced-zone scheme [19]

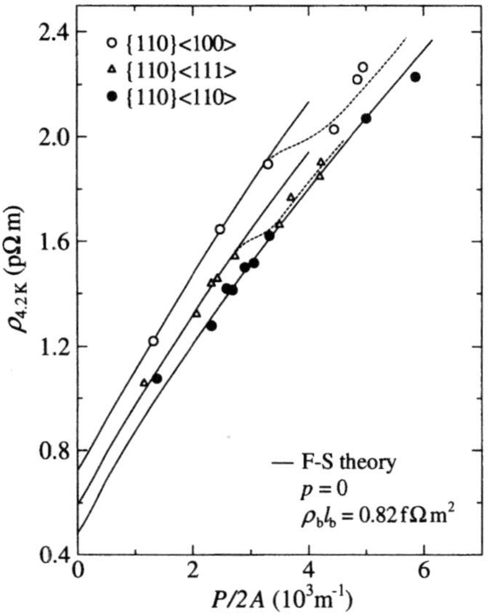

Fig. 9.7. Experimental values of $\rho_{4.2\text{K}}$ of ultra-high-purity aluminum single crystals (RRR $\simeq$ 50 000), plotted as a function of $P/2A$. The broken curves are drawn to guide the eye

On inspection of the geometry of the Fermi surface, especially around the [001] direction, it does appear that the relative number of conduction electrons moving in the [001] direction is larger than that on the free-electron Fermi sphere. Thus, if the specimen surface is parallel to the (001) plane, such electrons are effectively scattered out of their states by the surface scattering. This increases the number of unoccupied states into which an electron travelling in the forward direction will be scattered, leading to an enhancement of the rates of scattering of the electrons. Therefore, the apparent resistivity becomes larger than the prediction based on the F–S theory; that is, $\rho_{4.2\,\text{K}}$ should increase rapidly as the specimen thickness is decreased to a value comparable with the mean free path of the electrons moving in the [001] direction. With a further decrease in thickness, the increase in $\rho_{4.2\,\text{K}}$ will soon become a slow change and will accord well with the F–S theory, since the zone boundaries have only a slight effect on the surface scattering rate of electrons that move in directions that make a large angle with the surface normal. In the [110] direction, on the other hand, the Fermi surface intersects with the zone boundaries, but their influence is relatively small. The velocity distribution in this direction is nearly free-electron-like. Therefore, it is expected that the F–S theory will give a good description for the size effect due to a {110} surface. In fact, there is satisfactory agreement between theory and experiment for the thick {110}⟨100⟩ specimens whose side surfaces have the same orientation, {110}, as the main surface, as shown in Fig. 9.7.

9.4.2 Dependence on the Axis Orientation (Direction of Current Flow)

Figure 9.7 shows $\rho_{4.2\,\mathrm{K}}$ for the $\{110\}\langle 100\rangle$, $\{110\}\langle 111\rangle$ and $\{110\}\langle 110\rangle$ specimens, plotted as a function of $P/2A$. The main feature of the results is that $\rho_{4.2\,\mathrm{K}}$ depends strongly on the axis orientation (the direction of current flow); that is, $\rho_{4.2\,\mathrm{K}}$ increases in the following order of the directions: $\langle 110\rangle$, $\langle 111\rangle$, $\langle 100\rangle$. The degree of this anisotropy, however, decreases in the thin specimens in which the rate of scattering of conduction electrons at the specimen surface is large. As the thickness decreases, the $\rho_{4.2\,\mathrm{K}}$ values of the $\{110\}\langle 111\rangle$ and $\{110\}\langle 100\rangle$ specimens rapidly approach the $\rho_{4.2\,\mathrm{K}}$ values of the $\{110\}\langle 110\rangle$ specimens around $P/2A \sim 3\times 10^3\,\mathrm{m}^{-1}$ and $\sim 4\times 10^3\,\mathrm{m}^{-1}$, respectively. This implies that the size effect is independent of the axis orientation within the range of the experimental error.

The anisotropy of $\rho_{4.2\,\mathrm{K}}$ with respect to the axis orientation also occurs for the $\{100\}\langle 100\rangle$ and $\{100\}\langle 110\rangle$ specimens in the region $P/2A < 2\times 10^3\,\mathrm{m}^{-1}$, but its degree is small compared with the results for the $\{110\}$ specimens shown in Fig. 9.7. These results are reasonable, since the rate of scattering at a $\{100\}$ surface is larger than that at a $\{110\}$ surface.

The disappearance of the anisotropy due to the increased surface scattering suggests that the appearance of anisotropy is primarily due to enlargement of the bulk mean free path ℓ_b with improvement in the specimen purity; $\ell_\mathrm{b} \simeq 1.7\,\mathrm{mm}$ at 4.2 K in the specimens with RRR $\simeq$ 50 000. In such a situation, it is reasonable to expect that the deviation of the electron distribution induced by the application of the electric field to the equilibrium distribution can no longer be regarded as being a first order infinitesimal. The deviation of the distribution in the steady state generally depends on the direction of the electric field, due to the anisotropic structure of the Fermi surface. This leads to the idea that, in ultra-high-purity aluminum, the electron distribution function does not always retain its cubic symmetry, even though the applied electric field is small. Therefore, we would expect the anisotropic dependence of $\rho_{4.2\,\mathrm{K}}$ on the axis orientation to be a reflection of the anisotropy of the bulk resistivity ρ_b; that is, ρ_b increases in the following order of the directions: $\langle 110\rangle$, $\langle 111\rangle$ and $\langle 100\rangle$. The anisotropy of ρ_b is noteworthy in relation to electronic transport phenomena in cubic metals [20, 21]; the details will be given in the next section.

9.5 Anisotropy of the dc Electrical Resistivity in Cubic Metals

Usually, in metals, the current density $\boldsymbol{j}$ in the steady state is proportional to the electric field strength $\boldsymbol{E}$ (Ohm's law):

$$j_i = \sum_{j=1}^{3} \sigma_{ij} E_j ts, \tag{9.5}$$

where j_i and E_i are the components of $\boldsymbol{j}$ and $\boldsymbol{E}$, respectively, along the x_i-axis ($i = 1$, 2 or 3) and σ_{ij} is the ij component of a symmetric tensor $\boldsymbol{\sigma}$ of rank 2, called the *conductivity tensor*. The number of independent components of the tensor σ_{ij} is reduced because of crystal symmetry. As an example, $\boldsymbol{\sigma}$ becomes a scalar for crystals with cubic symmetry – that is, $j_i = \sigma_0 E_i$ – as shown in the following development.

Let the coordinate axes coincide with the cubic axes. Reflections in the x_2–x_3 and x_3–x_1 planes signify the transformations $x_1 \to -x_1$, $x_2 \to x_2$, $x_3 \to x_3$, and $x_1 \to x_1$, $x_2 \to -x_2$, $x_3 \to x_3$. The components of a tensor are transformed as the products of the corresponding coordinates. It is therefore clear that, in the transformations mentioned, all components σ_{ij} $(i \neq j)$ will change sign, while the other components will remain unchanged. By the symmetry of the crystal, however, all quantities that characterize its properties must remain unchanged on reflection in the plane of symmetry; then, obviously, $\sigma_{ij} = 0$ for $i \neq j$. Furthermore, rotations through an angle $\pi/2$ about the cube axes impose a restriction that $\sigma_{11} = \sigma_{22} = \sigma_{33}$. Hence $\sigma_{ij} = \sigma_0 \delta_{ij}$ (where δ_{ij} is the Kronecker delta) in crystals with cubic symmetry; that is, the conductivity tensor $\boldsymbol{\sigma}$ becomes a scalar, as in an isotropic material.

Following the symmetry argument, the dc conductivity σ_0 of a cubic metal is usually written [1]

$$\sigma_0 = \frac{\boldsymbol{j}}{\boldsymbol{E}} = \frac{e^2}{12\pi^3\hbar} S_{\mathrm{F}} \bar{\ell}\,, \tag{9.6}$$

irrespective of the detailed shape of the Fermi surface, where e is the electronic charge, $\hbar$ is Planck's constant, S_{F} is the total area of the Fermi surface and $\bar{\ell}$ $(= \ell_{\mathrm{b}})$ is the average mean free path over the Fermi surface.

In recent years, however, we have identified anisotropy of ρ_{b} at 4.2 K with respect to the current direction in single crystals of ultra-high-purity aluminum [22–25]. This anisotropy strongly correlates with the detailed structure of the Fermi surface, as will be described below.

The specimens were rectangular single-crystal strips with a $(1\bar{1}0)$ main surface. The bulk residual resistivity ρ_{b} was calculated from measured resistivity $\rho_{4.2\,\mathrm{K}}$ according to (9.3), with $p = 0$, using $\rho_{\mathrm{b}}\ell_{\mathrm{b}} = 0.82\,\mathrm{f\Omega\,m^2}$.

9.5.1 Anisotropy of the Bulk Residual Resistivity ρ_{b} at 4.2 K

Figure 9.8 shows $\rho_{\mathrm{b}}/\rho_{\mathrm{b}}^{110}$ at 4.2 K, plotted as a function of the angle θ between the crystallographic orientation [001] and the axis direction in the $(1\bar{1}0)$ plane. Here, ρ_{b}^{110} is the bulk residual resistivity along the [110] direction. As expected, the degree of anisotropy of ρ_{b} increases markedly with an increasing purity level. The characteristic features of the anisotropy of ρ_{b} are given as follows for each specimen purity level.

1. In the specimen with RRR $\simeq$ 10 000, the anisotropy of ρ_{b} is relatively weak, but full of variations: as θ increases, ρ_{b} rises around $\theta = 7$, 24 and

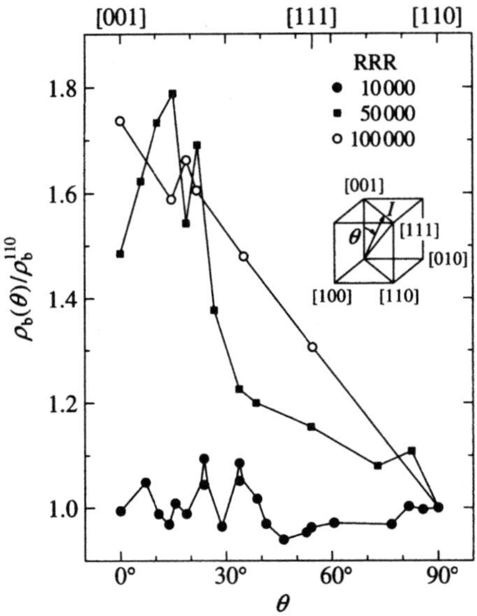

Fig. 9.8. Experimental values of $\rho_b(\theta)/\rho_b^{110}$, plotted as a function of θ for high-purity aluminum single crystals. Here, ρ_b^{110} = 2.529, 0.473 and 0.247 pΩ m for RRR ≃ 10 000, 50 000 and 100 000, respectively. The inset explains the angle θ

34°, and then gradually increases until θ reaches 90°, showing a small step-wise increase at $\theta \sim 80°$. The value along the [111] direction ($\theta = 54.7°$) is slightly smaller than those along the [001] and [110] directions.

2. In the specimens with RRR ≃ 50 000, the anisotropic behavior of ρ_b shows a salient structure over the range $0 < \theta < 30°$: as θ increases, ρ_b increases still more from the value along [001], showing a dip around $\theta = 20°$, and then rapidly decreases. With further increase in θ, the resistivity ρ_b decreases monotonically, except for a small increase around $\theta = 85°$. The values along [001] and [111] are about 50% and 20% larger than that along [110], respectively.
3. In the specimen with RRR ≃ 100 000, the θ dependence of ρ_b is approximately monotonic. The values along [001] and [111] are about 75% and 30% larger than that along [110], respectively.

The θ dependence of ρ_b correlates with the detailed structure of the Fermi surface of aluminum. To emphasize this correlation, polar plots of $\rho_b(\theta)/\rho_b^{110}$, fixed on the central $(1\bar{1}0)$ section of the Fermi surface in the extended-zone scheme, are shown in Fig. 9.9.

The strong correlation is clear, especially in the specimens with RRR ≃ 10 000. The ρ_b value rises when θ is nearly in the current direction in which the Fermi surface intersects the Brillouin zone boundary. This is surprising, since the resistivity is determined from an average mean free path over the entire Fermi surface. As the purity level of the specimens increases, the range

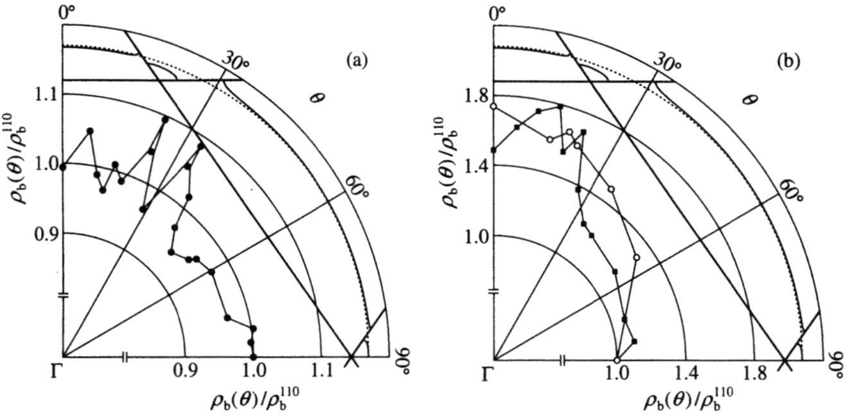

Fig. 9.9. Polar plots of $\rho_b(\theta)/\rho_b^{110}$, fixed on the central $(1\bar{1}0)$ section of the Fermi surface of aluminum in the extended-zone scheme: (**a**) RRR $\simeq$ 10 000; (**b**) RRR $\simeq$ 50 000 and 100 000. The thick lines represent the Brillouin zone boundaries and the dotted curves show the free-electron Fermi sphere. The calculations of the Fermi surface were made in the 4-OPW approximation using the pseudopotentials of Joss and Monnier [26]

of θ over which ρ_b increases markedly extends in the direction away from the intersection between the Brillouin zone boundaries and the second-zone Fermi surface. This tendency is pronounced around the [001] direction, in which the second-zone Fermi surface is nearest to the zone boundary, corresponding to the anisotropy such that $\rho_b^{110} < \rho_b^{111} < \rho_b^{001}$. These results suggest the following as a possible explanation for the anisotropy of ρ_b.

When an electric field is applied, the electron distribution is shifted along the electric field, and deviates from the equilibrium state. The deviation in the steady state generally depends on the direction of the electric field because of the anisotropic structure of the Fermi surface.

In aluminum, the structure of the Fermi surface is highly anisotropic near the Brillouin zone boundary, especially around the intersection between the Fermi surface and the zone boundary. Since the Bragg reflection occurs at this intersection, the Fermi surface must meet the zone boundary at right angles, resulting in a departure from the free-electron Fermi sphere. To satisfy this severe requirement, even a small deviation of the electron distribution near the zone boundaries, induced by the electric field, will enhance the scattering probability over the region where the influence of the zone boundaries is large. This influence should be significant around the [001] direction, as shown in Fig. 9.9.

In highly pure specimens, the electrons are still scattered by phonons at 4.2 K. However, as the phonon wave vector is small, this only gives rise to small-angle scattering, restricted to a small region of the Fermi surface. In this situation, no compensation occurs for the increase in the scattering probability due to the deviation of the electron distribution along the electric field, leading to the anisotropy of ρ_b as observed in the specimens with RRR $\simeq$ 100 000. As the purity level of the specimens decreases, impurity-scattering effects become sufficient to restore the electrons to their equilibrium distribution. Thus the region over which the scattering probability is increased decreases in size toward the intersection between the Fermi surface and the zone boundaries, and is finally limited to the intersection. This tendency leads to a steep rise in the bulk resistivity ρ_b, as observed in the specimens with RRR $\simeq$ 10 000.

9.5.2 The Temperature Dependence of the Anisotropy

Figure 9.10 shows the ratios $\rho_p^{110}(T)/\rho_p^{111}(T)$ and $\rho_p^{001}(T)/\rho_p^{111}(T)$ as a function of temperatue [23], where $\rho_p^{hkl}(T)$ means $\rho_p(T)$ $(= \rho(T) - \rho_0)$ of a specimen with the [*hkl*] axis. The deviation of the ratios from 1.0 represents the anisotropy of the temperature dependence of resistivity.

The behavior of the anisotropy of $\rho_p(T)$ shows a good correspondence with that of $\rho_{4.2\,K}$ as a function of $P/2A$ in Fig. 9.7. At low temperatures, where the rate of electron–phonon scattering is small, the anisotropy of $\rho_p(T)$ is the same as that of $\rho_{4.2\,K}$ in the region $P/2A < 3\times10^3\,m^{-1}$: $\rho_p^{110}(T) < \rho_p^{111}(T) < \rho_p^{001}(T)$. As the temperature increases to 25 K, the ratio $\rho_p^{110}(T)/\rho_p^{111}(T)$ rapidly approaches 1.0, while $\rho_p^{001}(T)/\rho_p^{111}(T)$ crosses 1.0 at about 25 K after showing a shoulder near 10 K. This implies that $\rho_p^{111}(T)$ approaches $\rho_p^{110}(T)$

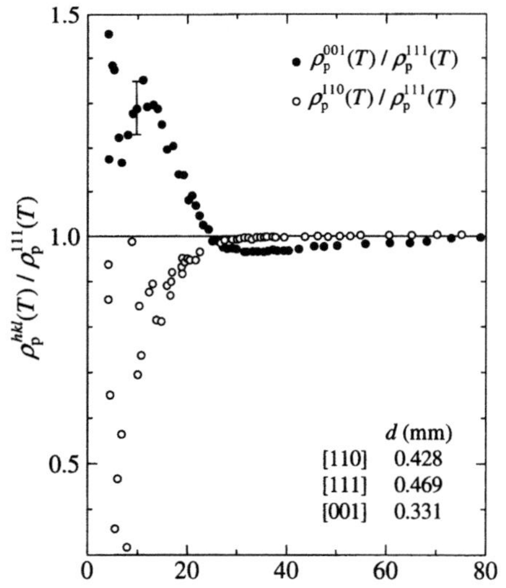

Fig. 9.10. The temperature dependence of $\rho_p^{hkl}(T)/\rho_p^{111}(T)$ for aluminum single crystals (RRR $\simeq$ 50 000) with a $(1\bar{1}0)$ surface

more rapidly than $\rho_{\mathrm{p}}^{001}(T)$ does, similar to the behavior of $\rho_{4.2\,\mathrm{K}}$ in the region $3 \times 10^3\,\mathrm{m}^{-1} < P/2A < 4 \times 10^3\,\mathrm{m}^{-1}$. With a further increase in temperature, $\rho_{\mathrm{p}}^{001}(T)/\rho_{\mathrm{p}}^{111}(T)$ takes the minimum value between 30 and 35 K and gradually approaches 1.0, corresponding to the disappearance of the anisotropy of $\rho_{4.2\,\mathrm{K}}$ for $P/2A > 4 \times 10^3\,\mathrm{m}^{-1}$.

These results, together with the purity dependence shown in Fig. 9.8, confirm that anisotropy of the bulk resistivity does not occur when scattering effects become sufficient to restore electrons to their equilibrium state. When the temperature is raised to over 20 K, normal electron–phonon scattering is more effective than anisotropic electron–phonon scattering at low temperatures and electron–electron scattering, as shown in Fig. 9.3. The change in the process of electron–phonon scattering will be a cause of the detailed behavior of $\rho_{\mathrm{p}}^{001}(T)/\rho_{\mathrm{p}}^{111}(T)$, that shows a minimum between 30 and 35 K.

In the preceeding, we have shown the details of the anisotropy of the bulk resistivity in ultra-high-purity aluminum. However, the question of which mechanisms correspond to the enhancement of the scattering probability, induced by the anisotropic deviation of the electron distribution near the zone boundary, still remains open.

9.6 Nonlinear *I*–*V* Characteristics in Ultra-High-Purity Metals

In some electronic transport problems in metals, it is usually assumed that the steady state distribution of electrons in the electric field does not depart very far from the equilibrium distribution; that is, the deviation from equilibrium is regarded as being a first order infinitesimal. In fact, in the experiments on ultra-high-purity aluminum, deviation from the equilibrium state of a conduction electron, induced by the application of the electric field, is of the order of 10^{-9} eV at most, being much smaller than the Fermi energy of ∼10 eV. However, such a small deviation may cause "new phenomena" in electronic transport when the electron mean free path is enlarged; an example is the anisotropy of ρ_{b}, as mentioned in the previous section. It is therefore necessary to measure the effect of electrons in nonequilibrium states more directly in order to confirm the "new phenomena". One possible way is to measure *I*–*V* characteristics under the action of a nonuniform electric field in ultra-high-purity metals.

In the following, we briefly describe the nonlinear *I*–*V* characteristics observed in ultra-high-purity aluminum [27].

Specimens were prepared from single-crystal plates, 0.5 mm thick, of ultra-high-purity aluminum with RRR ≃ 140 000. Tapered strips whose longitudinal sides were inclined at an angle ϕ to one another, shown in Fig. 9.11, were cut to serve as the specimens. The potential probes were placed sufficiently far inside to avoid end effects. The voltage difference V_{ij} between potential probes i and j was measured at 4.2 K as a function of current I for both the

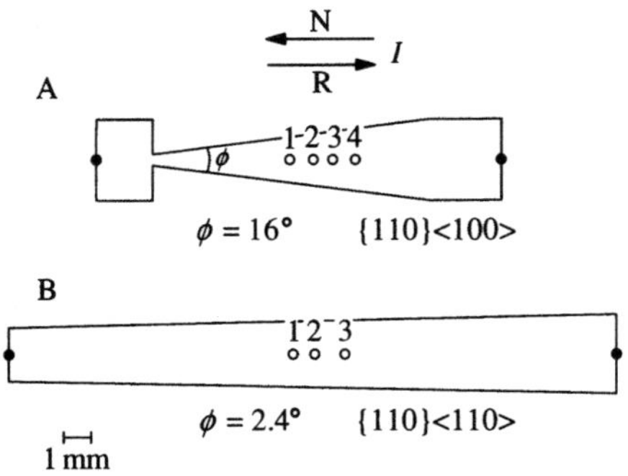

Fig. 9.11. Tapered strips A and B of ultra-high-purity aluminum single crystals with RRR ≃ 140 000. N and R indicate the current direction. Current contacts (•) and potential probes (○) are arranged along the specimen axis

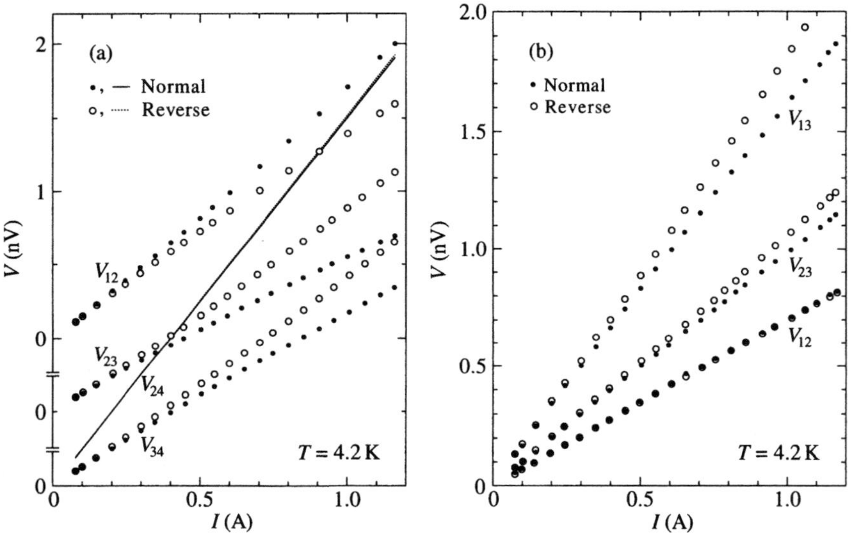

Fig. 9.12. The I–V characteristics of (**a**) a tapered strip of $\phi = 16°$, A, and (**b**) that of $\phi = 2.4°$, B. V_{ij} is the voltage difference between the potential probes i and j, measured by a standard four-probe method

normal (N) and reverse (R) current directions. The specimen was shielded with a superconducting Pb can, to avoid the influence of the Earth's field. The measuring current I was set at 0.0–1.2 A in the light of the results shown in Fig. 9.4.

Figure 9.12 shows the measured I–V characteristics of tapered strips A and B, in which the applied electric field $\boldsymbol{E}$ varies along the specimen axis. The salient feature of the results is the nonlinear behavior, which depends strongly on the position of the potential probes and on the direction of current flow. The Joule heat, the thermal EMF, the Earth's field and the magnetoresistance due to the measuring current I are not the causes of this nonlinear behavior. Therefore, the results show that Ohm's law, $\boldsymbol{j} = \boldsymbol{\sigma} \cdot \boldsymbol{E}$, does not hold for any portion of the specimen when a nonuniform electric field is applied. The details of the nonlinear behavior suggest that the redistribution of the electron density occurs to conserve the total current I flowing across any cross-sectional area of the specimen. The nonlinear response of the redistribution to the current I is a possible cause of the nonlinear I–V characteristics.

Under the action of a uniform electric field, Ohm's law apparently holds even when the electron mean free path is enlarged, since the current density is essentially uniform. In fact, the I–V characteristics of the rectangular strips, shown in Fig. 9.4, are linear within the range of the experimental error.

The nonlinear I–V characteristics as shown in Fig. 9.12 directly show the effect of electrons in nonequilibrium states. This confirms the anisotropy of the bulk resistivity in ultra-high-purity aluminum.

9.7 Ballistic Electron Transport Characteristics in Ultra-High-Purity Metals: Negative Bend Resistance

Three kinds of length, the Fermi wavelength λ_F, the phase coherence length L_φ and the mean free path ℓ_b, characterize the transport of electrons in a conductor. Here, λ_F is the de Broglie wavelength of an electron on the Fermi surface of the conductor, L_φ is the length through which the electron wave can propagate with its phase coherence and ℓ_b is the length of the linear motion of the electron conserving its initial momentum. In general, the relation between these quantities is as follows: $\lambda_F < L_\varphi < \ell_b$. The conduction system of electrons is categorized into three types according to the relative length between this characteristic length, x, and the size of the system, L, as shown in Table 9.3. The mean free path ℓ_b, which has also been discussed in previous sections, is the important index that characterizes the ballistic region where ℓ_b is larger than the system size.

Several works have been reported on electron conduction in the ballistic region. These are concerned with the bend resistance and the Hall effect in a conductor in the shape of a cross, the electron focusing in a magnetic field and the lens effect on an electron wave. All of these previous works

Table 9.3. The sizes of electron conduction systems and their characteristics

x	$L < x$	$L > x$
λ_F	Low-dimensional	Three-dimensional
L_φ	Mesoscopic	Macroscopic
ℓ_b	Ballistic	Diffusive

have been performed on one- or two-dimensional systems fabricated in the neighborhood of the heterojunction of GaAs/AlGaAs [28,29]. The mean free path ℓ_b is about 50 μm at most, even for a good heterostructure, and the size of the conduction system for the experiment is 0.1–100 μm. In contrast to this way of preparing a microstructure to realize the ballistic region, another way is to purify the material thoroughly to obtain a long mean free path ℓ_b.

In aluminum with RRR $\simeq$ 100 000, ℓ_b is already 3 mm at 4.2 K. This value is about 1000 times larger than that obtained in the heterostructure. This situation means that we can use a high-quality single crystal of the order of millimeter in size as the ballistic conduction system, where $L < \ell_b$. A sufficiently large single-crystal system assures us of a well-defined electronic structure, such as the Fermi surface, the Fermi velocity, and so on. At sufficiently low temperatures, where electron–phonon scattering is fully suppressed and $L < L_\varphi$, a mesoscopic system that is macroscopic in size will be available.

In the following, we are mainly concerned with the features of the ballistic conduction of electrons at 4.2 K, especially with the negative bend resistance observed in ultra-high-purity aluminum single crystals [30].

The specimen in the shape shown in Fig. 9.13 was prepared from a single-crystal plate of high-purity aluminum with RRR $\simeq$ 100 000 in the bulk. The orientation of the main surface of the specimen was (001). Each of the four arms a, b, c and d of the specimen was directed along ⟨110⟩. The thickness and the width of the arms were 0.5 mm and 0.4 mm, respectively. High-purity aluminum wires, 0.3 mm in diameter and 15 mm in length, were spot-welded on to each end of the four arms as current contacts and potential probes.

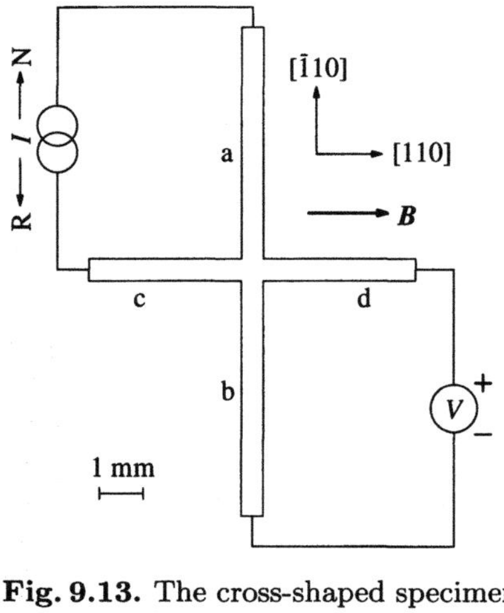

Fig. 9.13. The cross-shaped specimen of high-purity aluminum single crystal with RRR $\simeq$ 100 000. The orientation of the main surface is (001) and its thickness is 0.5 mm. N and R indicate the normal current direction and the reverse current direction, respectively

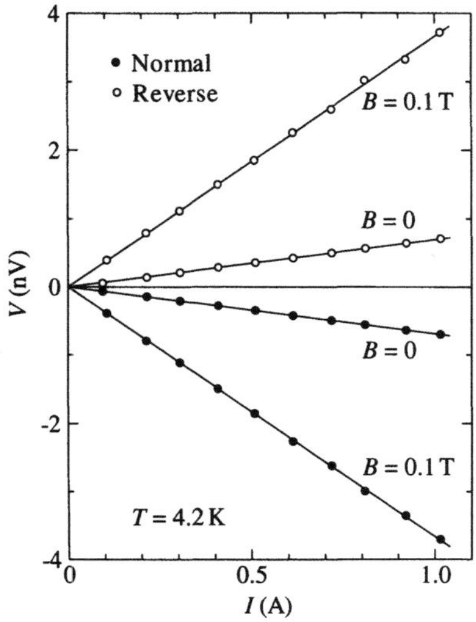

Fig. 9.14. The I–V characteristics of the cross-shaped specimen. A magnetic field B was applied parallel to the [110] arms c and d

The I–V characteristics of the cross-shaped specimen were measured at 4.2 K using the superconducting chopper amplifier. The polarity of this voltmeter was fixed, as shown in Fig. 9.13. A dc current of up to 1 A was fed through the specimen with a constant-current-regulated power supply.

Figure 9.14 shows the I–V characteristics of the cross-shaped specimen shown in Fig. 9.13 at 4.2 K, where the bulk mean free path ℓ_b of conduction electrons reaches about 3 mm in the specimen. A linear relationship between I and V, but a negative resistance, was clearly obtained for both the current directions N and R (see the plots of $B = 0$). All of the data lie on the solid lines that start from the origin, which were fitted by the least mean squares method. The thermal EMF gives $V \neq 0$ at $I = 0$. The Joule heat and the magnetoresistance due to a weak magnetic field generated by the current I cause an increase in V that is nearly proportional to I^2. Therefore, the linear behavior of the I–V characteristics in Fig. 9.14 assures us that the admixture of the thermal EMF, the influence of the Joule heat and the increase in magnetoresistance due to a self-magnetic field of the measuring current are all negligible in this experiment.

The width of the intersection of the specimen is 0.4 mm, while ℓ_b is about 3 mm. Some of the electrons shot from the current arm travel ballistically across the intersection and directly enter into the opposite arm for potential measurement without scattering. This potential arm is negatively charged. Consequently, a negative potential difference is detected for N, or vice versa for R, as pointed out in [28,29,31], if the intersection resistance is sufficiently small. The remaining electrons collide with the surface of the current arm and the intersection, or they are deflected to the crossing arms.

If this situation is true, a more effective shot of electrons into the potential arm on the opposite side is expected upon the application of a magnetic field parallel to the face of the specimen, by virtue of the cyclotron motion of electrons. Under the condition that the cyclotron radius r_c is sufficiently smaller than the thickness of the specimen, electrons are prevented from colliding with the surface. When the magnetic field is applied parallel to one of the current arms, the electrons with a screw motion go forward through this arm and the intersection without colliding with the surface. Then the negative resistance will be enhanced. This effect is also expected for the magnetic field perpendicular to the current arm. In this configuration, most of the electrons that have a trochoidal motion will enter the potential arm on the opposite side through the intersection, since only those electrons that possess a large velocity component along the magnetic field collide with the side surface perpendicular to the field.

To test this idea, we applied a weak magnetic field B of 0.1 T parallel to the [110] arms c and d, using a superconducting solenoid, as shown in Fig. 9.13. The magnetic field of 0.1 T corresponds to an r_c of 0.1 mm in aluminum on the free electron model. The magnetic field was reversed in order to eliminate a mixed Hall voltage. A linear I–V relationship and an increased negative resistance were also obtained, as shown in Fig. 9.14. The measured values of V for $B = 0.1$ T increased five times more than those for $B = 0$. These results clearly confirm the above description.

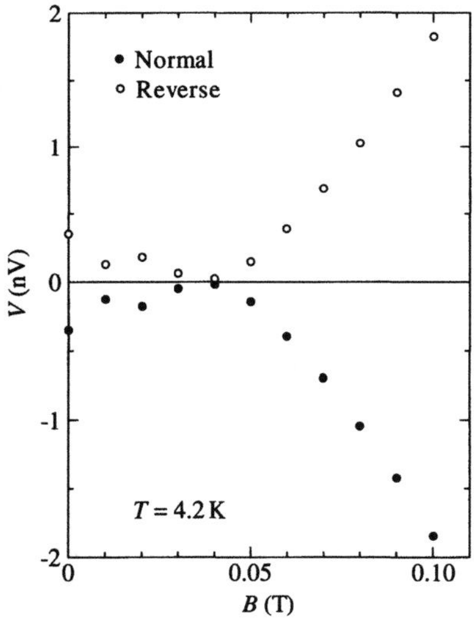

Fig. 9.15. Variations of the potential difference V with a magnetic field B parallel to [110] arms c and d of the specimen for a fixed current $I = 0.50$ A

In order to examine in more detail the effect of a magnetic field – that is, the cyclotron radius r_c – on the negative resistance observed in Fig. 9.14, we measured the potential difference V for a fixed current $I = 0.50\,\mathrm{A}$ in an increasing magnetic field B up to 0.1 T. The results are shown in Fig. 9.15.

The negative resistance is initially weakened for both current direction N and R in the low-magnetic-field region in Fig. 9.15. This is attributed to the galvanomagnetic size effect in the longitudinal configuration ($\boldsymbol{B} \parallel \boldsymbol{I}$) [32] and the transverse configuration ($\boldsymbol{B} \perp \boldsymbol{I}$) [33]. In a low magnetic field where r_c is larger than the thickness of a specimen, electrons are deflected from the initial orbit that they would take in the case of no magnetic field, and some of them collide with the specimen surface before scattering at impurities. This leads to an increase in resistivity in a low magnetic field in general. In these experiments, it results in a decrease in the number of electrons that can reach the potential arm through the intersection. Thus, the negative resistance is weakened. On the other hand, in $B > 0.05\,\mathrm{T}$, where the value of $2r_c$ is less than the size of the cross-section of the arms, the negative resistance is enhanced for both N and R. These results confirm the qualitative explanation given in the case of Fig. 9.14.

9.8 Magnetic Breakdown Oscillation in Magnetoresistance

New phenomena are also expected to appear in galvanomagnetic effects in ultra-high-purity metals. In this section, as an example, we present a new phenomenon observed on the oscillations in magnetoresistance due to magnetic breakdown.

In a magnetic field, the conduction electron in a metal is forced, by the Lorentz force, to move on a circle perpendicular to the field. This cyclotron motion is quantized due to the wave character of the electron, leading to various types of oscillatory phenomena. The well-known effects are the de Haas–van Alphen effect and the Shubnikov–de Haas effect [34]. These phenomena are controlled by the electronic structure intrinsic to the metal, and are only visible when the electron makes many cyclotron revolutions before it is scattered at impurities, crystal defects or phonons.

Another oscillatory phenomenon is the *magnetic breakdown oscillation* of the transverse magnetoresistance in a single-crystal specimen in high magnetic fields directed in the crystallographic orientation of high symmetry [35]. This phenomenon arises from the oscillation of the magnetic breakdown probability due to the requirement of phase coherence of the wavefunction on the involved electron orbits. Thus, in an ultra-high-purity metal in which the electron mean free path exceeds a few millimeters, new or prominent effects are expected to emerge in this oscillatory phenomenon.

We measured the transverse magnetoresistance at 4.2 K in ultra-high-purity aluminum single crystals with RRR $\simeq$ 100 000 in the bulk. The sizes,

Table 9.4. The specification of the specimens used for the magnetoresistance measurement

Sp.	Size (mm)	Orientation	$\rho_{4.2\,\mathrm{K}}$ (pΩ m)
AL-110	$0.4 \times 0.9 \times 16.5$	$\{110\}\langle 110\rangle$	1.44
AL-100	$2.0 \times 2.0 \times 18.2$	$\{100\}\langle 100\rangle$	0.85

the crystallographic orientations and the residual resistivities of the specimens are listed in Table 9.4.

The specimens were rigidly attached to a duralumin holder [36] at both the ends with varnish, GE 7031, so as not to move under the Lorentz force. The magnetoresistance measurements were performed at 4.2 K. Magnetic fields up to $B = 4.0\,\mathrm{T}$ were applied with a superconducting solenoid 26.0 mm in bore diameter, with the field inhomogeneity less than 0.43% within a sphere 26.0 mm in diameter. The voltage difference across specimen AL-110 was measured with the dc comparator potentiometer (Guildline, Model-9930). For specimen AL-100, the superconducting chopper amplifier [12] was used. The influence of stress was not found after the cooling cycles. The reproducibility of the resistance measurement was within the measurement error of the apparatus in a repeated run.

The magnetic fields $\boldsymbol{B}$ were directed parallel to the crystallographic orientation [001] for both specimens. The transverse magnetoresistivity $\rho(B)$ as a function of B is shown in Figs. 9.16 and 9.17 for AL-110 with the [110] current direction and Al-100 with the [100] current direction, respectively. Both sets of results display large oscillations due to magnetic breakdown.

In specimen AL-110, the oscillations with monotonically increasing amplitude are superimposed on the linear high-field magnetoresistivity, as previously observed for the [110] current direction [37, 38] and also for the [100] current direction [37–41]. The period of the oscillations, $\Delta(1/B) = (2.14 \pm 0.02) \times 10^{-2}\,\mathrm{T}^{-1}$, agrees well with previous values obtained from magnetoresistance data [38–41], and with those from de Haas–van Alphen data for the β orbit on the third-zone Fermi surface of aluminum [42, 43].

On the other hand, the $\rho(B)$ of specimen AL-100 exhibits a strange dependence on the magnetic field, as shown in Fig. 9.17. The bottoms and the tops of the oscillations increase stepwise every three oscillations with increasing magnetic fields, as guided by the lower and the upper dotted curves, respectively. The period $\Delta(1/B)$ observed for specimen AL-100 is $(2.15 \pm 0.05) \times 10^{-2}\,\mathrm{T}^{-1}$. The mean value agrees well with that obtained for specimen AL-110, but the standard deviation is large. This large deviation seems arising from the variation of period $\Delta(1/B)$ around the stepwise change of $\rho(B)$.

Here, we mention the appearance of stepwise oscillations due to magnetic breakdown in specimen AL-100 with a [100] current direction and their disappearance in specimen AL-110 with a [110] current direction. Open orbits

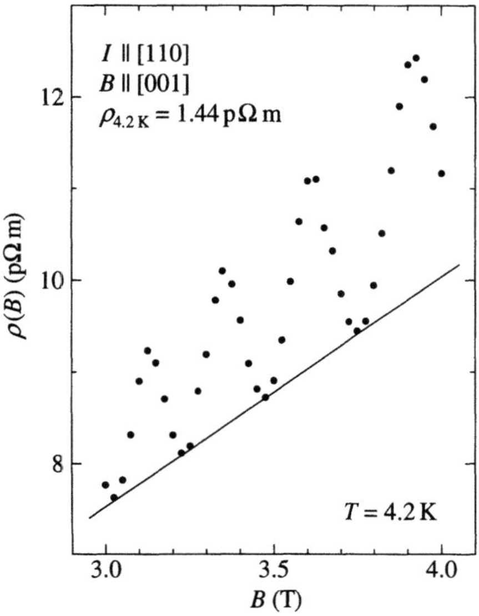

Fig. 9.16. The transverse magnetoresistivity $\rho(B)$ of specimen AL-110 with a [110] current direction. The magnetic field B is parallel to [001]

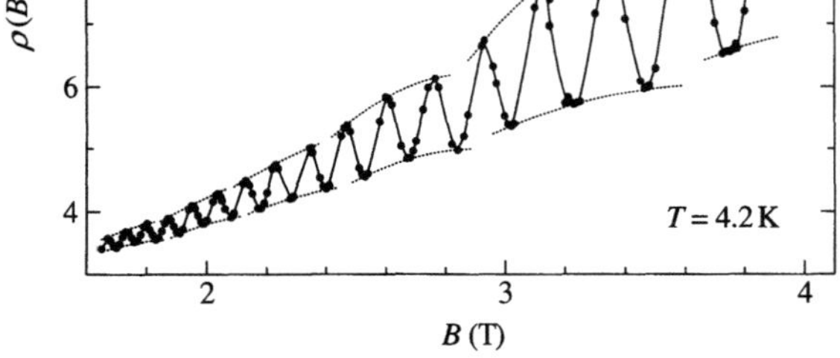

Fig. 9.17. The transverse magnetoresistivity $\rho(B)$ of specimen AL-100 with a [100] current direction. The magnetic field B is parallel to [001]. The dotted curves are drawn to illustrate the stepwise variation of $\rho(B)$

created by magnetic breakdown run in $\boldsymbol{k}$-space in the [100] and [110] directions as a whole in specimens AL-110 and AL-100, respectively [37, 40]. The [100] open orbit has only one β orbit per period as a bridge, as shown in Fig. 9.18. The junction points between the second-zone and the third-zone Fermi surfaces on this open orbit align in the [100] direction. Under this rel-

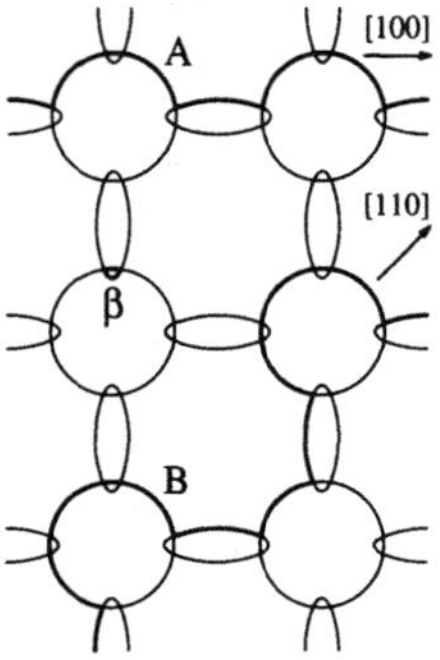

Fig. 9.18. Open orbits in the (001) section of the Fermi surface of aluminum. A and B denote [100] and [110] open orbits, respectively. The β orbit is exaggerated for clarity

atively simple condition, the phase coherence of electron waves will be persistently maintained for a sufficiently pure specimen. Thus, a new quantum interference effect could be observed in the form of the stepwise oscillations in the sufficiently pure specimen used in this experiment.

On the other hand, the [110] open orbit has two bridges perpendicular to each other in a cycle. Unless an applied magnetic field $\boldsymbol{B}$ lies on the (110) plane, the junction points of two bridges will have different characteristics, since these junction points are not placed at identical points relative to a plane perpendicular to $\boldsymbol{B}$. Additionally, the path that electrons take on the [110] open orbit in a cycle is longer than that on the [100] open orbit. As a result, the phase coherence of the electron waves must be lost, and stepwise oscillations could not be observed in the specimen with a [110] current direction.

9.9 Summary

On the way to realizing truly pure aluminum, we have reached the purity of its bulk residual resistance ratio of 150 000. In this material, the mean free path of a conduction electron becomes about 4.5 mm at low temperatures. In single-crystal specimens prepared from such ultra-high-purity aluminum, we have shown new peculiar phenomena in electronic transport: the dependence of the bulk resistivity and the size effect on the crystallographic orientation, the nonlinear current–voltage characteristics, the negative bend resistance, the stepwise variation of magnetic breakdown oscillations, and so on. The deviation from the equilibrium state of an electron is of the order of neV at most, compared with the Fermi energy of the order of 10 eV. Thus, it is often concluded that nothing special happens. However, the experimental results presented above show that unexpected phenomena may also appear in "macroscopic" metals when they enter in the state of the ballistic region or the mesoscopic system. Ultra-high-purity metals are still unknown objects.

References

1. J.M. Ziman: *Principles of the theory of solids*, 2nd edn. (Cambridge University Press, Cambridge 1972)
2. J. Bass: Adv. Phys. **21**, 431 (1972)
3. E. Hashimoto, Y. Ueda: Mater. Trans. JIM **35**, 262 (1994)
4. S. Kawata, T. Kino: J. Phys. Soc. Japan **39**, 684 (1975)
5. R.J.M. van Vucht, H. van Kempen, P. Wyder: Rep. Prog. Phys. **48**, 853 (1985)
6. G. Grimvall: *The Electron–phonon interaction in metals* (North-Holland, Amsterdam 1981)
7. J. de Launay: *Solid state physics*, Vol. 2 (Academic Press, New York 1956) p. 233
8. R. Stedman, L. Almqvist, G. Nilson: Phys. Rev. **162**, 549 (1967)
9. C. Kittel: *Introduction to solid state physics*, 7th edn. (John Wiley, New York 1996)
10. F.R. Fickett: Cryogenics **11**, 349 (1971)
11. J.R. Sambles, K.C. Elsom, G. Sharp-Dent: J. Phys. F: Metal Phys. **11**, 1075 (1981)
12. Y. Ueda, H. Hosoda, T. Kino: J. Phys. Soc. Japan **57**, 3896 (1988)
13. E.H. Sondheimer: Adv. Phys. **1**, 1 (1952)
14. R. Risnes, V. Sollien: Phil. Mag. **20**, 895 (1969)
15. E. Hashimoto, Y. Ueda: J. Phys.: Condens. Matter **10**, 6727 (1998)
16. J.R. Sambles, K.C. Elsom: J. Phys.: Met. Phys. **10**, 1487 (1980)
17. S.B. Soffer: J. Appl. Phys. **38**, 1710 (1967)
18. H. Yamasaki, Y. Ueda, E. Hashimoto: J. Phys.: Condens. Matter **8**, L399 (1996)
19. N.W. Ashcroft: Phil. Mag. **8**, 2055 (1963)
20. W.A. Harrison: *Solid state theory* (McGraw-Hill, New York 1970)
21. P.L. Rossiter: *The electrical resistivity of metals and alloys* (Cambridge University Press, Cambridge 1986)
22. E. Hashimoto, Y. Ueda, H. Tamura, T. Kino: J. Phys. Soc. Japan **62**, 4178 (1993)
23. Y. Ueda, H. Tamura, E. Hashimoto: J. Phys.: Condens. Matter **6**, L637 (1994)
24. Y. Ueda, H. Tamura, E. Hashimoto: J. Phys.: Condens. Matter **7**, L361 (1995)
25. E. Hashimoto, Y. Ueda, H. Yamasaki: J. Phys. Soc. Japan **66**, 1253 (1997)
26. W. Joss, R. Monnier: J. Phys. F: Metal Phys. **10**, 9 (1980)
27. Y. Ueda, E. Hashimoto: Kotai Butsuri (Solid State Physics) **32**, 656 (1997) (in Japanese)
28. Y. Takagaki, K. Gamo, S. Namba, S. Ishida, S. Takaoka, K. Murase, K. Ishibashi, Y. Aoyagi: Solid State Communs. **68**, 1051 (1988)
29. Y. Takagaki, K. Gamo, S. Namba, S. Takaoka, K. Murase, S. Ishida, K. Ishibashi, Y. Aoyagi: Solid State Communs. **69**, 811 (1989)
30. Y. Ueda, K. Sakoh, E. Hashimoto: J. Phys. Soc. Japan **65**, 3442 (1996)
31. M. Büttiker: IBM J. Res. Develop. **32**, 317 (1988)
32. Yi-Han Kao: Phys. Rev. **138**, A1412 (1965)
33. D.K.C. MacDonald, K. Sarginson: Proc. R. Soc. Lond., Ser. A **203**, 223 (1950)
34. N.W. Ashcroft, N.D. Mermin: *Solid state physics* (Holt, Rinehart and Winston, New York 1976)

35. A.B. Pippard: *Magnetoresistance in metals* (Cambridge University Press, Cambridge 1989)
36. Y. Ueda, H. Hosoda, T. Kino: J. Phys. Soc. Japan **54**, 3858 (1985)
37. W. Kesternich, C. Papastaikoudis: J. Phys. F: Metal Phys. **7**, 837 (1977)
38. H. Hosoda, Y. Ueda, T. Kino: J. Phys. Soc. Japan **56**, 2858 (1987)
39. R.J. Balcombe, R. A. Parker: Phil. Mag. **21**, 533 (1970)
40. H. Sato, K. Shimizu, K. Yonemitsu: J. Low Temp. Phys. **35**, 41 (1979)
41. R. Fletcher: Phys. Rev. B **28**, 1721 (1983)
42. C.O. Larson, W.L. Gordon: Phys. Rev. **156**, 703 (1967)
43. P.T. Coleridge, P.M. Holtham: J. Phys. F: Metal Phys. **7**, 1891 (1977)

in low-alloy steels that contain residual impurities such as phosphorus, tin and antimony [3–10]. In order to elucidate the mechanism of temper embrittlement, in which intergranular fractures are observed along prior austenitic boundaries, much effort has been made to conduct compositional analysis of these boundaries. It has been successfully proved that the impurities are segregated to prior austenitic boundaries, and the influence of segregation of the impurities on mechanical properties has been intensively investigated by means of a combination of mechanical tests and grain boundary analysis. Besides temper embrittlement in low-alloy steels, grain boundary segregation is considered to play a very important role in the microstructural evolution of metals and alloys, although this role is not clarified sufficiently yet.

In addition to the establishment of preparation techniques for high-purity metals and well-controlled alloys, surface analytic methods such as X-ray photoelectron spectroscopy (XPS) and Auger electron spectroscopy (AES) have advanced remarkably in the past few decades [14, 15]. They have enhanced systematic studies of surface and grain boundary segregation of impurities. XPS is mainly used to analyze the chemical composition and state of the solid surface, whereas AES enables us to observe microstructures and to analyze the composition of local areas in metals and alloys. These methods have been employed to provide definite evidence of surface and grain boundary segregation of impurities in metals and alloys, although AES is applicable only to intergranular planes of a sample fractured in an ultra-high vacuum. This work has indicated that the experimental results include some factors that are not taken into account in simple thermodynamic models, but may result from interactions between impurities, the effect of microstructures on segregation in polycrystalline samples and so on. In order to understand the parameters that influence practical surface and grain boundary segregation, systematic experimental studies have been performed in polycrystalline high-purity iron-based alloys. Studies of grain boundary segregation in Fe–Si bicrystals with specific misorientations have provided important information about the grain boundary structure.

With this background of studies on surface and interfacial phenomen a in mind, this chapter gives an overview of experimental and theoretical studies on surface segregation and grain boundary segregation of impurities, which have mainly been made on high-purity iron-based alloys by means of careful surface analyses. The characteristic features of impurities segregated at the surface and interface are comprehensively discussed. In addition, a native oxide layer formed on the metal surface is also described, to consider the effect of surface composition on surface stability.

10.2 Thermodynamic Features of Surfaces and Interfaces

The thermodynamic properties of surfaces and interfaces are represented by free energy, which is related to a measure of the unsatisfied atomic bonding in a relatively disordered structure. The free energy depends on the elemental composition at the surface and interface; that is, the positive or negative segregation of solute atoms. This section briefly describes the relationship between the free energy and segregation, in order to interpret experimental results thermodynamically.

10.2.1 Interfacial Energy

The thermodynamic relationship between the surface free energy and the chemical composition of the surface in a multi-component solid is given by the Gibbs adsorption isotherm. It is obtained by replacing the surface tension by the interfacial free energy γ. In the dilute case of a bulk solute molar concentration $X_c \ll 1$, this is expressed by

$$\left(\frac{\mathrm{d}\gamma}{\mathrm{d}\ \ln X_c}\right)_T = -\mathbf{R}T\Gamma\,, \tag{10.1}$$

where Γ is the interfacial excess of the solute species, expressed in $\mathrm{mol\,m^{-2}}$, $\mathbf{R}$ is the gas constant and T is temperature. This has been used experimentally to relate changes in composition of the bulk to the change in both the free energy and the surface excess composition for surfaces or interfaces in solids and liquids. In practice, it has been shown for liquid and solid surfaces and grain boundaries in pure iron that the interfacial free energy varies as a function of the bulk content [3–16]. The data given for these interfaces are analyzed by application of (10.1) to the results pertaining to the specific temperature of the experiment. The amount of interfacial excess expressed by Γ is transformed into equivalent monoatomic layers, and the surface composition is shown as a function of the bulk content. Based on the Gibbs adsorption isotherm, the dependence of the interfacial free energy and the surface and grain boundary excesses have been discussed in the case of iron containing dilute solutes [3–16].

10.2.2 Interfacial Segregation

The interfacial excesses mentioned above correspond to segregation of solute atoms observed in experiments. Segregation is classified into equilibrium segregation and nonequilibrium segregation. Equilibrium segregation takes place due to inhomogeneities in a solid, where there is a higher probability that solute atoms will occupy the places with a lower free energy. Since the interfacial structure is different from that of the matrix, there is a difference in free energy between these two locations, which is the driving force for segregation. At a given temperature, there is a unique concentration of solute atoms for each of these sites at equilibrium. On the other hand, non equilibrium segregation depends on rate processes, and generally disappears as time passes. This type of segregation may occur in different processes. In practice, equilibrium segregation is the most fundamental process in dealing with interfacial segregation of solute atoms. The expression for equilibrium segregation of solute atoms at grain boundaries, which is representative of the interface, was derived by McLean by statistical mechanics [17]. It is given by

$$\frac{X_I^{\phi}}{X_I^{0\phi} - X_I^{\phi}} = \frac{X_I}{1 - X_I} \exp\left(\frac{-\Delta G_{\mathrm{gb}}}{kT}\right), \tag{10.2}$$

where k is the Boltzmann constant, T is the equilibrium temperature, and ΔG_{gb} is the free energy of segregation per mole of solute. In the notation of Hondros and Seah [4], X_I is the bulk solute molar fraction, X_I^{ϕ} is the fraction covered with segregated atoms, and $X_I^{0\phi}$ is the fraction of the grain boundary monolayer available for segregated atoms at saturation. As this expression is similar to that for adsorption given by Langmuir, it is often referred to as the Langmuir–McLean equation. From this relation, it is understood that the segregated amount increases as the bulk solute content, X_I, increases, or as the temperature, T, decreases. Therefore, these two factors are considered to easily control the segregated amount of solute atoms.

Figure 10.1 schematically represents a relative relationship of the free energies of equilibrium segregation versus solid solution, together with precipitation and free surface segregation. Here, ΔG_{sol}, ΔG_{surf} and ΔG_{ads} are the free energies of precipitation, surface segregation and adsorption of a solute atom, respectively. ΔG_{sol} can be obtained by measuring the temperature dependence of the solubility limit of solute atoms. ΔG_{surf} and ΔG_{ads}

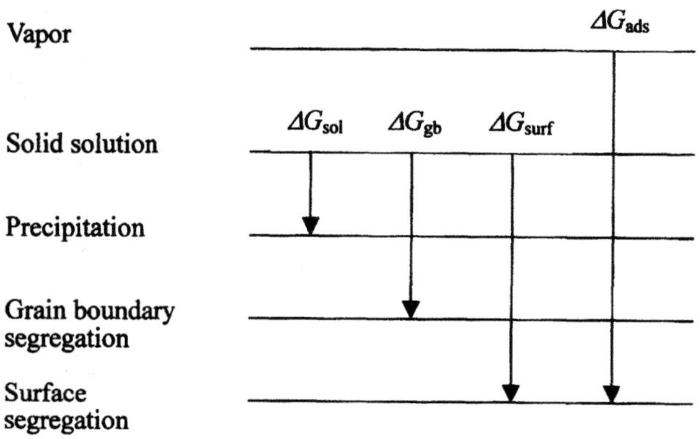

Fig. 10.1. A free energy diagram for solute atoms segregated at surface and grain boundaries

are parameters associated with surface behavior, which is estimated from equilibrium experiments in an ultra-high vacuum.

Guttmann expressed interfacial segregation in a multi-component system, considering interactions between different foreign atoms [18–21]. He applied a regular solution model to grain boundary segregation in a multi-component system, as given by

$$\frac{X_I^{\phi}}{X_I^{0\phi} - \Sigma_J^{M-1} X_J^{\phi}} = \frac{X_I}{1 - \Sigma_J^{M-1} X_J^{\phi}} \cdot \exp\left(\frac{-\Delta G_I}{\mathbf{R}T}\right), \tag{10.3}$$

where X_I^{ϕ} and X_I are the molar fractions of elements at the interface and in the bulk, respectively. The free energy of segregation, ΔG_I, for the multi-component system is given by

$$\Delta G_I = \Delta H_I^0 - T\Delta S_I^0 - 2\alpha_{MI}(X_I^{\phi} - X_I) + \sum\nolimits_{J \neq I}^{M-1} \alpha_{IJ}{}'(X_I^{\phi} - X_I), \tag{10.4}$$

where ΔH_I^0 and ΔS_I^0 are the standard molar enthalpy and entropy of segregation in a dilute binary M–I system, and α_{MI} and α'_{IJ} are the parameters for binary interaction between atoms I in matrix M, and for ternary interaction between atoms I and J in the same matrix, respectively. Some thermodynamic expressions, taking into account the interaction of impurities at high segregation level and so on, have also been proposed [4]. The above models are applicable to experimental results on interfacial segregation.

10.2.3 Interfacial Structure

Up to now, it has sometimes been recognized that interfacial phenomena differ from interface to interface in polycrystalline metals and alloys. This suggests that interfacial phenomena depend on the interfacial atomic structure, and therefore the possible atomic structures should be implicitly considered when the thermodynamic models mentioned above are discussed.

Figure 10.2 illustrates the atomic heterogeneity found on the surface of a single crystal. The practical solid surface may contain many kinds of defects, such as a surface vacancy, a kink, an adatom on a terrace, and a step adatom. There are a number of surface crystallographic orientations for grains in polycrystalline metals, and the density of the defects is considered to depend on this orientation. It is to be noted that the variation in the structure on the surface influences the total thermodynamics of segregation of solute atoms, although it is difficult to evaluate all surface structures with segregated atoms.

The atomic structure of grain boundaries is more complicated than that of surfaces, since boundary planes include parameters of misorientation between two grains, plane orientation and so on. Nevertheless, during the past few decades, studies on the structure of grain boundaries have progressed remarkably due to the provision of a large number of atomistic models and

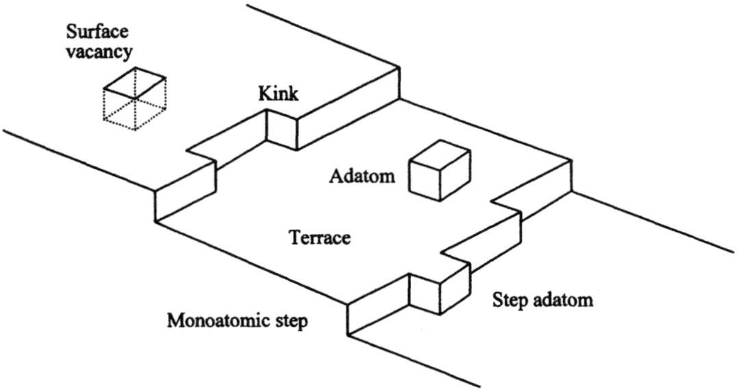

Fig. 10.2. A schematic representation of the atomic heterogeneity found on surfaces

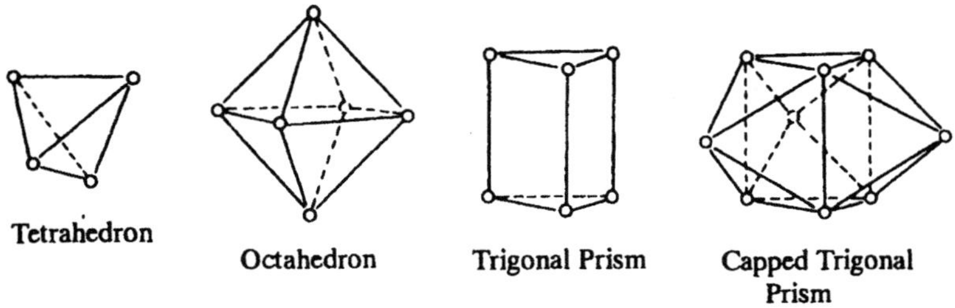

Fig. 10.3. Compact polyhedra of atoms found in grain boundaries

simulations of grain boundaries [22–26]. Indeed, it has been shown that there are a variety of types of atomistic configurations; for example, several compact polyhedra in polycrystalline materials, as illustrated in Fig. 10.3. Therefore, it is readily thought from the structural models that a large number of segregation structures exist in grain boundaries of polycrystalline metals.

10.3 Measurement of Surface and Interface Compositions

The interfacial energy and the amount of interfacial excess – that is, the segregated amount at the surface or grain boundaries cited above – are evaluated experimentally. With regard to impurity behavior on surface and grain boundaries, the effect of impurities on surface energy has been investigated in earlier work [16], where the surface energy γ_s was measured by the zero creep method. Furthermore, the grain boundary tension γ_b obtained by the method is given by

$$\gamma_b = 2\gamma_s \cos\left(\frac{\theta}{2}\right) , \tag{10.5}$$

where θ is the dihedral angle formed by the groove between two grains [16]. Thus, the surface and grain boundary energies were estimated as a function of the solute atoms, and segregation at the interfaces was discussed on the basis of these results.

On the other hand, surface segregation and grain boundary segregation are directly evaluated by analyzing the surface prepared in an ultra-high vacuum (UHV) and intergranular planes on a sample fractured in UHV to avoid adsorption and contamination from the atmosphere. AES and XPS have been mainly used to detect interfacial segregation. In order to discuss the practical data obtained in AES and XPS measurements, a few results are shown here.

Figures 10.4a–c show Auger spectra from an Ar^+ ion sputtered surface, a silicon segregated surface and a surface covered by tin on the (011) surface of an Fe–3 mass% Si with a small amount of tin, respectively [27]. The sample was heated up to 723 K under UHV. Clear segregation of silicon and tin on the surface is found upon heating. AES is often combined with ion sputtering, to obtain depth information on a surface layer. This information indicates that tin covers the silicon segregated surface of the sample given in Fig. 10.4c. However, clear depth profiles are not obtained by means of the combination of

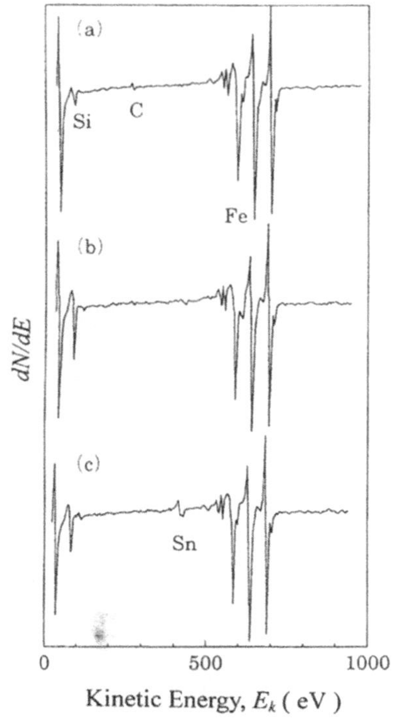

Fig. 10.4. Auger spectra from (**a**) a sputtered surface, (**b**) a silicon segregated surface and (**c**) a tin segregated surface in the iron–silicon [27]

AES and ion sputtering when the thickness of the surface layer is comparable to the escape depth of the Auger electrons. It should be remarked that ion sputtering may induce preferential sputtering of constituent elements and embedding of ions.

In order to overcome the problem of preferential sputtering, a nondestructive method is required. Angle-resolved XPS (AR-XPS) is an effective method for nondestructively characterizing very thin layers on the solid surface. X-ray photoelectrons are detected by changing the take-off angle against the specimen surface in this method. For instance, Figs. 10.5a,b show the take-off angle dependence of the concentration of iron and chromium obtained by AR-XPS in Fe–13 mass%Cr and Fe–25 mass%Cr, heated to 973 K, respectively [28]. The concentration was estimated from the measured XPS intensity data coupled with the sensitivity of the factors of each element. The experimental results are denoted by symbols. The decrease in the concentration of chromium with an increasing take-off angle indicates that chromium is concentrated into the surface region by heating. The dependences on the take-off angle of the concentrations for chromium and iron, calculated using a model of the layered structure of segregation, are plotted as lines, as shown in Figs. 10.5a,b. This good fit enables us to provide an effective thickness

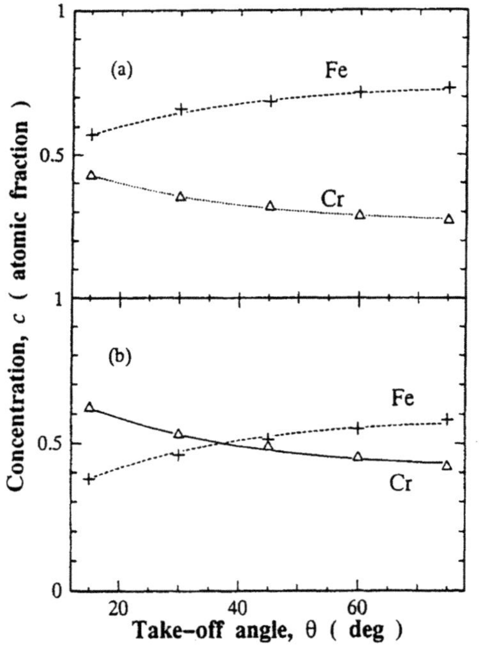

Fig. 10.5. The concentration of iron and chromium versus the take-off angle for the chromium surface segregated (**a**) Fe–13 mass% Cr and (**b**) Fe–25 mass% Cr alloys [28]

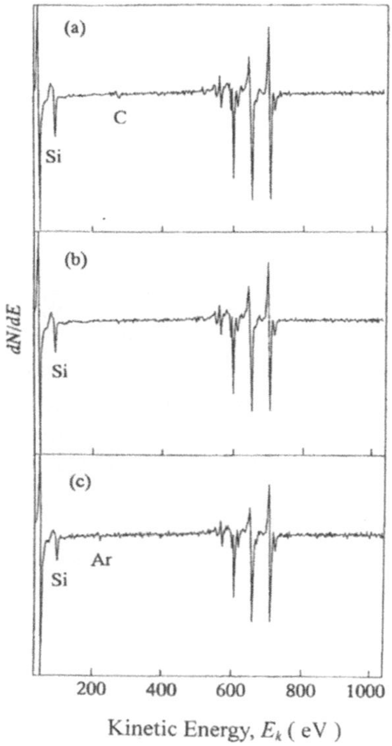

Fig. 10.6. Auger spectra from (**a**) an intergranular plane, (**b**) a transgranular plane and (**c**) an intergranular plane with Ar^+ ions in the iron–silicon with a $\Sigma 9$ boundary [29]

and concentration of the segregated layer. It is estimated that the effective thickness of the segregated layer does not exceed 1 nm, and that the surface segregated amount of chromium is a few times more than the bulk concentration. The results may indicate that the enriched chromium atoms may actually be distributed within not one atomic layer (ca. 0.25 nm) but a few atomic layers.

Contrary to surface segregation, grain boundary segregation is investigated by analyzing the intergranular fracture surface. Figures 10.6a–c demonstrate Auger spectra from an intergranular fracture plane, a transgranular fracture plane and intergranular fracture plane sputtered by Ar^+ ions in an Fe–3 mass% Si with a $\Sigma 9$ boundary, respectively [29]. From a comparison of intergranular and transgranular fractured plane spectra, segregation of carbon and some silicon at boundaries is found. It is also suggested that preferential sputtering of silicon on the surface occurs by applying Ar^+ ion bombardment. When several grain boundaries are contained in polycrystalline metals and alloys, different segregated amounts are often observed.

Thus, surface analytic methods are carefully employed to detect segregated atoms at surfaces or interfaces, paying particular attention to contamination by residual gases in UHV. These methods have given us the most useful information on the surface and grain boundary segregation in different samples, which has been discussed by comparing thermodynamic models and their atomic structures.

10.4 Surface Behavior

10.4.1 Surface Segregation

Surface segregation has been investigated mainly in a lot of iron-based alloys: iron–nonmetallic element systems, such as Fe–C [30–32], Fe–N [32, 33], Fe–Si [34–38], Fe–P [39], Fe–Sn [40–44] and Fe–Sb [45–47]; and iron–metallic element systems, such as Fe–Ti [48], Fe–Cr [48–58] and Fe–Mo [48]. AES or XPS has been mainly used in these experiments, where some interesting results have been obtained by finding systematic compositional changes on the specimen surface. However, the amount of surface segregation for an

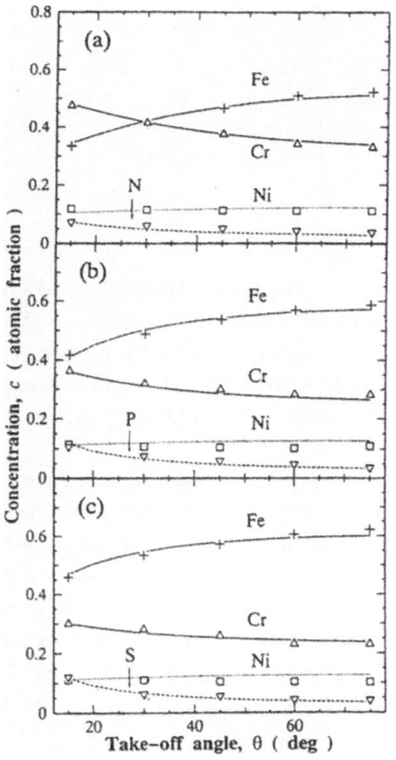

Fig. 10.7. The concentration of iron, chromium, nitrogen, phosphorus and sulfur versus the take-off angle for Fe–19 mass% Cr–15 mass% Ni specimens with (**a**) nitrogen, (**b**) phosphorus and (**c**) sulfur [71]

element has often seemed to be affected by other coexisting impurities, since the amount and the state of impurities contained in iron or steel samples were not controlled sufficiently. With regard to passive films formed on the iron surface, surface oxide layers have been also investigated by surface analytic techniques [59–66]. In these studies, the experimental results appear to be very sensitive to the surface preparation of the samples used.

Thus, in order to understand the intrinsic behavior of each element on the metal surface, characterization is required with respect to surface segregation in well-defined samples. Several AR-XPS studies were made on the surface behavior of impurity and alloying elements in high-purity iron-based alloys and well-controlled iron alloys [67–71], as demonstrated in the Fe–Cr system shown in Fig. 10.5. Figures 10.7a–c show the variation of the take-off angle as a function of the concentration for iron, chromium, nickel, nitrogen, phosphorus and sulfur in Fe–19 mass%Cr–15 mass%Ni alloys with 0.06 mass%N, 0.08 mass%P and 0.007 mass%S, which were heated at 973 K [71]. The decrease in concentration of nitrogen, phosphorus, sulfur and chromium with increasing take-off angle indicates that these nonmetallic elements, as well as chromium, concentrate on a surface layer. It may be stressed that the dependence of the concentration of phosphorus and sulfur on take-off angle is larger than that for chromium and nitrogen. This suggests that phosphorus and sulfur are segregated in a top surface layer, relative to chromium and nitrogen.

The fit of calculated curves to the experimental results in Fig. 10.7 provides the effective thickness and concentration of the segregated layer, as shown in Fig. 10.8. The effective thickness of the segregated layer of chromium is comparable to that for the high-purity iron–chromium steels shown in Fig. 10.5 [28], and nitrogen seems to be correlated with the chromium segregation. Phosphorus and sulfur are clearly segregated in the top surface. It is interesting to note in Fig. 10.8 that the degree of chromium segregation is strongly affected by the surface segregation of phosphorus and sulfur. This may be attributed to the replacement of chromium with phosphorus or sulfur in the surface layer; this is referred to as site competition on the specimen surface. It is also suggested in Fig. 10.8 that the segregated layers of phosphorus or sulfur are thinner than that of chromium, and they are considered to be more surface-active than chromium. Thus, the decrease in chromium concentration shown in Figs. 10.8b,c, as compared to the result shown in Fig. 10.8a, may take place through this surface-active behavior of sulfur and phosphorus. These results may include the effect of different surface orientations in polycrystalline specimens on the surface segregated layers. Nevertheless, it may be safely said that the present semi-quantitative estimation of the effective thickness and concentration is fairly acceptable for expressing the characteristic features of the surface behavior of these elements.

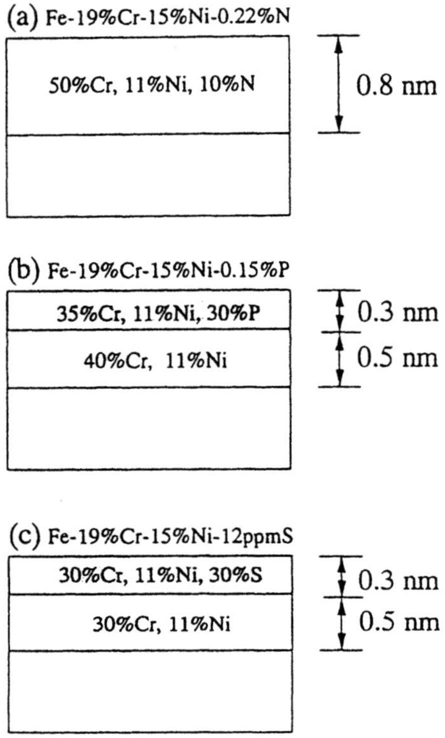

Fig. 10.8. A schematic diagram for the layered structures of a surface segregated layer of Fe–19 mass% Cr–15 mass% Ni specimens with (**a**) nitrogen, (**b**) phosphorus and (**c**) sulfur. The thickness of the segregated layers formed on the surface and the effective concentration (in at% or atppm) are denoted in the figures [71]

10.4.2 Native Oxide Layer

When a metal surface is exposed to reactive gases such as oxygen and air, a thin surface layer – for example, oxide layer – forms on the surface. Since the surface stability of metals and alloys against such oxidation is controlled by surface conditions, the relationship between the native oxide layers and the surface conditions has been systematically characterized by AR-XPS [72,73]. A few examples of the formation of the native oxide layers are shown below.

Figure 10.9 shows the thickness values of the oxide layers and contaminated layers formed on ultra-high-purity iron and copper, which were sequentially obtained by AR-XPS, as a function of exposure time [73]. The oxide layer thickness for ultra-high-purity iron is found to be almost unchanged by the exposure time, whereas the oxide layer in ultra-high-purity copper becomes considerably thicker under the present exposure conditions. These results are consistent with the fact that spectral changes in XPS by air exposure are remarkable in ultra-high-purity copper. In addition, the thickness

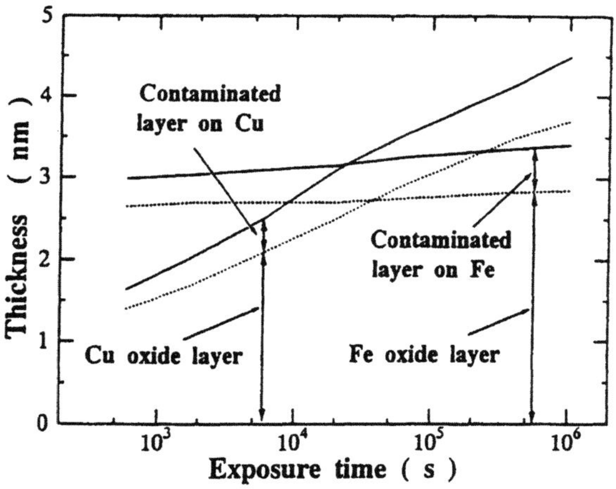

Fig. 10.9. The thickness of the oxide layers and the contaminated layers formed on the surface of ultra-high-purity iron and copper, as a function of air exposure time [73]

of contaminated layers increases with increasing exposure time in both ultra-high-purity iron and copper. The thickness values of the contaminated layers for ultra-high-purity iron and copper appear to be almost independent of the species of the substrate, although there is a difference in detail. The kinetics of these oxidation processes appear to follow a logarithmic relationship, which generally prevails at low temperature [74], rather than a linear and parabolic relationship. Also, the experimental results indicate that these metals are rapidly oxidized by a short exposure to air at room temperature, to form a stable film a few nanometers thick, as observed in the initial stage of slow oxidation.

The addition of chromium to iron and steel is well known to be effective in preventing their oxidation. AR-XPS is convenient for obtaining systematic information on the effect of bulk chromium and surface segregated chromium on native oxide layers formed on the surface of iron–chromium alloys. Figure 10.10 summarizes the thickness of oxide layers formed on the iron–chromium alloy surface versus the bulk chromium concentration. All samples were oxidized by air exposure at room temperature for 600 s. When the surface concentration is the same as that in the bulk, no surface segregation of chromium is detected. The thickness of an oxide layer decreases with increasing concentration of the bulk chromium. In addition, even if the bulk chromium concentration is low, enrichment of chromium by surface segregation appears to induce the decrease in thickness of the oxide layer. More effective reduction of the native oxide formation is found in Fe–Cr–Ni–N systems [69, 71]. This may result from the mutual segregation of chromium and nitrogen on the surface.

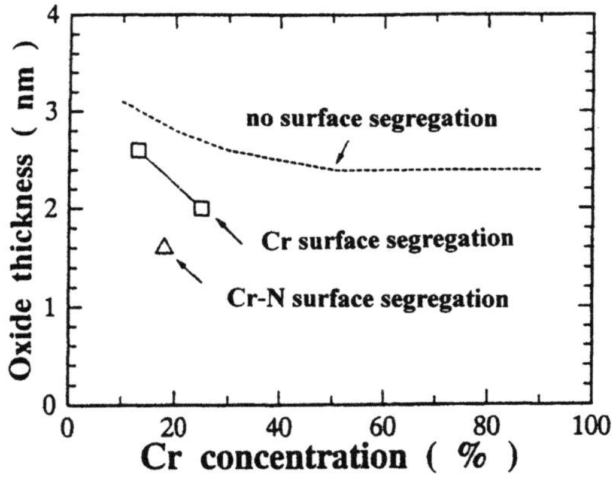

Fig. 10.10. The oxide thickness formed on the surface as a function of the bulk chromium concentration mass% in iron–chromium alloys [72]

The effect of chromium surface segregation on the reduction in thickness of the native oxide layers is also observed in the XPS spectra, as shown in Figs. 10.11 and 10.12. AR-XPS spectra of Fe 2p were measured for an Fe–18 mass%Cr–12 mass%Ni–0.08 mass%N alloy exposed to air after ion sputtering and the identical alloy exposed to air after surface segregation. It was shown in the results that the relative intensities of the metallic peaks (707 and 721 eV) to the oxide peaks (711 and 725 eV) in the Fe 2p spectra increase with the take-off angle. Since XPS signals measured at higher take-off angles are known to arise mainly from a deep part of the specimens, these results indicate that the matrix of the specimens is considered to be metallic. They also mean that the thickness of the oxide layer formed on the surface is comparable to the mean free path of X-ray photoelectrons. Thus, it is suggested that the surface properties of iron–chromium alloys can be controlled by changing the surface concentration near the surface region and, for such purposes, it would be preferable to modify the process conditions for the sample surface, such as atmospheric conditions and annealing temperature.

10.5 Interfacial Behavior

Impurity segregation at grain boundaries in polycrystalline metals and alloys often shows a complicated behavior, which cannot be explained by simple thermodynamic models. It may be attributed to a characteristic feature of the segregation of each solute atom, the interaction between solute atoms, grain boundary structure and so on. In order to discuss this behavior, several experiments on segregation of solute atoms in polycrystalline high-purity iron-based alloys and iron–silicon bicrystals with impurities are described in the present section.

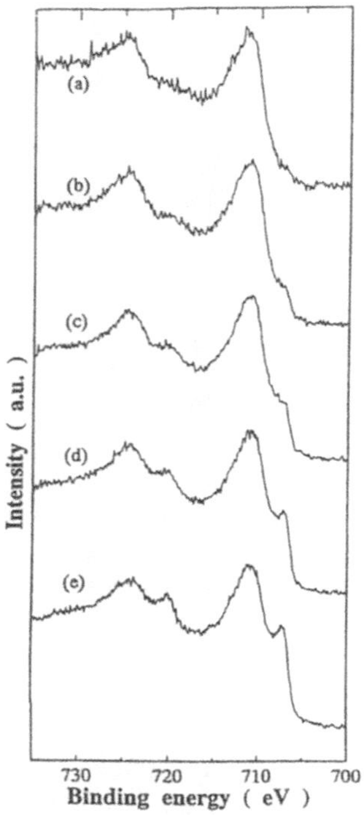

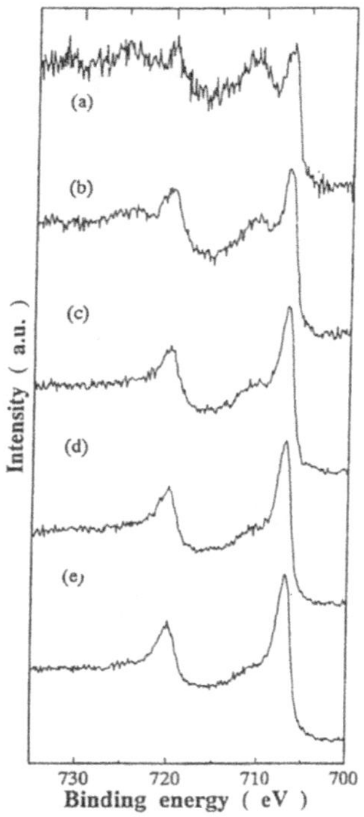

Fig. 10.11. (*left*) Fe 2p AR-XPS spectra obtained in the take-off angle of (**a**) 15°, (**b**) 30°, (**c**) 45°, (**d**) 60° and (**e**) 75° for Fe–18 mass% Cr–12 mass% Ni–N exposed to air after sputtering [72]

Fig. 10.12. (*right*) Fe 2p AR-XPS spectra obtained in the same take-off angle as in the case of Fig. 10.11 for Fe–18 mass% Cr–12 mass% Ni–N exposed to air after surface segregation [72]

10.5.1 Grain Boundary Segregation in Polycrystals

Grain boundary segregation of solute atoms has been surveyed not only in low-alloy steels but also in iron-based alloys which are in a ferritic phase with a relatively simple microstructure. Typically, the equilibrium grain boundary segregation of phosphorus in the ferritic phase was investigated by analyzing intergranular planes in iron–phosphorus binary alloys [75, 76], in which the grain boundary concentration of phosphorus was evaluated as a function of concentration and temperature. These resullts indicated that there is a saturation level of segregation and a temperature dependence of the grain boundary segregation of phosphorus. The theoretical models do not seem to correspond to all of the results for the scattering of the experimental data, but the free energy of segregation of phosphorus in iron was estimated from an equilibrium segregation experiment [76, 77].

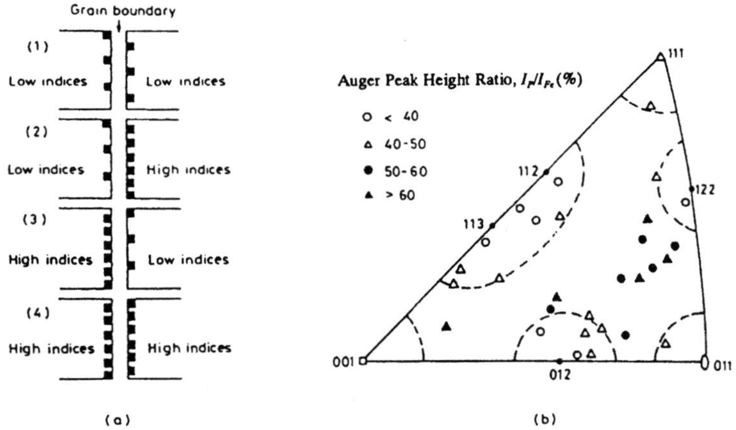

Fig. 10.13. The anisotropy of phosphorus grain boundary segregation in Fe–1 mass% P, showing (**a**) the segregated atoms (dark squares) in relation to the fracture process and (**b**) the level of segregation in relation to the relative orientation of each grain to the grain boundary [78]

In practical analyses, a variation of segregation on intergranular fracture planes is very often observed, if many boundary planes are analyzed in a polycrystalline sample. Since the variation of experimental data is not negligible in a thermodynamic analysis, the reason for the occurrence of scattering should be taken into account. Figure 10.13 shows the dependence of the orientation of each fracture plane on the segregated amount of phosphorus in a polycrystalline Fe–1 mass%P alloy, in which the phosphorus segregation is nearly saturated [78]. This result indicates the anisotropy of the grain boundary segregation of phosphorus, which is illustrated as segregated phosphorus (dark squares) in relation to the fracture process. Thus, it is suggested that the segregated phosphorus is not always equally distributed in two fracture planes, which will be discussed later, using results of bicrystals. The variation of the phosphorus segregation at many grain boundaries is considered to be related to the grain boundary structure, which was, for instance, explained in terms of free volumes of grain boundaries [79].

The above results on phosphorus segregation also suggest that the density of a segregation site or the saturation level of the segregation differs from boundary to boundary in polycrystalline metals. In order to eliminate the effect of the density of a segregation site on the evaluation of segregation in a polycrystalline sample, a number of intergranular fracture planes are analyzed. When experimental results of samples prepared under different conditions are compared, the average value of the segregated amount obtained at each plane is used. In other words, the models of (10.2) and (10.3) implicitly assume that the density of a segregation site at grain boundaries is constant. However, even if the average segregation amount is considered

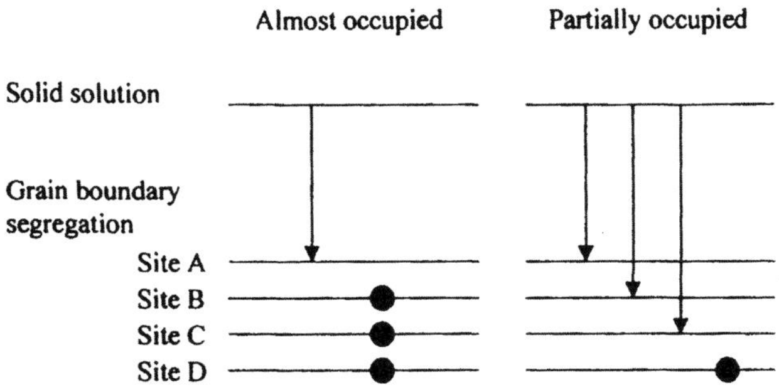

Fig. 10.14. The free energy diagram for solute atoms segregated at different grain boundary sites

as a representative value of all grain boundaries, the value may not always be useful for thermodynamic analysis. This is because there are a variety of types of atomistic configurations in the grain boundary; for example, compact polyhedra in polycrystalline metals, as illustrated in Fig. 10.3, which reveal different free energies of segregation. Therefore, the influence of this distribution of the free energy on the thermodynamic evaluation for different segregation sites should be considered [80, 81], as shown in Fig. 10.14.

The effect of a certain distribution of the free energy of segregation on the measured amount of segregation can be expressed mathematically. The average segregated amount, $X_I^{\phi \mathrm{av}}$, is obtained from many fractured planes consisting of various segregation sites with individual free energies of segregation, and is given by

$$X_I^{\phi\ \mathrm{av}} = \int X_I^{\phi} f(\Delta G_i) \mathrm{d}\Delta G_i \,, \tag{10.6}$$

where X_I^{ϕ} is the fraction occupied by segregated atoms with the free energy of segregation ΔG_i, and $f(\Delta G_i)$ is a distribution function of the free energy of segregation. The temperature dependence of the average segregated amount can be calculated by assuming a mean free energy of segregation and a specified distribution. For example, if the mean free enthalpy for grain boundary segregation in a Gaussian distribution is 50 kJ/mol, and if the standard deviation of the free enthalpy is 30%, the calculated values are shown in Fig. 10.15 [81]. The dashed and solid lines in Fig. 10.15 show the temperature dependence of the segregated amount calculated by the Langmuir–McLean relation (10.2) and the calculated relation taking into account the energy distribution (10.6), respectively. The influence of the bulk concentration on segregation is also shown in Fig. 10.15. The temperature dependence of the segregation can be semi-quantitatively understood by referring to the idea of Fig. 10.14. Segregation sites with lower segregation energies are preferentially occupied at equilibrium. Higher absolute values of the free energies of

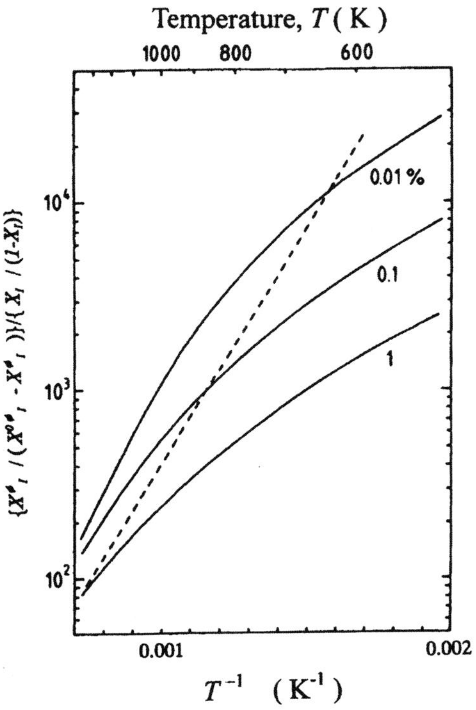

Fig. 10.15. The temperature dependence of the segregated amount calculated by the Langmuir–McLean relation (*dashed lines*) and the calculated relation taking into account the enthalpy distribution (*solid lines*) [81]

segregation (i.e. a steeper slope of the curves in Fig. 10.15) are observed for low occupation of segregation sites, which corresponds to the results obtained in samples that have a low bulk concentration at higher temperatures. On the other hand, lower absolute values of the free energies of segregation are experimentally obtained in the case of almost completely occupied boundaries, which is segregation in samples having a high concentration at low temperatures. Thus, the effects of the distribution of free energies and the bulk concentration on grain boundary segregation are crucial in analyzing grain boundary segregation thermodynamically.

The addition of a third element to a binary-alloy system influences the grain boundary segregation of solute atoms to some degree. A large number of studies have been performed in order to examine the influence of third elements on the segregation of solute atoms in iron-based alloys [82–98]. When different solute atoms are segregated in grain boundaries, they are distributed to occupy given segregation sites on grain boundaries according to the degree of free energy of segregation, as shown in (10.2). Therefore, the segregated amounts of different atoms mutually influence each other even if there is

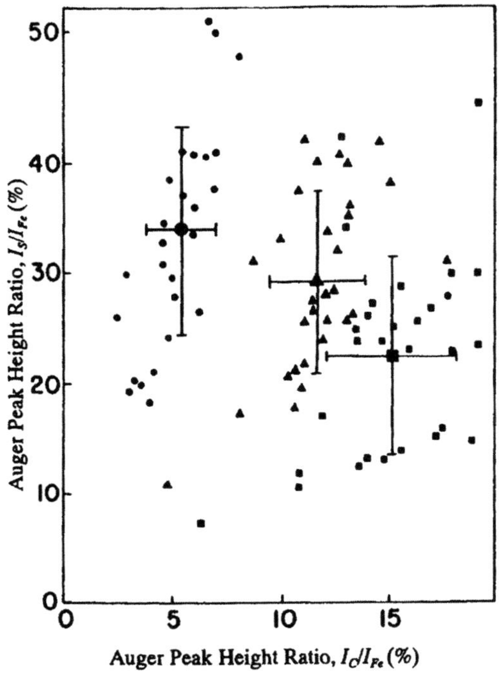

Fig. 10.16. The Auger peak height ratio of sulfur to iron versus that of carbon to iron for Fe–100 mass ppm S alloys containing 12 (•), 40 (▲) and 75 (■) mass ppm of carbon. Each small point was taken at one grain boundary facet, and the large points with error bars attached thereto represent the average degree of segregation in each specimen with the standard deviation [91]

no chemical interaction between two elements. This behavior is called site competition, and is reported in several iron-based alloys that contain different metalloid elements.

Figure 10.16 shows the Auger peak height ratio of sulfur to iron versus that of carbon to iron for Fe–100 mass ppm S alloys containing 12, 40 and 75 mass ppm carbon [91]. Each point in Fig. 10.16 shows the result on a grain boundary fracture plane, and the averages and standard deviations are also denoted. It should be noted that the segregated amount of sulfur decreases with carbon segregation. Similar behavior was observed in phosphorus and carbon in the Fe–P–C system [86]. This behavior reveals that site competition apparently occurs between two elements in iron-based ternary alloys, although the structure of the segregation sites is not specified for each element. Site competition seems to be a general phenomenon for different metalloid elements with relatively low free energies of segregation.

On the other hand, the segregation behavior found in an Fe–Cr–P ternary system is shown in Fig. 10.17, where the Auger peak height ratio of phospho-

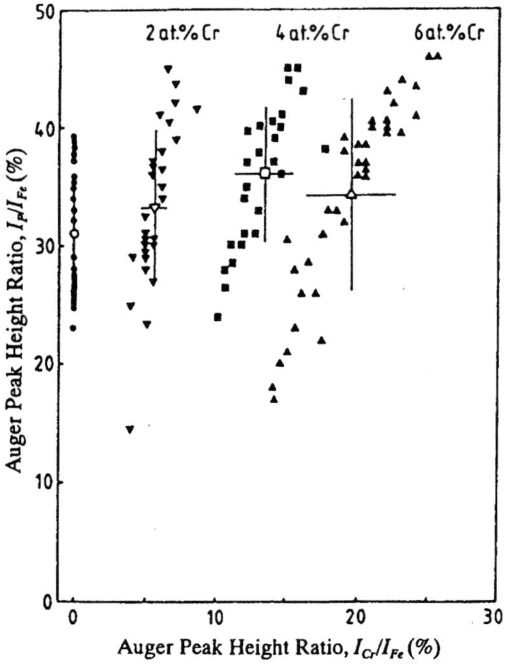

Fig. 10.17. The Auger peak height ratio of phosphorus to iron versus that of chromium to iron for Fe–0.5 mass% P alloys containing 2 (▼), 4(■) and 6 (▲) mass% Cr [97]

rus to iron versus that of chromium to iron is plotted for Fe–0.5 mass%P alloys, containing 2, 4 and 6 mass% of chromium [97]. The segregation behavior of solute atoms in the Fe–Cr–P ternary system appears to be contrary to that for the Fe–S–C ternary system. The addition of chromium does not seem to significantly affect the grain boundary segregation of phosphorus, while chromium segregation occurs slightly. Since chromium behaves as a metallic element in iron-based alloys, these results suggest that chromium does not occupy segregation sites as metalloid elements but occupies iron sites at grain boundaries. Thus, the chemical character of segregated solute atoms is responsible for a variation in the segregation behavior in ternary systems.

Furthermore, each point plotted for each sample, as shown in Figs. 10.16 and 10.17, demonstrates a characteristic correlation between the segregation of sulfur and carbon, and such a correlation between the segregation of chromium and phosphorus in these samples, which was observed in the Fe–P–C system [86]. Also, such a correlation was obtained in plots between chromium and phosphorus segregation in a Cr–Mo steel [82] and between molybdenum and phosphorus segregation in a Mo steel [95]. This relation has sometimes been interpreted as being due to the attractive interaction between chromium and phosphorus. However, in the Fe–P–C and Fe–S–C systems, the interactions between phosphorus and carbon, and between sulfur and carbon, are thermodynamically repulsive. Rather, the correlation indicates that the segregated amounts of two kinds of solute atoms change at individual grain boundary fracture planes in a similar way, probably due to the grain boundary structure.

It is interesting to estimate the tendency of solute atoms to segregate, based on the results of experiments on segregation in multi-component systems of iron-based solid solution alloys. Several experiments show that a pronounced segregation of boron [96], carbon [86, 91] and sulfur [91] takes place under the condition that they are in solution. The segregation of phosphorus [86] and tin [98] reveals an intermediate tendency, while metallic elements are slightly segregated in grain boundaries. Thus, it may be concluded that the tendency of solute atoms to segregate may be estimated roughly as B > C > S > P > metallic solutes.

In addition, it should be noted that the quenching rate of samples is also an important factor in estimating thermodynamic parameters, because the diffusivity of solute atoms strongly affects the segregated amount of solute atoms. The segregation of solute atoms with high mobility can take place at a low quenching rate. If such solute atoms migrate to grain boundaries during quenching, the measurable change of segregation with a change in equilibrium temperatures is underestimated. This process is a kind of nonequilibrium segregation. Actually, the influence of quench-induced nonequilibrium

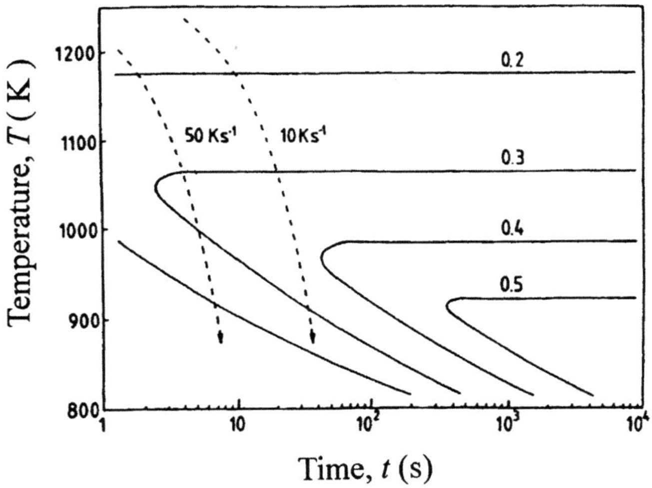

Fig. 10.18. A temperature–time–segregation diagram for Sn grain boundary segregation. The numbers in the figure denote the respective degree of the segregation, and different quenching rate are plotted in the figure [98]

segregation of a solute atom was found in sputtering depth profiles near the fracture surface in Ni–In bicrystals [99].

As a typical example, a temperature–time–segregation diagram for tin segregation in iron is shown in Fig. 10.18 [98]. The numbers in the figure denote the respective degree of segregation, and the different quench rates are plotted. The segregated amount of tin in samples quenched at a low rate is increased by additional segregation during quenching. However, if the quenching rate is more than 1000 K/s, this effect is small. Furthermore, if the mobility of the solute atoms is high, or the heat conductivity of a sample is low, the effect of the quenching rate on segregation will become apparent, resulting in an underestimation of the free enthalpy of segregation.

10.5.2 Grain Boundary Segregation in Bicrystals

The use of bicrystals with a specific lattice misorientation and boundary orientation plane is adequate to investigate the effect of grain boundary structure on grain boundary segregation. Some attempts were made to study the effect of misorientation on grain boundary segregation in bicrystals of iron–silicon [100] and iron–tin [101] alloys that were cut from polycrystals with coarse grains. The results showed that the segregated amount of silicon and tin increases with increasing misorientation angle, although the orientation of samples used in those experiments was not intensively controlled. This misorientation dependence of segregation seems to be consistent with the dislocation model, in which the density of dislocations corresponding to segregation sites increases with an increasing misorientation angle.

Furthermore, a systematic study of grain boundary segregation at intergranular fracture planes was performed on carefully prepared bicrystals of an Fe–Si system containing some impurities [102–110], in which the misorientation and boundary orientation were changed. The results of the experiments on grain boundary segregation may be related to the grain boundary structure. Some essential points obtained in these experiments are described here.

Grain boundaries are sometimes classified as symmetric and asymmetric boundaries. Since segregation behavior is discussed with respect to data for segregation on one of the fracture planes measured by AES, it is very important to certify if symmetric segregation in two fracture planes is observed in AES. In order to study the effect of symmetry of boundaries on segregation, the amounts of the segregated elements were measured on two boundary planes, the (001) fracture plane and the (011) fracture plane, in 45° [100] tilt iron–3.55 at% silicon bicrystals containing a small amount of phosphorus and carbon [110]. The results show that the two fracture planes reveal different segregated amounts of silicon, phosphorus and carbon. The results obtained in measurements of one side of the fracture planes have suggested that there is a correlation between phosphorus segregation and orientation of fracture plane in iron–phosphorus alloys, as shown in Fig. 10.13. Thus, asymmetry of

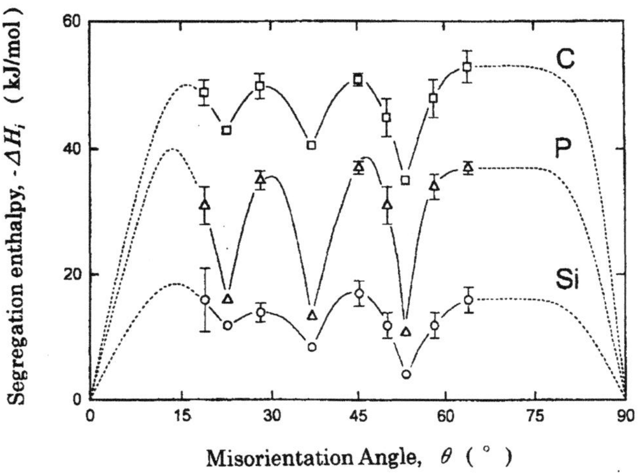

Fig. 10.19. A schematic diagram of the segregation enthalpies of different elements at the symmetrical tilt boundary versus the misorientation angle [11]

segregation on fracture planes seems to be a general form of behavior, since most of the grain boundaries are asymmetric.

Since the orientation of a boundary plane is an important factor in discussing grain boundary structure, segregation has been studied in boundary planes with different orientations for a given lattice misorientation. The grain boundary segregation of carbon, phosphorus and silicon was surveyed in bicrystals with a (001)/(011) asymmetrical boundary and a {0kl} symmetric boundary in 45° [100] tilt iron–3.55 at% silicon bicrystals [110]. The results for the latter are obtained for symmetrical boundaries, and show that that the number of segregated atoms on one boundary surface is equal to that on the other boundary surface. Thermodynamic parameters for segregation are calculated from this temperature dependence of the segregated amounts at grain boundaries. The segregation enthalpies of these elements at the (001)/(011) asymmetrical boundary were shown to be different from those at the {0kl} symmetrical boundary, which suggests that grain boundary segregation depends on the orientation of the boundary plane.

Since the above results concern segregation in bicrystals with a specific lattice misorientation but with differently oriented boundary planes, segregation in bicrystals with different misorientations is discussed here. For this experiment, bicrystals with symmetrical boundaries have mainly been used to avoid the effect of asymmetry of the segregated amount due to fracture paths. Figure 10.19 shows the orientation dependence of the segregation enthalpies of carbon, phosphorus and silicon, respectively, at the symmetrical tilt grain boundaries of an Fe–Si–P–C alloy [11]. The segregation of each element represents the characteristic orientation dependence, which is related to the grain boundary structure; that is, to the density of segregation sites. These plots demonstrate that the absolute values of the segregation enthalpy for carbon and phosphorus are high. They also reveal that the enthalpy for each element depends on misorientation; in particular, the plot exhibits significant cusp 22.6° [100], {015}, 36.9° [100], {013} and 53.1° [100], {012} tilt grain boundaries. Similar cusps were also observed at various asymmetric boundaries. The segregation enthalpy of an element was related to the solid solubility, as proposed in earlier work [4]. Recently, this segregation enthalpy has also been expressed as a function of bulk solid solubility and grain boundary structures, and the experimental grain boundary segregation diagram was constructed for the [100] tilt grain boundaries in α-Fe [11].

10.6 Further Approaches for Attaining Surface and Interface Characterization

We have emphasized studies on surface and grain boundary segregation of impurities investigated by XPS and AES, which may be applied to general metallurgical studies. However, the experimental results obtained by these methods may not always represent all the features of the distribution of impurities in metals and alloys with microstructures, and therefore the more realistic features of the impurity distribution should be examined by further studies. For instance, micro-focused X-ray photoelectron spectroscopy has been developed to analyze segregation in local areas [110]. The method must be very effective to analyze segregation of surface and intergranular fracture planes. In addition, methods for analyzing grain boundaries without intergranular fractures are required to carry out studies on grain boundary segregation. Micro-focused secondary-ion spectrometry and atom probe field-ion microscopy may provide useful information in the future, by utilizing the high sensitivity of ion analysis. Cross-sectional analysis of grain boundaries has been attempted using nanoprobe electron microscopy with an energy-dispersive X-ray detector and an energy-loss spectrometer [111], which is applicable to analysis of the interface in thin films. Characteristic X-ray spectra and electron energy-loss spectra from grain boundary regions reveal the signals of segregated elements, although there is still a problem in conducting a quantitative analysis of the distribution of electron beams in a sample. Thus, these nanoprobe analysis methods must prompt further studies on surface and interface layers, particularly the characteristic segregation behavior at individual surfaces and boundaries. Nevertheless, the information on surface and grain boundary segregation obtained up to now is indispensable for discussing the segregation of impurities in metals and alloys.

10.7 Summary

The present chapter has mainly described the characteristic features of the segregation of impurity atoms at surfaces and interfaces in polycrystalline high-purity iron-based alloys, on the basis of the results of experiments by XPS and AES. In particular, the metallurgical factors in the experimental results have been discussed, in order to understand the realistic behavior of impurity and alloying elements in metals and alloys.

XPS has been systematically used for investigating the surface segregation of metallic and nonmetallic elements in high-purity iron-based alloys. It has also been applied to characterize oxide layers formed on the surface of these alloys, with and without segregation at room temperature. Metallic elements such as chromium are clearly segregated to the iron surface by in situ heating at high temperatures. The effective thickness of the segregated layers is estimated to be less than 1 nm. Nonmetallic elements such as nitrogen, phosphorus and sulfur are easily segregated to the iron surface. While nitrogen is mutually segregated with chromium in the surface layer, phosphorus and sulfur appear to replace chromium segregation at the top surface. The surface segregation of chromium reduces the thickness of the native oxide layer formed on iron–chromium alloys, although the segregated layer is thinner than the oxide layer.

In polycrystalline high-purity iron-based alloys, the effects of third elements, microstructures and heat treatment on measured segregation have been reviewed. It has been pointed out that grain boundaries with different structures or with different free energies of segregation are responsible for deviations from predictions made on the basis of simple thermodynamic models. From comparisons of the segregation behavior of various solute atoms at grain boundaries, it has been suggested that nonmetallic solute atoms are effectively segregated at grain boundaries. On the basis of systematic studies of grain boundary segregation of solute atoms in Fe–Si bicrystals with various misorientations, it has been shown that the grain boundary structure plays a crucial role in grain boundary segregation.

Acknowledgments

I am very grateful to Professor Y. Waseda and Professor M. Isshiki for their continuous support and encouragement. I am also indebted to Professor Hofmann and Dr P. Lejcek for having stimulating discussions with me about interfacial segregation.

References

1. K. Sugimoto, S. Matsuda, M. Isshiki, T. Ejima, K. Igaki: J. Japan Inst. Metals **46**, 155 (1982)
2. K. Sugimoto, N. Hara, M. Isshiki, T. Ejima, K. Igaki: J. Japan Inst. Metals **46**, 703 (1982)
3. E.D. Hondros, M.P. Seah: Int. Metall. Rev. **22**, 867 (1977)
4. E.D. Hondros, M.P. Seah: Metall. Trans. **8A**, 1363 (1977)
5. M.P. Seah: J. Phys. F: Metal Phys. **10**, 1043 (1980)
6. S. Hofmann: J. Chim. Phys. **84**, 141 (1987)
7. H.J. Grabke: Steel Res. **58**, 1013 (1987)
8. H.J. Grabke: Surf. Interface Anal. **14**, 686 (1989)
9. C.L. Briant: Metall. Trans. **21A**, 2339 (1990)
10. C.J. McMahon, Jr.: Mater. Character. **26**, 269 (1991)
11. P. Lejcek, S. Hofmann: Crit. Rev. Solied State Mater. Sci. **20**, 1 (1995)
12. E.D. Hondros, M.P. Seah, S. Hofmann, P. Lejcek: In *Physical metallurgy*, ed. by R.W. Cahn, P. Haasen, (North-Holland, Amsterdam 1996) p. 1201
13. S. Hofmann, P. Lejcek: Interface Sci. **3**, 241 (1996)
14. L.C. Feldman, J.W. Mayer: *Fundamentals of surface and thin film analysis* (North-Holland, Amsterdam 1986)
15. M.P. Seah: In *Practical surface analysis*, Vol. 1, ed. by D. Briggs, M.P. Seah (John Wiley, Chichester 1990) p. 107
16. E.D. Hondros, D. McLean: In *Surface phenomena of metals* (Society for Chemical Industry, London 1968) p. 39
17. D. McLean: *Grain boundaries in metals* (Oxford University Press, London 1957) p. 116
18. M. Guttmann: Surface Sci. **53**, 213 (1975)
19. M. Guttmann: Metal Sci. **10**, 337 (1976)
20. M. Guttmann, Ph. Guttmann, M. Waymen: Metall. Trans. **13A**, 1183 (1976)
21. M. Guttmann, D. McLean: In *Interfacial segregation*, ed. by W.C. Johnson, J.M. Bionkely (American Society for Metals, Metals Park, Ohio 1979) p. 261
22. R.W. Baluffi: In *Interfacial segregation*, ed. by W.C. Johnson, J.M. Blakely (American Society for Metals, Metals Park, Ohio 1979) p. 193
23. M.F. Ashby, F. Spaepen, S. Williams: Acta Metall. **26**, 1647 (1978)
24. V. Vitek, A.P. Sutton, D.A. Smith, R.C. Pond: In *Grain boundary structure and kinetics* (American Society for Metals, Milwaukee 1979) p. 115
25. C.L. Briant, R.P. Messer: Acta Metall. **30**, 1811 (1982)
26. V. Paidar: Phil. Mag. **A66**, 41 (1992)
27. S. Suzuki: Mater. Trans. JIM **35**, 35 (1994)
28. S. Suzuki,T. Kosaka,H. Inoue, M. Isshiki, Y. Waseda: Appl. Surface Sci. **103**, 495 (1996)
29. Y. Yoshitomi, S. Suzuki, T. Ueda, S. Tsurekawa, N. Nakashima, H. Yoshinaga: Scr. Metall. Mater. **32**, 1067 (1995)
30. H.J. Grabke, G.Tauber, H. Viefhaus: Scr. Metall. **9**, 1181 (1975)
31. H.J. Grabke, W. Paulitsche, G. Tauber, H. Viefhaus: Surface Sci. **3**, 377 (1977)
32. H.J. Grabke, H. Viefhaus, G. Tauber: Arch. Eisenhüttenw. **49**, 391 (1978)
33. H.J. Grabke: Mater. Sci. Eng. **42**, 91 (1980)
34. N. Egert, H.J. Grabke, Y. Sakisaka, T.N. Rhodin: Surface Sci. **141**, 397 (1984)

35. H. de Rugy, H.J. Grabke: Surface Sci. **109**, 418 (1986)
36. L. Marchut, C.J. McMahon, Jr:, Metall. Trans., **12A**, 1135 (1981)
37. J. du Plessis, P.E. Viljoen: Surface Sci. **131**, 321 (1983)
38. F. Bezuidenhout, J. du Plessis, P.E. Viljoen: Surface Sci. **171**, 397 (1986)
39. B. Egert, G. Panzner: Surface Sci. **118**, 345 (1985)
40. M.P. Seah, C. Lea: Phil. Mag. **31**, 627 (1975)
41. C. Lea, M.P. Seah: Phil. Mag. **31**, 213 (1977)
42. H. Viefhaus, M. Rusenberg: Surface Sci. **159**, 1 (1985)
43. Y.X. Zhou, R. Hsiao, C.J. McMahon, Jr., E.W. Plummer: Surface Sci. **179**, 209 (1987)
44. L. Marchut, T.M. Buck, C.J. McMahon, Jr., G.H.Wheatley, W.M. Augstyniak: Surface Sci. **180**, 252 (1987)
45. L.Q. Clayton, G.T. Burstein: Metall. Sci. **13**, 530 (1979)
46. M. Oku, K. Hitokawa, H. Kimura, S. Suzuki: Surface Interface Anal. **8**, 67 (1986)
47. M. Rusenberg, H. Viefhaus: Surface Sci. **172**, 615 (1986)
48. P. Doumoulin, H. Guttmann: Mater. Sci. Eng. **42**, 249 (1980)
49. C. Leygraf, G. Hultquist, S. Ekelund, J.C. Ericksson: Surface Sci. **46**, 157 (1974)
50. C. Leygraf, G. Hultquist, S. Ekelund: Surface Sci. **51**, 409 (1975)
51. C. Leygraf, G. Hultquist: Surface Sci. **61**, 60 (1976)
52. C. Leygraf, G. Hultquist: Surface Sci. **61**, 69 (1976)
53. W. Yu-Quing, C.J. McMahon, Jr.: Scr. Metall. **20**, 19 (1986)
54. C.L. Briant: Surface Interface Anal. **13**, 209 (1988)
55. C. Uebing, H. Viefhaus, H.J. Grabke: Appl. Surface Sci. **32**, 363 (1988)
56. H.J. Grabke: Surface Interface Anal. **14**, 209 (1989)
57. C. Uebing: Surface Sci. **225**, 97 (1990)
58. S. Suzuki: Mater. Trans. JIM **31**, 1085 (1990)
59. I. Olefjord: Corros. Sci. **15**, 687 (1975)
60. G. Hultquist, C. Leygraf: Corros. Sci. **22**, 331 (1982)
61. T.G. Hultquist, M. Seo, N. Sato: Oxid. Metals **25**, 363 (1986)
62. H.J. Mathieu, D. Landolt: Appl. Surface Sci. **3**, 348 (1979)
63. J.E. Castle, R. Ke, J.F. Watts: Corros. Sci. **30**, 771 (1990)
64. H.J. Mathieu, D. Landolt: Surface Interface Anal. **14**, 744 (1989)
65. C. Palcio, H.J. Mathieu, D. Landolt: Surface Sci. **214**, 493 (1989)
66. C. Palcio, H.J. Mathieu, V. Stambouli, D. Landolt: Surface Sci. **295**, 547 (1993)
67. S. Suzuki, T. Kosaka, H. Inoue, Y. Waseda: Mater. Trans. JIM **36**, 1379 (1995)
68. S. Suzuki, T. Kosaka, H. Inoue, Y. Waseda: ISIJ International **36**, 433 (1996)
69. S. Suzuki, T. Nakazawa, Y. Waseda: ISIJ International **36**, 1273 (1996)
70. S. Suzuki, Y. Waseda: Scr. Mater. **36**, 915 (1997)
71. S. Suzuki, U. Tsu, T. Ohtsuka, T. Shindo, M. Tanino, Y. Waseda: In *ECASIA'97* (John Wiley, Chichester 1997) p. 639
72. S. Suzuki: High Temp. Mater. Process. **17**, 1 (1998)
73. S. Suzuki, Y. Ishikawa, M. Isshiki, Y. Waseda: Mater. Trans. JIM **38**, 1004 (1997)
74. O. Kubaschewski, B.E. Hopkins: *Oxidation of metals and alloys* (Butterworths, London 1962) p. 38

75. T. Matsuyama, H. Suto: Trans. Japan Inst. Metals **20**, 44 (1979)
76. H. Erhart, H.J. Grabke: Metal Sci. **15**, 401 (1981)
77. H.J. Grabke: ISIJ International **29**, 529 (1989)
78. S. Suzuki, K. Abiko, H. Kimura: Scr. Metall. **15**, 1139 (1981)
79. T. Ogura, M. Koiwa: Scr. Metall. **16**, 1273 (1982)
80. C.L. White, D.F. Stein: Metall. Trans. **9A**, 13 (1978)
81. S. Suzuki: ISIJ International **30**, 1000 (1990)
82. J. Yu, C.J. McMahon, Jr: Metall. Trans. **11A**, 277 (1980)
83. P. Gas, M. Guttmann, J. Bernardini: Acta Metall. **30**, 1309 (1982)
84. P. Dumoulin, M. Guttmann, Ph. Maynier, P. Chevalier: Metal Sci. **17**, 1013 (1983)
85. C.L. Briant: Acta Metall. **31**, 257 (1983)
86. S. Suzuki, M. Obata, K. Abiko, H. Kimura: Scr. Metall. **17**, 1325 (1983)
87. S. Suzuki, M. Obata, K. Abiko, H. Kimura: Trans. Iron Steel Inst. Japan **25**, 62 (1985)
88. H. Kimura, K. Abiko, S. Suzuki, M. Obata, J. Kumagai, H. Kimura: In *Grain boundary structure and related phenomena*, Proc. JIMIS-4, Suppl. Trans. JIM (Japan Inst. Metals, Sendai 1986) p. 53
89. W. Yu-Qing, C.J. McMahon, Jr: Mater. Sci. Technol. **3**, 207 (1987)
90. H.J. Grabke, R. Moeller, H. Erhart, S.S. Brenner: Surface Interface Anal. **10**, 202 (1987)
91. S. Suzuki, S. Tanii, K. Abiko, H. Kimura: Metall. Trans. **18A**, 1109 (1987)
92. C.L. Briant: Mater. Sci. Technol. **4**, 956 (1988)
93. N. Saito, K. Abiko, H. Kimura: Mater. Sci. Eng. A **102**, 169 (1988)
94. M. Paju, H.J. Grabke: Steel Res. **60**, 41 (1989)
95. M. Menyhard, C.J. McMahon, Jr: Acta Metall. **37**, 2287 (1989)
96. C.M. Liu, T. Nagoya, K. Abiko, H. Kimura: Metall. Trans. **23A**, 263 (1992)
97. S. Suzuki : Z. Metallkde. **83**, 153 (1992)
98. S. Suzuki, K. Kuroki, H. Kobayashi, N. Takahashi: Mater. Trans. JIM **33**, 1068 (1992)
99. T. Muschik, W. Gust, S. Hofmann, B. Predel: Acta Metall. **37**, 1989 (1989)
100. T. Watanabe, T. Murakami, S. Karashima: Scr. Metall. **12**, 361 (1978)
101. T. Watanabe, S. Kitamura, S. Karashima: Acta Metall. **28**, 1980 (1980)
102. P. Lejcek, S. Hofmann: Surface Interface Anal. **16**, 546 (1990)
103. P. Lejcek, S. Hofmann: Acta Metall. Mater. **39**, 2469 (1991)
104. S. Hofmann, P. Lejcek: Scr. Metall. Mater. **25**, 2259 (1991)
105. P. Lejcek, J. Adamek, S. Hofmann: Surface Sci. **264**, 449 (1992)
106. S. Hofmann, J. Adamek, P. Lejcek: Surface Interface Anal. **19**, 601 (1992)
107. P. Lejcek, S. Hofmann: Interface Sci. **1**, 163 (1993)
108. P. Lejcek, S. Hofmann: Surface Sci. **307–309**, 798 (1994)
109. P. Lejcek: Anal. Chim. Acta **297**, 165 (1994)
110. P. Lejcek, V. Paidar, J. Adamek, S. Hofmann: Acta Mater. **45**, 3915 (1997)
111. K.R. Hallam, R.K. Wild: Surface Interface Anal. **23**, 133 (1995)
112. L.A. Vatter, J.M. Titchmarsh: Surface Interface Anal. **25**, 760 (1997)

11 Mechanical Properties

L. Hollang

11.1 Introduction

As early as the 1920s single crystals of cadmium and zinc, with remarkably high purity, had already been produced by Boas and Schmid [1, 2], shortly before the concept of dislocations was introduced in 1934 by Orowan [3], Polanyi [4] and Taylor [5], based on Prandtl's "Vernier defects" [6] and Dehlinger's *"Verhakungen"* [7]. Accordingly, the early work on the mechanical properties of metal crystals was concentrated on the plasticity of the hexagonal close-packed (hcp) metals cadmium and zinc, followed by investigations on face-centered cubic (fcc) metals and alloys. Most of the basic findings on the plasticity of metals were established by investigations on metals with a close-packed structure: the essentially athermal nature of the deformation of crystals by slip, the crystallography of slip and twinning, and Schmid's law [8] for the onset of slip. (For a review up to 1934, see Schmid and Boas [9]; for the work up to 1958, see [10].)

However, by the second half of the 1920s Taylor [11–13] had already realized, on the basis of experiments on α-iron [11] and β-brass [13], that in the body-centered cubic (bcc) structure the slip geometry was different from that of the above-mentioned metal structures. He coined the expression "pencil glide" for a slip mode in which the crystallography determines the *slip direction* but not the *slip planes*. However, doubts were cast on the correctness of this idea by subsequent work on α-iron [14–16], which came to the conclusion that in α-iron the slip planes were crystallographically determined too, but that, in contrast to the fcc metals, there were several of them.

In the course of time it turned out that there was an even more striking difference between the plasticity of bcc metals on the one hand and that of cadmium, zinc and the fcc metals on the other. While the *critical resolved shear stress*[1], σ_0, of cadmium and zinc crystals and of fcc metals increased only moderately when the deformation temperature was lowered below room temperature, for bcc metals σ_0 increases strongly below a critical temperature, later called the "knee temperature" T_K [17] (cf. Fig. 11.1). Moreover,

[1] The *critical resolved shear stress*, σ_0, is defined as the minimum shear stress in the slip system under consideration at which large-scale dislocation movements are possible; that is the shear stress at which *macroscopic* plasticity sets in (cf. Sect. 11.2).

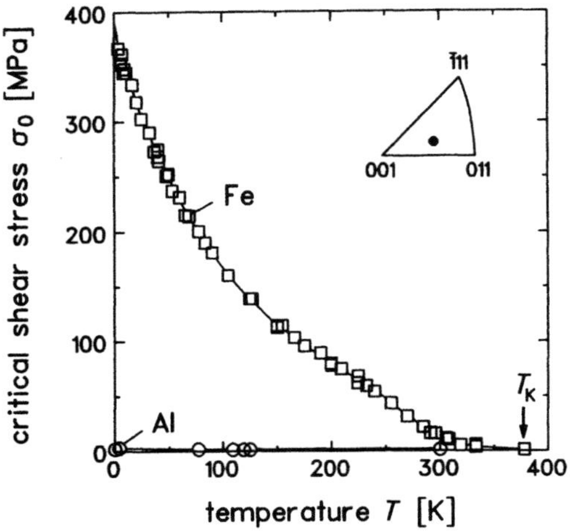

Fig. 11.1. The critical shear stress, $\sigma_0(T)$, of fcc aluminum [18] (cf. Fig. 11.7) and of bcc α-iron [19]. The high-purity single crystals, oriented for single slip, were deformed in tension. σ_0 was calculated for the $[\bar{1}01](111)$ slip system of aluminum and the $[111]$ $(\bar{1}01)$ slip system of α-iron, respectively. The knee temperature, T_K, refers to α-iron

whereas in the hcp and fcc metals σ_0 obeys Schmid's law to a good approximation, but depends on the perfection of the crystal, for bcc metals the exceptional high values of σ_0 below T_K appeared to be an intrinsic property, but showed strong violations of Schmid's law. Furthermore, in tensile tests at low temperatures the bcc metals usually failed by brittle fracture, clearly a problem of great technological relevance.

It was known early on that the mechanical properties of bcc metals were rather sensitive to the presence of interstitially dissolved impurities. However, only in the 1970s did it become possible, by zone-melting inside an ultra-high-vacuum system, to reduce the content of interstitially dissolved impurities such as N, O and C to such an extent that an extrapolation of the mechanical properties of, for example, tantalum single crystals to zero concentration of "heavy" (in contrast to hydrogen) foreign interstitial atoms became meaningful (Fig. 11.2). In this way [20–22] it could be shown that the strong temperature dependence of the critical shear stress is an intrinsic property of tantalum, and presumably of the other bcc metals as well. Since similar features are also observed in intermetallic compounds of great commercial relevance (e.g. in NiAl [23–25]), it is of general interest to achieve a full understanding of the mechanisms that determine the plastic behavior of pure bcc metals. (For a review up to 1978, see Šesták and Seeger [26]; and up to 1983, see Christian [27].)

Obviously, for the bcc metals the standard characterization of dislocations by their line direction and their Burgers vector is not sufficient. Since the line direction and the Burgers vector (plus elastic constants) determine the long-

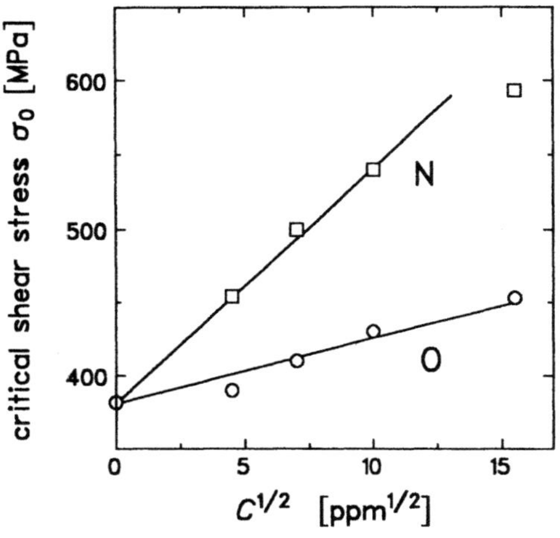

Fig. 11.2. The critical shear stress, extrapolated to zero temperature, σ_0 (0 K), of tantalum single crystals with orientations in the middle of the stereographic triangle, as a function of the atomic fraction, C, of nitrogen and oxygen [20]

range strain fields of dislocations, and since there is no doubt that these are correctly described by elasticity theory, it was concluded that it must be special properties of the *dislocation cores* that are responsible for the particular features of the mechanical properties of bcc metals. In 1960, Hirsch [28] introduced a concept to explain the plastic behavior, which was supported later by results of transmission electron microscopy (TEM), according to which it became evident that the *flow stress*[2], σ, of bcc metals below T_K is governed by the motion of screw dislocations with the Burgers vector $\boldsymbol{b} = a_0$<111>/2 (a_0 is the edge length of the elementary cube). The observation of long screw-dislocation segments with an obviously reduced mobility was ascribed to their sessile nature, which in turn is a consequence of the three-fold symmetry of the <111> directions in the bcc lattice.

A decisive step forward to a quantitative theory was made in 1981 by the suggestion (first quantitatively expressed by Seeger [17]) that the flow stress of metals with a bcc structure is determined by the thermally activated *formation of kink pairs* on screw dislocations and their subsequent migration along the dislocation line. The kink model was originally introduced to explain the so-called Bordoni maximum in the internal friction of cold-worked fcc metals [29,30]. The application of this concept to account for the temperature dependence of the flow stress of bcc metals permits us to describe their macroscopic plastic behavior in terms of a few kink parameters that can be determined directly from a deformation experiment. However, to avoid the

[2] The *flow stress*, σ, is the resolved shear stress measured in an uniaxial straining experiment needed to maintain plastic deformation at the externally prescribed strain rate.

above-mentioned strong influences of interstitial impurities on the mechanical properties, the measurements have to be performed on single crystals of the highest achievable purity.

This chapter concentrates on the mechanical properties of bcc metals, especially their plasticity. Nevertheless, after a short discussion concerning the onset of slip in metal crystals in Sect. 11.2, in Sect. 11.3 some of the essential properties of fcc metals are briefly summarized as far as needed to emphasize the differences between fcc and bcc metals. Afterwards, in Sect. 11.4, the plastic behavior of pure bcc metals will be discussed on the basis of recent measurements on single crystals of high-purity α-iron, niobium, tantalum, molybdenum and tungsten. After a brief discussion of the slip geometry in bcc metals in Sect. 11.4.1 and of work-hardening behavior and dislocation microstructure, the subject of Sect. 11.4.2, in Sect. 11.4.3 attention is focused on the flow stress of bcc metals, especially its strong and non-uniform temperature and strain-rate dependence. Furthermore, in Sects. 11.4.4, 11.4.5, and 11.4.6 three special features, observed mainly in bcc metals of high purity, are treated in more detail: namely, the marked tension–compression asymmetry of the flow stress, the occurence of anomalous slip, and the phenomenon of softening caused by atomic lattice defects, respectively. These particularities may be understood within the framework of kink theory, which will be first sketched in Sect. 11.4.7 and then applied to the data in Sects. 11.4.8 and 11.4.9.

11.2 The Onset of Slip

Among the first quantitative laws on crystal plasticity was the "shear-stress law" of Schmid [8]. For uniaxial deformation of crystals of different crystallographic orientations at a given temperature, T, and rate of resolved plastic shear strain, $\dot{\varepsilon}_{\mathrm{pl}}$, Schmid's law states that extensive plastic deformation by slip sets in at the *axial stress*, σ_{A} (tensile stress or compression stress), at which the resolved shear stress,

$$\sigma = (\cos\kappa \cos\xi)\sigma_{\mathrm{A}} = \mu_{(hkl)}\sigma_{\mathrm{A}} \,, \tag{11.1}$$

in the active slip system reaches a critical value, called the *critical resolved shear stress*, σ_0, which is *independent* of the crystallographic orientation of the straining axis. In (11.1), κ and ξ denote the angles between the stress axis and the slip-plane normal or the slip direction, respectively (cf. Fig. 11.3). The quantity $\mu_{(hkl)} \equiv \mu$ is the Schmid factor of the slip system under consideration for a given slip direction, characterized by the Miller indices of the slip planes (hkl). It follows from elementary continuum mechanics that the Schmid factor obeys the inequalities $0 \leq \mu \leq 1/2$. A corollary of Schmid's law is that in a set of crystallographic equivalent slip systems – for example, $<110>\{111\}$ in fcc crystals or $<111>\{110\}$ in bcc crystals – the slip system operating first in an uniaxial straining test is that with the Schmid factor closest to 1/2, which

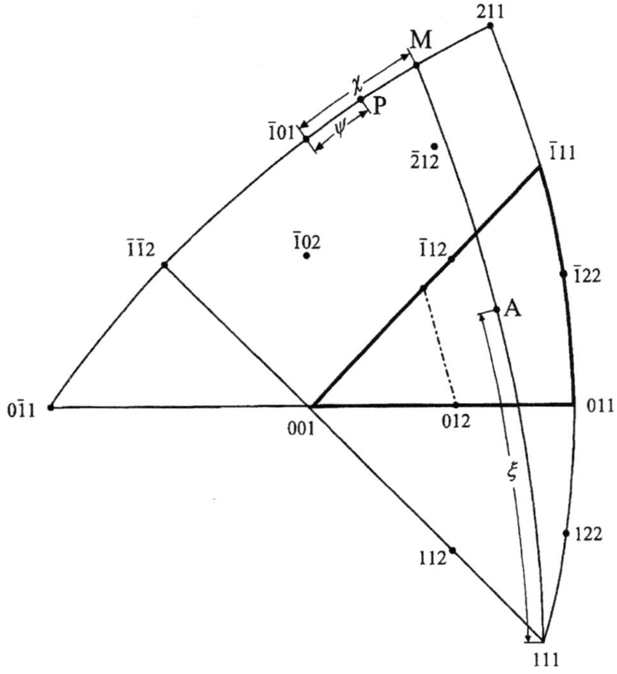

Fig. 11.3. The stereographic projection of a cubic lattice. The axis of uniaxial stress is denoted by A. If A lies in the interior of the triangle marked by bold lines (the so-called standard triangle), in the fcc lattice, the <110>{111} slip system with the largest Schmid factor is $[\bar{1}01](111)$ and in the bcc lattice the corresponding <111>{110} slip system is $[111]$ $(\bar{1}01)$. The angles refer to the bcc lattice: ξ denotes the angle between the axis A of uniaxial stress and the slip direction $[111]$. The angle κ in (11.1) (not shown) between A and the slip-plane normal P may be substituted by the Taylor angles ψ and χ, where χ is the angle between the $(\bar{1}01)$ plane and the plane of maximum resolved shear stress with normal M, and ψ is the angle between $(\bar{1}01)$ and the slip plane with normal P. The interrupted line denotes (approximately) the dividing line at which in the bcc lattice the <111>{211} slip system with the largest Schmid factor changes from $[111](\bar{2}11)$ to $[\bar{1}11](1\bar{1}2)$

is therefore called the *primary system*[3]. The other slip systems with smaller values of μ are *secondary systems.* Schmid's law has been extensively verified

[3] In order to permit a detailed comparison with dislocation theory, the experimental quantities flow stress, σ, plastic shear strain, $\varepsilon_{\mathrm{pl}}$, and plastic shear strain rate, $\dot{\varepsilon}_{\mathrm{pl}}$, used in this chapter have been calculated with the Schmid factor of the favored slip system relative to the axis of uniaxial stress (see, for example, [10]). For the often chosen "middle orientation" $[\bar{1}49]$ of the stress axis, where $\kappa = \xi = 45°$ in the standard stereographic triangle, the quantities σ, ε and $\dot{\varepsilon}_{\mathrm{pl}}$ refer to the primary $[\bar{1}01](111)$ slip system of fcc metals or the $[111](\bar{1}01)$ slip system of bcc metals, respectively. In both cases, the Schmid factor assumes its maximum value of $\mu = 1/2$.

for the <110>{111} slip systems of fcc metals and the basal slip systems in the hcp metals zinc, cadmium and magnesium, with the possible exception of the so-called almost-symmetric orientations, at which there are two or more slip systems with approximately the same large Schmid factor. In bcc metals the situation has turned out to be more complicated and therefore those will be discussed in more detail in Sect. 11.4.1. Schmid's law found its dislocation-theory explanation in the result that the force component per unit length acting on a dislocation in its slip plane is given by $b\sigma$. If only dislocations with a given modulus of the Burgers vector, b, have to be considered, Schmid's law thus means that in order to induce dislocation movements leading to *macroscopic* slip, the force per unit dislocation length must reach the critical value $b\sigma_0$.

When the concept of dislocations was introduced it had already become clear that the motion of a dislocation in its slip plane may be impeded by an *intrinsic* lattice effect, leading to a nonvanishing critical resolved shear stress, σ_0, even in an otherwise perfect crystal. Because of the atomistic structure of the slip planes, the line energy, $U(u)$, of a straight dislocation running parallel to a low-index crystallographic direction is a periodic function of its position on its slip plane, u, with a period that is determined by the crystal structure. This means that such a dislocation may occupy a sequence of energetically

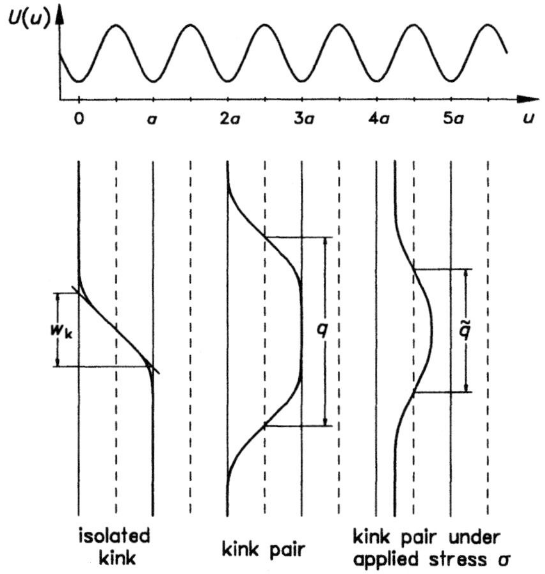

Fig. 11.4. A schematic illustration of the Peierls potential, $U(u)$, felt by a straight dislocation as a function of its displacement, u, and various configurations of kinks in dislocations. The kinks and kink pairs are characterized geometrically by the kink height, a, the kink width, w_{k}, and the separation, q, of the two kinks of a pair, respectively. $\tilde{q}$ is the separation in the configuration of unstable equilibrium

favored positions, with separations of the order of magnitude of and commensurate with the lattice parameter. Hence, the shape of the resulting so-called *Peierls potential of the first kind* (called the *Peierls potential* for short) depends strongly on the atomic arrangement on both sides of the slip plane and reflects the symmetry of that arrangement (Fig. 11.4). The maximum slope of the periodic energy function $U(u)$ gives the Peierls stress, σ_P, required in order to lift a straight dislocation out of its low-energy *Peierls valley* over the *Peierls barrier* in the absence of thermal or quantum-mechanical fluctuations. The original estimate of Peierls [31] has been corrected and improved by Nabarro [32] and various other authors (see, for example, [33]). For a given Burgers vector, σ_P depends on the crystallographic direction of the dislocation line; that is, on the dislocation character. For slip on close-packed planes (i.e. for fcc metals and for basal slip in hcp metals) the calculated Peierls stresses are of the order of magnitude of 10^{-4} of the shear modulus, M.

Later, it turned out that moving dislocation lines have to be considered as being flexible [34]. At finite temperatures, the Peierls potential may easily be overcome by the formation of *kink pairs* on dislocations. Kinks are short segments of a dislocation line, which connect parts of the otherwise straight dislocations in different Peierls valleys (cf. Fig. 11.4). Accordingly, the dislocation can move from one Peierls valley to the next one by the migration of the two kinks of a pair along the Peierls ridge in the opposite direction. Kink pairs may form in thermal equilibrium and reduce the stress required to move an otherwise straight dislocation through the crystal. As a natural consequence of the atomic discreteness of the crystal lattice, a kink moving along a dislocation line feels small periodic fluctuations of its potential energy, which define the *Peierls potential of the second kind* [35]. Thus, we have a hierarchy of two potentials, which have to be overcome in order to initiate macroscopic slip. (For a comprehensive survey about the history of the Peierls model, see Nabarro [36].)

11.3 The Plastic Behavior of fcc Metals

As already mentioned, some of the early basic experiments on metal plasticity were performed on hcp materials with high purity; for example, those on cadmium single crystals of "Kahlbaum" quality, with 99.996% purity [1]. The pioneering work of Boas et al. [1,2] and Meissner, Polanyi and Schmid [37] on cadmium single crystals of this purity and on zinc single crystals (involving measurements at 4.2 K and below!) lead to a surprising result: in uniaxial tests at constant temperature and strain rate, the flow stress, σ, increases strongly with increasing strain (thus exhibiting the phenomenon of work-hardening) but the critical shear stress, σ_0, was very low (compared with the shear modulus), and not only at room temperature – it was not even drastically higher at liquid-helium temperatures. Moreover, σ_0 at and below 4.2 K was found to be $< 10^{-5} M$; that is, at least a power of ten lower than the theoretical estimates mentioned in the section above. Later, it turned out that this behavior is not limited to the hcp metals zinc and cadmium, but is also found in pure fcc metals [38]. However, for the fcc metals the very successful interpretation of Bordoni relaxation, in terms of the stress-assisted thermally activated overcoming of the Peierls barriers by kink-pair formation, implies that σ_P is of the order of $10^{-4}M$. The question as to why the low-temperature critical shear stress, σ_0, of fcc meals as determined experimentally is substantially lower than the calculated Peierls stress, σ_P, may be answered qualitatively as follows [39, 40]:

(a) At low temperatures, quantum processes become important and permit a sufficiently large rate of kink-pair generation for the plastic deformation to be maintained even under moderate applied stresses (see, for example, [41]).
(b) The already low effective low-temperature Peierls barriers may be reduced further by the interaction of slipping dislocations with imperfections such as surfaces, impurities and, presumably more important, other dislocations (respectively, their stress fields).

Disregarding the differences between hcp metals slipping on the base plane[4] and fcc metals, the most important part of item (b), namely the interaction of dislocations and their influence on the flow stress, will be discussed briefly in the following example on pure fcc metals.

11.3.1 The Work-Hardening Behavior and Microstructure of fcc Metals

Figure 11.5 shows typical work-hardening curves[5] of as-grown single crystals of pure nickel, with the orientations of the crystal axis in the middle of the standard stereographic triangle. The curves display a subdivision into three stages. As introduced by Diehl [43], the stages are denoted with increasing strain by the roman numerals I, II and III. At the beginning of plastic deformation, the curves show a nearly linear Stage I with a relatively small slope. Stage II is characterized by a linear increase too, but with a significantly larger slope, whereas in Stage III the increase is reduced and is nonlinear.

This characteristic three-stage work-hardening behavior is found to be qualitatively the same in all fcc metals. However, the extension of the stages,

[4] Nonbasal slip systems are observed in hcp metals such as titanium, zirconium and beryllium with subnormal c/a ratios. These metals deform easily by slip on prismatic planes.

[5] Work-hardening curves (or uniaxial stress–strain curves) of metal crystals are obtained from uniaxial tests at constant temperature, T, and plastic strain rate, $\dot{\varepsilon}_{\mathrm{pl}}$, by plotting, as a function of the plastic strain, $\varepsilon_{\mathrm{pl}}$, the flow stress, σ, required to maintain the deformation with the given plastic strain rate.

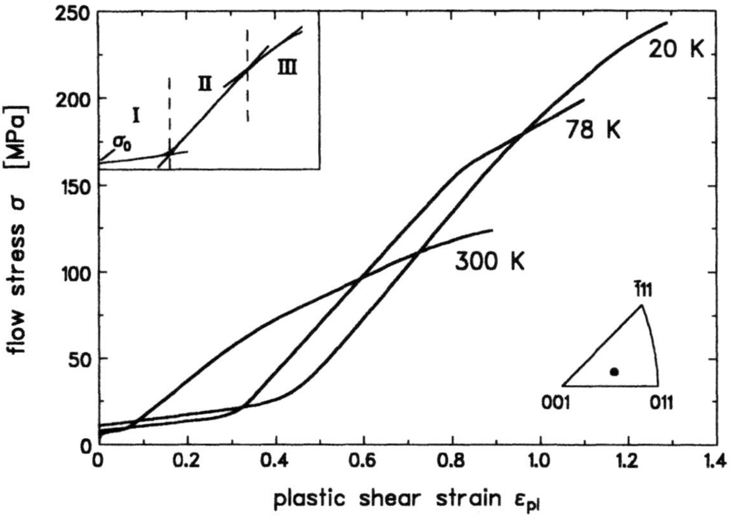

Fig. 11.5. Work-hardening curves of high-purity nickel single crystals, oriented for single slip in the $[\bar{1}01](111)$ slip system, at three different deformation temperatures [42]. The inset displays the subdivision of the curves into three stages schematically

as well as the work-hardening rates $\Theta = \mathrm{d}\sigma/\mathrm{d}\varepsilon_{\mathrm{pl}}$ in the stages, exhibit different pronounced dependences on the deformation temperature, the plastic strain rate, the orientation of the crystal axis and the impurity content. Generally, Stage I of fcc metals is much more sensitive to the abovementioned conditions than Stage II. Therefore, the extension of Stage I decreases strongly toward higher temperatures, whereas the work-hardening rate Θ_{I} increases. Raising the impurity content leads to an opposite effect: the extension of Stage I increases and Θ_{I} decreases with increasing temperature. In complete contrast, the work-hardening rate Θ_{II} in Stage II is found to be not only much less dependent on the deformation temperature and the impurity content, but also (if normalized by the shear modulus, M) nearly equal for all fcc metals investigated. However, the extension of Stage II decreases with increasing temperature and decreasing impurity content.

The characteristic three-stage hardening behavior discussed above is typical for crystals with orientations in the middle of the standard stereographic triangle; that is, for single slip orientations where the $[\bar{1}01](111)$ slip system is predominant (cf. Fig. 11.3). Getting closer to the boundary of the stereographic triangle, where multiple slip occurs more easily, the shape of the curves changes in such a way that Stage I becomes shorter, and even disappears for pure boundary orientations [43].

The three stages in the stress–strain curves are reflected by the development of the dislocation microstructure. Observations of the slip-line pattern on the surface of deformed copper single crystals in Stage I [44, 45] show fine, long and straight slip lines. TEM investigations of the stress-applied

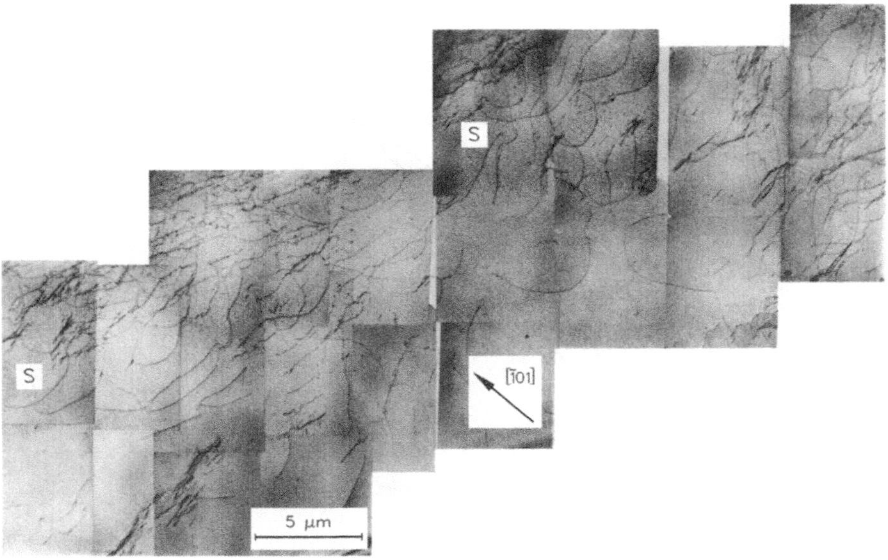

Fig. 11.6. A TEM image of 99.999% pure copper, deformed at 293 K to the transition from Stage I to Stage II. The foil is parallel to the primary slip plane (111). The dislocations were pinned in the unloaded state by irradiation with fast neutrons [53]. The dislocation sources, marked by "S", have generated outbowed glissile dislocation segments

state of copper single crystals, obtained by pinning the dislocations by neutron irradiation [46], have shown that the dislocation structure in Stage I is dominated by long curved primary dislocations and a few edge-dislocation dipoles (cf. Fig. 11.6). Groups of primary dislocations of the same sign, piled up against obstacles, were observed at higher strains. The dislocation structures of crystals that have been strained up to Stage II consist primarily of bundles and networks of primary and secondary dislocations and their reaction products, which are arranged roughly parallel to the primary slip plane [47]. These layer-like structures grow by the successive incorporation of other dislocation groups. In the stress-applied state, some pile-ups of primary dislocations on the same slip plane have been found, bowing out of the bundles between stable anchoring points [48]. In accordance with the development of the dislocation arrangement in the bulk, surface observations show in Stage II an increasingly inhomogenous and structured slip [44]. At the transition to Stage III, the slip line pattern is characterized by the appearance of sharp slip bands with cross-slip traces [44]. As TEM images have shown, in Stage III the dislocation structure is characterized by dense dislocation layers alternating with nearly dislocation-free areas, still parallel to the primary slip plane, and dislocation walls perpendicular to the primary slip direction [47, 49]. (A comprehensive survey of the work up to 1979 is given by Basinski and Basinski [50].)

A general result of the investigations mentioned above is that in fcc metals both edge and screw dislocations are of comparable high mobility. In fcc crystals with a stress axis in the middle of the stereographic triangle, the *macroscopic* slip is regularly crystallographic in all three stages of the work-hardening curve; that is, the slip occurs mainly on the primary slip plane (111) in the primary slip direction $[\bar{1}01]$. This was confirmed by investigations of the rotation of the tensile axis [51, 52] and by shape-change experiments [52] on aluminum and silver single crystals, respectively, which show that the ratio of secondary to primary slip is only of the order of 10^{-2}.

Based on the investigations discussed above, the work-hardening behavior of pure fcc metals may be explained qualitatively as follows [53, 54]. In Stage I the pre-existing dislocation forest remains practically unchanged, and the slight increase in the flow stress results mainly from an increase in the dislocation density. The beginning of Stage II is then determined by the flow stress at which secondary slip systems become active on a macroscopic scale. As a result, the rapid hardening in Stage II is mainly due to the formation and growth of dislocation bundles and layers, leading on the one hand to a reduction of the free slip distances in the primary system, due to the formation of immobile dislocation arrangements (e.g. Lomer–Cottrell dislocations [55, 56]), and on the other to an increasing number of dislocation sources (cf. Fig. 11.6). The beginning of Stage III is attributed to the thermally activated cross-slip at the heads of piled-up screw dislocation groups throughout the crystal. The decrease of the work-hardening rate in Stage III becomes possible by surmounting, or circumvention of, the slip barriers in the primary slip plane through the cross-slip or climb (at higher temperatures) of the screw and edge dislocations, respectively. Moreover, in Stage III dislocations of opposite sign may increasingly annihilate, which leads to a decrease in the work-hardening rate due to dynamic recovery.

11.3.2 Thermally Activated Dislocation Motion in fcc Metals

Due to the importance of plasticity for technological applications of metals, an enormous amount of effort has been put into their experimental and theoretical investigation. (The early models were critically reviewed by Nabarro, Basinski and Holt [57], by Mughrabi [54], and by Basinski and Basinski [50]; for more recent work see, for example, [58–60].) However, from the TEM observations discussed in the preceding section it is evident that the dislocation structures in deformed fcc metals are generally too complicated to permit the description of plastic deformation and work-hardening by any one simple mechanism developed from first principles.

A theoretical approach to explain the flow stress of fcc metals, which has stood the test of time, was introduced by Seeger and co-workers [10, 61–66] and then developed further by Kocks, Argon and Ashby [67] and Schoeck [68]. The basic physical concept is that the dislocations in a given slip plane give rise to long-range stress fields. The increase in flow stress during plastic deformation is mainly due to the increase of the "passing stress"; that is, the stress necessary to push a gliding dislocation through the elastic stress fields due to dislocations on the parallel planes. However, the flow stress at the beginning of plastic deformation will be determined not only by the interactions between the first dislocations generated on the chosen slip system but also by the interaction of these dislocations with those that are present in the as-grown state. These grown-in dislocations are assumed to be randomly distributed, and many of them pierce the primary slip planes. Usually, this dislocation microstructure is referred to as the "dislocation forest". Thus, the critical resolved shear stress, σ_0, and its dependence on temperature and strain rate, are strongly influenced by the initial dislocation density of the crystals.

The *long-range internal stress model* sketched above is well suited to explain the plastic behavior of as-grown single crystals at the beginning of plastic deformation. According to Seeger [61], the flow stress in the primary slip system, which is predominant in Stage I, may be decomposed into two components:[6]

$$\sigma = \sigma^* + \sigma_M . \tag{11.2}$$

The athermal stress contribution, σ_M, results from elastic interactions of the primary dislocations, mainly with dislocations lying in planes parallel to the slip plane and partly with dislocations intersecting the slip plane. Since the wavelength of the internal stress fields resulting from these interactions will be approximately equal to the dislocation spacing (viz. 10^{-6} m in an as-grown crystal), thermal fluctuations should not make any significant contribution to assist the applied stress to move the dislocations. Thus, the temperature dependence of σ_M should be entirely due to that of the elastic moduli, M, as indicated by the subscript. As at low enough temperatures self-diffusion is negligibly small, the other part of the flow stress in (11.2), σ^*, arises from primary gliding dislocations intersecting, with the help of thermal activation, secondary dislocations piercing the slip plane and acting as *short-range obstacles*. As in the fcc lattice both edge and screw dislocations can dissociate into partial dislocations, the temperature and strain-rate dependence of σ^* can be attributed partly to the fact that a gliding dislocation must be constricted to permit such intersection processes. Accordingly, σ^* depends on the separation, η, of the partials or on the stacking-fault energy, γ, of the crystal, respectively (see Table 11.1). However, most pure elements have stacking-fault energies that are so high that their equilibrium partial separation is of the order of a few lattice spacings and it would require stresses in excess of $10^{-2}M$ to dissociate them. Such stresses occur at most at the head of dislocation pile-ups, which are observed only at later stages

[6] Seeger [61] originally introduced the components σ_S for "*Schneidspannung*" (intersection stress) and σ_G for the stress contribution that depends on temperature as the "*Schubmodul G*" (the shear modulus; here denoted by M).

Table 11.1. The stacking-fault energies, γ, normalized by the shear modulus, M, and the modulus of the Burgers vector, b, and the equilibrium separation, η, of partials in edge and screw dislocations [69]

	$\gamma/Mb[10^{-3}]$	η_{screw}/b	η_{edge}/b
Some elements with allotropic transformations, many solid solutions	<1	>10	>50
Ag	2	7	30
Cu	4	5	15
Au	6	3	10
Ni	13	2	4
Al	27	1	2

of deformation. On the other hand, most solution-hardened alloys and some elements have rather extended dislocations, and therefore the constriction of the partials has to be considered.

As an example, Fig. 11.7 shows the influence of short-range obstacles on the dislocation motion and hence on the temperature dependence of the critical resolved shear stress, $\sigma_0(T)$, of pure and alloyed aluminum.[7] A characteristic feature of both curves in Fig. 11.7, and most of the other fcc metals and alloys, is the existence of a characteristic temperature, T_C, which is low with respect to the melting point. According to Seeger [10, 70], the linear increase of σ_0 below T_C is attributed to the formation of jogs in edge dislocations during intersection, assisted by the thermal component of the applied stress, σ^*. Then, T_C characterizes the temperature at which intersection processes become possible without the aid of the applied stress. Consequently, the constant values of σ_0 above T_C correspond entirely to the athermal component, σ_M. Whereas in pure aluminum the stacking-fault energy, γ, is so high that intersection processes do not involve the constriction of partial dislocations, in an aluminum–magnesium alloy[8] with a much lower γ, constrictions play a dominant role (cf. Table 11.1). Accordingly, as can be seen from Fig. 11.7, alloying leads not only to a higher athermal component of the critical shear stress, σ_M, (solid-solution hardening) but also to a stronger increase of the thermal component, σ^*, below T_C.

[7] For the fcc and hcp metals, σ_0 is usually obtained by back-extrapolation of the linear part of the stress–strain curves in Stage I to zero strain (cf. Fig. 11.5). This procedure is permitted since in fcc and hcp metals both edge *and* screw dislocations already contribute to the flow stress at the beginning of plastic deformation; that is, *macroscopic* plasticity is indeed observed.

[8] In the case of aluminum single crystals containing 5.52at.% magnesium the notation "impurity" as used in this chapter does, of course, not apply any more. Nevertheless, the curve in Fig. 11.7 that was obtained for this solution-hardened alloy gives a qualitative description of the influence of both substitutional and interstitial impurities on the critical resolved shear stress of fcc metals (see, e.g., [41]).

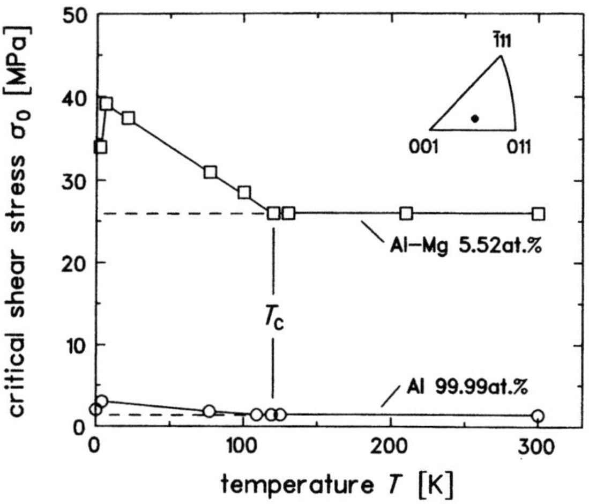

Fig. 11.7. The temperature dependence of the critical resolved shear stress, σ_0, of pure aluminum and of an aluminum–magnesium alloy; single crystals oriented for slip in the $[\bar{1}01](111)$ slip system [18]. The athermal stress component, σ_M, is marked by the dashed horizontal lines

The stress-assisted thermally activated overcoming of short-range obstacles may be described quantitatively within the framework of Kramers' theory of diffusion-controlled reactions [71], which deals with the escape of particles over potential barriers. In this picture, a dislocation has to wait in front of an obstacle until a successful thermal fluctuation helps to overcome it, and it then moves on to the next obstacle. Accordingly, the dislocation motion occurs in a jerky way. The plastic shear-strain rate, $\dot{\varepsilon}_{\mathrm{pl}}$, is given by the Orowan equation [72],

$$\dot{\varepsilon}_{\mathrm{pl}} = bN_{\mathrm{d}}\bar{v}_{\mathrm{d}}(\sigma^*, T)\,, \tag{11.3}$$

where b is the modulus of the Burgers vector, N_{d} denotes the density, and $\bar{v}_{\mathrm{d}}(\sigma^*, T)$ is the mean velocity of the dislocations. $\bar{v}_{\mathrm{d}}$ is determined essentially by the rate of occurrence of successful thermal fluctuations, $\Gamma_{\mathrm{d}}(\sigma^*, T)$, according to

$$\bar{v}_{\mathrm{d}} = ZA\Gamma_{\mathrm{d}}(\sigma^*, T)\,. \tag{11.4}$$

In (11.4), Z denotes the number of thermally activated processes per unit dislocation length and A is the area swept over by the dislocation per successful activation event. Under the assumption that the time of flight between the obstacles is small compared with the waiting time in front of the obstacles (the transition-state approximation), Γ_{d} can be obtained from general arguments as an Arrhenius equation [71],

$$\Gamma_{\mathrm{d}} = \nu_{\mathrm{d}} \exp\left\{-\frac{H(\sigma^*)}{k_{\mathrm{B}}T}\right\}\,, \tag{11.5}$$

with ν_{d} as the attempt frequency of the dislocation segment to overcome the obstacle. The enthalpy, $H(\sigma^*)$, of the activation process may be calculated for various kinds and distributions of obstacles. For short-range obstacles, which are especially sensitive to thermal activation, a useful phenomenologic expression is given by Kocks et al. [67]:[9]

$$H(\sigma^*) = H_0 \left[1 - \left(\frac{\sigma^*}{\hat{\sigma}}\right)^p\right]^q . \tag{11.6}$$

Here, $\hat{\sigma}$ is the obstacle strength – that is the stress needed to surmount the obstacles without thermal activation – and H_0 is the enthalpy necessary to overcome the obstacles without the aid of an applied stress and, therefore, characterizes an obstacle. The activation enthalpy $H(\sigma^*)$ decreases from H_0 to zero as σ^* is raised from zero to its maximum value $\hat{\sigma}$ at zero temperature. The empirical parameters p and q are found to be in the range $0 \leq p \leq 1$ and $1 \leq q \leq 2$, respectively, depending on the shape of the "obstacle profile"; that is on the force–distance curve of the obstacle [67].

11.4 The Plastic Behaviour of Pure bcc Metals

Pure bcc metals exhibit particular mechanical properties compared to those of hcp and fcc metals:

1. A strongly temperature-dependent shape of the stress–strain curves, with two distinct hardening regimes separated by a transition temperature, T_0.
2. The existence of a critical temperature, T_{K}, below which the flow stress needed to initiate *macroplastic* deformation increases very strongly with decreasing deformation temperature and increasing plastic strain rate.
3. A strong dependence of the flow stress on the crystallographic orientation of the stress axis, and a pronounced asymmetry for loading in tension and compression.
4. The occurrence of "anomalous" slip on the stress-unfavored $(0\bar{1}1)$ planes at low temperatures.
5. The striking effect of small concentrations of atomic lattice defects on the mechanical properties.

Some of these pecularities will be discussed in detail while analyzing recent deformation experiments on high-purity[10] single crystals of niobium, tantalum, α-iron, molybdenum and tungsten. It is interesting that most of the

[9] In [67] the Gibbs free enthalpy $G = H - TS$ of thermal activation is used, but within the framework of this section the activation entropy, S, is assumed to be negligibly small.

[10] In particular, the refractory metals molybdenum and tungsten can be produced in ultra-high purity. For example, in molybdenum the total concentration of dissolved interstitial impurities may be decreased to values of 0.02 at.ppm [73].

following also holds for the bcc alkali metal potassium [74], which, in contrast to the neighboring alkali metals lithium and natrium, does not undergo a low-temperature phase transition.

11.4.1 The Slip Geometry of Pure bcc Metals

The origin of the above-mentioned peculiarities of pure bcc metals is the fact that, in complete contrast to fcc metals, in the bcc lattice only the <111> slip directions are determined clearly crystallographically, but not the slip planes. The slip planes may be low-index planes of type {110}, {211}, and possibly {321}, depending on temperature, crystal orientation, purity and so on. This makes it difficult to calculate the resolved shear stress, σ, from the axial stress, σ_A, according to (11.1). Moreover, Schmid's law, if valid at all, is not easily to apply, since the critical shear stress, σ_0, could not be expected to be the same for different types of slip planes. As outlined for the fcc metals, an experimental determination of the operating slip planes is relatively simple if long and straight slip lines can be observed at the specimen surface. In contrast, for high-purity bcc metals deformed at higher temperatures (e.g. at room temperature) the slip lines, if clearly visible, are always wavy. On the average, they often do not follow the traces of a plane with low Miller indices (*hkl*), but coincide with or are close to the traces of the plane of maximum resolved shear stress of the operating <111> zone. At low temperatures, the situation is radically different. Here, the slip lines are found to be long and straight, and there is no doubt that the slip occurs on the closest-packed {110} planes. However, when high-purity specimens became available, it turned out by slip-line observations that in the bcc metals Schmid's law is not even obeyed for {110} slip planes.

In order to describe the complex slip geometry of bcc metals quantitatively, the angle κ in (11.1) has been replaced by the Taylor angles ψ and χ, which are more suited to treat the problem. In this notationing the Schmid factor is given by

$$\mu = \sin\xi \cos\xi \cos(\chi - \psi) , \tag{11.7}$$

where ξ is again (cf. (11.1)) the angle between the slip direction (chosen as [111]) and the stress axis (in Fig. 11.3 denoted by A), ψ is the angle between the {110} plane with the largest Schmid factor (chosen as $(\bar{1}01)$) and the observed slip plane (denoted by P), and χ is the angle between $(\bar{1}01)$ and the plane of maximum resolved shear stress (called the mrss plane, denoted by M, for short). Inside the basic stereographic triangle, indicated by bold lines in Fig. 11.3, these two angles may vary between $\pm 30°$. By convention, they are taken as positive if P and M lie on the side of the [111]–$[\bar{1}11]$ great circle and negative if they lie on the side of [111]–[001]. Using this description, the dependence of the observed slip plane on the orientation of the stress axis is usually plotted as a $\psi(\chi)$ curve.

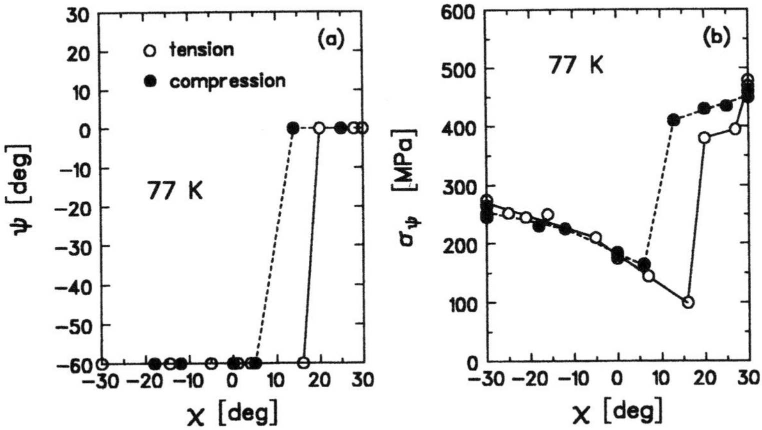

Fig. 11.8. The orientation dependence of slip in high-purity molybdenum single crystals deformed at 77 K up to $\varepsilon_{\mathrm{pl}} = 0.005$ [75]: (**a**) the ψ–χ relationship; (**b**) the χ dependence of the flow stress resolved on the observed slip plane, σ_ψ

Figure 11.8 shows how dramatically Schmid's law may be violated in bcc metals, using the example of high-purity molybdenum single crystals deformed slightly at liquid-nitrogen temperatures [75]. As can be seen from Fig. 11.8a, slip on the primary $\{110\}$ plane $(\bar{1}01)$ with $\psi = 0°$ was observed only for $\chi \geq 20°$ in tension and $\chi \geq 15°$ in compression, respectively. At smaller angles χ, the slip shifts to $\psi = -60°$; that is, to the stress-unfavored $(0\bar{1}1)$ planes with very low Schmid factors of about $\mu_{(0\bar{1}1)} = 0.2$ for $[\bar{1}49]$ crystals. For that reason, this slip mode is called *anomalous slip*. In the present case of molybdenum, it is limited to small strains. Figure 11.8b displays the strong orientation dependence of the corresponding flow stress resolved on the observed slip plane, σ_ψ, which shows, moreover, a marked *tension–compression asymmetry* at large positive χ. Both phenomena – the tension–compression asymmetry and the anomalous slip – will be discussed in more detail in Sects. 11.4.4 and 11.4.5, respectively.

11.4.2 The Work-Hardening Behaviour and Microstructure of bcc Metals

In following, the discussion of work-hardening behavior and dislocation microstructure is limited to high-purity molybdenum single crystals[11] with axis orientation $[\bar{1}49]$. As can be seen from the stress–strain curves in Fig. 11.9 at the transition temperature, T_0 (370 K in molybdenum), bcc metals undergo a change of deformation mechanism from a three-stage hardening behavior

[11] For detailed information about the purification and characterization of the ultra-high-purity molybdenum single crystals produced and investigated in Dresden [76–82], see Köthe [83].

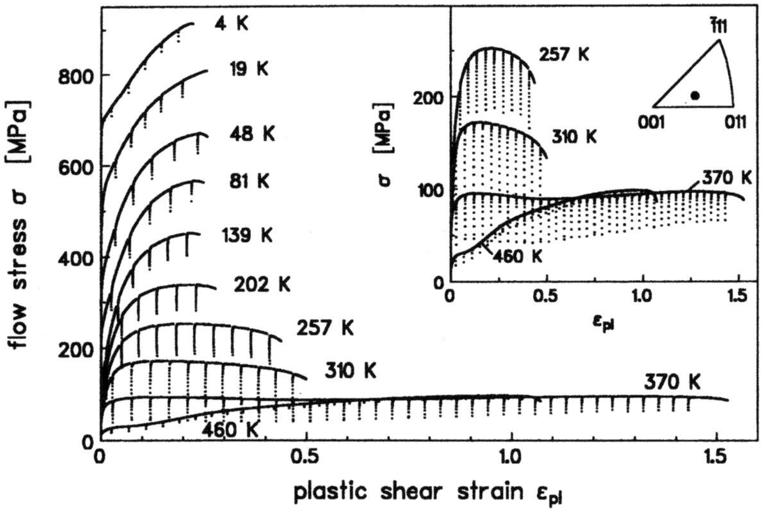

Fig. 11.9. Work-hardening curves of ultra-high-purity molybdenum single crystals with stress axis $[\bar{1}49]$ deformed in tension at $\dot{\varepsilon}_{pl} = 8.4 \times 10^{-4}\,s^{-1}$ [40, 84]. Note the exceptional high ductility of the crystal deformed at the transition temperature $T_0 = 370\,K$. The interruptions are caused by stress-relaxation experiments

at higher temperatures, which is similar to that of fcc metals, to a strongly parabolic[12] hardening behavior at lower temperatures.

11.4.2.1 Temperatures Below the Transition Temperature. If the temperature is not too low – that is, $120\,K \leq T \leq 370\,K$ in molybdenum – the extreme increase of the flow stress at the beginning of plastic deformation is followed at higher strains by a nearly constant flow-stress level, which is maintained until specimen failure. These two different parts of the stress–strain curves are reflected by the development of both the slip-line pattern and the dislocation microstructure, investigated mainly at room temperature [76–82,86–89]. As can be seen from the slip-line pattern, at the beginning of plastic deformation several slip systems are active [80–82]. The dislocation microstructure in the bulk is characterized by short dislocation segments, mostly of edge type, and small dislocation loops. Long screw dislocations were observed only rarely. Upon further straining the slip concentrates on the primary system $[111](\bar{1}01)$ and the conjugate system[13] $[\bar{1}11](101)$, and the slip lines become increasingly wavy, indicating extensive cross-slip. TEM investigations confirm the surface observations. Long straight primary and sec-

[12] α-iron, however, is an exception. Here the work-hardening curves are found to be nonparabolic, with an approximately horizontal course of flow stress versus strain from the beginning [85].

[13] The conjugated slip system $[\bar{1}11](101)$ is the secondary <111>{110} system with the highest Schmid factor; for example, $\mu_{(101)} = 0.47$ for crystals with stress axis $[\bar{1}49]$.

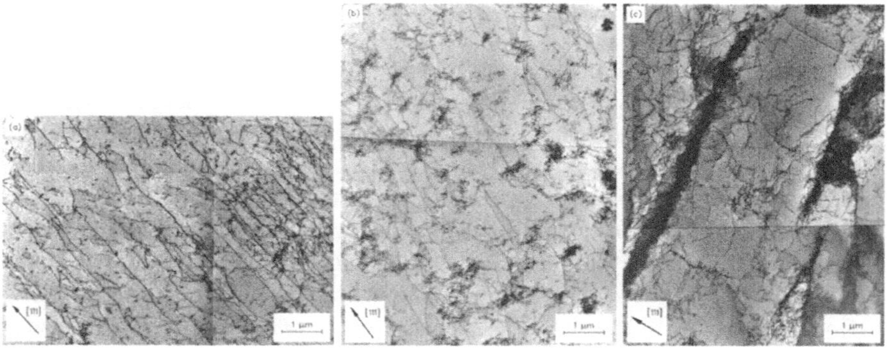

Fig. 11.10. The dislocation arrangement in high-purity niobium single crystals after deformation at (**a**) 77 K, (**b**) 195 K, and (**c**) 470 K. The dislocations were fixed in the unloaded state by irradiation with fast neutrons. Foils were parallel to the primary $(\bar{1}01)$ slip plane; primary screw dislocations were parallel to the [111] direction [90]. The images are representative for molybdenum single crystals, too

ondary screw dislocations are rather uniformly distributed (cf. Fig. 11.10b). Moreover, the motion of the long screw dislocations leaves behind a high amount of debris in the form of elongated edge dipole loops and small prismatic loops [78, 79]. With increasing strain, the densities of the long screw dislocations and of the debris increase. Where the flow stress approaches its constant level, besides a further uniform distribution of screws and debris, the TEM images show dense dislocation tangles of varying sizes and shapes (cf. Fig. 11.10b) [78, 79, 87–89]. At higher strains, the main change in the dislocation structure consists of continued growth and the compaction of existing tangles, whereas the number of tangles increases only slowly with strain. The density of the long screw dislocations remains nearly constant. At very high strains, most of the total dislocation length is localized within large tangles [78, 79, 88].

At very low temperatures, $T < 120$,K in molybdenum, work-hardening occurs until the specimen fails by necking. Fine, straight slip lines on the surface of molybdenum single crystals deformed at 4.2 K [91, 92] and 77 K [76, 92] appear already at the beginning of plastic deformation. The slip lines belong mainly to the primary system $[111](\bar{1}01)$, but also to secondary systems.[14] Accordingly, the dislocation microstructure in the bulk [86–88] is characterized by straight and uniformly distributed primary and secondary screw dislocations (cf. Fig. 11.10a). This dislocation arrangement remains qualitatively unchanged during further straining, whereas the average dislocation density increases. In contrast to the behavior at higher temperatures, dislocation tangles were not observed at and below 77 K.

From the observations reported above, we may conclude that the work-hardening behavior of bcc metals below T_0 is governed by the low mobility

[14] Anomalous slip can be observed in molybdenum at 77 K [75] (cf. Sect. 11.4.1).

of the screw dislocations. Immediately below T_0, the onset of so-called *microplastic* deformation (also called *microyielding*) at relatively small stresses is caused by the motion of grown-in edge dislocation segments [93]. The highly mobile edge dislocation segments move over large distances through the crystal, trailing behind pairs of long screw dislocations of opposite sign. Obviously, the strong increment of the flow stress at the beginning of the plastic deformation can be closely correlated with the increasing density of long immobile screw dislocations and the exhaustion of glissile edge dislocations. With increasing stress, the screw dislocations become more and more mobile. This leads over to the *macroplastic* regime, where both edge and screw dislocations contribute to the slip. The jogged appearance of the screws indicates that they do not slip continuously as a whole. Rather, the various segments move intermittently over short distances on different slip planes, creating steps of edge type. The development of dislocation tangles seems to be responsible for the flow-stress plateau. The growth of tangles is probably the result of their interaction with slipping dislocations [79]. Tangles are impenetrable barriers for dislocations and incorporate permanently new dislocation segments. Tangles are also involved in the process of dislocation generation, as appropriate edge segments break away and trail behind pairs of screws. However, the formation and growth of large tangles even at constant flow stress requires the operation of cross-slip and climb. Therefore, the tendency for dislocation tangling reduces as the deformation temperature is decreased. Obviously, the absence of a flow-stress plateau at very low temperatures is caused by the suppression of dislocation tangling.

11.4.2.2 Temperatures Above the Transition Temperature. Above the transition temperature T_0, bcc metals show remarkable similarities with fcc metals, such as high ductility, a low initial flow stress and a (quasi-) three-stage hardening behavior. Similar to fcc metals, the stress–strain curves can be separated into several stages, as shown in Fig. 11.5. At the beginning of the plastic deformation a pronounced microplastic region is also observed, known from temperatures below T_0. Analogous to the behavior at lower temperatures in this early stage of plastic flow, several slip systems are active with nearly equal weights [80]. In Stage I the slip-line patterns suggest that only the stress-favored primary $[111](\bar{1}01)$ slip system operates [80,86]. This observation has been confirmed by TEM investigations which showed that the majority of the dislocations were primary edge dislocations [78,89]. Long screw dislocations were observed only rarely. In contrast to pure fcc metals, in the bcc metals the work-hardening Stage II has degenerated to an inflection point leading over to Stage III [26]. At the beginning of Stage II/III the slip-line pattern shows evidence of secondary slip, which increases strongly with increasing strain. From TEM studies, it can be concluded that a dislocation cell structure with dislocation layers parallel to the primary slip plane and dense dislocation walls perpendicular to the primary slip direction develops in the work-hardening Stage II/III (cf. Fig. 11.10c). The cell boundaries surround virtually dislocation-free areas. Between neighboring cells there normally exist misorientations [78]. In spite of the distinct increase of secondary slip in Stage II/III, the tensile axis rotates clearly toward the primary Burgers vector [80]. As for the fcc metals, this leads to the conclusion that the primary slip is predominant.

The preceding may be summarized as follows. At temperatures above T_0, the work-hardening behavior of bcc metals is determined by the increasing mobility of the screw dislocations with increasing temperature. The almost complete absence of long screw dislocations in Stage I can be explained by their easy cross-slip and subsequent annihilation. This is indicated by the waviness of the slip lines, which is already evident at the beginning of the deformation. The end of Stage I is marked by the exhaustion of primary dislocation sources and the beginning of large-scale cross-slip of the screw dislocations. The activation of secondary systems leads to the formation of the dislocation layers. The long-range internal stresses originating from these layers are responsible for the large increase in work-hardening during the transition to Stage II/III. Obviously, macroscopic slip occurs in the primary slip system between the layers, where the stress distribution is very inhomogenous.

11.4.3 The Effective Flow Stress of bcc Metals

From the preceding section, we may conclude that a quantitative description of the temperature and strain-rate dependence of the flow stress of bcc metals below T_K is tantamount to a description of the temperature- and strain-rate-dependent mobility of the $a_0\langle 111\rangle/2$ screw dislocations in the bcc lattice – at least at high strains. For that purpose it is necessary to distinguish between the influence of the decreasing mobility of the screw dislocations on the flow stress and the effect of the dislocation arrangement. Following the suggestion of Seeger [17], the *macroscopic* flow stress, σ, of bcc metals may be decomposed according to (11.2) into a thermal and an athermal component. However, in complete contrast to the behavior in hcp and fcc metals, the pronounced temperature and strain-rate dependence of the so-called *effective flow stress*, σ^*, is an intrinsic property of pure bcc metals and can be entirely ascribed to the thermally activated motion of the $a_0\langle 111\rangle/2$ screw dislocations in the bcc lattice.

A very successful method to determine the temperature and strain-rate dependence of σ^* was developed by Ackermann, Mughrabi and Seeger [94]. The procedure is based on the cyclic-deformation technique, which permits us to obtain a large set of flow-stress data covering a wide range of temperatures and plastic strain rates from measurements on one single crystal, thus avoiding the reproducibility problems connected with the use of different samples in more standard tests (e.g. the tensile tests shown in Fig. 11.9). The basic

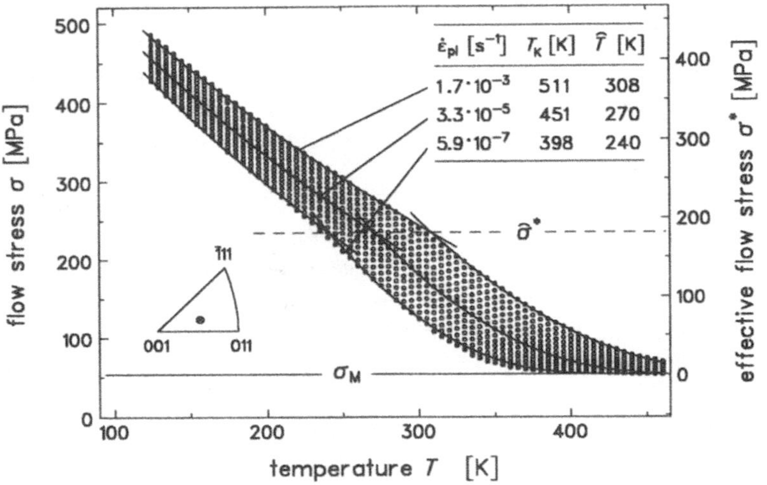

Fig. 11.11. The temperature dependence of the flow stress, $\sigma(T)$, and of the effective flow stress, $\sigma^*(T)$, respectively, for 15 different strain rates measured at *one* molybdenum single crystal [95]. The solid lines represent the fits for the three emphasized $\sigma^*(T)$ curves. The corresponding strain rates, $\dot{\varepsilon}_{\mathrm{pl}}$, knee temperatures, T_{K}, and upper-bend temperatures, $\widehat{T}$, are collected in the table. The athermal component of the flow stress, σ_{M}, and the upper-bend stress, $\widehat{\sigma}^*$, are marked by the full and dashed horizontal lines, respectively

idea is to produce a stable microstructure in the specimen bulk by preconditioning it at an elevated temperature, until a saturation-state is reached where the maximum resolved shear stress is independent of the number of cycles. Dislocation cell structures obtained by pre-conditioning the specimens well above T_{K} have proved to be very stable at lower temperatures during a few cycles involving changes of temperature and strain rate. After temperature and strain-rate changes, the dislocation microstructure can be checked and restored (if needed) by repeated cycling at the preconditioning temperature. An example is shown in Fig. 11.11: the complete set of flow-stress data was obtained by cyclic-deformation experiments on *one* molybdenum single crystal, pre-conditioned at 530 K [95]. At high temperatures and low strain rates, the flow stress, σ, assumed a strain-rate-independent value. This strain-rate independence indicates that under these conditions thermally activated processes no longer contributed to the flow stress; hence σ_{M} can be determined. The above-mentioned experimental conditions permit us to keep this σ_{M} value constant even at lower temperatures. Thus, in Fig. 11.11 the effective flow stress σ^* is given by subtracting the constant value σ_{M} (disregarding the small temperature dependence of M) from the measured flow stress, σ, in the entire temperature range investigated.

The experimental determination of σ^* from stress–strain curves of as-grown specimens – for example, that shown in Fig. 11.9 – is usually not

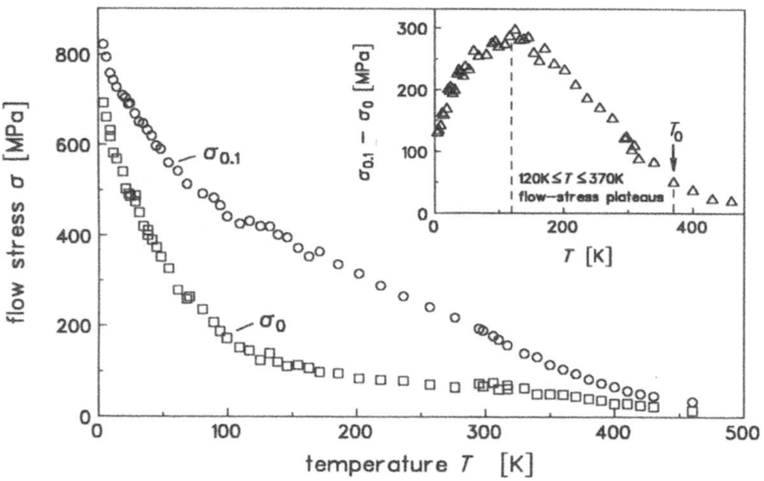

Fig. 11.12. The temperature dependence of the flow stress, σ, of molybdenum single crystals, obtained from the work-hardening curves in Fig. 11.9 at $\varepsilon_{\mathrm{pl}} = 0(\sigma_0)$ and $\varepsilon_{\mathrm{pl}} = 0.1(\sigma_{0.1})$. The inset displays the temperrature dependence of work-hardening, represented by the difference $\sigma_{0.1} - \sigma_0$. In the region $120\,\mathrm{K} \leq T \leq 370\,\mathrm{K}$, emphasized by the dashed lines, flow-stress plateaus are observed at high strains (cf. Fig. 11.9)

straightforward. As emphasized in the preceding section, the onset of *microplastic* deformation at relatively small stresses is caused by the motion of grown-in edge dislocation segments. Only at higher stresses, in the *macroplastic* regime, do both edge and screw dislocations contribute comparably to the slip. It is difficult to decide at which point of the stress–strain curves the influence of the screw dislocations on the flow stress becomes predominant (cf. Fig. 11.12), but this is important for comparison with theory. Both TEM investigations [79] and stress-relaxation experiments [96] have shown that, for tensile tests on molybdenum single crystals, the density of long screw dislocations reaches its maximum value at $\varepsilon_{\mathrm{pl}} \approx 0.1$ and remains nearly constant at higher strains (cf. Fig. 11.13b–d). Therefore, the corresponding values $\sigma_{0.1} := \sigma(\varepsilon_{\mathrm{pl}} = 0.1)$ may be tentatively identified as the effective flow stresses, σ^*, characteristic for the chosen deformation temperature, T, and plastic strain rate, $\dot{\varepsilon}_{\mathrm{pl}}$. As shown in Fig. 11.13a, this interpretation is supported by temperature-lowering tests on samples that have been pre-deformed at high enough temperatures [96].

Bcc metals show pronounced nonuniformities of the effective flow stress – temperature curves. A remarkable feature of the $\sigma^*(T)$ curves at constant strain rate in Fig. 11.11 is that, after a first increase below T_{K}, the slope $\partial\sigma^*/\partial T$ decreases to lower temperatures over a relatively narrow temperature interval. In this interval, the $\sigma^*(T)$ curves are convex from above. This first characteristic nonuniformity in the temperature dependence of the effective

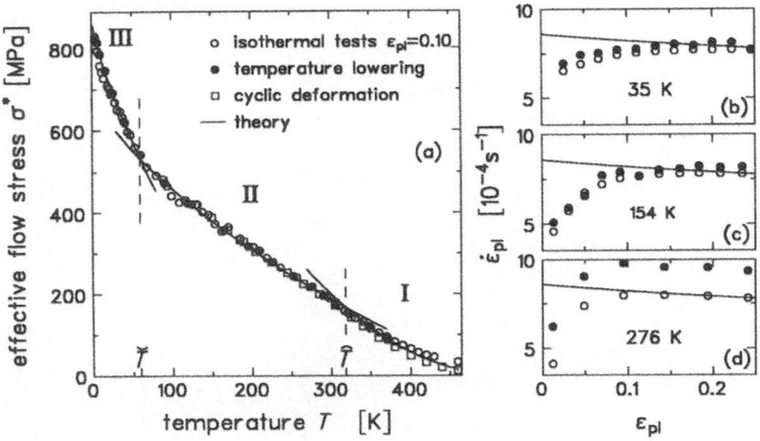

Fig. 11.13. The evaluation of the molybdenum data. (**a**) The temperature dependence of the effective flow stress, $\sigma^*(T)$: a comparison of the data obtained from the isothermal tests in Fig. 11.9 with the data of temperature-lowering tests [96] and cyclic-deformation tests (Fig. 11.11). (**b–d**) Three examples of the dependence of the plastic strain rate, $\dot\varepsilon_{\rm pl}$, on the accumulated plastic strain, $\varepsilon_{\rm pl}$. The open and full circles refer to different procedures for determining $\dot\varepsilon_{\rm pl}$. The straight lines give $\dot\varepsilon_{\rm pl}$ as calculated from the uniaxial straining rate [96]

flow stress, which is called the "upper bend"[15] [40], was observed early on for various bcc metals (α-Fe [97–101], Nb [102–104], Ta [22, 105] and Mo [106–108]). As Fig. 11.11 shows, the upper bend occurs at a flow stress $\widehat{\sigma}^*$ that is independent of the strain rate, but the corresponding "upper-bend temperature", $\widehat{T}$, decreases with decreasing strain rate.

A second characteristic nonuniformity in the temperature dependence of the flow stress at low temperatures was first found on α-Fe single crystals at about 100 K [19]. It is characterized by the fact that the extrapolation of the flow-stress data, $\sigma^*(T)$, from the regime well below the nonuniformity to higher temperatures, gives flow-stress values that lie significantly below those measured at temperatures above the nonuniformity (cf. Fig. 11.13). This entails a rather rapid variation of the slope $\partial\sigma^*/\partial T$ at constant plastic strain rate over a fairly narrow temperature interval. In molybdenum this so-called "lower bend" appears at about 60 K. In order to emphasize that at the lower bend the sign of the change $\partial\sigma^*/\partial T$ is opposite to that of the upper bend – that is, at the lower bend $\sigma^*(T)$ remains concave from above – the corresponding temperature is denoted by $\breve{T}$ and the flow stress by $\breve{\sigma}$ [40]. According to Brunner and Diehl [19, 85, 109], the three typical regimes in the temperature dependence of the flow stress of high-purity bcc metals were denominated by roman numerals I, II, and III.[16] As indicated in Fig. 11.13,

[15] In the literature, the "upper bend" is sometimes referred to as "the hump".

regime I lies between T_{K} and $\widehat{T}$, regime II between $\widehat{T}$ and $\breve{T}$, and regime III below $\breve{T}$.

11.4.4 The Flow-Stress Asymmetry of bcc Metals

A striking effect, which distinguishes high-purity bcc metals from pure fcc and hcp metals, is a marked difference of the flow stress, σ, measured in tension and in compression, respectively. This is not surprising. Since in the bcc lattice the {111} planes perpendicular to the dominant slip directions <111> are *not* mirror planes of the crystal structure, there is – in contrast to the fcc metals[17] – no general reason why the flow stress of a bcc crystal should remain the same when the sign of the resolved shear stress in the slip system is reversed. Therefore, in uniaxial tests, bcc metals should in general show an asymmetry,

$$\Delta\sigma := \sigma_{\mathrm{t}} + \sigma_{\mathrm{c}} \tag{11.8}$$

of the flow stress measured in tension, σ_{t} (being positive), and in compression, σ_{c} (being negative) – denoted as *flow-stress asymmetry.* An exception occurs if the slip planes happen to be {110} planes. In this case, the fact that {110} planes are mirror planes of the bcc structure too brings in an additional symmetry element compared with {211} or {321} planes. This picture is supported by the fact that no flow-stress asymmetry is observed in pure bcc metals if the slip occurs entirely on the anomalous $(0\bar{1}1)$ slip planes (e.g. at $\chi < 15°$ in Fig. 11.8b).

The fact that, in bcc metals, the stress value measured in an uniaxial deformation experiment depends on the sign of the load was observed early on by numerous authors (e.g. for molybdenum by Aono, Kuramoto, Brunner and Diehl [75] and by Guiu and his associates [106,111–115]). However, their results are contradictory and hard to compare, since it is rather difficult to deduce a relatively small difference in the flow stress from measurements on different specimens using different experimental set-ups for the tension and compression tests, respectively. The cyclic-deformation technique described in Sect. 11.4.3 permits us to avoid those difficulties, and is therefore well suited for precise measurements of the flow-stress asymmetry. Figure 11.14 shows the $\Delta\sigma(T)$ of high-purity molybdenum single crystals measured in cyclic deformation [95, 116]. For the three specimens investigated, at high temperatures the flow stress in tension equals the flow stress in compression; that is, $\Delta\sigma$ vanishes. At a strain-rate-dependent temperature, the sign of the flow stress asymmetry begins to change – for the specimens with $\chi < 30°$

[16] Although the notation is similar to that of the three hardening stages in fcc metals (cf. Fig. 11.5), there is no connection.

[17] In fcc metals the slip directions *are* perpendicular to mirror planes, viz. {110}. Hence the reversal of the resolved shear stress should not change the flow stress, in agreement with numerous observations (see, e.g., [110]).

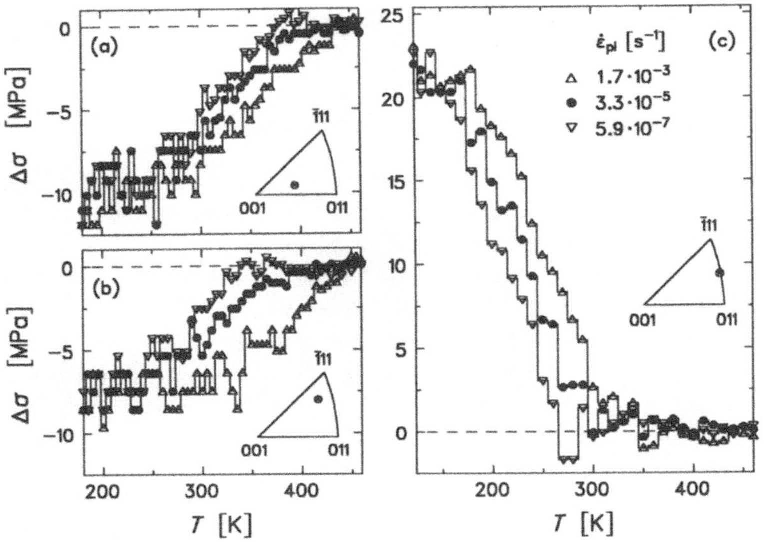

Fig. 11.14. The flow-stress asymmetry, $\Delta\sigma$, as a function of the temperature and plastic strain rate measured in cyclic deformation on three molybdenum single crystals with different orientations of the stress axis [116]: (**a**) $[\bar{1}49]$ with $\chi = 0°$, $\xi = 45°$ and $\mu_{(\bar{1}01)} = 0.50$; (**b**) $[\bar{5}79]$ with $\chi = +21°$, $\xi = 60°$ and $\mu_{(\bar{2}11)} = 0.43$; and (**c**) $[\bar{1}22]$ with $\chi = +29°$, $\xi = 54°$ and $\mu_{(\bar{2}11)} = 0.48$

to moderate negative values and for the specimen with $\chi \approx 30°$ to large positive values. The asymmetry rises sharply over a temperature interval of about 150 K. The temperatures at the center of this interval lie about 130 K below T_K and show a similar strain-rate dependence as T_K (cf. Fig. 11.11). At lower temperatures the flow-stress asymmetry seems to saturate for all three specimens. From the point of view of crystal symmetry, it is highly remarkable that in the symmetric crystal orientation with $\chi = 0°$ the asymmetry is not only different from zero but even more pronounced than in the orientation with $\chi = 21°$. Moreover, the sign of the asymmetry of the specimen with $\chi = 21°$ is not in agreement with the hypothesis that in bcc metals slip by dislocation motion should be easier in the *twinning* direction than in the *anti-twinning* direction. Only the specimen with $\chi \approx 30°$ meets this expectation, which will be briefly discussed in the following.

The flow-stress asymmetry observed in high-purity bcc metals is closely correlated with the geometry of twinning. Deformation twinning in the bcc structure takes place by shifting neighboring {211} lattice planes over each other in <111> directions by the amount $s = +3^{-1/2}a_0/2$; in this way, the mirror image (a "twin") of the original structure is obtained. Shifts of $s = -3^{-1/2}a_0/2$, which have not been observed, would not result in twins but in a crystal structure of very high energy. Hence the two opposite directions of <111> shifts are distinguished as the "twinning direction" and the "anti-

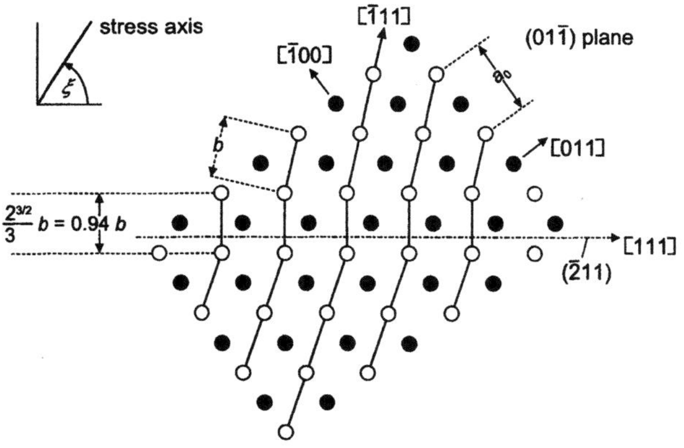

Fig. 11.15. An $[01\bar{1}]$ projection of the bcc structure with a stacking fault on $(\bar{2}11)$. The full and empty circles denote the unrelaxed atom positions on alternating $(01\bar{1})$ planes

twinning direction". Twinning in the system with the largest Schmid factor (which is identical with that of the corresponding <111>{211} slip system) shortens the specimen in the stress-axis direction if $\chi > 0$ and lengthens it if $\chi < 0$. This may be seen in Fig. 11.15, which shows a twin with a thickness of two atomic layers. The atoms adjacent to the $(\bar{2}11)$ plane (indicated by the interrupted line) are in mirror positions with respect to that plane. Barring exceptional situations such as cracking, in tensile tests on bcc metals twinning has indeed been observed only if $\chi < 0$, and in compression tests only if $\chi > 0$.

The asymmetry of the energy–shear-strain relationship for the simple shear leading to twinning is correlated with the Peierls potential $U(u)$ on {211} planes. $U(u)$ is softer if a dislocation is moved away from its position of minimal energy at $u = 0$ in the twinning direction than if it is moved in the opposite direction. The asymmetry of $U(u)$ should be reflected in the flow stress if slip takes place on {211} planes. In the early work on the flow-stress asymmetry of the bcc metals, this led to the view that the flow-stress asymmetry as defined by (11.8) should have the same sign as χ (the so-called twinning–anti-twinning rule). Irrespective of what was observed, the twinning direction was often referred to the "soft direction" and the anti-twinning direction as the "hard direction". For $\chi = 0°$, the conclusion that $\Delta\sigma$ should vanish agrees with the symmetry that is obtained if the stress axis lies in the $(\bar{1}01)$ plane. The observation of a small, but significant negative flow-stress asymmetry in ultra-high-pure molybdenum with $\chi = 0°$ (Fig. 11.14a) is clearly at variance with the above symmetry considerations. Moreover, at large positive χ, where the twinning–anti-twinning rule predicted large *positive* $\Delta\sigma$, a *negative* asymmetry with even smaller $\Delta\sigma$ (Fig. 11.14b) was observed. Since the direction of the χ-dependence is in agreement with the twinning–anti-twinning rule, these observations were interpreted as indicating the compensation of the "twinning–anti-twinning asymmetry" by an effect with an opposite but weak χ-dependence. This effect, called the "straining asymmetry" [116], will be discussed in Sect. 11.4.9. The assumption that the twinning–anti-twinning asymmetry is predominant at $\chi \approx 30°$ is confirmed by the measurements shown in Fig. 11.14c, where a large *positive* effect was indeed observed.

11.4.5 Anomalous Slip in bcc Metals

As already mentioned in Sect. 11.4.1 in high-purity bcc metals the so-called anomalous slip on $(0\bar{1}1)$ planes can be observed at low temperatures (cf. Fig. 11.8). Above all, anomalous slip is pronounced in the group Va metals vanadium [117], niobium [118] and tantalum [119]. Anomalous slip can be observed in both tensile and compression tests. It occurs preferentially for stress axes in the [001] corner of the standard stereographic triangle [118]; roughly speaking, to the left of the interrupted line in Fig. 11.3. Extensive studies of anomalous slip in middle-oriented niobium and tantalum single crystals of high purity have been performed recently by Wasserbäch [120] in the temperature region between 77 K and 250 K. It turned out that anomalous slip already occurs at about 200 K in niobium and only at 77 K in tantalum, respectively. In this case the slip-line patterns consist mainly of coarse and crystallographic slip bands that are visible around the whole circumference of the specimen and that correspond to the anomalous $(0\bar{1}1)$ planes. Investigation of the change of the crystallographic orientation of the specimens showed that in the case where anomalous slip lines occurred on the surface, the tensile axes rotated approximately toward the [011] direction. The [011] direction is the sum of the primary [111] and the conjugate $[\bar{1}11]$ slip directions (cf. Fig. 11.3). Therefore the rotation of the tensile axis can be regarded as the result of simultaneous slip in the $[111](0\bar{1}1)$ and the $[\bar{1}11](0\bar{1}1)$ slip system to almost equal extents. This picture is supported by TEM investigations, which show in the $(0\bar{1}1)$ planes a regular cross-grid of primary and conjugate screw dislocations of almost equal densities. The presence of the conjugate $a_0[\bar{1}11]/2$ screw dislocations will be explained in Sect. 11.4.9. The reason why the primary and conjugate screw dislocations (if already present) prefer the anomalous $(0\bar{1}1)$ planes is as follows [121]: when moving under the influence of the applied stress, the primary and conjugate screw dislocations will intersect each other unless they move on their common slip plane, which is the $(0\bar{1}1)$ plane. The intersection of the screw dislocations would lead to the formation of jogs in them. The dragging of these jogs by the moving dislocations would result in the formation of intrinsic atomic defects – in particular, of self-interstitials. Unless the temperature is high enough to contribute substantially to overcoming the energy barriers associated with these processes, the prescribed strain rate will be maintained by both types of screw dislocations cross-slipping to the anomalous $(0\bar{1}1)$ plane, where they can move conservatively without jog formation.

The phenomenon of anomalous slip is an example of the striking effect of small concentrations of impurities on the mechanical properties of bcc metals. A comparison of the above-mentioned results with earlier work performed on less pure specimens [122] shows that both the temperature below which anomalous slip sets in and the maximum angle χ where anomalous slip is still observed are strongly dependent on purity. Generally, the purer the specimen, the more the T and χ regions of anomalous slip are extended. On the other hand, in high-purity single crystals of vanadium, niobium and tantalum, anomalous slip may be substantially decreased and may finally completely be suppressed by doping the crystals with increasing amounts of interstitials (e.g. hydrogen in tantalum [120]).

11.4.6 Softening of bcc Metals by Atomic Defects

The fact that atomic defects influence the mechanical properties of bcc metals is understandable, since the perturbation of the crystal structure accompanying the incorporation of atomic defects – be they *intrinsic* (e.g. vacancies or self-interstitials) or *extrinsic* (foreign atoms) – interferes with the motion of dislocations. Generally, atomic defects induce hardening; that is, the flow stress of bcc metals increases (partly very strong) with increasing defect content. This leads to the well-known and thoroughly studied hardening phenomena, such as alloy hardening, irradiation hardening and quench hardening. Since this behavior is discussed in detail elsewhere (see, for example, [123]), this section will be focused on an unexpected phenomenon observed only in ultra-high-purity bcc metals; namely, softening caused by small amounts of atomic lattice defects. This effect, which was first noticed in α-iron alloys [124], will be discussed in the following by means of investigations on high-purity α-iron, especially those carried out by Brunner, Diehl and associates [125–128].

The occurrence of softening depends strongly on the species of the defects and on the deformation temperature [129, 130]. While, for example, interstitially dissolved foreign atoms soften α-iron in the temperature regime 80–270 K only (solid solution or alloy softening), atomic lattice defects induced by low-temperature irradiation with energetic particles cause softening at even lower temperatures (irradiation softening). It has turned out recently that softening in bcc metals can be properly analyzed phenomenologically and interpreted thoroughly only on the basis of the nonuniformities of the low-temperature flow stress of the pure metal (cf. Sect. 11.4.3). Since this phenomenon is only fully developed in materials of ultra-high purity, a reasonable study of softening requires ultra-high-purity starting materials.

Extrinsic atomic lattice defects of both kind, substitutional or interstitial, may lead to solution softening in high-purity bcc metals. However, whereas, for example, substitutional impurities in α-iron, such as silicon, cause only a weak softening [26], small amounts of interstitials may already have a strong

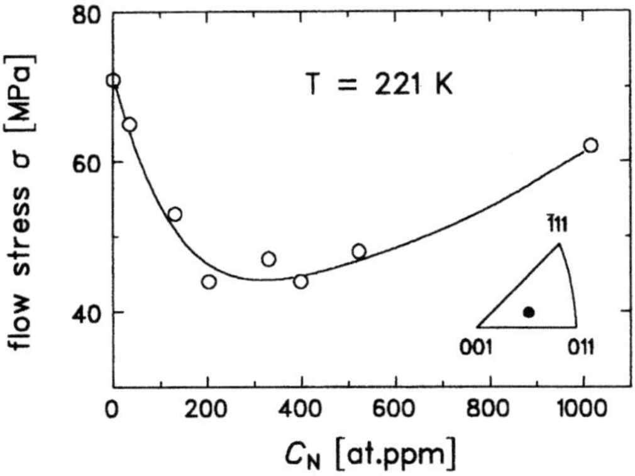

Fig. 11.16. The flow stress, σ, of α-iron single crystals, oriented for $[111](\bar{1}01)$ slip, as a function of the nitrogen concentration, C_N [126]

effect. For example, Fig. 11.16 shows the influence of a range of nitrogen concentrations on the flow stress of α-iron [126]. Figure 11.17a shows the temperature dependence of the flow stress of the specimens containing 330 at.ppm nitrogen (maximum of alloy softening in Fig. 11.6) and, for comparison, that for undoped high-purity crystals of the same orientation. It is evident that the low-temperature regime III in the notation of Brunner and Diehl appears not only to be unaffected by nitrogen doping but is also extended to higher temperatures. Consequently, after doping the transition to regime II changes; and regime I seems not to be present. A qualitative interpretation of this behavior will be given in Sect. 11.4.9.

Intrinsic atomic lattice defects cause softening effects in high-purity bcc metals which are much more complicated than those due to foreign atoms. On the basis of flow-stress measurements on α-iron, irradiated at the liquid-nitrogen temperature 77 K with 2.5 MeV electrons, it was shown that one must distinguish between *primary* and *secondary* irradiation softening [125] (Fig. 11.17b). Primary irradiation softening (PIS) is observed below $\breve{T}$ – that is, in the temperature range that remained unaffected by doping with nitrogen – provided that the irradiated sample was not annealed at temperatures of about 145 K or above. Secondary irradiation softening (SIS) occurs if the α-iron samples are annealed in the temperature range between 145 K and 220 K. Secondary irradiation softening is confined to temperatures above the lower bend; that is, it resembles alloy softening in every respect, except for the fact that it recovers at about 220 K.

11.4.7 The Kink Theory of bcc Metals

In 1971, Seeger and Šesták [131] proposed that the so-called alpha and gamma relaxations of plastically deformed high-purity bcc metals, which had been studied in detail by Chambers [132] and others, had an explanation ana-

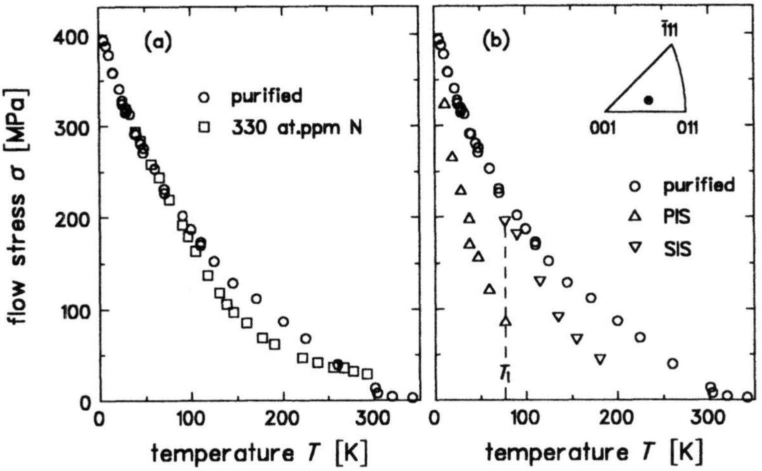

Fig. 11.17. The temperature dependence of the flow stress, σ, of purified α-iron single crystals compared with (**a**) crystals doped with 330 at.ppm nitrogen, and (**b**) crystals irradiated with an electron dose of $5.5 \times 10^{22}\,\mathrm{m}^{-2}$ in liquid nitrogen ($T_\mathrm{i} = 77\,\mathrm{K}$) and then (i) deformed at lower temperatures (PIS) or (ii) after prestraining at 77 K and subsequent annealing (20 min at 185 K), deformed at higher temperatures (SIS) [126, 127]

loguous to that of Bordoni relaxation in fcc metals [29]. Specifically, they suggested that the gamma relaxation, which in bcc metals such as niobium, tantalum or α-iron occurs near or somewhat above room temperature, is caused by thermally activated kink-pair formation in screw dislocations. In 1981, Seeger [17] extended the kink model to explain the low-temperature flow stress of bcc metals. Since that time, the kink theory has proved to be very successful in describing the particularities of the plastic behavior of pure bcc metals quantitatively.

11.4.7.1 The Flow Stress of bcc Metals. We assume that the effective flow stress, σ^*, of bcc metals below the knee temperature is determined by the properties of one type of dislocations only; namely, screw dislocations with a Burgers vector $\boldsymbol{b} = a_0\langle 111\rangle/2$. Then the macroscopic plastic strain rate, $\dot{\varepsilon}_\mathrm{pl}$, may be expressed in terms of the density, N_d, and the mean velocity, $\bar{v}_\mathrm{d}$, of these dislocations by the Orowan equation (11.3). The migration of the kinks along the dislocation line is limited by two different mechanisms. Kinks may either annihilate when they meet kinks of opposite sign on the same dislocation line, after having travelled a mean distance $\bar{x}_\mathrm{k}$, or they may become immobilized by being trapped at insurmountable obstacles separated by a mean distance $\bar{L}_\mathrm{d}$. For common deformation experiments, which involve high densities of dislocations and therefore many obstacles, we may assume that $\bar{L}_\mathrm{d} \ll \bar{x}_\mathrm{k}$ [40]. Then, the mean dislocation velocity is given by

$$\bar{v}_\mathrm{d} = a\bar{L}_\mathrm{d}\Gamma_\mathrm{k}(\sigma^*, T)\,, \tag{11.9}$$

where the kink height a is the distance between the Peierls valleys connected by a kink in a dislocation (cf. Fig. 11.4), and $\Gamma_\mathrm{k}(\sigma^*, T)$ is the net rate of kink-pair formation per unit dislocation length.

The overcoming of the Peierls barrier by kink-pair formation may be described by the diffusion and drift of the kinks over an energy barrier (a saddle point in a many-dimensional configuration space) resulting from the superposition of the attraction between kinks of opposite sign and the work performed on kinks by the effective flow stress, σ^*, which tends to drive the two kinks apart [17, 133]. Under these conditions, the net rate of kink-pair formation, which is the key quantity in the theory, may be calculated by an extension of Kramers' theory of diffusion-controlled reactions [71]. This leads to

$$\Gamma_\mathrm{k} = \frac{1}{\sqrt{\pi m_\mathrm{k} k_\mathrm{B} T}} \frac{2\pi H_\mathrm{k}}{w_\mathrm{k}^2} \tanh\frac{\Phi}{2} \exp\left\{-\frac{H_\mathrm{kp}(\sigma^*)}{k_\mathrm{B}T}\right\} \tag{11.10}$$

with

$$\sinh\Phi :\equiv 2\sqrt{2m_\mathrm{k}}\lambda(\sigma^*)\mu_\mathrm{k}(T) \equiv: x\,, \tag{11.11}$$

where $2H_\mathrm{k}$ denotes the formation enthalpy of two isolated kinks at zero stress, $H_\mathrm{kp}(\sigma^*)$ is the kink-pair formation enthalpy under the action of the effective flow stress σ^*, a is the height, w_k is the width (cf. Fig. 11.4), m_k is the mass and μ_k is the mobility of the kinks. The quantity $\lambda(\sigma^*)$ is the decay rate of the kink pairs in unstable equilibrium; it is determined by the shape of the energy barrier [133]. Kramers' theory contains, as limiting case $x \gg 1$, the so-called transition-state theory, in which the kink mobility no longer appears. (In this case, $\tanh(\Phi/2)$ is replaced by unity.) As a detailed discussion [133] shows, in the kink-formation problem in pure bcc metals, the transition-state theory is applicable except for small σ^*, where the full expression (11.11) must be used.

11.4.7.2 The Kink-Pair Formation Enthalpy. For calculating the kink-pair formation enthalpy, $H_\mathrm{kp}(\sigma^*)$, two different models may be used, depending on the separation q between the kinks in the configuration of unstable equilibrium at the energy barrier.

1) The Elastic-Interaction Model. At small σ^* and, therefore, large separations q between the two kinks of a kink pair (cf. Fig. 11.4), the kink–kink interaction is governed by the elastic long-range interaction between the kinks, which obeys a Coulomb-type law, $-a^2\gamma_0/2q$. The force exerted on the kinks by the effective stress σ^* leads to a repulsive potential $-abq\sigma^*$. The resulting total potential reads

$$U(\sigma^*, q) = -abq\sigma^* - \frac{a^2\gamma_0}{2q}\,. \tag{11.12}$$

The quantity γ_0 denotes the pre-logarithmic factor of the elastic contribution to the dislocation line tension $\gamma_d = \gamma_0 \ln(R/r_0)$, where R is the outer and r_0 the inner cut-off radius of the stress-field contribution to the dislocation line energy. γ_0 may be calculated unambigously using the second order elastic constants [134]. From (11.12), the kink-pair formation enthalpy follows as

$$H_{\mathrm{kp}}^{\mathrm{EI}}(\sigma^*) = 2H_{\mathrm{k}} - 2\alpha\sigma^{*1/2}\,, \tag{11.13}$$

with

$$\alpha = (a^3 b\gamma_0/2)^{1/2}\,. \tag{11.14}$$

2) The Line-Tension Model. Equation (11.12) fails at large stresses and hence smaller separations q of the two kinks of a pair. In this case, the kink-pair formation enthalpy may be calculated from the so-called line-tension model. In this model the self-interaction of a curved dislocation line is described in terms of the dislocation line tension γ_d and the potential energy $U(u)$ per unit dislocation length, as a function of its displacement u from the bottom of the Peierls valley. The kink-pair formation enthalpy is given by the difference between the energy of a kinked dislocation under the action of the effective stress, σ^*, and a straight dislocation under the same stress. Although $H_{\mathrm{kp}}(\sigma^*)$ may be calculated for any potential $U(u)$, for simplicity the so-called Eshelby potential [135],

$$U(u) = \frac{a^2 U''(a)}{2}\left(\frac{u}{a}\right)^2\left(1-\frac{u}{a}\right)^2\,, \tag{11.15}$$

is often employed, where $U''(a)$ is the second derivative of the Peierls potential in its minima. Since the Eshelby potential is even in u, it cannot describe the tension–compression asymmetry discussed in Sect. 11.4.4. Moreover, it does not possess the spatial periodicity of the crystal. It does reflect, however, the main features involved in the formation of a single kink pair, as long as we disregard the tension–compression asymmetry, which is usually a relatively small effect (cf. Fig. 11.14). The Eshelby potential has the advantage that an analytic expression for the kink-pair formation enthalpy may be derived:

$$H_{\mathrm{kp}}^{\mathrm{LT}}(\sigma^*) = 2H_{\mathrm{k}}\left[\beta - \frac{6b}{aU''(a)}\mathrm{arctg}\left(\frac{3^{1/2}\beta}{(1+2\beta^2)^{1/2}}\right)\right]\,. \tag{11.16}$$

The parameter β is related to a, b and $U''(a)$ by

$$(1-\beta^2)(1+2\beta^2)^{1/2} = 3^{1/2}\frac{6b}{a\sigma^*}U''(a)\,. \tag{11.17}$$

For $\sigma^* \ll aU''(0)/b$, (11.16) may be simplified and written as

$$H_{\mathrm{kp}}^{\mathrm{LT}}(\sigma^*) = 2H_{\mathrm{k}} - ab\sigma^*\left[\frac{\gamma_d}{U''(a)}\right]^{1/2}\left[1+\ln\left(\frac{\tilde{\sigma}}{\sigma^*}\right)\right] + O(\sigma^{*2})\,, \tag{11.18}$$

with $\tilde{\sigma} = 2aU''(a)/b$. However, (11.18) holds not only for the Eshelby potential (11.15) but for any physically acceptable $U(u)$. The general expression for $\tilde{\sigma}$ reads

$$\tilde{\sigma} = \frac{4aU''(a)}{2b} \exp\left\{ \int_{u=0}^{a} \left[\frac{u}{a} \left(\frac{2U''(a)}{U(u) - U(0)} \right)^{1/2} - \frac{2}{a-u} \right] du \right\} . \quad (11.19)$$

The argument of the exponential function in (11.19) is independent of the kink height a and the strength of the Peierls potential. If (11.15) is exact, $U''(a)$ can be expressed in terms of the Peierls stress, σ_{P}. We then have

$$U''(a) = \frac{2b\sigma_{\mathrm{p}} 3^{3/2}}{a} \quad (11.20)$$

and

$$\tilde{\sigma} = 12\sigma_{\mathrm{p}} 3^{1/2} . \quad (11.21)$$

It should be noted again, however, that in the present case, in which $U(u)$ is known to be asymmetric (cf. Sect. 11.4.4), its description in terms of the Eshelby potential is just a book-keeping device.

11.4.8 Elementary Slip Steps in High-Purity bcc Metals

From the above calculations, we may derive the following relationships between the macroscopic measurable quantities plastic strain rate, $\dot{\varepsilon}_{\mathrm{pl}}$, effective flow stress, σ^*, and temperature, T, on the one hand, and the parameters that describe the kinks and the dislocation arrangement on the other:

$$\dot{\varepsilon}_{\mathrm{pl}} = \frac{abN_{\mathrm{d}}\bar{L}_{\mathrm{d}}}{\sqrt{\pi m_{\mathrm{k}} k_{\mathrm{B}} T}} \frac{2\pi H_{\mathrm{k}}}{\omega_{\mathrm{k}}^2} \frac{x}{\sqrt{1+x^2}+1} \exp\left\{ -\frac{H_{\mathrm{kp}}(\sigma^*)}{k_{\mathrm{B}}T} \right\} , \quad (11.22)$$

$$x = 4(b\sigma^*)^{3/4} \left(\frac{2am_{\mathrm{k}}^2}{\gamma_0} \right)^{1/4} \frac{D_{\mathrm{k}}}{k_{\mathrm{B}}T} . \quad (11.23)$$

Equation (11.23) is obtained from (11.11) by using the Einstein relation $D_{\mathrm{k}} = \mu_{\mathrm{k}} k_{\mathrm{B}} T$ (where D_{k} is the kink diffusivity) and the curvature λ of the energy barrier, given by (11.12), at the saddle-point separation, $\tilde{q}$, at which a kink pair is in unstable equilibrium (cf. Fig. 11.4). For the kink-pair formation enthalpy $H_{\mathrm{kp}}(\sigma^*)$, the expressions according to the elastic-interaction model (11.13) or according to the line-tension model (11.18) must be used.

The cyclic-deformation experiments on ultra-high-purity single crystals of bcc metals in particular have confirmed the predictions of the theory to a very high degree [94, 95, 136, 137]. One of the critical tests of the theory is that the temperature dependence of the effective flow stress, $\sigma^*(T)$, at higher temperatures should show a clearly discernable bend (the "upper bend"), at an effective flow stress level $\widehat{\sigma}^*$ that is *independent* of the chosen strain rate,

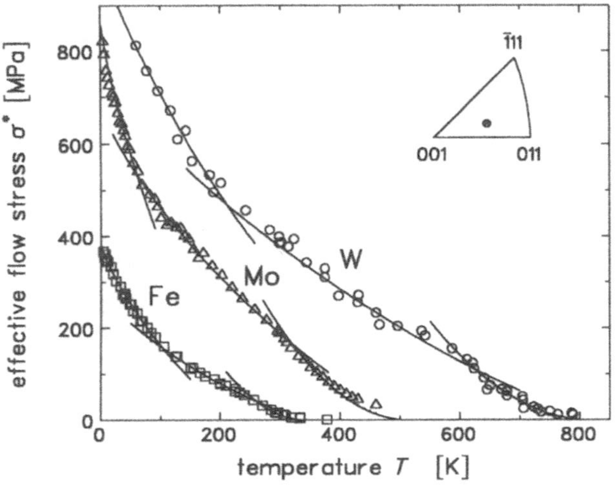

Fig. 11.18. The effective flow stress versus temperature of α-iron [109], molybdenum [96] and tungsten [138]. The single crystals were measured in tension at $\dot{\varepsilon}_{\mathrm{pl}} = 8.6 \times 10^{-4}\,\mathrm{s}^{-1}$. The full lines represent the calculated curves (cf. Table. 11.2)

$\dot{\varepsilon}_{\mathrm{pl}}$ (cf. Fig. 11.11). The physical meaning of $\widehat{\sigma}^*$ is that of the effective flow stress, σ^*, at which the kink-pair formation enthalpies $H_{\mathrm{kp}}(\sigma^*)$ as calculated by the line-tension model or by the elastic-interaction model intersect. At this intersection, the slope $\mathrm{d}H_{\mathrm{kp}}/\mathrm{d}\sigma^*$ predicted by the line-tension model is larger than that predicted by the elastic-interaction model, by a factor between 1.59 and 1.69. For example, for ultra-high-purity molybdenum, the ratio is found to be 1.63 [95], in agreement with the prediction.

The physical quantity of main interest is the kink height a. It is defined as the distance between the two Peierls valleys connected by the kinks (cf. Sect. 11.2). For a dislocation line in the bcc structure running along <111>, the Peierls potential has the period

$$a_{\{110\}} = (2/3)^{1/2} a_0, \tag{11.24}$$

$$a_{\{211\}} = 2^{1/2} a_0 = 3^{1/2} a_{\{110\}}, \tag{11.25}$$

$$a_{\{321\}} = (8/3)^{1/2} a_0 = 2 a_{\{110\}}, \tag{11.26}$$

for the slip planes $\{110\}$, $\{211\}$ and $\{321\}$, respectively. From a naive point of view, one expects a to agree with (11.24); in other words: a should equal the separation of the Peierls valleys on the $\{110\}$ slip planes. This appears natural, since $a_{\{110\}}$ is the shortest kink height in the bcc structure and since, for a given height of the Peierls barrier, the kink-pair formation enthalpy increases with $a^{3/2}$ [cf. (11.13)]. Above and below $\widehat{T}$ – that is, in regimes I and II in the notation of Brunner and Diehl – a can be deduced with exceptionally high accuracy by fitting the flow-stress data available from cyclic-

Table 11.2. A summary of the main kink parameters $2H_k$ and a (expressed in terms of $a_{\{110\}}$) and of the characteristic stresses $\hat{\sigma}^*$, $\breve{\sigma}^*$ and σ_p for the three metals in Fig. 11.18. As an exception, for α-Fe Brunner and Diehl [109] found that in regime II both $2H_k$ and a are lower than in regime I

	III			II			I		
Metal	σ_P [MPa]	$2H_k$ [eV]	a [$a_{\{110\}}$]	$\breve{\sigma}^*$ [MPa]	$2H_k$ [eV]	a [$a_{\{110\}}$]	$\hat{\sigma}^*$ [MPa]	$2H_k$ [eV]	a [$a_{\{110\}}$]
α-Fe	395	0.60	1.12	160	0.74	1.12	45	0.93	1.73
Mo	870	0.63	0.98	530	1.27	1.67	166	1.27	1.67
W	982	1.27	1.22	470	2.06	1.92	110	2.06	1.80

deformation experiments [94,95,136,137] to the elastic-interaction model and the line-tension model, respectively (cf. Fig. 11.11). It is somewhat surprising that the experiments on all high-purity bcc metals investigated so far are in disagreement with the above-mentioned expectation of {110} slip steps, particularly since, in all specimens investigated, the [111]($\bar{1}$01) slip system was strongly favored. However, there is an excellent agreement with (11.25); that is, with elementary slip steps on {211} planes. This result also holds for tensile tests [84, 85, 96, 109, 138]. As a consequence, the upper bend is *not* the result of a change in the slip mechanism, but an intrinsic feature of the kink-pair mechanism. The observation that the slip lines at intermediate temperatures are wavy and closely correlated to the plane of maximum resolved shear stress (cf. Sect. 11.4.1) can be explained without contradiction by the hypothesis of elementary {211} slip steps and easy cross-slip of the screw dislocations in the temperature range between $\breve{T}$ and T_K.

At temperatures below $\breve{T}$ – that is, in regime III according to Brunner and Diehl – determining the alternative "{110} or {211} slip planes" is more difficult. It has been demonstrated on high-purity α-Fe [109, 139], on ultrapure molybdenum [96], and on high-purity tungsten [138] that the enthalpy of formation a pair of isolated kinks, $2H_k$, comes out much smaller if determined by the extrapolation $\sigma \to 0$ from below $\breve{T}$ than from above $\breve{T}$ (cf. Table 11.2). This is a most unusual result, since it means that high-temperature slip has a higher activation energy than the process that controls slip at low temperatures (cf. Fig. 11.19). The explanation is as follows [40]. The core configuration of lowest energy of the a_0<111>/2 screw dislocations is such that the dislocations can slip by the kink-pair mechanism on {110} planes. This accounts for the finding that the kink height below $\breve{T}$ is in agreement with (11.24) – that is, with slip steps on {110} planes – and for the observation that at very low temperatures the slip lines of bcc metals are straight and compatible with <111>{110} slip. The screw dislocations are postulated to have a second, metastable core configuration which has, due to symmetry breaking, a well defined {211} slip plane. The Gibbs free enthalpies of the

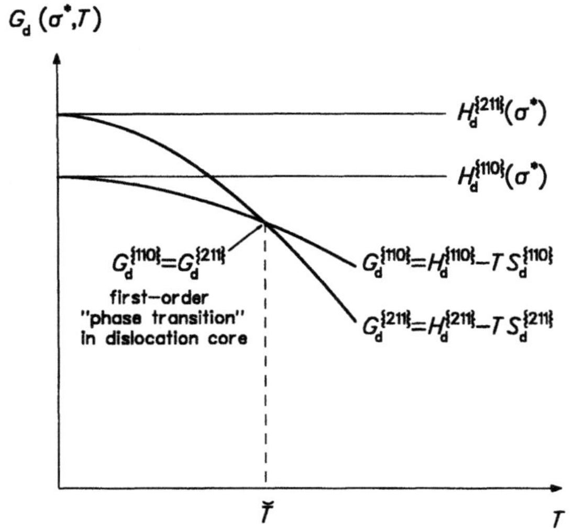

Fig. 11.19. A cut through the proposed surface of the Gibbs free dislocation line enthalpy, $G_\mathrm{d}(\sigma^*, T)$, for a a_0<111>/2 screw dislocation with either {110} or {211} slip planes [40]. The cut follows the temperature variation $\sigma^*(T)$ of the effective flow stress for a given plastic strain rate, $\dot{\varepsilon}_\mathrm{pl}$; hence it depends on σ^*. The intersection of $G_\mathrm{d}^{\{110\}}$ and $G_\mathrm{d}^{\{211\}}$ gives the temperature $\breve{T}$ of the lower bend

different core configurations are given by

$$G_\mathrm{d}^{(hkl)}(\sigma^*, T) = H_\mathrm{d}^{(hkl)}(\sigma^*, T) - TS_\mathrm{d}^{(hkl)}(\sigma^*, T) \,. \tag{11.27}$$

In the usual way, in (11.27), H_d refers to the dislocation-line enthalpy, and S_d to the dislocation-line entropy. The lower bend of the flow-stress–temperature relationship, $\sigma^*(T)$, is then identified with the point $\breve{\sigma}^*$, $\breve{T}$ at which the Gibbs free enthalpies,

$$G_\mathrm{d}^{\{110\}}(\breve{\sigma}^*, \breve{T}^*) = G_\mathrm{d}^{\{211\}}(\breve{\sigma}^*, \breve{T}) \,, \tag{11.28}$$

of the {110} and the {211} core configurations become equal (cf. Fig. 11.19). According to Seeger [40], here the cores of the a_0<111>/2 screw dislocations undergo a first order "phase transition" from their low-temperature configuration, capable of slipping on {110} planes, to their high-temperature configuration, slipping on one of the {211} planes.

11.4.9 Conclusions for Slip in bcc Metals

The existence of two core configurations of a_0<111>/2 screw dislocations, allowing for slip on the {110} or {211} planes, respectively, accounts for the unknown contribution to the *flow-stress asymmetry* discussed in Sect. 11.4.4. A screw dislocation with its core configuration allowing {211} slip must be extended on its slip plane, albeit by a small amount, since otherwise during slipping the dislocation would pass through an unstable configuration, from which it would fall back into the low-energy {110} slip configuration [40]. It is natural to assume that this extension bears some relationship to {211} stacking faults. In the bcc structure, the {211} planes appear to be the only ones on which stacking faults are mechanically stable; however, in most bcc metals and alloys their specific fault energy, γ, is too high to lead to a splitting of a_0<111>/2 dislocations into partial dislocations that are comparable with that which is familar in the fcc structure (cf. Sect. 11.3.2). A likely reason for this difference may be seen in Fig. 11.15. Whereas in the fcc structure a stacking fault produced by a rigid shift on a {111} plane leaves all the nearest-neighbor distances of the atoms unchanged, this is not so in the present case. Here, the distance between the atoms facing each other across the fault along the $[\bar{2}11]$ axis is about 6% less than the nearest-neighbor distance in the perfect lattice, which gives rise to a large strain contribution to the stacking-fault energy that is absent in the fcc structure. This contribution is lowered, and hence γ is reduced, if there is a tensile strain along the $[\bar{2}11]$ axis. In Fig. 11.15, the plane of the drawing, $(01\bar{1})$, corresponds to $\chi = 30°$. It is easily seen that, for realistic Poisson ratios, uniaxial tensile stress with an axis inside the stereographic triangle and χ near 30° leads to positive strain along $[\bar{2}11]$; the more the closer ξ approaches 90°. The argument may be extended to most of the interior of the basic stereographic triangle. Then it may be argued that a tensile stress along the specimen axis reduces the energy difference between the {211} and the {110} configurations, and hence the height of the Peierls barrier for {211} slip, whereas compression has the opposite effect. This explains the existence and the sign of the flow-stress asymmetry at $\chi = 0°$. At larger χ, the "straining asymmetry" is gradually compensated by the "twinning-anti–twinning asymmetry", which is the dominant effect at $\chi = 30°$.

Furthermore, the change of the elementary slip steps from {110} planes below $\breve{T}$ to {211} planes above allows us to explain why and under what circumstances in tensile or compression tests there should be a significant density of $a_0[\bar{1}11]/2$ screw dislocations that is sufficient to produce *anomalous slip*, as discussed in Sect. 11.4.5. The key is to recognize that if the restriction to the [111] slip direction is dropped, the <111>{211} slip system[18] with the largest Schmid factor on the [001] side in the standard stereographic triangle (cf. Fig. 11.3) is no longer $[111](\bar{1}\bar{1}2)$ but $[\bar{1}11](1\bar{1}2)$ (ref. [121]). Below $\breve{T}$, where the slip is confined to {110} planes, the motion of conjugate $a_0[\bar{1}11]/2$ screw dislocations ceases after exhaustion of the edge segments,

[18] In cubic crystals, the <111>{211} slip system with the largest Schmid factor is not the same inside the entire standard stereographic triangle. For the slip direction [111], the two {211} slip planes with the largest Schmid factors inside the triangle are $(\bar{2}11)$ on the $[011]$–$[\bar{1}11]$ side and $(\bar{1}\bar{1}2)$ on the [001] side, respectively (cf. Fig. 11.3).

owing to their small Schmid factor – except, of course, in the trivial case of the stress axis at or very close to [001]. Hence, anomalous slip should not occur. Above $\breve{T}$, however, for part of the time $[\bar{1}11]a_0/2$ screw dislocations possess $(1\bar{1}2)$ slip planes, since in the left-hand part of the standard stereographic triangle the Schmid factor of this configuration exceeds that of all three {211} configurations of $a_0[111]/2$ dislocations. These $a_0[\bar{1}11]/2$ screw dislocations on $(1\bar{1}2)$ planes act as "mother" dislocations for subsequent slip on the anomalous $(0\bar{1}1)$ plane, where both the primary and the conjugate screw dislocations can move conservatively.

Finally, it should be pointed out that the assumption of a first order phase transition in the cores of $a_0<111>/2$ screw dislocations at the transition from regime III to regime II offers a straightforward way of explaining the phenomena of *alloy softening and irradiation softening* at $T \geq \breve{T}$, introduced in Sect. 11.4.6. It may be postulated [40] that certain atomic defects can reach the dislocation cores and cause them to "overheat"; that is, to shift the phase transition from the low-temperature {110} configuration to the high-temperature {211} configuration toward higher temperatures. Particularly good candidates for producing this overheating effect are the "crowdions" created by irradiation below $\breve{T}$ [140]. These intrinsic atomic defects are mobile at $\breve{T}$ and can therefore reach the dislocation cores. However, as shown in Fig. 11.17a, other atomic defects – for example, "heavy" interstitial foreign atoms such as nitrogen – can be incorporated in the cores of $a_0<111>/2$ screw dislocations of the refractory bcc metals by suitable heat treatments too, and can give rise to alloy softening. We may thus conclude that both the secondary irradiation softening and the solid-solution softening are due to the overheating of the transition from the {110} configuration to the {211} configuration of the screw dislocation core.

11.5 Summary

The early experiments on pure cadmium single crystals have already shown that the critical resolved shear stress of metals with a close-packed structure is very low at both room temperature and liquid-helium temperature. The observed three-stage work-hardening behavior was realized as typical for hcp and fcc metal crystals with orientations in the middle of the stereographic triangle. It turned out that in metals with a close-packed structure the slip is regularly crystallographic in all three stages of the work-hardening curve; that is, the slip takes place on stress-favored densely packed planes in the direction of the shortest lattice vector. On the basis of TEM investigations, it became clear that the deformation behavior of hcp and fcc metals is determined by the fact that both edge and screw dislocations are in principle highly mobile even at the lowest temperatures. Work-hardening during straining was found to be governed by the increasing number of dislocations, which interact with one another in a complex manner over long distances through their stress fields and over short distances by intersection processes. Since the latter process depends strongly on the stacking-fault energy of the crystal, the strong influence of atomic lattice defects on the mechanical properties of hcp and fcc metals can be attributed, among other things, to their influence on the stacking-fault energy.

When measurements on high-purity refractory bcc metals became available, both their parabolic work-hardening behavior and the rapid increase of the flow stress with decreasing temperature and increasing strain rate were established as intrinsic properties. This behavior was soon ascribed to the sessile nature of the $a_0<111>/2$ screw dislocations in the bcc lattice which leads to a strong decrease in the dislocation mobility below the knee temperature. Accordingly, exhaustion of the more mobile nonscrew dislocations was found to be the origin of the extended microplastic region observed in most bcc metals at the beginning of plastic deformation. It turned out that, contrary to hcp and fcc metals, in bcc metals deformed below the knee temperature the influence of dislocation interactions on the flow stress is small compared with the resistance offered by the exceptionally high intrinsic Peierls barriers to the motion of the screw dislocations.

At finite temperatures, high Peierls barriers may be surmounted by the stress-assisted thermally activated formation of kink-pairs in dislocations. This process is indeed present in all metals, and was found to be the origin of Bordoni relaxation in fcc metals and gamma relaxation in bcc metals, respectively. Moreover, it was realized that the formation of kink-pairs on $a_0<111>/2$ screw dislocations and their subsequent migration along the dislocation line governs the low-temperature flow stress of pure bcc metals almost entirely.

Today, advanced experimental techniques allow precise measurements of the macroscopic effective flow-stress component that helps the dislocations to overcome the Peierls barriers. This was done very successfully for the screw dislocations in the bcc metals α-iron, niobium, tantalum, molybdenum and tungsten. The application of the kink theory to the flow-stress data gives detailed information about the kink parameters, especially the formation enthalpy of the kinks and the kink height. The latter permits us to identify the slip planes on which the elementary slip steps occur. The result is that in the bcc metals the elementary slip steps occur only at very low temperatures at the densely packed $\{110\}$ planes. When the temperature is increased, a transition from $\{110\}$ slip to $\{211\}$ slip takes place, which corresponds to the "lower bend" in the flow-stress–temperature relationship. Microscopically, it can be described as a first order phase transition in the atomic structure of the screw-dislocation cores. The fact that a first order phase transition may be undercooled gives an important hint to an explanation of the unusual effects of solution and secondary irradiation softening observed in pure bcc metals. The occurrence of the "upper bend" in the flow-stress–temperature relationship within the range of $\{211\}$ slip steps can be described within the framework of the kink theory. It may be understood by a transition from the line-tension model to the elastic-interaction model that describes the kink–kink interaction at lower and higher temperatures, respectively. Moreover, the occurrence of $\{211\}$ slip steps at intermediate temperatures permits us to explain the flow-stress asymmetry and the occurrence of anomalous slip in pure bcc metals.

Acknowledgments

The author is grateful to Professor A. Seeger and to Drs. D. Brunner, U. Essmann and W. Wasserbäch for valuable suggestions, and for the permission to use their results. My particular thanks are directed to Drs. A. Vojta and T. Vojta for carefully reading the manuscript.

References

1. W. Boas, E. Schmid: Z. Phys. **54**, 16 (1929)
2. W. Boas, E. Schmid: Z. Phys. **57**, 575 (1929), **61**, 767 (1930)
3. E. Orowan: Z. Phys. **89**, 634 (1934)
4. M. Polanyi: Z. Phys. **89**, 660 (1934)
5. G.I. Taylor: Proc. Roy. Soc. Lond. A **145**, 362 (1934)
6. L. Prandtl: Z. Angew. Math. Mech. **8**, 85 (1928)
7. U. Dehlinger: Ann. Phys. Lpzg. **2**, 749 (1929)
8. E. Schmid, In: *Proceedings of the International Congress for Applied Mechanics* (Delft, 1924), p. 342
9. E. Schmid, W. Boas: *Kristallplastizität unter besonderer Berücksichtigung der Metalle* (Springer, Berlin 1935)
10. A. Seeger: In *Encyclopedia of physics*, ed. by S. Flügge (Springer, Berlin 1958) Vol. VII/2, p. 1
11. G.I. Taylor, C.F. Elam: Proc. Roy. Soc. Lond. A **112**, 337 (1926)
12. G.I. Taylor: In *Proceedings of the 2nd International Congress for Applied Mechanics* (Orell Füssli, Zürich 1927) p. 45
13. G.I. Taylor: Proc. Roy. Soc. Lond. A **118**, 1 (1928)
14. H.J. Gough: Proc. Roy. Soc. Lond. A **118**, 498 (1928)
15. W. Fahrenhorst, E. Schmid: Z. Phys. **78**, 383 (1932)
16. F. Sauerwald, H.I. Sossinka: Z. Phys. **78**, 634 (1932)
17. A. Seeger: Z. Metallkde. **72**, 369 (1981)
18. T. Parkhomenko: Fiz. Nizk. Temp. **3**, 74 (1977); Sov. J. Low Temp. Phys. **3**, 37 (1977)
19. D. Brunner, J. Diehl: Phys. Stat. Sol. (a) **104**, 145 (1987)
20. R.L. Simalek, T.E. Mitchell: Phil. Mag. **22**, 1105 (1970)
21. R. Lachenmann, H. Schultz: Scr. Metall. **6**, 731 (1972)
22. R. Lachenmann, H. Schultz: Z. Metallkde. **66**, 443 (1975)
23. D.B. Miracle: Acta Metall. Mater. **41**, 649 (1993)
24. R.D. Noebe, R.R. Bowman, M.V. Nathal: Int. Mater Rev. **38**, 193 (1993)
25. B. Brunner, P. Gumbsch: to be published in Mater Sci. Eng. A
26. B. Šesták, A. Seeger: Z. Metallkde. **69**, 195, 355, 425 (1978)

27. J.W. Christian: Metall. Trans. A **14**, 1237 (1983)
28. P.B. Hirsch: *Discussion, 5th Int. Conf. on Crystallography* (Cambridge 1960)
29. P.G. Bordoni: J. Acoust. Soc. Am. **26**, 495 (1954)
30. A. Seeger: Phil. Mag. **1**, 651 (1956)
31. R. Peierls: Proc. Roy. Soc. Lond. **52**, 34 (1940)
32. F.R.N. Nabarro: Proc. Roy. Soc. Lond. **59**, 236 (1947)
33. A. Seeger, P. Schiller: In *Physical acoustics*, ed. by W. P. Mason (Academic Press, New York London 1966) Vol. IIIA, p. 361
34. W. Shockley: Trans. Amer. Inst. Min. Metall. Engrs. **194**, 829 (1952)
35. G. Schottky: Phys. Stat. Sol. **5**, 697 (1964)
36. F.R.N. Nabarro: Mater Sci. Eng. A **234-236**, 67 (1997)
37. W. Meissner, M. Polanyi, E. Schmid: Z. Phys. **66**, 477 (1930)
38. T.H. Blewitt, R.R. Coltman, J.K. Redman: In *Report on a conference on defects in solids, held in Bristol 1954* (The Physical Society, London 1965) p. 349
39. A. Seeger: In *Selected topics on condensed matter physics,* ed. by W. E. Alnaser (University of Bahrein, Omega Scientific, Wallingford, U.K. 1993) p. 192
40. A. Seeger: J. Physique (IV) **5**, C7-45 (1995)
41. V.I. Startsev: In *Dislocations in solids*, ed. by F.R.N. Nabarro (North-Holland, Amsterdam 1983) Vol. 6, p. 143
42. P. Haasen: Phil. Mag. **3**, 384 (1958)
43. J. Diehl: Z. Metallkde. **47**, 331, 411 (1956)
44. S. Mader: Z. Phys. **149**, 73 (1957)
45. A. Seeger, H. Kronmüller, S. Mader, H. Träuble: Phil. Mag. **6**, 639 (1961)
46. H. Mughrabi: Phil. Mag. **18**, 1211 (1968)
47. U. Essmann: Phys. Stat. Sol. **12**, 707, 723 (1965)
48. H. Mughrabi: Phil. Mag. **23**, 869, 897, 931 (1971)
49. U. Essmann: Phys. Stat.Sol. **3**, 932 (1963)
50. S.J. Basinski, Z.S. Basinski: In *Dislocations in solids*, ed. by F.R.N. Nabarro (North-Holland, Amsterdam 1979) Vol. 4, p. 261
51. M. Ahlers, P. Haasen: Acta Metall. **10**, 977 (1962)
52. U.F. Kocks: Trans. AIME **230**, 1160 (1964)
53. A. Seeger, M. Wilkens: In *Realstruktur und Eigenschaften von Reinststoffen*, ed. by E. Rexer (Akademie-Verlag, Berlin 1967) Vol. 3, p. 29
54. H. Mughrabi: In *Constitutive equations in plasticity* ed. by A. S. Argon (MIT Press, Cambridge, Mass. 1975) p. 199
55. W.M. Lomer: Phil. Mag. **42**, 1327 (1951)
56. A.H. Cottrell: Phil. Mag. **43**, 645 (1952)
57. F.R.N. Nabarro, Z.S. Basinski, D.B. Holt: Adv. in Phys. **13**, 193 (1964)
58. P. Lukác, J Balík: In *Key engineering materials*, ed. by P. Lukác, ISPMA 6 (Trans Tech Publications, Switzerland 1994) Vols. 97–98, p. 307
59. E. Nes: Prog. Mater. Sci. **41**, 129 (1998)
60. D. Kuhlmann-Wilsdorf: Phil. Mag. **79**, 955 (1999)
61. A. Seeger: Phil. Mag. **45**, 771 (1954)
62. A. Seeger: In *Dislocations and mechanical properties of crystals*, ed. by. J.C. Fisher, W.G. Johnston, R. Thomson, T. Vreeland, Jr. (John Wiley, New York 1957) p. 243
63. A. Seeger, J. Diehl, S. Mader, H. Rebstock: Phil. Mag. **2**, 323 (1957)
64. H. Kronmüller: In *Moderne Probleme der Metallphysik* ed. by A. Seeger (Springer, Berlin 1965) Vol. 1, p. 126

65. H. Kronmüller: Can. J. Phys. **45**, 631 (1967)
66. A. Seeger: In *Work hardening*, ed. by J.P. Hirth, J. Weertman (Gordon & Breach, New York 1968) p. 27
67. U.F. Kocks, A.S. Argon, M.F. Ashby: Prog. Mater. Sci. **19**, 1 (1975)
68. G. Schoeck: In *Dislocations in solids*, ed. by F.R.N. Nabarro (North-Holland, Amsterdam 1980) Vol. 3, p. 63
69. P.C.J. Gallagher: Metall. Trans. **1**, 2429 (1970)
70. A. Seeger: Phil. Mag. **46**, 1194 (1955)
71. H.A. Kramers: Physica **7**, 284 (1940)
72. E. Orowan: Proc. Roy. Soc. Lond. **52**, 8 (1940)
73. D. Elefant: Dr. rer. nat. thesis, Akademie der Wissenschaften der DDR/ Zentralistitut für Festkörperphysik und Werkstofforschung Dresden (1971)
74. W. Pichl, M. Krystian: Phys. Stat. Sol. (a) **160**, 373 (1997)
75. Y. Aono, E. Kuramoto, D. Brunner, J. Diehl: In *Strength of metals and alloys*, ed. by P.O. Kettunen, T.K. Lepistö, M.E. Lehtonen ICSMA 8 (Pergamon Press, Oxford 1988) Vol. 1, p. 271
76. L. Kaun, A. Luft, J. Richter, D. Schulze: In *Realstruktur und Eigenschaften von Reinststoffen*, ed. by E. Rexer (Akademie-Verlag, Berlin 1967) Vol. 3, p. 489
77. L. Kaun, A. Luft, J. Richter, D. Schulze: Phys. Stat. Sol. (a) **26**, 485 (1968)
78. A. Luft, L. Kaun: Phys. Stat. Sol. (a) **37**, 781 (1970)
79. A. Luft: Phys. Stat. Sol. (a) **42**, 429 (1970)
80. J. Richter: Phys. Stat. Sol. (a) **46**, 203 (1971)
81. A. Luft, J. Richter, W. Reitzenstein, P. Finke: In *3. Internationales Symposium Reinststoffe in Wissenschaft und Technik*, ed. by M. Balarin (Akademie-Verlag, Berlin 1972) Vol. 4, p. 285
82. A. Luft, L. Kaun: Phys. Stat. Sol. (a) **18**, 109 (1973)
83. A. Köthe: In *Ultra-high-purity base metals*, ed. by K. Abiko, K. Hirokawa, S. Takaki, UHPM 96 (The Japan Institute of Metals, Sendai 1995) p. 291
84. L. Hollang: Dr. rer. nat. thesis, Universität Stuttgart, Stuttgart (1995)
85. D. Brunner, J. Diehl, A. Seeger: In *The structure and the properties of crystal defects*, ed. by V. Paidar, L. Lejcek (Elsevier, Amsterdam New York 1984) p. 175
86. A. Lawley, H.L. Gaigher: Phil. Mag. **10**, 15 (1964)
87. H.W. Loesch, F.R. Brotzen: J. Less-Common Metals **13**, 565 (1967)
88. H. Matsui, H. Kimura: Trans. Jpn. Inst. Metals **22**, 481 (1981)
89. D. Vesely: Phil. Mag. **27**, 607 (1973)
90. W. Wasserbäch: Phil. Mag. A **38**, 401 (1978)
91. K. Kitajima, Y. Aono, H. Abe, E. Kuramoto: Scr. Metall. **13**, 1033 (1979)
92. K. Kitajima, Y. Aono, E. Kuramoto: Scr. Metall. **15**, 919 (1981)
93. C.J. McMahon, Jr.: In *Advances in materials research*, ed. by H. Herman (Interscience, New York 1968) Vol. 2, p. 121
94. F. Ackermann, H. Mughrabi, A. Seeger: Acta Metall. **31**, 1353 (1983)
95. L. Hollang, M. Hommel, A. Seeger: Phys. Stat. Sol. (a) **160**, 329 (1997)
96. L. Hollang, D. Brunner, A. Seeger: to be published in Mater. Sci. Eng. A
97. D.J. Quesnel, A. Sato, M. Meshii: Mater. Sci. Eng. **18**, 208 (1975)
98. J. Diehl, M. Schreiner, S. Staiger, S. Zwiesele: Scr. Metall. **10**, 949 (1976)
99. D. Tseng, K. Tangri: Scr. Metall. **11**, 719 (1977)
100. H. Matsui, S. Moriya, S. Takaki, H. Kimura: Trans. Jpn. Inst. Metals **19**, 163 (1978)

101. E. Kuramoto, Y. Aono, K. Kitajima: Scr. Metall. **13**, 1039 (1979)
102. C.D. Statham, D.A. Koss, J.W. Christian: Phil. Mag. **26**, 1080 (1972)
103. A.J. Garratt-Reed, G. Taylor: Phil. Mag. A **33**, 577 (1976)
104. J. Nakagawa, M. Meshii: Phil. Mag. A **44**,1165 (1981)
105. S. Takeuchi, K. Maeda: Acta Metall. **25**, 1485 (1977)
106. F. Guiu, P.L. Pratt: Phys. Stat. Sol. **15**, 539 (1966)
107. Y. Aono, E. Kuramoto, K. Kitajima: In *Strength of metals and alloys*, ed. by R.C. Gifkins, ICSMA 6 (Pergamon Press, Oxford 1982) Vol. 1, p. 9
108. H.J. Kaufmann: Phys. Stat. Sol. (a) **65**, 53 (1981)
109. D. Brunner, J. Diehl: Phys. Stat. Sol. (a) **124**, 155, 455 (1991), **125**, 203 (1991)
110. G.I. Taylor: J. Inst. Metals **62**, 307 (1938)
111. P.J. Sherwood, F. Guiu, H.C. Kim, P.L. Pratt: Can. J. Phys. **45**, 1075 (1967)
112. F. Guiu: Scr. Metall. **3**, 449 (1969)
113. G.J. Irwing, F. Guiu, P.L. Pratt: Phys. Stat. Sol. (a) **22**, 685 (1974)
114. M. Anglada, B. Etemad, J.A. Planell, F. Guiu: Scr. Metall. **14**, 1319 (1980)
115. J.A. Planell, F. Guiu: Phil. Mag. **54**, 325 (1986)
116. A. Seeger, L. Hollang: Mater. Trans. JIM **41**, 141 (2000)
117. G. Taylor, R. Bajaj, O.N. Carlson: Phil. Mag. **28**, 1035 (1973)
118. M.S. Duesberry, R.A. Foxall: Phil. Mag. **20**, 719 (1969)
119. S. Takeuchi, E. Kuramoto, T Suzuki: Acta Metall. **20**, 909 (1972)
120. W. Wasserbäch: Phys. Stat. Sol. (a) **147**, 417 (1995)
121. A. Seeger: to be published in Mater. Sci. Eng. A
122. A.J. Garratt-Reed, G. Taylor: Phil. Mag. A **39**, 597 (1979)
123. H. Suzuki: In *Dislocations in solids*, ed. by F.R.N. Nabarro (North-Holland, Amsterdam 1979) Vol. 4, p. 191
124. E. Pink: Z. Metallkde. **64**, 871 (1973)
125. Y. Aono, D. Brunner, J. Diehl: In *Strength of metals and alloys*, ed. by P.O. Kettunen, T.K. Lepistö, M.E. Lehtonen ICSMA 8 (Pergamon Press, Oxford 1988) Vol. 1, p. 509
126. D. Brunner, J. Diehl: Mater. Sci. Eng. A **164**, 350 (1993)
127. D. Brunner, J. Diehl: In *Strength of metals and alloys*, ed. by K. Oikawa, K. Maruyama, S. Tacheuchiu, M. Yamaguchi ICSMA 10 (The Japan Institute of Metals, Sendai 1994) Vol. 1, p. 141
128. D. Brunner, J. Diehl: Phys. Stat. Sol. (a) **160**, 355 (1997)
129. A. Sato, T. Mifune, M. Meshii: Phys. Stat. Sol. (a) **18**, 699 (1973)
130. E. Pink, R.J. Arsenault: Prog. Mater. Sci. **24**, 1 (1979)
131. A. Seeger, B. Šesták: Scr. Metall. **5**, 875 (1971)
132. R.H. Chambers: In *Physical acoustics*, ed. by W.P. Mason (Academic Press, New York London 1966) Vol. IIIA, p. 123
133. A. Seeger: In *Dislocations 1984*, ed. by P. Veyssière, L. Kubin, J. Castaing (Editions C.N.R.S., Paris 1984) p. 141
134. Y.T. Chou, G.T. Sha: Metall. Trans. **3**, 2857 (1972)
135. J.D. Eshelby: Proc. Roy. Soc. A **266**, 222 (1962)
136. M. Werner: Phys. Stat. Sol. (a) **104**, 63 (1987)
137. U. Holzwarth, A. Seeger: In *Strength of metals and alloys*, ed. by D.G. Brandon, R. Chaim, A. Rosen, ICSMA 9 (Freund, London 1991) Vol. 1, p. 577
138. D. Brunner, V. Glebovski: Mater. Lett. **42**, 290 (2000)
139. D. Brunner, J. Diehl: Z. Metallkde. **83**, 828 (1992)
140. W. Frank, A. Seeger: Mater. Sci. Forum **15–18**, 57 (1987)

12 Atomic Defects in High-Purity Metals: Fundamentals and Equilibrium Concentrations

A. Seeger

12.1 What Is an Atomic Defect?

A *crystal defect* is any *permanent* perturbation of the crystal structure. In crystals, the interatomic distances and the size of the elementary cells provide us with a natural length scale. This allows us to base a first classification of crystal defects on the number of dimensions in which the perturbation extends over distances that are *large* compared with the chosen length scale. For metals and most other inorganic materials it does not matter whether we choose the distances between nearest-neighbour atoms or the size of the elementary cells as the length scale, since they either coincide (in the case of Bravais lattices) or differ only by a factor of the order of magnitude unity. The situation is different in molecular crystals built up from large molecules, such as crystalline polymers. This brings in features that go beyond the above classification.

Inclusions of second phases and *voids* are *three-dimensional* defects. Examples of *two-dimensional* defects are the surfaces of crystals, extended stacking faults, and the interfaces between crystallites, the *grain boundaries*. The most important *one-dimensional* defects are *dislocations*. *Zero-dimensional* defects, the subject of the present chapter, extend in all three space dimensions over atomic distances only. Therefore, a more descriptive name is *'atomic defects'*. It will be used throughout this chapter in preference to the frequently encountered name 'point defects', which is not really appropriate. Atomic defects possess important properties that are incompatible with 'points'. For example, we shall have to deal extensively with the *volume* of atomic defects. Some of them may take up different *crystallographic orientations*.

An important distinction is that between *intrinsic* and *extrinsic* defects. *Intrinsic* defects involve only atom species that are present in the *perfect* crystals. They can be generated in an otherwise perfect crystal environment without participation of *foreign atoms*. An easy-to-visualize example of an intrinsic atomic defect in a crystal is a *Frenkel pair*. Here an atom is removed from its site in the unperturbed crystal structure (in the following called 'lattice site' for simplicity) and inserted into a not-too-distant interstice of the crystal structure, leaving behind an empty lattice site (for short called 'vacancy'). An atom located in an interstice is called and *interstitial atom* or, for short, *interstitial.* In the Frenkel-pair example we deal with *self-interstitials*, in contrast to *foreign interstitials*, such as hydrogen, carbon, nitrogen, or oxygen atoms accommodated in interstices of crystal structures.[1]

Extrinsic atomic defects involve *foreign atoms*, which may have been introduced during crystal growth, by intended or unintended in-diffusion in the solid state, by ion implantation, or by transmutation induced by irradiation with neutrons or energetic charged particles.[2] Thermodynamics tells us that foreign atoms are always present in macroscopic samples, even if these are of the highest achievable purity. Since foreign atoms may interact strongly with intrinsic defects, unintentionally introduced foreign atoms cannot a priori be disregarded in the study of intrinsic atomic defects in high-purity materials, although most research strategies attempt to minimize the influence of these 'impurities'. In discussing the interactions between intrinsic and extrinsic atomic defects, it is important to distinguish between *foreign interstitials* and *substitutional atoms* (i.e., foreign atoms that have replaced host atoms on lattice sites). The main reason is that foreign interstitials are much more mobile than substitutional atoms. Over a wide temperature range foreign interstitials may migrate towards intrinsic defects and interact with them. By contrast, interactions between intrinsic atomic defects and substitutional atoms usually come about when during their migration through the crystal the intrinsic defects encounter the less mobile substitutional foreign atoms.

The restriction to *permanent perturbations* in the definition of crystal defects given in the introductory paragraph aims primarily at excluding perturbations such as phonons that can be accommodated within the perfect crystal structure by considering time averages over a few lattice vibration periods. Beyond this, the restriction is also relevant for atomic defects as defined above, however. Consider, as an example, *Frenkel pairs.* A crystal containing a Frenkel pair has a higher energy than a perfect crystal consisting of the same number of atoms; it may hence be considered as an excited state of the crystal. Owing to its high degeneracy, at high temperatures this excited state may be populated in spite of the large energy of a Frenkel pair compared with the temperature equivalent, $k_{\mathrm{B}}T$ (cf. Sect. 12.2). The 'thermal' Frenkel pairs may be looked upon as thermal fluctuations, their majority disappearing rapidly by self-interstitial–vacancy recombination. Whether a given experimental technique 'sees' the Frenkel pairs or not depends on whether the lifetime of the excited state is longer or shorter than the time required for the detection of the defects. Section 12.7 takes up again this question, to which not always sufficient attention was paid in the past.

[1] Note that the *Frenkel pairs* as defined above are a more restricted concept than the *Frenkel defects* in the theory of ionic crystals, which are equal numbers of ions in interstices and vacant sites irrespective of their origin.

[2] For stacking faults the distinction "intrinsic–extrinsic" does not involve foreign atoms and differs, therefore, from the above usage.

12.2 How Do We Characterize Intrinsic Atomic Defects?

Intrinsic atomic defects may be characterized by a three-tier description of increasing sophistication. The first tier is a purely *geometric* description, such as the one used in Sect. 12.1 to define a Frenkel pair. On this level of description, for an isolated vacancy (from now on called *monovacancy*, abbreviated V_1) it suffices to say that an atom has been removed from its lattice site and to state the position of the ensuing monovacancy, viz. that of the emptied lattice site. A *vacancy pair* (henceforth called *divacancy*, abbreviated V_2) consists of two neighbouring vacant lattice sites. In this case, within the first-tier description a classification arises depending on whether the two members of the pair are **n**earest **n**eighbours, **n**ext-**n**earest **n**eighbours, etc. If several species occur simultaneously, we have to attach modifiers such as nn, nnn etc. to the divacancy symbol. The description may be condensed into considering the midpoint between the pair as the divacancy location (which may or may not be a lattice site) and ascribing to it a crystallographic orientation. In a co-ordination structure with z nearest neighbours, an nn-divacancy may take up $z/2$ crystallographically equivalent orientations. (Vacancies are indistinguishable, so that interchanging the two member vacancies does not give us a new defect.) The number of possible configurations increases very rapidly if we consider *multiple vacancies* consisting of three, four, or more vacant sites. In Sect. 12.4, the first-tier description will be used to calculate the *configuration entropy* of atomic defects in crystals.

Let us illustrate the preceding by a few examples. In the face-centred cubic (fcc) lattice nn-divacancies may be orientated along 6 distinct $\langle 110 \rangle$ axes, nnn-divacancies along 3 different $\langle 100 \rangle$ axes. In the body-centred cubic (bcc) lattice there are 4 different $\langle 111 \rangle$ axes of nn-divacancies, and again 3 different $\langle 100 \rangle$ axes for the nnn-divacancies. In the hexagonal close-packed (hcp) structure we must distinguish between two kinds of nn-divacancies, depending on whether the two vacant nearest-neighbour sites lie on the same basal plane or on adjacent ones. Clearly, the number of possible configurations increases very rapidly if we consider *multiple vacancies* consisting of three, four, or more vacant sites.

For *self-interstitials* the need to consider more than one possible configuration arises already for single interstitials (generically denoted by I_1). The fcc and hcp structures may be envisaged as being made up of two types of interstices, octahedral and tetrahedral. Octahedral interstitial sites are surrounded by six atoms, tetrahedral sites by four. On the present simple description level, self-interstitials may be accommodated in either type of interstice, hence we have to distinguish between at least two interstitial configurations. However, as we shall see presently, in reality there are further and much more serious complications.

In the first-tier description the reaction of the surroundings to the removal of atoms from their lattice sites or their insertion into interstices is disregarded. Taking this reaction into account, i.e. relaxing the assumption that the crystal structure is *rigid*, constitutes the *second tier* of sophistication. For qualitative estimates of what to expect it suffices to picture the metallic bonding as a balance between the electrostatic attraction between the positively charged ion cores and the conduction electrons on the one hand and the Fermi pressure of the electron gas and the repulsion between the ion cores on the other hand. It is true that d-electrons give a further contribution to the bonding, as the large binding energies of the noble metals (Cu, Ag, Au) and, even more so, of the transition metals show. Taking this into account would hardly alter the following *qualitative* discussions, however.

As a particularly simple example, we consider *monovacancies* first. When an atom is removed from a lattice site, the balance of forces is disturbed in such a way that the adjacent ion cores move towards the vacant site ("inward relaxation"). In closed-packed structures of elements with large ion cores, rather small displacements (of the order of magnitude 10^{-2} of the interatomic distance) suffice to reach a new equilibrium position because the strong repulsion between neighbouring cores prevents them from getting too close. In less densely packed structures, e.g. in the bcc lattice, the inward relaxation is expected to be larger, particularly so in the bcc *alkali metals*, in which the diameters of the ion cores are much smaller than the nearest-neighbour distances. Irrespective of the sophistication of the underlying model, so far all computations indicate that in the cubic Bravais lattices the relaxation around monovacancies maintains the octahedral point symmetry of the lattice sites, in agreement with the first-tier description introduced above. This means that in these cases the lattice relaxation is characterized completely by the local volume change. The relationship of the local volume change to the monovacancy formation volume, V_{1V}^{F}, will be discussed in Sect. 12.6.4. The available experimental information on V_{1V}^{F} is in agreement with the preceding qualitative arguments.[3]

For *self-interstitials*, going from the first to the second tier, i.e. giving up the rigidity hypothesis, has much more serious consequences. The reason is that in the metal structures the interstices are too small to accommodate an atom, hence the neighbours of an inserted atom – with the possible exception of interstitial hydrogen – will undergo quite large *outward displacements*. This gives rise to a strain energy that contributes decisively to the formation energy of the self-interstitials. The tendency to minimize the strain energy determines to a large extent which interstitial configurations are mechanically stable. This has several consequences: *Intuitive arguments* do not nearly go as far as in the vacancy case. The *anisotropy* of the elastic constants may play

[3] The monovacancy formation volume may be obtained from the dependence of the monovacancy formation enthalpy on the hydrostatic pressure p [cf. Eq. (12.1)]. Such measurements are rare, however. The above statement is mainly based on measurements of the pressure dependence of *self-diffusion*, which in the materials discussed in this chapter provides us with approximate values of V_{1V}^{F}.

a major role. In order to be useful, computations have to be performed on *realistic* models. It does not suffice to insert the interstitial atoms into *high-symmetry* sites and to admit relaxations compatible with this symmetry. Configurations of *lower symmetry* have to be considered as well.

Among the important low-symmetry configurations are those obtained by squeezing the interstitial atom between two nearest-neighbour atoms and letting the strain energy relax by displacements that extend rather far out along the particular nearest-neighbour axis. The resulting interstitial configurations have been termed 'crowdions'. Other possibilities with lower-than-cubic symmetry are the *dumbbell* configurations. Here an atom is removed from its lattice site, and two atoms are inserted symmetrically with respect to the vacant site. If in the cubic Bravais lattices this is done along one of the symmetry axes $\langle 100 \rangle$, $\langle 110 \rangle$ or $\langle 111 \rangle$, and if during the relaxation process the symmetry is maintained, configurations of mechanical equilibrium may be obtained (cf., for example, [1]). Of course, the majority of the configurations which are in mechanical equilibrium by symmetry are not mechanically *stable.* If in the multi-dimensional space of the relative displacements of the atoms there is just one *unstable* mode, the particular equilibrium configuration is a *saddle-point configuration.* The self-interstitials may have to pass through or near to this configuration during their migration. The existence of more than one unstable mode in a configuration of mechanical equilibrium usually indicates that the configuration will relax towards a stable or saddle-point configuration of lower symmetry.

In view of the multitude of possibilities just outlined, it is not surprising that the majority of modern model computations support the expectation that in a given metal isolated self-interstitials can exist in more than one configuration. As a rule, in the cubic metals these configurations are predicted to have lower-than-cubic symmetry, which means that in analogy to the divacancies discussed above they may take up different crystallographic orientations. An exception is an atom located in the centre of an octahedral interstice in fcc metals (briefly called *octahedral interstitial*), which by some models is predicted to be mechanically stable and which has the same point symmetry as a regular lattice site.

Knowing the configuration of an atomic defect is a prerequisite for the third tier of the description of atomic defects. This level of description aims at making quantitative predictions of measurable physical quantities and thus at testing the validity and reliability of the underlying models by comparing the predictions with experimental data. A rather simple example of such a quantity, the monovacancy formation volume $V^{\mathrm{F}}_{1\mathrm{V}}$, has already been mentioned. By well-known thermodynamic arguments the formation volume of any atomic defect, V^{F}, is related to the *enthalpy of formation* of the defect, $H^{\mathrm{F}} = H^{\mathrm{F}}(p)$, according to

$$V^{\mathrm{F}} = \partial H^{\mathrm{F}} / \partial p \,|_S \,, \qquad (12.1)$$

where the subscript S means that the differentiation has to be performed under the condition of constant entropy S. The precise definition of H^{F} and its companion quantity, the entropy of defect formation S^{F}, will be given in Sect. 12.3. From H^{F} and S^{F} we may calculate the *free enthalpy* (i.e. the Gibbs free energy) of *defect formation*,

$$G^{\mathrm{F}} = H^{\mathrm{F}} - TS^{\mathrm{F}} , \tag{12.2}$$

where T denotes the absolute temperature. In terms of the dependence of $G^{\mathrm{F}} = G^{\mathrm{F}}(p)$ on the hydrostatic pressure p the defect formation volume may be written as

$$V^{\mathrm{F}} = \partial G^{\mathrm{F}} / \partial p \,|_T \,. \tag{12.3}$$

Making use of the basic thermodynamic relationship

$$S^{\mathrm{F}} = -\partial G^{\mathrm{F}} / \partial T \,|_p \tag{12.4}$$

we find by differentiating (12.2) with respect to T

$$\partial H^{\mathrm{F}} / \partial T \,|_p = T \partial S^{\mathrm{F}} / \partial T \,|_p \,. \tag{12.5}$$

Equation (12.5), to which we shall return in Sect. 12.4, shows that the temperature variations of the enthalpies and entropies of formation of atomic defects are coupled. The absolute values of H^{F} and S^{F}, e.g. their values $H^{\mathrm{F}}(T_0)$ and $S^{\mathrm{F}}(T_0)$ at some reference temperature T_0, are, of course, independent of each other.

In addition to the four *thermodynamic quantities* characterizing a given kind J of point defect, $G^{\mathrm{F}}_{\mathrm{J}}, H^{\mathrm{F}}_{\mathrm{J}}, S^{\mathrm{F}}_{\mathrm{J}}$, and $V^{\mathrm{F}}_{\mathrm{J}}$, there are quite a few other physical quantities that are important for the identification of point defects and for studying their properties. Among them are the contribution of a unit atomic concentration of the defects to the electrical resistivity of the metal, ρ_{J}, the amplitudes of X-ray and neutron scattering by the defect, the rates of trapping of thermalized positrons or positive muons by a unit concentration of defects (cf. Sects. 12.6.4 and 12.6.5), the electric field gradients induced by the defects at neighbouring nuclei, and – particularly important but also particularly difficult to calculate – the enthalpies and entropies of migration, $H^{\mathrm{M}}_{\mathrm{J}}$ and $S^{\mathrm{M}}_{\mathrm{J}}$. Predicting these quantities reliably for the various defect configurations can be a rather demanding task and is usually not possible without introducing advanced, tailor-made models. Thus, the third tier of our description of atomic defects necessarily comprises a range of different approaches.

12.3 How Are Atomic Defects Generated?

The answer to the above question is closely related to the necessity to provide unambiguous definitions for the four thermodynamic characterizations introduced in Sect. 12.2. This is straightforward for Frenkel pairs (characterized by the subscript FP), since they can be generated in the interior of a crystal in an otherwise perfect environment. Hence the definitions of $G^{\mathrm{F}}_{\mathrm{FP}}$, $H^{\mathrm{F}}_{\mathrm{FP}}$, $S^{\mathrm{F}}_{\mathrm{FP}}$, and $V^{\mathrm{F}}_{\mathrm{FP}}$ do not present problems. For example, $H^{\mathrm{F}}_{\mathrm{FP}}$ is simply the difference of the enthalpies of two crystals consisting of the same very large number of atoms and kept under the same hydrostatic pressure p. The first crystal contains precisely one Frenkel pair, whereas the second crystal is perfect.

The definition of the basic thermodynamic quantities of the partner defects, monovacancies, V_1, and isolated self-interstitials, I_1, called *elementary intrinsic atomic defects*, requires some care, as may be seen from the fact that in textbooks erroneous statements may be found. As in the case of the Frenkel pairs, we wish to compare a defect-containing crystal with a perfect crystal consisting of the *same* number of atoms. Using the monovacancy formation enthalpy $H^{\mathrm{F}}_{1\mathrm{V}}$ as an example, we illustrate the essential considerations in the next paragraph.

When in a *Gedankenexperiment* ('thought experiment') aimed at 'determining' $H^{\mathrm{F}}_{1\mathrm{V}}$ we remove an atom from its lattice site, we have to reinsert it into the crystal in such a way that the energetics far away from the vacant site are not altered. This requires that the insertion be done in a *wiederholbarer Schritt (repeatable step)*, i.e. in such a way that an unlimited number of additional atoms can be removed and reinserted without changing the energetics at the insertion sites. This condition is *not* fulfilled if an atom is put onto an otherwise smooth crystal surface (giving us an 'ad-atom'), since a second atom put next to the first ad-atom finds itself in an energetically different situation from the first ad-atom. The appropriate procedure is to put the atom into a *Halbkristallage* (somewhat clumsily translated as *half-crystal position*). If we think of a crystal as being held together by bonds between atoms, in a *Halbkristallage* half the bonds of a regular site in the bulk are present whereas the other half are missing. Examples of *Halbkristallagen* are kinks in ledges on crystal surfaces and elementary jogs in edge dislocations or in dislocations with well-developed edge components. (Jogs connect two segments of a dislocation line lying on different parallel slip planes. A jog is called *elementary* if it is as short as possible, i.e. if it connects segments on adjacent slip planes.) Comparing the numbers of interatomic interactions (which need neither be central nor two-body interactions) shows that in large enough crystals they are the same in the two crystals except in the neighbourhood of the vacancy. If we then let the relaxation processes in this neighbourhood take place as described in Sect. 12.2, the ensuing difference in the total enthalpies of the crystals is the appropriate measure of the monovacancy formation enthalpy; hence we identify this energy difference with $H^{\mathrm{F}}_{1\mathrm{V}}$.

The analogous procedure (removing an atom from a *Halbkristallage*, inserting it into an interstice in the bulk, letting the neighbouring atoms relax, and subtracting from the enthalpy of the crystal with the self-interstitial that of the perfect crystal) gives us $H^{\mathrm{F}}_{1\mathrm{I}}$. From the definitions it follows immediately that the relationship

$$H^{\mathrm{F}}_{\mathrm{FP}} = H^{\mathrm{F}}_{\mathrm{1V}} + H^{\mathrm{F}}_{\mathrm{1I}} \tag{12.6}$$

holds, where the left-hand side refers to Frenkel pairs with partners that are separated widely enough for their interaction energy to be neglected.

The same *Gedankenexperiment* gives us also the other three basic thermodynamic quantities characterizing the elementary atomic defects. We note that in the first step, inserting an atom into a *Halbkristallage* or removing it from there, the crystal volume increases or decreases by one atomic volume, V_{A}. Thus, the monovacancy formation volume $V^{\mathrm{F}}_{\mathrm{1V}}$ is the *sum* of the (negative) local volume change discussed in Sect. 12.2 and of V_{A}; hence we expect

$$0 < V^{\mathrm{F}}_{\mathrm{1V}} < V_{\mathrm{A}} \,. \tag{12.7}$$

In order to obtain the self-interstitial formation volume $V^{\mathrm{F}}_{\mathrm{1I}}$, we have to *subtract* the atomic volume from the local volume change due to the relaxation of the neighbouring atoms. Computations based on models as well as intuitive arguments tell us that at least for metals with d-electron cores the relaxation term exceeds V_{A}, so that in these cases we expect $V^{\mathrm{F}}_{\mathrm{J}}$ to be positive for self-interstitials, too. For Frenkel pairs the contributions $\pm V_{\mathrm{A}}$ cancel, hence with the same proviso as in (12.6) the relationship

$$V^{\mathrm{F}}_{\mathrm{FP}} = V^{\mathrm{F}}_{\mathrm{1V}} + V^{\mathrm{F}}_{\mathrm{1I}} \tag{12.8}$$

holds.

In applying the preceding considerations to the *entropies of defect formation*, $S^{\mathrm{F}}_{\mathrm{1V}}$ and $S^{\mathrm{F}}_{\mathrm{1I}}$, we note that these quantities refer to the entropy per defect at a fixed location and, in the case of defects capable of taking up different crystallographic orientations, of fixed orientation. Included in $S^{\mathrm{F}}_{\mathrm{J}}$ are the differences in the spectra of the phononic, electronic, and magnetic excitations of the defect-containing and the perfect crystal. Apart from rare exceptions, the dominating contributions are due to phonons. If we describe these in terms of harmonic oscillators with circular frequencies ω'_j in the perturbed and ω_j in the unperturbed crystal, the well-known expression for the free enthalpy of a set of harmonic oscillators gives us, in conjunction with (12.4),

$$S^{\mathrm{F}}_{\mathrm{J}} = -k_{\mathrm{B}} \sum_{j=1}^{3N-6} \left\{ \ln \frac{1-\exp(-\hbar\omega'_j/k_{\mathrm{B}}T)}{1-\exp(-\hbar\omega_j/k_{\mathrm{B}}T)} - \frac{\hbar}{k_{\mathrm{B}}T} \left[\frac{\omega'_j}{\exp(\hbar\omega'_j/k_{\mathrm{B}}T)-1} - \frac{\omega_j}{\exp(\hbar\omega_j/k_{\mathrm{B}}T)-1} \right] \right\}, \tag{12.9}$$

where $\hbar = 1.055 \times 10^{-34}\,\mathrm{J\,s}$ denotes Planck's constant divided by 2π, and $k_{\mathrm{B}} = 1.381 \times 10^{-23}\,\mathrm{J\,K^{-1}} = 8.617 \times 10^{-5}\,\mathrm{eV\,K^{-1}}$ is Boltzmann's constant. In (12.9) the summation extends over the $3N-6$ vibrational degrees of freedom of a crystal containing N atoms.

In accordance with the third law of thermodynamics, $S_{\mathrm{J}}^{\mathrm{F}}$ goes to zero if $T \to 0$. At high enough temperatures (12.9) simplifies to

$$S_{\mathrm{J}}^{\mathrm{F}} = -k_{\mathrm{B}} \sum_{j=1}^{3N-6} \ln(\omega_j'/\omega_j)\,. \tag{12.10}$$

Usually the temperature-independent expression (12.10) is meant when the entropy of an atomic defect is stated. We see that Boltzmann's constant, k_{B}, is the natural unit in which to express the formation entropies of atomic defects.

The phonon frequencies ω_j' and ω_j may be obtained by solving the equations of motion of the atoms, say in the Born–Oppenheimer approximation. Since sufficiently far away from the defects the equations of motion are the same in the perturbed and in the unperturbed crystal, it suffices to evaluate expressions such as (12.9) or (12.10) for suitably chosen clusters of atoms that have been cut out of the full crystal. It is intuitively clear that the introduction of a vacancy makes such a cluster 'softer' and, therefore, lowers its phonon frequencies. We thus expect the phonon contribution to the entropies of vacancy formation to be always *positive*.

The extension of the preceding considerations to multiple atomic defects is straightforward; we shall come back to it in Sect. 12.4.

The *Gedankenexperiment* invoked in defining the four basic thermodynamic quantities characterizing atomic defects is often not too far from reality. By electron irradiation with e^- energies of the order of magnitude 1 MeV enough momentum may be transferred to atoms on lattice sites for them to be displaced permanently. A displaced atom may land in an interstice and form a Frenkel pair together with the vacant lattice site it left behind. Alternatively, the displaced atom may collide with other atoms and displace them, itself occupying a lattice site that has been emptied by such a displacement. By both types of process a Frenkel pair is formed. However, the minimum energy that must be transferred by the incoming radiation in order to generate a Frenkel pair, called *displacement threshold energy* E_{d}, is substantially larger than $H_{\mathrm{FP}}^{\mathrm{F}}$, typically by a factor between 5 and 10. The reason is that the displacement of a lattice atom resulting from the energy transfer is an adiabatic process, whereas $H_{\mathrm{FP}}^{\mathrm{F}}$ is an equilibrium property (as are all the thermodynamic quantities introduced above).

For a given material, the displacement threshold energy depends on both the temperature and the crystallographic direction of the transferred momentum, $\boldsymbol{\Omega}$. Measurements of the displacement cross section as a function of the electron energy allow us to determine $E_{\mathrm{d}}(\boldsymbol{\Omega},T)$ [2]. From this we may deduce information on the ion–ion interaction and on the elementary radiation damage processes [3]. Electron beams in the energy range required for such investigations are available in electrostatic accelerators such as Van de Graaff generators or pelletrons, in dynamitrons, and – with particularly high flux densities and therefore particularly well suited for detailed studies – in high-voltage electron microscopes.

In the absence of external agents producing intrinsic atomic defects, such as irradiation with energetic particles or plastic deformation, in small crystals the kinks in surface ledges are the most important sources and sinks of intrinsic atomic defects. In larger, *strongly dislocated* crystals the main sources and sinks are the jogs in dislocation lines, referred to in Sect. 12.2. In both cases the considerations involved in the *Gedankenexperiment* are fully applicable. By contrast, for vacancy supersaturations introduced by, say, quenching from high temperatures into samples that are both large and almost dislocation-free, the sinks just mentioned may be too rare to allow the complete removal of the vacancies. In such cases vacancies may condense in clusters and form voids, stacking-fault tetrahedra, or dislocation loops.

12.4 Intrinsic Atomic Defects in Thermal Equilibrium

Although crystals containing defects have higher energies than defect-free crystals, intrinsic atomic defects may, nevertheless, be present in *thermal equilibrium* in concentrations that are accessible to experimental investigation. The reason for this is, of course, that it is the *free enthalpy* of the crystals and *not* their *enthalpy* or energy that is minimal in thermal equilibrium. At elevated temperatures, the fact that defects of the same kind (e.g. vacancies) are indistinguishable and that a given number of defects may be arranged in many different ways gives a strong negative entropy contribution to the free enthalpy of a defect-containing crystal.

Let us consider a crystal consisting of N atoms that contains n_1 monovacancies and n_2 divacancies; hence the total number of sites (irrespective of whether they are occupied or not) is $N + n_1 + 2n_2$. The number of energetically equivalent but crystallographically different orientations of the divacancies is denoted by $z/2$ (cf. Sect. 12.2; in the present context z need not be the coordination number, however). We assume that the interaction between the vacancies may be disregarded unless divacancies are formed. Then the interaction is described by the free enthalpy of divacancy binding,

$$G^{\mathrm{B}}_{2\mathrm{V}} = 2G^{\mathrm{F}}_{1\mathrm{V}} - G^{\mathrm{F}}_{2\mathrm{V}} \,. \tag{12.11}$$

$G^{\mathrm{B}}_{2\mathrm{V}}$ is the gain in free enthalpy if two distant and therefore non-interacting monovacancies are brought together to form a divacancy. It may be subdivided in the usual way into a binding enthalpy, $H^{\mathrm{B}}_{2\mathrm{V}}$, and a binding entropy, $S^{\mathrm{B}}_{2\mathrm{V}}$, according to

$$G^{\mathrm{B}}_{2\mathrm{V}} = H^{\mathrm{B}}_{2\mathrm{V}} - TS^{\mathrm{B}}_{2\mathrm{V}} \,. \tag{12.12}$$

The simplification in the preceding description of the vacancy–vacancy interaction is less serious than it may seem, since owing to the highly effective screening of electrical charges in metals, vacancies interact with each other only over small distances. (The elastic interaction between vacancies is usually negligibly small.) Furthermore, in metals the concentrations of vacancies

Table 12.1. Melting temperatures T_m (in energy equivalents), total vacancy concentrations at T_m, $C_V^{eq}(T_m)$, according to differential-dilatometry measurements, and effective vacancy formation enthalpy H_{eff}^F [eV] of various metals. The last entry is based on $H_{eff}^F = k_B T_m[1 - \ln C_V^{eq}(T_m)]$

Metal	Cu	Ag	Au	Al	Pb	Mg	Zn	Cd
$k_B T_m$ [meV]	117	106	115	80	52	80	60	51
$10^4 \times C_V^{eq}(T_m)$	5.2	7.6	13	9.0	1.7	7.2	3.0	4.5
H_{eff}^F [eV]	1.00	0.87	0.88	0.64	0.50	0.66	0.55	0.44

in thermal equilibrium even close to their melting points are fairly small (at the most about 10^{-3} in atom fractions, see Table 12.1); hence the formation of trivacancies etc. is rare. Occasional attempts to attribute the melting of metals to a catastrophic increase in the concentration of multiple vacancies as the melting temperature is approached have so far lacked experimental support.

Under the above assumptions the monovacancies may be distributed on $N + n_1$ sites. (Note that the sites occupied by divacancies are inaccessible to monovacancies and that the restriction to mono- and divacancies implies that we disregard the complications that arise if a monovacancy occupies a site next to a divacancy, i.e. if a trivacancy is formed.) Similarly, we have $(zN/2 + n_2)$ possibilities to arrange divacancies in the crystal. According to Boltzmann's statistical interpretation of entropy, the configuration entropy of n indistinguishable objects that may be distributed over N' equivalent sites is given by

$$S(n) = k_B \ln[N'!/(N' - n)!n!] \,. \tag{12.13}$$

In the present case the configuration entropy reads

$$S(n_1, n_2) = k_B\{\ln[(N + n_1)!/N!n_1!] + \ln[(zN/2 + n_2)!/N!n_2!]\} \,. \tag{12.14}$$

The difference in free energy of the crystals with and without vacancies at the temperature T may thus be written as

$$\Delta G(n_1, n_2; T) = n_1 G_{1V}^F + n_2 G_{2V}^F - TS(n_1, n_2) \,. \tag{12.15}$$

After simplifying (12.14) by means of Stirling's asymptotic formula for large arguments of the factorial function and solving the conditions for minimal ΔG,

$$\partial \Delta G/\partial n_j = 0 \quad (j = 1, 2) \,, \tag{12.16}$$

we obtain

$$\begin{aligned} G_{1V}^F + \ln[n_1/(N + n_1)] &= 0 + O(1/n_1) \,, \\ G_{2V}^F + \ln[n_2/(zN/2 + n_2)] &= 0 + O(1/n_2) \,. \end{aligned} \tag{12.17}$$

Apart from the certainly justified use of Stirling's formula, (12.17) are exact deductions from our model. Making use of the fact that in all metals so far investigated in thermal equilibrium $n_1, n_2 \ll N$ holds, we obtain for the *atomic thermal-equilibrium concentrations*, defined as[4]

$$C_j^{\mathrm{eq}} := n_j/(N + jn_j)\,, \quad (j = 1, 2)\,, \tag{12.18}$$

$$C_{\mathrm{1V}}^{\mathrm{eq}} = \exp(-G_{\mathrm{1V}}^{\mathrm{F}}/k_{\mathrm{B}}T) = \exp(S_{\mathrm{1V}}^{\mathrm{F}}/k_{\mathrm{B}})\exp(-H_{\mathrm{1V}}^{\mathrm{F}}/k_{\mathrm{B}}T) \tag{12.19}$$

and

$$\begin{aligned} C_{\mathrm{2V}}^{\mathrm{eq}} &= (z/2)\exp(-G_{\mathrm{2V}}^{\mathrm{F}}/k_{\mathrm{B}}T) \\ &= (z/2)\exp(S_{\mathrm{2V}}^{\mathrm{F}}/k_{\mathrm{B}})\exp(-H_{\mathrm{2V}}^{\mathrm{F}}/k_{\mathrm{B}}T)\,. \end{aligned} \tag{12.20}$$

Equations (12.19) and (12.20) obey the law of mass action for the reaction

$$2\mathrm{V}_1 \leftrightarrow \mathrm{V}_2\,, \tag{12.21}$$

$$\begin{aligned} C_{\mathrm{2V}}^{\mathrm{eq}} &= z\exp(G_{\mathrm{2V}}^{\mathrm{B}}/k_{\mathrm{B}}T)(C_{\mathrm{1V}}^{\mathrm{eq}})^2/2 \\ &= z\exp(-S_{\mathrm{2V}}^{\mathrm{B}}/k_{\mathrm{B}})\exp(H_{\mathrm{2V}}^{\mathrm{B}}/k_{\mathrm{B}}T)(C_{\mathrm{1V}}^{\mathrm{eq}})^2/2\,. \end{aligned} \tag{12.22}$$

The following physical picture indicates that in metals $H_{\mathrm{2V}}^{\mathrm{B}}$ is positive. The generation of a monovacancy may be thought of as removing a positively charged ion core (not a neutral atom!) from a lattice site, putting it into a *Halbkristallage* as described in Sect. 12.3, letting the electron gas of the metal expand into the increased (by the monovacancy formation volume $V_{\mathrm{1V}}^{\mathrm{F}}$) crystal volume, and screening the effective negative charge at the vacant site by repulsing electrons near that site. The last step increases the kinetic energy of the electrons, whereas the preceding step decreased it [4]. If we bring together two vacant sites to form a divacancy, the electronic kinetic energy connected with the screening decreases, since the "screening cloud" of repelled electrons common to the two vacancies is now further extended, hence the kinetic energy of the electrons is smaller than the sum of the corresponding energies of the isolated vacancies. This simple picture can be used to estimate $H_{\mathrm{2V}}^{\mathrm{B}}$ [5]. The estimates $0.1\,\mathrm{eV} \le H_{\mathrm{2V}}^{\mathrm{B}} \le 0.2\,\mathrm{eV}$ for the noble metals are in agreement with the experimental data.

For the sign of

$$S_{\mathrm{2V}}^{\mathrm{B}} = 2S_{\mathrm{1V}}^{\mathrm{F}} - S_{\mathrm{2V}}^{\mathrm{F}} \tag{12.23}$$

a general argument may be given, too. In the cluster picture introduced in Sect. 12.3, it is clear that bringing together two vacancies softens the cluster, hence lowers the phonon frequencies, and increases the entropy. As a consequence, the *divacancy binding entropy* $S_{\mathrm{2V}}^{\mathrm{B}}$ is predicted to be *negative*. It is therefore convenient to use the association entropy $\Delta S_{\mathrm{2V}} := -S_{\mathrm{2V}}^{\mathrm{B}} > 0$

[4] For remarks on the definitions (12.18) see Sect. 12.6.3.

in equations such as (12.22). Since ΔS_{2V} may reach a few k_B and since in metals z is typically of the order of magnitude 10^1 (= 12 in the fcc and hcp structures, and 14 in the bcc structure if nearest and next-nearest neighbours are considered together), the pre-exponential factor $z\exp(\Delta S^B_{2V}/k_B)/2$ on the right-hand side of (12.22) may be quite large. Hence in those metals in which, in thermal equilibrium, the total concentrations of vacancies,

$$C_V = \sum_j jC_{jV}\,, \tag{12.24}$$

reaches the order of magnitude 10^{-3} near the melting point, the fraction C^{eq}_{2V}/C^{eq}_{1V} should not be neglected without explicit justification.

The considerations leading to (12.19) and (12.20) hold, mutatis mutandis, also for self-interstitials. Indeed, the literature contains several attempts to attribute high-temperature observations on metals to self-interstitials in thermal equilibrium. However, the only defects contributing appreciably to the thermal-equilibrium population are those with the smallest formation enthalpies, unless there is a further one with an exceptionally large entropy factor $\exp(S^F_J/k_B)$. All computations of formation energies in metals have yielded self-interstitial formation energies that are so much larger than the vacancy formation enthalpies that this difference cannot be compensated by entropy terms. Hence the phenomena attributed to thermal-equilibrium self-interstitial concentrations that are comparable with C^{eq}_V must have other explanations. (Owing to the smallness of its ion cores, the smallest ratio H^F_{1I}/H^F_{1V} is expected for Li, for which a rather reliable ab initio calculation [6] has given $H^F_{1I}/H^F_{1V} = 1.4$ for the two self-interstitial configurations with the lowest formation enthalpies, viz. the $\langle 111\rangle$ and $\langle 110\rangle$ dumbbells.)

In the derivation of (12.19) and (12.20), T was just a parameter; hence these equations are valid irrespective of the temperature variations of S^F_J and H^F_J. From the fact that these variations are coupled according to (12.5) it follows that the important relationship

$$\partial \ln C^{eq}_J(T,p)/\partial(1/k_B T)\,|_p = -H^F_J(T,p) \tag{12.25}$$

holds for *any* temperature variation of $H^F_J(T,p)$ and any species J. The widespread belief that a *linear* variation of H^F_J with temperature may be absorbed into the entropy S^F_J and, therefore, gives the same functional form of $C^{eq}(T,p)$ as temperature-independent formation enthalpies is *wrong*. Rather, the temperature law

$$H^F(T) = H^F(T_0) + \alpha k_B(T - T_0)\,, \tag{12.26}$$

which may be considered as the beginning of a Taylor expansion of the enthalpy in powers of $(T - T_0)$, leads to

$$S^F(T) = S^F(T_0) + \alpha k_B \ln(T/T_0)\,, \tag{12.27}$$

and hence to

$$C^{\mathrm{eq}} = (T/T_0)^{\alpha} \exp\{[S^{\mathrm{F}}(T_0)/k_{\mathrm{B}}] - \alpha\} \exp\{-[H^{\mathrm{F}}(T_0) - \alpha k_{\mathrm{B}} T_0]/k_{\mathrm{B}}T\} . \tag{12.28}$$

Thus, the temperature dependence (12.26) modifies Boltzmann laws such as (12.19) and (12.20) by a temperature-dependent pre-exponential factor. The choice of the reference temperature T_0 is a matter of expediency as long as it lies within the range of validity of the approximation (12.10). For fitting experimental data by (12.28), a convenient choice is often the midpoint, on a T^{-1} scale, of the interval covered by the data.

12.5 The Influence of Impurities on the Vacancy Concentration in Thermal Equilibrium

From the remarks on self-interstitials in Sect. 12.4 it follows that high-temperature measurements allow us to investigate vacancies without the need to separate off the influence of *self-interstitials*. However, even in high-purity metals we may have to take into account the influence of *impurities* if their interaction with vacancies is attractive and thus enhances the total vacancy population. An estimate of this effect may be obtained from a very simple model, called *Lomer model* [7]. It assumes that an impurity, present in the atomic concentration C_{i} (defined as the ratio of impurity atoms to the sum of host and impurity atoms), may capture at most one monovacancy at one of z_{i} neighbouring sites and that the free enthalpy of binding of a vacancy at these sites is $G^{\mathrm{B}}_{\mathrm{Vi}}$. Taking into account that the formation of such a vacancy–impurity complex reduces the number of sites available for unbound vacancies by $z_{\mathrm{i}}C_{\mathrm{i}}$, we obtain as equilibrium concentration of unbound vacancies

$$C^{\mathrm{eq}}_{1\mathrm{V}} = (1 - z_{\mathrm{i}}C_{\mathrm{i}}) \exp(-G^{\mathrm{F}}_{1\mathrm{V}}/k_{\mathrm{B}}T) . \tag{12.29}$$

By following the procedure of Sect. 12.4, the concentration of the bound vacancies is found to be

$$C^{\mathrm{eq}}_{\mathrm{Vi}} = \frac{z_{\mathrm{i}}C_{\mathrm{i}} \exp[-(G^{\mathrm{F}}_{1\mathrm{V}} - G^{\mathrm{B}}_{\mathrm{Vi}})/k_{\mathrm{B}}T]}{\{1 + z_{\mathrm{i}} \exp[-(G^{\mathrm{F}}_{1\mathrm{V}} - G^{\mathrm{B}}_{\mathrm{Vi}}/k_{\mathrm{B}}T)]\}} . \tag{12.30}$$

The enhancement of the vacancy concentration normalized to that of the impurity-free metal is given by [8]

$$\begin{aligned} \Delta C_{\mathrm{V}} \exp(G^{\mathrm{F}}_{1\mathrm{V}}/k_{\mathrm{B}}T) &= (C^{\mathrm{eq}}_{1\mathrm{V}} + C^{\mathrm{eq}}_{\mathrm{Vi}}) \exp(-G^{\mathrm{F}}_{1\mathrm{V}}/k_{\mathrm{B}}T) - 1 \\ &= \frac{z_{\mathrm{i}}C_{\mathrm{i}}\{\exp(G^{\mathrm{B}}_{\mathrm{Vi}}/k_{\mathrm{B}}T)[1 - z_{\mathrm{i}} \exp(-G^{\mathrm{F}}_{1\mathrm{V}}/k_{\mathrm{B}}T)] - 1\}}{1 + z_{\mathrm{i}} \exp[-(G^{\mathrm{F}}_{1\mathrm{V}} - G^{\mathrm{B}}_{\mathrm{Vi}})/k_{\mathrm{B}}T]} . \end{aligned} \tag{12.31}$$

It increases rapidly with decreasing temperature. If

$$\exp(G^{\mathrm{F}}_{1\mathrm{V}}/k_{\mathrm{B}}T) \gg \exp(G^{\mathrm{B}}_{\mathrm{Vi}}/k_{\mathrm{B}}T) \gg 1$$

holds, (12.31) may be replaced by

$$\Delta C_{\mathrm{V}} \exp(G^{\mathrm{F}}_{1\mathrm{V}}/k_{\mathrm{B}}T) = zC_{\mathrm{i}} \exp(G^{\mathrm{B}}_{\mathrm{Vi}}/k_{\mathrm{B}}T) . \tag{12.32}$$

12.6 Experimental Studies of Thermal-Equilibrium Defects

12.6.1 General Remarks

As already emphasized, from measurements of the vacancy concentration in thermal equilibrium over a sufficiently wide temperature range the enthalpy and entropy of monovacancy formation, $H^{\mathrm{F}}_{1\mathrm{V}}$ and $S^{\mathrm{F}}_{1\mathrm{V}}$, may be deduced without interference from self-interstitials. For several reasons, $H^{\mathrm{F}}_{1\mathrm{V}}$ is a key quantity for the understanding of intrinsic defects in metals.

1. In conjunction with self-diffusion data, from which, in many metals, rather accurate values of $H^{\mathrm{F}}_{1\mathrm{V}}+H^{\mathrm{M}}_{1\mathrm{V}}$ may be derived (see, for example, [9]), the knowledge of $H^{\mathrm{F}}_{1\mathrm{V}}$ allows us to deduce the enthalpy of *monovacancy migration*, $H^{\mathrm{M}}_{1\mathrm{V}}$. Arguably, the enthalpies of migration constitute the most important 'fingerprints' when we wish to identify atomic defects in complex situations, such as in irradiated or plastically deformed metals.
2. Since monovacancies are the simplest crystal defects, comparisons of ab initio computations of $H^{\mathrm{F}}_{1\mathrm{V}}$ with reliable experimental determinations are well suited for judging the reliability and accuracy of the underlying theoretical framework and the computational techniques.
3. Experimental $H^{\mathrm{F}}_{1\mathrm{V}}$ values are often used to adjust parameters of semi-empirical potentials employed in computations of other defect properties. Unreliable input data would clearly invalidate the conclusions from such computations.

Since monovacancies interact with impurities as discussed in Sect. 12.5, for the experimental determination of $H^{\mathrm{F}}_{1\mathrm{V}}$ the purest materials available should be used. Generally speaking, the measurements least affected by residual impurities are those performed at high temperatures. Since in 'pure' metals there are many more sites at which the vacancies do not interact with the impurities than sites with strong vacancy–impurity interactions, with increasing temperature the configuration entropy shifts the free-enthalpy balance towards 'free' vacancies (cf. Sect. 12.5).

Unfortunately, circumstances make it often quite difficult to obtain the desired high-temperature information. The effects of the atomic defects on the physical quantity under investigation are superimposed on those of the unperturbed crystal. In thermal equilibrium, it is not possible to determine the background due to the unperturbed crystal separately. Save for one exception (see Sect. 12.6.3), it must be obtained by extrapolation from lower temperatures at which the vacancy contribution is negligible. Unavoidably, this brings in uncertainties. Since in metals $C^{\mathrm{eq}}_{\mathrm{V}}(T)$ is small compared to unity even close to the melting point, and since it decreases exponentially with decreasing temperature, in all cases very sensitive measuring techniques are required. Furthermore, at high temperatures we have to take into account the possibility of divacancy contributions to the measured effects (Sect. 12.4).

Lowering the temperature reduces this complication in the determination of $H^{\rm F}_{\rm 1V}$ but enhances the unwanted effects of residual impurities.

In principle, any physical property P that responds to the presence of vacant lattice sites and that can be measured with sufficient accuracy at high temperatures may be used to investigate $C^{\rm eq}_{\rm V}(T)$. A broad survey of such measurements on metals, containing more than 400 references, has recently been published by Y. Kraftmakher [10]. In the following we consider only the most important techniques. In line with what has been said in Sect. 12.4, the interpretation of high-temperature measurements in terms of intrinsic atomic defects may be confined to mono- and divacancies. Their contributions per unit concentration will be denoted by P_1 and P_2, respectively, and the background due to the unperturbed crystal by $P_{\rm lattice} = P_{\rm lattice}(T)$. Thus, the temperature variation of the property P reads

$$
\begin{aligned}
P(T) &= P_1 \exp(S^{\rm F}_{\rm 1V}/k_{\rm B}) \exp(-H^{\rm F}_{\rm 1V}/k_{\rm B}T) \\
&\quad + P_2 (z/2) \exp(S^{\rm F}_{\rm 2V}/k_{\rm B}) \exp(-H^{\rm F}_{\rm 2V}/k_{\rm B}T) + P_{\rm lattice} \, . \qquad (12.33)
\end{aligned}
$$

From (12.33) we see immediately that, in general, from high-temperature measurements alone it is not possible to factorize the products $P_j \exp(S^{\rm F}_{j\rm V}/k_{\rm B})$. Important exceptions are calorimetric measurements (cf. Sect. 12.6.2), since here the contributions per unit defect concentration are related to the enthalpies of formation of the defects. For example, if P is the enthalpy per atom, $P_1 = H^{\rm F}_{\rm 1V}$, $P_2 = H^{\rm F}_{\rm 2V}$ holds. The reduction of the number of adjustable parameters in (12.33) resulting from this permits, at least in principle, the experimental determination of $S^{\rm F}_{j\rm V}$ and thus, together with $H^{\rm F}_{j\rm V}$, the calculation of the equilibrium concentrations using (12.19) and (12.20).

12.6.2 Calorimetry

Calorimetric measurements have been employed by many workers, most notably by Y. Kraftmakher, to whose recent review [10] the reader is referred for full coverage and detailed references. There are essentially two different experimental approaches, *drop calorimetry* and *high-temperature calorimetry*.

Drop calorimetry is basically a quenching technique.[5] The specimen is rapidly dropped into a calorimeter bath. From the temperature increase of

[5] In quenching experiments the sample is equilibrated at a high temperature, T, and then cooled rapidly, usually by dropping it into an iced brine or a cryogenic fluid. Ideally, the quench is so fast that the atomic defects in thermal equilibrium at T are 'frozen in'. Since owing to different migration enthalpies (in fcc metals: $H^{\rm M}_{\rm 1V} > H^{\rm M}_{\rm 2V}$) different atomic defects migrate and recover at different annealing temperatures, they may subsequently be separated by successive annealing in a sort of 'fractional distillation'. However, it is virtually impossible to achieve an 'ideal quench'. The thermal stresses accompanying very fast quenches will lead to plastic deformation and thus to the introduction of dislocations. The dislocations act as sinks for the atomic defects and thus enhance the annealing out during the quench (cf. Sect. 12.3). Reducing the quenching rate in order to avoid damaging

the bath or the heat release, the heat content of the specimen is determined as a function of the quenching temperature.

High-temperature calorimetry comprises several techniques. The basic idea of *adiabatic calorimetry* is to minimize the heat exchange between the sample and the environment (which is detrimental to accurate measurements) by keeping a thermal shield surrounding the specimen at the sample temperature during rapid heating. The heat capacity of the sample is derived from the relationship between the sample temperature and the heat supplied to the sample, usually by electric heating. In *modulation calorimetry*, which goes back to O.M. Corbino [13,14] and was later on extensively used for heat-capacity measurements by several workers, the specimen is heated by an alternating electrical current. From the amplitude of the temperature variations, measured either directly or, say, through the electrical resistance of the sample, the heat capacity may be deduced. *Pulse* and *rapid-heating techniques* minimize the heat exchange between sample and environment by performing the measurements as fast as possible. *Pulse calorimetry* achieves this by employing small temperature increments; hence the specific heats measured are averages over narrow temperature intervals. By *rapid-heating calorimetry* the entire heat-capacity–temperature curve up to the melting point and above may be measured within about a microsecond or less.

If c_{lattice} denotes the heat capacity per atom (for briefness in the following called *specific heat*) of the unperturbed crystal and if we restrict ourselves again to mono- and divacancies, the specific heat of the specimen at constant pressure is given by

$$c_p(T) = \{H_{1V}^{F} \exp(S_{1V}^{F}/k_B) \exp(-H_{1V}^{F}/k_B T) \\ + z H_{2V}^{F} \exp(S_{2V}^{F}/k_B) \exp(-H_{2V}^{F}/k_B T)\}/z k_B T^2 + c_{\text{lattice}} \,. \quad (12.34)$$

the sample gives the defects time to react with each other, e.g. by forming multiple vacancies and eventually dislocation loops, voids, and stacking-fault tetrahedra, and with impurities.

Owing to their basic simplicity, quenching experiments played an important role in the early studies of point defects in metals and contributed valuable information. (See, for example, the proceedings of the 1964 conference on quenching of metals [11]. The notation used in the present chapter is based on a convention agreed upon at that conference.) Notwithstanding the advances in high-temperature techniques discussed in the present chapter, rapid quenching from high temperatures has remained a powerful technique for the determination of both formation and migration enthalpies of vacancy-type defects in metals and alloys. While it is true that the complications mentioned above necessitate discernment in the quantitative interpretation of quenching experiments, disregarding their results summarily – as has been done in a recent data collection ([12], Sect. 2.3.2) – is unjustified. Regrettably, a lack of space does not allow us to discuss the H_{1V}^{F} determinations by quenching experiments in the present chapter in any detail.

The interpretation of the high-temperature measurements in terms of mono- and divacancy properties suffers from the difficulty that at high temperatures anharmonicities contribute to c_{lattice}, and that these are not well known. We shall return to this particular aspect of the background problem in Sect. 12.7. Drop calorimetry does not have to face this problem, since the degrees of freedom contributing to c_{lattice} adjust themselves to the instantaneous temperature in times that are short compared with the times involved in the experiments.

12.6.3 Differential Dilatometry

The 'background problem' plaguing most high-temperature techniques for studying thermal-equilibrium defects may be avoided by comparing the macroscopically measured crystal volume, which contains a contribution from the defects present in thermal equilibrium (cf. Sect. 12.3), with that calculated from lattice-parameter measurements. Moreover, this method shares with the calorimetric measurements the advantage that the contribution per defect is not one of the parameters to be adjusted, so that, at least in principle, absolute defect concentations and, after having determined the formation energies, defect formation entropies may be deduced. Disregarding for the moment the considerable experimental demands, *differential dilatometry* is thus the ideal method for studying atomic defects in thermal equilibrium.

According to T. Hehenkamp [15], the basic idea goes back to J. Beyer and C. Wagner [16]. Since on most metals pycnometric measurements are not feasible, macroscopic dilatometric measurements are compared with the temperature variation of the lattice parameter as measured by X-ray or neutron diffraction, hence the name *differential dilatometry.* The first recorded application of the technique, by A. Gott on aluminium [17], did not give a positive result, nor did measurements by J. van Dujn and J. van Galen [18] on lead. In both attempts the experimental accuracy was presumably insufficient. The first successful measurements were those by R. Feder and A.S. Nowick [19] on aluminium. Subsequently, the technique was used by many groups, most extensively by those of R.O. Simmons and R.W. Balluffi [20] and of T. Hehenkamp [15]. The metal with the highest melting temperature so far reported in the literature is Cu [21–23].

In order to keep the arguments as simple as possible, we derive the basic equations under the assumption that we may ascribe to each atom the same atomic volume V_{A}. (This covers all Bravais lattices but also the hcp structure and β–Sn; the generalization to more complex non-Bravais structures is straightforward.) The difference between the macroscopically measured volume of a crystal into which n_{V} vacancies and n_{I} self-interstitials have been introduced and that of a defect-free reference crystal containing the same number N of atoms may be written as

$$\delta v_{\mathrm{macr}} = N\delta V_{\mathrm{A}} + n_{\mathrm{V}}V_{\mathrm{V}}^{\mathrm{F}} + n_{\mathrm{I}}V_{\mathrm{I}}^{\mathrm{F}} \,. \tag{12.35}$$

In (12.35) δV_{A} denotes the increase of the atomic volume between the temperature of the reference crystal and the temperature T at which we wish to determine the concentration of atomic defects. Eshelby [24] showed that the macroscopic volume and the 'diffraction volume' as calculated from the size of the lattice cells are equally affected not only by the thermal expansion but also by *inhomogeneous* lattice strains. The latter result allows us to disregard the distinction between monovacancies, divacancies, etc. and to work with average values $V_{\mathrm{V}}^{\mathrm{F}}$ and $V_{\mathrm{I}}^{\mathrm{F}}$. Thus, the difference in the *diffraction volumes* of the defect-containing and the reference crystal reads

$$\delta v_{\mathrm{diff}} = N\delta V_{\mathrm{A}} + n_{\mathrm{V}}[V_{\mathrm{V}}^{\mathrm{F}} - V_{\mathrm{A}}(T)] + n_{\mathrm{I}}[V_{\mathrm{I}}^{\mathrm{F}} + V_{\mathrm{A}}(T)]\,. \tag{12.36}$$

Subtracting (12.36) from (12.35) gives us the exact result

$$\delta v_{\mathrm{macr}} - \delta v_{\mathrm{diff}} = (n_{\mathrm{V}} - n_{\mathrm{I}})V_{\mathrm{A}}(T)\,. \tag{12.37}$$

As argued in Sect. 12.4, in high-temperature measurements on metals n_{I} may be neglected. Hence under these circumstances (12.37) simplifies to

$$\delta v_{\mathrm{macr}} - \delta v_{\mathrm{diff}} = n_{\mathrm{V}}(T)V_{\mathrm{A}}(T)\,. \tag{12.38}$$

The macroscopic volume at the temperature T, $v_{\mathrm{macr}}(T)$, is given by

$$v_{\mathrm{macr}}(T) = (N_{\mathrm{V}} + n_{\mathrm{V}})V_{\mathrm{A}}(T) + n_{\mathrm{V}}(T)[V_{\mathrm{V}}^{\mathrm{F}} - V_{\mathrm{A}}(T)]\,. \tag{12.39}$$

Since in metals the second term on the right-hand side of (12.39) is negligibly small compared with the first one, we obtain, making use of (12.17),

$$\begin{aligned}(\delta v_{\mathrm{macr}} - \delta v_{\mathrm{diff}})/v_{\mathrm{macr}}(T) &= \exp(-G_{\mathrm{V}}^{\mathrm{F}}/k_{\mathrm{B}}T)\\ &= \exp(S_{\mathrm{V}}^{\mathrm{F}}/k_{\mathrm{B}})\exp(-H_{\mathrm{V}}^{\mathrm{F}}/k_{\mathrm{B}}T)\,.\end{aligned} \tag{12.40}$$

In (12.40) the subscript V indicates that the thermodynamic quantities are *effective* quantities that must be interpreted in terms of monovacancies, divacancies, etc.

In order to determine the left-hand side of (12.40) experimentally, for cubic metals it suffices to measure the thermal expansion $\delta l/l_0$ along any linear dimension of the sample and to determine the relative change of the lattice parameter, $\delta a/a_0$, starting from a defect-free state l_0, a_0. Hexagonal, tetragonal, and trigonal crystal structures require measurements in two directions that do not both lie in the basal plane. The prediction that the left-hand side of (12.40) is zero at low temperatures has been verified in many experiments.

The expansion of the left-hand side of (12.40) in powers of $\delta l/l_0$ and $\delta a/a_0$ gives us

$$3(\delta l/l_0 - \delta a/a_0)\{1 - 2\delta l/l_0 + \delta a/a_0\} = \exp(S_{\mathrm{V}}^{\mathrm{F}}/k_{\mathrm{B}})\exp(-H_{\mathrm{V}}^{\mathrm{F}}/k_{\mathrm{B}}T)\,. \tag{12.41}$$

The term in the curly bracket in (12.41) is a correction factor that may play a role close to the melting point. The reason why it differs from the correction factor given earlier [8] is related to a certain ambiguity in defining *atomic* concentrations. The concentration of a substance is originally a macroscopic concept and refers to the amount of matter *per unit volume.* If we consider the *atomic* concentration of intrinsic defects, it is not a priori clear whether we should refer it to the number of atoms or to the number of unit cells. In other words: Should we take into account the increase in crystal volume due to the vacancy formation or not? We have indirectly answered the question by going back to Sect. 12.4 [e.g. to (12.17)], where it is shown that the description of equilibrium concentrations in terms of the free enthalpies of defect formation is simplest if the change in crystal volume due to the defects is allowed for in the definition of the atomic concentration.

In the literature the analysis of the dilatometric measurements on cubic crystals is usually based on

$$\delta l/l_0 - \delta a/a_0 = (C_{\mathrm{V}} - C_{\mathrm{I}})/3\,, \tag{12.42}$$

with $C_{\mathrm{I}} \equiv 0$ in the case of high-temperature measurements. (Note that the inclusion of C_{I} is essential, e.g., in investigations of the recovery of intrinsic defects *after particle irradiation.*) In the approximation (12.42) the question of the correction factor discussed in the preceding paragraph does not arise.

Although in theory differential-dilatometric measurements are background-free since in (12.37) the 'background' due to the thermal expansion of the perfect crystal has dropped out, in each single measurement this background is large compared with the effect due to the defects. Thus, the technique involves subtracting two large quantities from each other in order to obtain a small difference. Moreover, the 'background' varies with temperature; hence it is of paramount importance that the macroscopic and the diffraction measurements be performed at the same temperature within very narrow limits. Since the length measurements integrate over the sample, the use of small samples permitted by optical techniques (in the case of the Göttingen measurements to be discussed below [15] 18 mm in diameter and 22 mm in length) constitutes a definite advantage over the mechanical measurements of Simmons and Balluffi [20], who had to use much bigger samples.

The main weakness of differential dilatometry is that, as a consequence of the subtraction procedure, the *absolute* errors are approximately independent of the measured quantity, hence the *relative* errors *increase* with decreasing $\delta l/l_0 - \delta a/a_0$. In practice, this makes it difficult to determine $H^{\mathrm{F}}_{1\mathrm{V}}$ and $S^{\mathrm{F}}_{1\mathrm{V}}$ separately with high accuracy. Since, as we shall see presently, $H^{\mathrm{F}}_{1\mathrm{V}}$ is not very sensitive against the precise choice of $S^{\mathrm{F}}_{1\mathrm{V}}$, further ab-initio computations of $S^{\mathrm{F}}_{1\mathrm{V}}$ such as the one recently undertaken by V. Schott, M. Fähnle and P.A. Madden [25] on the alkali metals would help in deducing reliable $H^{\mathrm{F}}_{1\mathrm{V}}$ values from high-temperature differential dilatometry even if their accuracy is limited. These authors obtained $S^{\mathrm{F}}_{1\mathrm{V}} = 1.36k_{\mathrm{B}}$ for sodium and $S^{\mathrm{F}}_{1\mathrm{V}} = 0.78k_{\mathrm{B}}$ for potassium.

The most extensive high-temperature differential dilatometry has been performed on Al. Here the results of the various groups are in quite good agreement with each other. This is not so for metals with higher melting temperatures T_m. The measurements of the Göttingen group on Cu [23], Ag [26], and Au [27], which comprised also detailed studies on several dilute alloys (cf. [15, 28]), gave substantially higher vacancy concentrations than those of Simmons and Balluffi [21, 29, 30]. Since the Göttingen group used a superior technique (see above) and performed much more extensive measurements, their results deserve more confidence. The vacancy concentrations extrapolated to the melting point, $C_V(T_m)$, are listed in Table 12.1 together with the average value of four different measurements on Al [28,31–33] as well as those on Pb [34], Mg [35], Zn [36–38], and Cd [39, 40].[6] Table 12.1 gives also the effective activation enthalpies of vacancy formation at the melting point, $H_V^F(T_m)$, as calculated from

$$H_V^F(T_m) = -k_B T_m \ln C_V^{eq}(T_m) + T_m S_V^F(T_m) \tag{12.43}$$

with the assumption $S_V^F(T_m) = k_B$. The numerical values are not very sensitive against the choice of $S_V^F(T_m)$. For example, a change of S_V^F by $\pm 0.2 k_B$ changes H_V^F by $\pm 1.72 \times 10^{-2}(T_m/\text{K})$ meV.

What is the relationship between $H_V^F(T_m)$ and H_{1V}^F? The answer is simple if $C_V^{eq}(T_m)$ is small enough for V_2 to be neglected, as is the case in Pb, Zn, and Cd (cf. Table 12.1). Then $H_V^F(T_m)$ and H_{1V}^F coincide within experimental uncertainty, provided $S_V^F(T_m)$, which coincides with S_{1V}^F, has been chosen correctly. Since in these cases the temperature interval over which $C_V^{eq}(T)$ can be determined with good accuracy is narrow, it is advisable to ascertain that the determination of H_{1V}^F according to (12.25) leads to reasonable S_{1V}^F.

If $C_V^{eq}(T_m)$ is larger, as in Au, two correction terms have to be considered. For a given choice of the entropy term, the curly bracket in (12.41) increases $H_V^F(T_m)$ by a small amount. The second correction, that due to divacancies (cf. Sect. 12.4), goes in the same direction but is likely to be larger unless the free enthalpy of divacancy binding is very small (Fig. 12.1). The assumption that $C_{2V}^{eq}(T_m)/C_{1V}^{eq}(T_m) = 1/4$, i.e. that at the melting point there are twice as many vacant sites in monovacancies than in divacancies, leads to H_{1V}^F values that exceed the $H_V^F(T_m)$ of Table 12.1 by $k_B T_m \ln(3/2) = 3.5 \times 10^{-2}(T_m/\text{K})$ meV. In the specific case of Au, $H_{1V}^F - H_V^F(T_m) = 0.04$ eV appears to be a realistic estimate, hence $H_{1V}^F \approx 0.92$ eV. Making use of $H_{1V}^M = H_{1V}^{SD} - H_{1V}^F$, where H_{1V}^{SD} is the activation enthalpy of the monovacancy contribution to self-diffusion (in Au 1.74 eV [42]), we obtain for the migration enthalpy of monovacancies in Au $H_{1V}^M \approx 0.82$ eV. Even if we allow for the (small) uncertainty of this estimate, we may state that the value recommended in Sect. 2.3.2 of [12], $H_{1V}^M = (0.71 \pm 0.05)$ eV, is incompatible with the high-temperature data and therefore very unlikely to be correct.[7]

---

[6] A review of atomic defects in the hexagonal metals Zn and Cd by the present writer includes a detailed discussion of H_{1V}^F determinations in these metals [41].

[7] Disregarding a large body of experimental data, in particular those obtained in quenching experiments (cf. Footnote 5), contributed to the erroneous assignment.

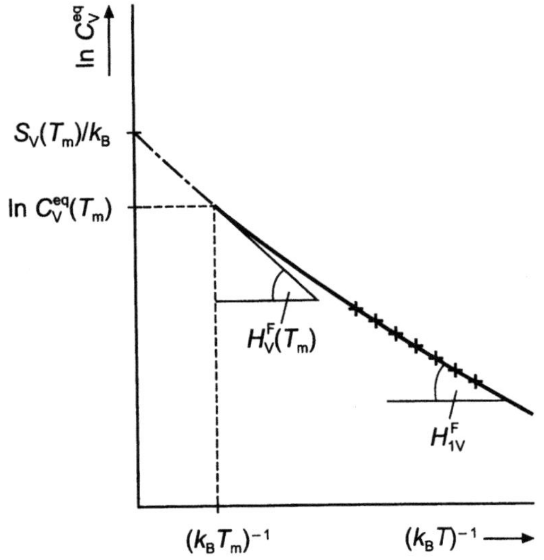

Fig. 12.1. Boltzmann plot of the equilibrium concentration of vacancies, $C_V^{eq}(T)$, as a function of the inverse temperature T^{-1}. The crosses indicate the temperature range of the 'concentration window' of positron annihilation. The slope at the low-temperature end of the plot gives the monovacancy formation enthalpy H_{1V}^F, the slope at the melting temperature T_m defines the effective vacancy formation enthalpy H_V^F and the effective formation entropy S_V^F

In metals with intermediate $C_V^{eq}(T_m)$, such as Al, Ag, Mg, and Cu, the correction term $H_{1V}^F - H_V^F(T_m)$ is presumably smaller than the above estimate for Au.

12.6.4 Positron Annihilation

General Background. Positrons (e^+) are the antiparticles of electrons (e^-), having the same mass, m_e, as e^- and carrying one positive elementary electric charge. They are stable particles that are readily available from the β^+-decay of radioactive nuclei. When implanted into metals, their vast majority annihilates with electrons of the host material according to the reaction

$$e^+ + e^- \to 2\gamma \tag{12.44}$$

at annihilation rates of the order of magnitude $10^{10}\,s^{-1}$. Owing to energy and momentum conservation, in the centre-of-mass system of the annihilating e^+e^--pairs the two gamma quanta (in the following referred to as *annihilation photons*) have the same energy, viz. $m_e c^2 = 511\,keV$, and opposite momenta $\pm m_e c$ (c = speed of light in vacuum). The interest of e^+ annihilation for defect studies lies in the information on the annihilation sites that is transmitted by the annihilation photons. This information may be subdivided into two main classes.

1. One of the annihilation photons may be used as the 'death signal' of an implanted positron. Together with a 'birth signal', the 'death signal' allows us to determine the *positron age*, i.e. the time an e^+ has spent in the sample before it annihilates. The 'birth signal' may be obtained either from a so-called prompt photon emitted by the radioactive mother nucleus at about the same time as the e^+ or from the passage of the e^+ through a scintillation counter placed between the sample and the e^+ source (which in this case usually is a monoenergetic beam in the MeV range). The first type of age determination is known as the $\gamma\gamma$-techniquegg-technique@$\gamma\gamma$-technique, the second one as the $\beta^+\gamma$-technique. The distribution function of the measured positron ages is the so-called *lifetime spectrum.* It contains information on the e^+ *annihilation rates* and the probabilities with which the positrons annihilate at definite rates. Since the annihilation rates are proportional to the electron densities at the e^+ annihilation sites, the analysis of the lifetime spectrum, known as *lifetime spectrometry*, allows us to deduce information on the sites at which the e^+ annihilate and on the probabilities with which they do so.
2. The annihilation radiation may be used to obtain information on the *momenta, p*, of the annihilating e^+e^--pairs. Since in metals the implanted positrons slow down to thermal energies within times that are short compared with their average lifetime and hence have negligible momenta when they annihilate, the e^+e^- momenta are determined entirely by the momentum distribution of the electrons at the annihilation sites. Thus, measurements of the momentum distribution, too, give us information on these sites. This information, however, is different from and therefore supplementary to that obtainable by lifetime spectrometry.
   The momentum measurements are made possible by the fact that in general the centre-of-mass systems of the annihilating e^+e^--pairs do not coincide with the laboratory system. This is exploited by two rather different experimental techniques, which supplement each other very well.
   (a) If the annihilation photons result from the annihilation of an e^+e^--pair with momentum component p_l longitudinal to their emission axis, their energies differ from $m_e c^2$ by the Doppler shifts
   $$\Delta E_\gamma = \pm c p_l / 2\,. \qquad (12.45)$$
   Measurements of the *energies* of the annihilation photons, usually performed with a Ge detector, give us the shape of the Doppler-broadened 511 keV photon line; therefore, the technique is referred to as *Doppler broadening* or the ΔE_γ-technique.
   (b) While collinear in the rest system of the annihilating e^+e^--pairs, in the laboratory system the emission directions of the two annihilation photons form an angle that differs from 180° by
   $$\theta = p_t / m_e c\,. \qquad (12.46)$$
   In (12.46), p_t denotes the momentum component transverse to the emission axis in the direction in which the angle θ is measured by

correlated detection of the emission directions of the two annihilation photons. The technique is referred to as ACAR (**a**ngular **c**orrelation of **a**nnihilation **r**adiation).

The relevance of the positron annihilation techniques to the study of atomic defects in metals rests on the following four facts:

1. As already mentioned, e^+ implanted into metals thermalize in times that are short compared with their lifetimes.
2. In the thermalized state, the e^+ diffuse very fast, with diffusivities D^+ that may be as large as $10^{-4}\,\mathrm{m^2\,s^{-1}}$ at elevated temperatures and even larger at low temperatures.
3. Positrons may be trapped by vacant lattice sites. If we assume that the trapping is diffusion-limited and ascribe to the traps a trapping radius r_0, the trapping rate per unit atomic concentration of the traps (henceforth called the *specific trapping rate*) is given by

$$\sigma = 4\pi r_0 D^+/V_A\,. \tag{12.47}$$

Considering bcc iron as and example ($V_A = a_0^3/2$; a_0=length of the elementary cube edge $= 2.87 \times 10^{-10}\,\mathrm{m}$) and assuming as the capture radius of monovacancies $r_0 = 2a_0/3$ and as the positron diffusivity $D^+ = 1 \times 10^{-4}\mathrm{m^2 s^{-1}}$, we obtain

$$\sigma = 16\pi D^+/3a_0^2 = 2.0 \times 10^{16}\mathrm{s^{-1}}\,. \tag{12.48}$$

With the positron lifetime of bcc Fe, $\tau = 115\,\mathrm{ps}$, (12.48) gives us $\sigma\tau = 2.3 \times 10^6$. The inverse of this number indicates the order of magnitude of the atomic fraction of monovacancies in which *a substantial fraction* of the implanted positrons will be trapped. The *detection limit* of trapped positrons is about one power of 10 lower, i.e. somewhat less than 10^{-7}. Similar estimates hold for other metals.
4. All three techniques mentioned above [1, 2(a), 2(b)] may be used to distinguish e^+ annihilation in traps from annihilation of '*free*' positrons. Together with fact 3, this makes positron annihilation a high-sensitivity tool for studying those defects in metals that provide e^+ traps.

The fact that different annihilation modes and, hence, different traps give *distinguishable signals* constitutes a definite advantage over the other high-sensitivity technique for studying atomic defects in metals, the measurement of the residual electrical resistance. This feature brings with it an important characteristic of positron-annihilation studies of defects, common to all techniques based on trapping, namely the appearance of *saturation effects*, to be discussed presently.

The existence of a *lower limit* of the defect concentrations C detectable by positron annihilation is, as mentioned in fact 3, a consequence of the limited lifetime τ of positrons in matter. If the product $\sigma\tau C$ is very small compared to unity, the fraction of positrons capable of reaching the defects is too small to give a detectable contribution to the annihilation signal. For determining the dependence of the trap concentration on an external parameter, say the temperature, there is also an *upper* limit. It arises in the following way. If $\sigma\tau C$ is much larger than unity, all positrons are trapped virtually instantaneously. Irrespective of the detection technique used, the signal 'saturates', becoming independent of the trap concentration. This feature was recognized by several workers [43–45] and formalized, in its simplest form, in the so-called *two-state trapping model* of Bergersen–Stott [46] and Connors–West [47]. The saturation effect has the practical consequence that positron-annihilation studies of the dependence of the trap concentration on an external parameter are confined to a certain *concentration window*, which typically extends over about three powers of 10. If the two-state trapping model is valid, the signal shows a sigmoidal dependence on concentration, varying between that of free positrons and the trapped-positron signal. The observation, on Al, In, Zn, and Cd, of such a sigmoidal variation of the mean e^+ lifetime with increasing temperature by I.K. MacKenzie, T.L. Khoo, A.B. McDonald, and B.T.A. McKee [48] in 1967 marked the beginning of the study of vacancies in thermal equilibrium by positron annihilation.

As an example, we consider the mean positron lifetime, $\overline{\tau}$, in a slight extension of the original two-state trapping model allowing for positron *detrapping* with an escape frequency ν_{esc}. If we denote the lifetime of the 'free' positrons by τ_{f} and that of the trapped positrons by τ_{t}, the mean lifetime is given by [49, 50]

$$\overline{\tau} = \tau_{\mathrm{f}} \frac{1 + \tau_{\mathrm{t}}(\sigma C + \nu_{\mathrm{esc}})}{1 + \tau_{\mathrm{f}}\sigma C + \tau_{\mathrm{t}}\nu_{\mathrm{esc}}} ts. \tag{12.49}$$

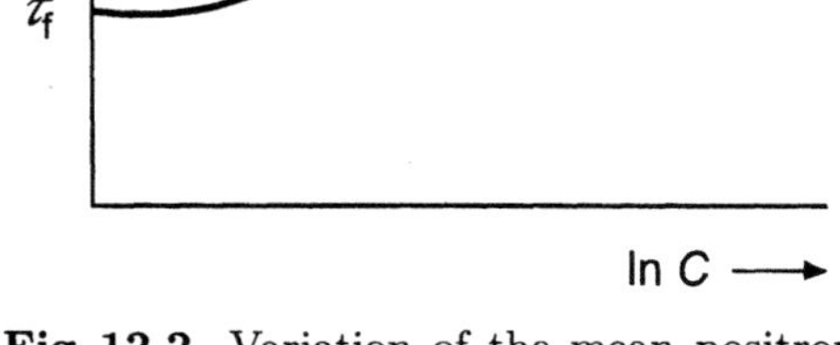

Fig. 12.2. Variation of the mean positron lifetime, $\overline{\tau}$, as a function of the logarithm of the trap concentration C or, provided (12.28) holds and the limetimes τ_{f} of 'free' and τ_{t} of trapped positrons are temperature-independent, of the inverse temperature T^{-1} · ν_{esc} denotes the escape frequency from the traps

Figure 12.2 illustrates the sigmoidal behaviour following from (12.49). For $C = 0$ the mean lifetime is $\overline{\tau} = \tau_{\mathrm{f}}$, whereas for large enough C we have $\overline{\tau} = \tau_{\mathrm{t}}$, irrespective of whether detrapping is negligible or not. If $\nu_{\mathrm{esc}} > 1$, (12.49) predicts that the mean lifetime still reaches the saturation value $\overline{\tau} = \tau_{\mathrm{t}}$, though only at large trap concentrations. Since the detrapping rate increases with temperature [cf. (12.62)], in the application to the study of $C_{1\mathrm{V}}^{\mathrm{eq}}(T)$ this may require temperatures beyond the melting temperature! In comparisons of high-temperature measurements with low-temperature $\tau_{1\mathrm{V}}$ determinations, e.g. on electron-irradiated samples, detrapping effects may have to be taken into account.

Expressions analogous to (12.49) hold also for ACAR and Doppler broadening. The generalization of (12.49) to the inclusion of a second type of trap (with negligible detrapping) may be found in [50].

Vacancies as Positron Traps. In this subsection we show that the electron-gas model used in Sect. 12.5 to explain the attractive interaction between neighbouring vacancies allows us not only to understand qualitatively the observation [48] that vacancies act as e^+ traps in most but not all metals but also provides us with an order-of-magnitude estimate of the energy $E_{1\mathrm{V}}^{+}$ with which a positron is bound to a vacant lattice site in a monovalent metal. In this model, the enthalpy of monovacancy formation may be written as the sum of three terms, viz.

$$H_{1\mathrm{V}}^{\mathrm{F}} = \Delta E_{\mathrm{screen}} + \Delta E_{\mathrm{relax}} + \Delta E_{\mathrm{expnd}} \,. \tag{12.50}$$

The first term (> 0) results from the redistribution of the electrons required to shield the effective negative electric charge produced by the removal of the positively charged ion core. The second term (< 0) represents the lowering of the energy of the system by the relaxation of the adjacent atoms into the vacant site. The third term (> 0) is the energy gained by the expansion of the electron gas into the increased crystal volume. From the dependence of the Fermi energy, E_{F}, on the $-(2/3)^{\mathrm{rd}}$ power of the specific volume, it follows that this term is given by

$$\Delta E_{\mathrm{expnd}} = -2V_{1\mathrm{V}}^{\mathrm{F}} E_{\mathrm{F}}/5V_{\mathrm{A}} \,. \tag{12.51}$$

The Fermi energy E_{F} may either be deduced from experimental data or estimated from

$$E_{\mathrm{F}} = \hbar^2 (3\pi^2/V_{\mathrm{el}})^{2/3}/2m^* \,, \tag{12.52}$$

where m^* denotes the effective mass of the conduction electrons (treated as 'quasi-free') and V_{el} the volume per electron. For monovalent metals, V_{el} is equal to the atomic volume $V_{\mathrm{A}} =: 4\pi r_{\mathrm{A}}^3/3$.

The energy $E_{1\mathrm{V}}^{+}$ with which a positron is bound to a vacancy is calculated in two steps (Fig. 12.3). First, move a 'free' $\mathrm{e}^+(1)$ into the vacancy and endow it with an infinitely large mass (2). This step lowers the energy of the system

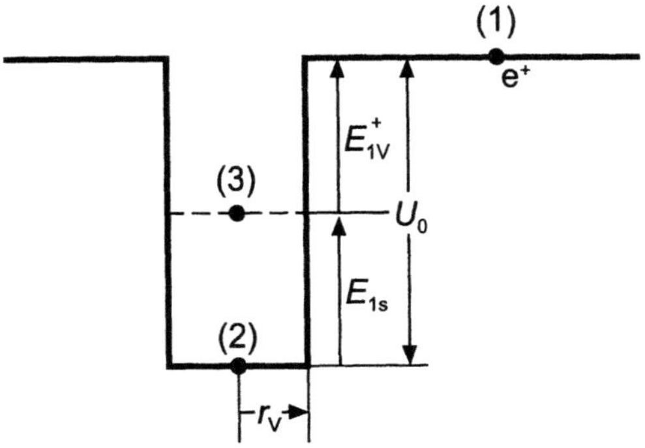

Fig. 12.3. Positron bound state in a potential well of radius r_{V} and depth U_0. (1) Potential energy of a 'free' positron. (2) Energy level of an elementary positive charge of infinite mass. (3) Ground-state energy level of a positron in the well (mass m_{e})

by U_0, the depth of the potential well provided by the vacancy for an elementary positive electric charge. Since now the charge compensates the effective negative charge of the vacant site, the need for the screening rearrangement of the electron gas is obviated and the screening energy $\Delta E_{\mathrm{screen}}$ is gained. This gives us, with (12.50), the relationship

$$U_0 = \Delta E_{\mathrm{screen}} = H^{\mathrm{F}}_{1\mathrm{V}} - \Delta E_{\mathrm{relax}} - \Delta E_{\mathrm{expnd}} \,. \tag{12.53}$$

In a second step the positron is given back its mass, m_{e}. The positron will then occupy the ground state in the well corresponding to m_{e}, with energy $E_{1\mathrm{s}}$ above the bottom of the well (3). The relaxation of the neighbours will be changed, too, but since $\Delta E_{\mathrm{relax}}$ is a small term, presumably not exceeding the inherent uncertainty of the model, we neglect it in the further estimates. We thus obtain

$$E^{+}_{1\mathrm{V}} = U_0 - E_{1\mathrm{s}} \approx H^{\mathrm{F}}_{1\mathrm{V}} - \Delta E_{\mathrm{expnd}} - E_{1\mathrm{s}} \,. \tag{12.54}$$

For simplicity we assume the potential well to be a so-called square well of radius r_{V} (see Fig. 12.3), for which the ground-state energy in three dimensions of a particle of mass m_{e} is known to be

$$E_{1\mathrm{s}} = (\hbar X_{1\mathrm{s}}/r_{\mathrm{V}})^2/2m_{\mathrm{e}} \,. \tag{12.55}$$

The dimensionless quantity $X_{1\mathrm{s}}$ is determined by the product $m_{\mathrm{e}}U_0r_{\mathrm{V}}^2$; in general, it must be found numerically [51]. It varies between $X_{1\mathrm{s}} = \pi$, reached in the limit $U_0 \to \infty$, and $X_{1\mathrm{s}} = \pi/2$, corresponding to the limiting case

$$E_{1\mathrm{s}} = U_0 = (\hbar\pi/r_{\mathrm{V}})^2/8m_{\mathrm{e}} \,, \tag{12.56}$$

in which there is no longer a bound state.

The volume of the cavity representing a monovacancy, $4\pi r_{\mathrm{V}}^3/3$, differs from the vacancy formation volume $V^{\mathrm{F}}_{1\mathrm{V}}$. This is due to the fact that, in order to keep the sample surface traction-free, the long-range elastic strains set up by the relaxation of the neighbouring atoms must be compensated by surface stresses which, in turn, give rise to a dilational (approximately homogeneous) strain. The relationship between the two volumes is [52]

$$V_{1V}^{F} = (1+\gamma_E) 4\pi r_V^3/3\,, \tag{12.57}$$

where the material-dependent parameter γ_E is called the Eshelby factor. In an elastically isotropic material with Poisson's ratio ν it is given by

$$\gamma_E = 3(1-\nu)/(1+\nu)\,. \tag{12.58}$$

For anisotropic materials, γ_E is obtained from the second-order elastic constants by numerical calculation [53]. An approximate formula for weak elastic anisotropy has been given elsewhere [54].

Using the preceding expressions including (12.52), we obtain for the difference between the positron *binding* energy and the *formation* enthalpy of a monovacancy in a monovalent metal

$$E_{1V}^{+} - H_{1V}^{F} = \frac{\hbar^2}{2m_e r_A^2}\left\{\frac{2}{5}\left(\frac{9\pi}{4}\right)^{2/3}\frac{m_e V_{1V}^{F}}{m^* V_A} - X_{1s}^2\left(\frac{(1+\gamma_E)V_A}{V_{1V}^{F}}\right)^{2/3}\right\}. \tag{12.59}$$

The order of magnitude of the right-hand side of (12.59) is determined by $\hbar^2/2m_e = 3.81 \times 10^{-20}\,\mathrm{eVm}^2$. The relationship between r_A and the edge length of the elementary cube, a_0, is $r_A/a_0 = 0.391$(fcc) or 0.492 (bcc). The numerical factor in (12.59) is $2(9\pi/4)^{2/3}/5 = 1.47$. Since $0 < V_{1V}^{F} \leq V_A$ and since for monovalent metals we may assume $0 < m_e/m^* \leq 1$, the curly bracket in (12.59) is always negative. This means that in monovalent metals

$$E_{1V}^{+} < H_{1V}^{F}\,, \tag{12.60}$$

or, with the approximate empirical relationship $H_{1V}^{F} \approx 10 k_B T_m$, that

$$\exp(-E_{1V}^{+}/k_B T_m) > 10^{-4.3} \tag{12.61}$$

should hold. If we write the attempt frequency for the escape of the positrons from their traps as

$$\nu_{esc} = \nu_{esc}^{0} \exp(-E_{1V}^{+}/k_B T_m)\,, \tag{12.62}$$

owing to the smallness of m_e we have to expect large attempt frequencies ν_{esc}^{0}, presumably lying between $10^{13}\mathrm{s}^{-1}$ and $10^{14}\mathrm{s}^{-1}$. Typical lifetimes for positrons in monovacancies in metals are $\tau_{1V} \approx 2 \times 10^{-10}\mathrm{s}^{-1}$. Thus, the preceding estimates indicate that in order for the escape to be negligible in high-temperature e^+ studies of vacancies, E_{1V}^{+} must not be much smaller than H_{1V}^{F}. According to (12.59), this means that the vacancy formation volume, V_{1V}^{F}, should be close to the atomic volume, V_A, or, in other words, that the relaxation of the neighbouring atoms discussed in Sect. 12.6.3 should be small. This condition is fulfilled in Cu, Ag, and Au. In agreement with this, there is no indication that in the high-temperature positron-annihilation experiments on these metals the e^+ escape from vacancies must be taken into account.

By rewriting (12.56) and dropping the first term on the right-hand side of (12.59), we find the following approximate condition for the existence of a positron bound state in a monovacancy:

$$V_{1V}^{F}/V_{A} > (\hbar^2/2m_e H_{1V}^{F} r_{A}^2)^{3/2}(2/\pi)^3(1+\gamma_E)\,. \tag{12.63}$$

For bcc Na [$H_{1V}^{F} = 0.36\,\mathrm{eV}$, $a_0 = 4.29 \times 10^{-10}\mathrm{m}$, $(1+\gamma_E)^{-1} = 0.728$] the right-hand side of (12.63) equals 1.3, whereas self-diffusivity measurements under hydrostatic pressure indicate that the left-hand side of (12.63) is only about 0.2–0.3. From the strong violation of condition (12.63) we conclude that positrons may *not* be trapped in monovacancies in Na. For Li ($H_{1V}^{F} \approx 0.4\,\mathrm{eV}$, $a_0 = 3.51 \times 10^{-10}\,\mathrm{m}$, $(1+\gamma_E)^{-1} = 0.715$) this conclusion holds a fortiori. Both conclusions are in agreement with the failure to detect the trapping of positrons by thermal-equilibrium vacancies [55] in the alkali metals and with the explanation given for this result [54].

For K, whose vacancy formation enthalpy is presumably not much smaller than that of Na but whose lattice parameter is substantially larger ($a_0 = 5.23 \times 10^{-10}\,\mathrm{m}$ at 5 K), the left-hand side of the inequality (12.63) is equal to 0.76 [with $(1+\gamma_E)^{-1} = 0.737$ and the assumption $H_{1V}^{F} = 0.34\,\mathrm{eV}$]. In view of the uncertainties of this estimate and the fact that in K the ratio V_{1V}^{F}/V_{A} is not well known, it cannot be excluded that there is an e^+ bound state in the monovacancies of body-centred cubic potassium. If this was so, the binding energy E_{1V}^{+} would be very small, however, so that *high-temperature* trapping should not be observable. Nevertheless, it appears worthwhile to look for e^+ trapping after *low-temperature* electron irradiation of potassium, in which the bcc structure is stable even at low temperatures (in contrast to Na and Li).

The preceding discussion shows that the simple model developed for positrons is fairly successful for monovalent metals. A general feature, valid for all metals, is the reduction of the positron–vacancy binding energy, E_{1V}^{+}, by strong inward relaxations of the neighbouring atoms. For multivalent *simple* metals (i.e. metals *without* unfilled d-shells) a further qualitative result may be inferred. In such metals the volume per valence or conduction electron, V_{el}, is smaller than in a monovalent metal with the same atomic volume, hence according to (12.51) and (12.52) the quantity $-E_{expnd}$ is larger. It is true that now a negative contribution resulting from the incomplete compensation of E_{screen} has to be added on the right-hand side of (12.54). However, as may be deduced from (12.50) and the fact that H_{1V}^{F} is positive, this contribution is smaller than the increase of $-E_{expnd}$, hence the ratio E_{1V}^{+}/H_{1V}^{F} is larger than in monovalent metals. We conclude that in high-temperature studies of monovalent simple metals with fairly dense packing such as Zn, Cd, Al, In, and Pb, we are allowed to disregard the detrapping problem. An example of a metal *with* an unfilled d-shell where this is not so (bcc Fe) will be discussed in Sect. 12.6.5.

Comparison of the Experimental Techniques. All three basic positron-annihilation techniques (Doppler broadening, ACAR, lifetime spectrometry) may be used to study crystal defects, including vacancies, provided the defects trap the positrons at least for a time comparable with the positron lifetime. The techniques and the data analysis required in defect studies have been reviewed repeatedly, with emphasis on different aspects [54,56–58]. The following is mainly devoted to a comparative discussion of the strengths and weaknesses of the various techniques. It will turn out that for defect studies the strengths are supplementary.

Good positron annihilation data require good measuring statistics, which usually means long measuring times. In high-temperature positron-annihilation measurements this is not a trivial issue because of the high vapour pressure of many metals near their melting points. In work on high-purity metals the danger of contamination increases with increasing measuring time. Furthermore, long measuring times may entail stability problems in the measuring electronics.

Of the three basic techniques mentioned above, the *Doppler-broadening technique* is by far the simplest and fastest, the reason being that it is *not* a *coincidence technique*. It is therefore well suited for data-taking as a function of one or more external parameters. For example, measurements as a function of temperature may indicate fairly quickly whether there is trapping by vacancies in thermal equilibrium or not. Similarly, it is well suited for measurements as a function of the annealing temperature. Moreover, by monitoring a γ-ray line close to the 511 keV annihilation line – e.g. the 514 keV line of ^{85}Sr –, stability of the energy measurement can be maintained for long periods. The main weakness of the method is its poor resolution. Because of this, Doppler-broadening measurements are *not* the best method to identify the nature of the traps or to separate different kind of traps, e.g. mono- and divacancies present at high temperatures (cf. Sect. 12.4).

For *identifying* the defects acting as traps, ACAR is the most powerful technique. Its high resolution (typically a power of 10 better than that of Doppler broadening) allows us to plot out the distribution of the momenta of the annihilating e^+e^- pairs (in 2D-ACAR in both directions transverse to the axis of the 2γ-emission) and to compare it with computed distributions. Provided the computations are reliable, in this way proposals for the nature of the traps may be tested rather stringently. The contributions of traps giving sufficiently different momentum distributions may be separated. The weakness of ACAR is that – as a coincidence technique – it requires long measuring times. For the reasons indicated above, it is therefore not well suited for high-temperature measurements. In order to reduce the measuring time, in investigations of the dependence on temperature or on another external parameter often only information obtainable with low statistics, e.g. the count rate in the peak region of the angular correlation curve, is made use of. Equivalent information can usually be obtained more simply by the Doppler-broadening technique, however.

Of the three basic techniques, *lifetime spectrometry* permits the most direct physical interpretation. The decomposition of the lifetime spectrum into a sum of exponentials gives us the decay constants $\tau_j (j = 0, 1, 2, \ldots; \tau_j < \tau_{j+1})$ and the relative intensities I_j associated with them ($\Sigma I_j = 1$). The trapping model allows us to deduce from these data the e^+ lifetimes in the various traps. As already emphasized, their inverses, the annihilation rates of the trapped e^+, are directly related to the electron densities at trapping sites, which in turn are accessible to physical intuition. As a simple example, let us consider the *two-state trapping model* [46, 47]. In the absence of detrapping, the model is characterized by three parameters, namely the rate κ with which the positrons are trapped and the lifetimes τ_f of the 'free' (i.e. untrapped) positrons and τ_t of the trapped positrons. Since at the trapping sites positive electric charge is missing and, as a consequence of the screening of the effective negative charge, the electron density is lower than in the perfect crystal, the lifetimes satisfy the inequality $\tau_f < \tau_t$. The decomposition of the lifetime spectrum into two exponential decay curves gives us three parameters, too, namely two decay constants, τ_0 and τ_1, and the intensity ratio I_1/I_0. The relationships between the two parameter sets are

$$\tau_f = \tau_0 + (\tau_1 - \tau_0)(I_1/I_0) , \qquad \tau_t = \tau_1 , \tag{12.64}$$

$$\kappa = [(\tau_1 - \tau_0)/\tau_0][\tau_1 - \tau_0 + \tau_0(I_0/I_1)]^{-1} . \tag{12.65}$$

If the monovacancies in thermal equilibrium are the only traps that have to be taken into account, (12.65) may be used to deduce from high-temperature positron annihilation lifetime measurements the trapping rate

$$\kappa = \sigma_{1V} C_{1V}^{eq}(T) \tag{12.66}$$

and the vacancy formation enthalpy

$$H_{1V}^F = -\mathrm{d} \ln \kappa / \mathrm{d}(1/k_B T) + \mathrm{d} \ln \sigma_{1V} / \mathrm{d}(1/k_B T) . \tag{12.67}$$

In (12.66) and (12.67) σ_{1V} denotes the specific trapping rate of monovacancies [59]. If (12.47) holds, i.e. if the trapping rate is diffusion-limited, and if the e^+ diffusivity is limited by the scattering by acoustic phonons, we have $\sigma_{1V} \sim T^{-1/2}$, hence $\mathrm{d} \ln \sigma_{1V} / \mathrm{d}(1/k_B T) = k_B T/2$ [60]. In this case the last term in (12.67) is only a small correction term. If the positrons diffuse so fast that the trapping rate is limited not by their diffusivity but by the capture in the bound state (i.e., by a quantum mechanical transition), σ_{1V} is temperature-independent. However, in some materials there is evidence that at high temperatures e^+ may occupy self-localized metastable states in which they move by thermally activated hopping [61]. In those cases the last term in

(12.67) may give a significant negative contribution to H_{1V}^{F}. This topic, which is of considerable practical importance, will be taken up again in Sect. 12.7.

Both techniques for measuring positron ages, the $\gamma\gamma$-technique and the $\beta^+\gamma$-technique, are *coincidence techniques*. With regard to the rate of data accumulation they are, therefore, intrinsically inferior to the ΔE_γ-technique. However, thanks to the introduction of BaF_2 γ-quantum detectors with a detection efficiency for the annihilation photons of about 0.2 and of the $\beta^+\gamma$-technique based on a relativistic positron beam (kinetic e^+ energy about 4 MeV), which gives a detection efficiency of the e^+ birth signal of unity, a lifetime spectrum with good statistics can now be obtained within about 2 h [62]. This makes lifetime spectrometry clearly superior to ACAR as a tool for studying the thermal-equilibrium concentration of vacancies. Compared with ACAR, a weakness of lifetime spectrometry, shared with Doppler broadening, is *limited resolution*. The widths of the time-resolution functions of even the most advanced lifetime spectrometers are still as large as about $2\cdot 10^{-10}$ s, i.e. comparable with the lifetimes of positrons in metals. Moreover, they have to be determined as part of the data fitting procedure. In investigating the influence of an external parameter on the lifetime spectrum, such as required for determining H_{1V}^{F} from the temperature variation of $C_{V}^{eq}(T)$, this is not too serious a problem. It does mean, however, that the *absolute values* of the lifetimes associated with the various positron annihilation modes in metals may be subject to considerable uncertainties. The fact that different authors report often the same positron lifetimes does not contradict this statement, since the same or at least very similar procedures and assumptions on the shape of the resolution function are being used. The present writer has recently proposed a different time-resolution function which has a better physical justification than the one in current use [63]. It remains to be seen whether the two approaches give significantly different results or not.

12.6.5 Spin Rotation of Positive Muons (μ^+SR)

From the point of view of elementary particle physics, muons are "heavy electrons". Positive (μ^+) and negative (μ^-) muons carry the same electric charge and the same spin (viz. $\hbar/2$) as positrons and electrons. The muon mass, $m_\mu = 207 m_e$, is much larger than the electron mass but still small compared with the proton mass $m_p \approx 9 m_\mu$. The behaviour of positive muons in metals is therefore intermediate between that of positrons and protons but closer to that of protons. In materials with the right magnetic properties (see below) the μ^+ behaviour may be studied by several techniques denoted by the acronym μ^+SR. Quite some time ago it was pointed out that the particular μ^+SR technique called 'spin **r**otation of **p**ositive **mu**ons' may be used to investigate crystal defects in much the same way as the various positron techniques [64, 65].

All μ^+SR techniques rest on the following basic facts (see, for example, [66,67]):

1. Beams of positive muons with rather high *spin polarization* may be obtained from the decay of positive π-mesons, π^+. Like positrons, positive muons may be implanted into any material.
2. In contrast to e^+ and e^-, muons are *not* stable particles. Positive muons *decay* with a mean lifetime $\tau_\mu = 2.2\,\mu s$ into a positron and two neutrinos. Therefore, in contrast to implanted positrons, their age distribution does not depend on the environment in which the decay takes place.
3. The decay of positive muons is mediated by the *Weak Interaction*. The *non-conservation of parity* has the consequence that the probability of observing a certain emission direction of the decay positron is correlated with the muon spin direction at the instant of decay.
4. The *magnetic moments* of the muons, which are smaller than those of positrons by the factor m_e/m_μ but larger than the proton moment by the factor 3.18, are parallel to the spin directions. Hence the rotation of the magnetic moments due to their interaction with magnetic fields, including the internal fields generated by the magnetic moments of nuclei and electrons, changes the spin polarization and, according to fact 3, the distribution of the e^+ emission directions.
5. The average energy of the decay positrons is very high ($\approx 36\,\mathrm{MeV}$). Therefore, high-efficiency recording of the μ^+SR signal, which contains information on the *'magnetic history'* of the μ^+ (i.e., on the magnetic fields felt by the muons between implantation into the sample and their decay) does not present problems even for very massive samples.

For the study of atomic defects by μ^+SR a *first prerequisite* is that a sufficiently large fraction of the implanted muons are trapped by the defects. The conditions for this to happen are similar but not quite the same as for positrons. Taking vacancies in monovalent metals as an example, owing to the larger μ^+ mass the term E_{1s} (cf. Sect. 12.6.4) is much smaller than for positrons and often negligible. Hence, in contrast to the e^+ case, (12.54) predicts muon binding energies that are slightly larger than H^F_{1V}. (Note that the term $-\Delta E_{\mathrm{expnd}}$ gives a small positive contribution.) Atomic defects without a *positron* bound state may have rather deep-lying bound states for *positive muons*.

Another feature that makes the conditions for μ^+ trapping less stringent than those for e^+ trapping is that the mean lifetime of μ^+ exceeds that of positrons in metals by a factor of the order of magnitude 10^4. Therefore, μ^+ have much more time to reach traps than e^+. However, as a rule this is counterbalanced by the fact that, owing to the higher μ^+ mass, the μ^+ diffusivity (in general determined by hopping processes – for a recently demonstrated exception see [68]) is usually smaller than the e^+ diffusivity (which under most circumstances in pure metals is limited by the scattering of e^+ Bloch waves by phonons [60]). The appropriate measure of the 'trapping capability' is the dimensionless product $\sigma\bar{\tau}$ of the specific capture rate and the mean lifetime. Which one of the two probes, e^+ or μ^+, has the larger trapping capacity depends on circumstances. An open crystal structure (as opposed to fcc and hcp) and high temperatures favour μ^+.

The *second prerequisite* for defect studies employing μ^+ is a suitable *magnetic structure* of the material. In the present context, two kinds of solid-state magnetism are of interest, viz. those due to *nuclear magnetic moments* or to *ferromagnetism*. The magnetic fields generated by the *nuclear magnetic moments* are small, and at the temperatures and the magnetic fields $\boldsymbol{B}_{\text{appl}}$ of interest, almost randomly orientated. This gives rise to a slight inhomogeneity of the fields $\boldsymbol{B}_\mu$ acting on the μ^+ magnetic moments and hence to a spread in the angular precession frequencies

$$\omega_\mu = \gamma_\mu B_\mu \tag{12.68}$$

of the muon spins, where $\gamma_\mu = 8.516 \times 10^8 \text{rad}\,\text{T}^{-1}\,\text{s}^{-1}$ is the gyromagnetic moment of the muons. The spread leads to a Gaussian damping of the time-differential spin-rotation signal with damping constant Γ_0 given by van Vleck's formula [69]

$$\Gamma_0^2 = \left(\frac{\mu_0}{4\pi}\right)^2 \gamma_\mu^2 \gamma_I^2 I(I+1)\frac{\hbar^2}{6}\sum_j \frac{(1-3\cos^3\theta_j)^2}{r_j^6}\,. \tag{12.69}$$

In (12.69), in which isotopic purity has been assumed for simplicity, the summation extends over all nuclei; $\boldsymbol{r}_j$ is the vector from the muon site to the position of nucleus j, θ_j the angle between $\boldsymbol{r}_j$ and $\boldsymbol{B}_{\text{appl}}$, γ_I the gyromagnetic moment of the nuclei, I the nuclear spin in units of $\hbar = 1.055 \times 10^{-34}$Js, and $\mu_0 \equiv 4\pi \times 10^{-7}\,\text{H}\,\text{m}^{-1}$.

The following numerical estimates for monovacancies in the bcc structure will be confined to the first three shells in the series of (12.69), assuming the $\boldsymbol{r}_j$ to be those of rigid crystals, and to the cases $\boldsymbol{B}_{\text{appl}} \parallel \langle 100\rangle$ and $\boldsymbol{B}_{\text{appl}} \parallel \langle 111\rangle$. In the first case the contribution of the nearest-neighbour shell vanishes. We obtain

$$\Sigma[1-\cos^2\theta_j]^2/r_j^6 = [0+12+9/4+\cdots]a_0^{-6} = (14.25+\cdots)a_0^{-6}\,. \tag{12.70}$$

In the case $\boldsymbol{B}_{\text{appl}} \parallel \langle 111\rangle$, in which the contribution of the next-nearest neighbours vanishes, we obtain

$$\Sigma[1-\cos^2\theta_j]^2/r_j^6 = [2^{11}/3^4+0+3/2+\cdots]a_0^{-6} = (26.78+\cdots)a_0^{-6}\,. \tag{12.71}$$

Since, as shown in Sect. 12.6.4, there is no chance to observe *positron* trapping by monovacancies in Li and Na, a natural question to ask is whether in these metals monovacancies may be detected by using *positive muons* as probes. The result $E^+_{1\text{V}} > H^{\text{F}}_{1\text{V}}$, derived above, guarantees that there is a deep-lying

μ^+ bound state. The abundant nuclides, ^{7}Li and ^{23}Na, have spin $(3/2)\hbar$ and fairly large gyromagnetic ratios. For spin-3/2 particles

$$6^{-1/2}(\mu_0/4\pi)\hbar\gamma_\mu I^{1/2}(I+1)^{1/2} = 1.421 \times 10^{-32}\text{rad}^{-1}\text{Tm}^3 \qquad (12.72)$$

holds. Specializing (12.69) and (12.71) to bcc Li with $a_0 = 3.51 \times 10^{-10}$ m, $\gamma_I(^7\text{Li}) = 1.039 \times 10^8$ rad T^{-1}s^{-1}, and taking into account that in natural Li the abundance of ^{7}Li is 93%, we find for $\boldsymbol{B}_{\text{appl}} \parallel \langle 111 \rangle$

$$\Gamma_0 = 1.7 \times 10^5\text{s}^{-1}, \quad \Gamma_0\tau_\mu = 0.36\,. \qquad (12.73)$$

The numerical result (12.73) is an underestimate since in (12.71) the inward displacements of the neighbouring nuclei have been neglected. If a vacancy is occupied by a positive muon (equivalent to a substitutional μ^+), these displacements are smaller than in an 'empty' vacancy. Their magnitude and the ensuing increase of Γ_0 have not yet been calculated. For Li, a reasonable estimate is $\Gamma_0\tau_\mu \approx 0.5$. The corresponding value for a monovacancy in Na is smaller because the 100% abundant ^{23}Na has a smaller gyromagnetic ratio ($\gamma_I = 0.71 \times 10^8$rad T^{-1}s^{-1}) than ^{7}Li and a larger atomic volume $\{V_\text{A}(\text{Na})/V_\text{A}(\text{Li}) = [a_0(\text{Na})/a_0(\text{Li})]^3 = 1.83\}$.

According to the above estimate it should be easy to detect the Gaussian damping of the μ^+SR signal of *positive muons* trapped by vacancies in bcc Li, provided that the muons can reach the vacancies before their majority has decayed and that the damping of the signal is not suppressed by motional averaging. The first condition is presumably satisfied even for fairly low vacancy concentrations since in low-Z metals with open crystal structure such as Li and Na we may expect the μ^+ to be highly mobile, hence $\sigma\tau_\mu$ will be large. The second condition requires that *substitutional muons* do not move too fast. Although the diffusivity of substitutional muons in Li has not yet been estimated by theory, it appears likely that in the temperature range of interest it is of the order of magnitude of, but somewhat larger than, the Li self-diffusivity.

Taken all together, in contrast to what has been said in Sect. 12.6.4 on positron annihilation, we expect the equilibrium concentration of vacancies in Li to be detectable by μ^+SR. To the writer's knowledge the experiment has not yet been undertaken. It would certainly be worthwhile since as the metal with the simplest electronic structure, Li is the ideal testing ground for ab initio calculations on defects in metals. Theoretical and experimental information on vacancies as well as on interstitial and substitutional μ^+ might be quantitatively compared with each other. Furthermore, owing to the fact that ^{7}Li has a nuclear quadrupole moment, van Vleck's formula (12.69) requires an extension at small applied magnetic fields. Comparison with μ^+SR measurements as a function of the strength and direction of $\boldsymbol{B}_{\text{appl}}$ will give information on the electrical field gradients at the neighbouring nuclei due to the perturbation of the cubic symmetry at their sites by the substitutional muon [70]. At moderately large applied fields, where this perturbation becomes unimportant, (12.69) predicts that the damping parameter Γ_0 for $\boldsymbol{B}_{\mathrm{appl}} \parallel \langle 100 \rangle$ is smaller by the factor $(14.25/26.8)^{1/2} \approx 0.73$ than that for $\boldsymbol{B}_{\mathrm{appl}} \parallel \langle 111 \rangle$. The angular dependence of Γ_0 permits us to test whether the observed damping is indeed due to μ^+ trapped by *monovacancies* and not by other defects. These would give rise to a different dependence of Γ_0 on the crystallographic direction of $\boldsymbol{B}_{\mathrm{appl}}$.

The magnetic fields in *ferromagnets* are electronic in origin and therefore much larger than those due to nuclear magnetism. For μ^+SR studies of vacancies in thermal equilibrium they can be used in those ferromagnetic metals in which the Curie temperature, T_{C}, is high enough for μ^+ to be trapped by vacancy concentrations of the order of magnitude $C_{\mathrm{V}}^{\mathrm{eq}}(T_{\mathrm{C}})$. The only candidates for this to happen are Fe and Co and their alloys. In bcc Fe the Curie temperature (1043 K) is lower than in hcp Co (1394 K), but this is partly compensated by the fact, already referred to, that in the more open bcc structure the μ^+ diffusivity is higher than in the close-packed metal structures. The situation is most favourable in bcc FeCo alloys near the composition $Fe_{45}Co_{55}$, at which T_{C} is as high as 1258 K. Measurements have so far been performed on Fe [71–74] and Co [72]. Those on Fe have recently been extensively discussed together with quenching and high-temperature positron-annihilation experiments on bcc Fe [75]. In the following we give a brief summary of the three μ^+SR techniques available as well as of the main results on Fe.

All three μ^+SR techniques make use of the following three facts:

1. In ferromagnetic metals the *conduction electrons* are spin-polarized.
2. This spin polarization gives rise to the so-called *contact interaction* between electron spins and muon spins (quantitatively described by the *Fermi* field $\boldsymbol{B}_{\mathrm{Fermi}}$).
3. The Fermi field acting on the magnetic moment of *substitutional* muons (i.e. muons trapped in monovacancies) is different from that acting on *interstitial* muons.

Owing to their coupling to the magnetization, in the absence of an applied magnetic field non-vanishing Fermi fields exist only below the Curie temperature. The simplest of the three μ^+SR techniques, the measurement of the *spin rotation in a zero applied field*, is therefore confined to temperatures below T_{C}. This means that in Fe, where the ratio of the Curie temperature to the melting temperature is only 0.58 and, hence, the accessible equilibrium concentration of vacancies is extremely low, the trapping of implanted μ^+ by these vacancies can at best be detected in a fairly narrow temperature interval below T_{C}. Because of the smallness of the spontaneous magnetization in this interval, the spin precession of the 'free' and the trapped μ^+ do not give rise to separate signals; rather, one observes only one spin precession frequency, from whose damping the fraction of trapped muons must be deduced. Since the μ^+ diffusivity is known (by extrapolation from measurements at lower temperatures), the major uncertainty in the quantitative interpretation of the measurements is the trapping radius r_0 [cf. (12.47)]. A good fit to the data on Fe [71, 72] has been obtained with $H^{\mathrm{F}}_{1\mathrm{V}} = 1.7\,\mathrm{eV}$, $S^{\mathrm{F}}_{1\mathrm{V}} = 2k_{\mathrm{B}}$, and $r_0 = 1.4a_0$. From the fact that the experimental data do not show evidence for detrapping, it may be concluded that the energy barrier for the escape of a muon trapped by a vacancy exceeds 1.1 eV.

Accuracy, sensitivity, and temperature range of the μ^+SR measurements may be increased by performing the measurements not in a zero applied field but in a magnetic field of the order of magnitude of a few Tesla. The magnetic field suppresses the order–disorder transition at T_{C} and hence the extreme sensitivity of the magnetization against temperature fluctuations near T_{C}, which reduces the temperature interval accessible to the zero-field measurements at its high-temperature end. This *high-field method* is applicable also above T_{C} (hence at higher vacancy concentrations) in a limited temperature interval whose extension increases with increasing $\boldsymbol{B}_{\mathrm{appl}}$. Furthermore, the uniform magnetization of the specimen in an external field results in a larger μ^+SR signal under otherwise equal conditions. The advantages of this second technique over the zero-field technique are quite substantial, so that it should supplan uture work.

The third method is an extension of nuclear magnetic resonance (NMR) and electron paramagnetic resonance (EPR) to positive muons. As spin-1/2 particles, in a magnetic field parallel or antiparallel to their spin polarization positive muons possess two Zeeman levels, corresponding to magnetic quantum numbers $m = \pm 1/2$. The energy difference between the two levels is given by $\hbar\gamma_\mu \boldsymbol{B}_\mu$, where $\boldsymbol{B}_\mu$ is the sum of the applied and the internal magnetic fields, including the Fermi field. Irradiation with a radio frequency matching this 'Zeeman splitting' introduces transitions between the two levels and reduces the original μ^+ spin polarization. This leads to a reduction of the asymmetry of the e^+ emission of the decaying muons and allows us, owing to the fact that the Zeeman splittings of 'free' and trapped muons differ, to determine the relative occupancies of the two muon sites. This *radio-frequency method* has in common with the *high-field method* that it works also in a temperature interval above T_{C} but is more demanding from the experimental point of view. It requires laminated specimens in order to permit the penetration of the radio-frequency radiation. The results on Fe [73, 74] agree with those of the zero-field technique and have been incorporated in the numerical value $H^{\mathrm{F}}_{1\mathrm{V}} = 1.7\,\mathrm{eV}$ quoted above.

12.7 A Comparison of the Different Techniques

The various experimental techniques discussed in Sect. 12.6 have their own strengths and weaknesses. In the literature these are not always made sufficiently clear, however. In order to obtain a reliable picture it is therefore important to compare the different techniques critically with each other.

From the point of view of the reliable interpretation of the experimental data, *differential dilatometry* is the 'cleanest' method. Most of the data may be interpreted in terms of the approximation (12.42). Furthermore, in the analysis of the high-temperature measurements the term C_{I} of (12.42) may safely be neglected since the enthalpies of formation of self-interstitials in metals obtained by *ab-initio* computations exceed the monovacancy formation enthalpy by a sufficiently large margin. Near the melting point, though, it may be necessary to allow for the correction terms discussed in Sect. 12.6.3 [cf. (12.41)]. All considered, differential dilatometry gives us the equilibrium concentration $C_{\mathrm{V}}^{\mathrm{eq}}(T)$ at high temperatures, though with a relative accuracy that decreases rapidly with decreasing temperature (cf. the remarks at the end of Sect. 12.6.3). As discussed in Sect. 12.6.3, so far the technique has been applied up to the melting temperature of Cu (1356 K). According to T. Hehenkamp [27], the extension to materials with higher melting points such as Fe, Ni, Co, and Si is feasible.

The values $H_{\mathrm{V}}^{\mathrm{F}}$ and $S_{\mathrm{V}}^{\mathrm{F}}$ obtainable by fitting the differential-dilatometry data according to (12.41) are *effective* formation enthalpies and entropies, since at the temperatures at which the experimental accuracy is highest (i.e. near T_{m}) divacancies may play a role. In principle, by analyzing the curvatures of the Boltzmann plots $\ln C_{\mathrm{V}}^{\mathrm{eq}}(T)$ vs. $1/k_{\mathrm{B}}T$, the mono- and divacancy contributions to $C_{\mathrm{V}}^{\mathrm{eq}}$ may be separated. In practice, however, the experimental accuracy well below T_{m} does not suffice to do this unambiguously, particularly so because the 'thermal expansion correction', α, is not a priori known and possibly not negligible. Seeger and Mehrer [76] proposed a procedure to achieve the monovacancy–divacancy separation, nevertheless, by introducing plausible assumptions for the formation entropies and making use of the following two generally valid results.

1. If in (12.41) $H_{\mathrm{V}}^{\mathrm{F}}$ and $S_{\mathrm{V}}^{\mathrm{F}}$ are treated as temperature-independent, they satisfy the inequalities
$$H_{1\mathrm{V}}^{\mathrm{F}} \leq H_{\mathrm{V}}^{\mathrm{F}}, \quad S_{1\mathrm{V}}^{\mathrm{F}} \leq S_{\mathrm{V}}^{\mathrm{F}}. \tag{12.74}$$
2. Since the free enthalpy of formation of divacancies is higher than that of monovacancies, at sufficiently low temperatures $C_{\mathrm{V}}^{\mathrm{eq}}(T)$ will be dominated by monovacancies.

The 'concentration windows' of the various *positron-annihilation techniques* (cf. Sect. 12.6.4) extend usually so far down in temperature that at least at their low-temperature ends divacancies may be neglected. Under these conditions, in *pure* metals the only atomic defects that have to be taken into account are *monovacancies*. However, this is the temperature regime in which attractive interactions between *impurities* and vacancies may become significant, resulting in an increase in the vacancy concentration over that of the 'pure' metal (cf. Sect. 12.5). Since such an impurity enhancement of the trap concentration leads to apparent formation enthalpies that are *lower* than the monovacancy formation enthalpies of the pure metals, it is essential to use high-purity material for the $H^{\mathrm{F}}_{1\mathrm{V}}$ determinations by positron annihilation or, if this is not possible, to perform systematic experiments with controlled purities.

A weakness of the study of atomic defects by positron annihilation is the uncertainty in the specific trapping rate σ. In the case of $H^{\mathrm{F}}_{1\mathrm{V}}$ studies the following three facts contribute to this.

1. The *trapping radius* r_0 of the vacancies is not well known. For geometrical reasons, r_0 is presumably not smaller than the nearest-neighbour distance, i.e. than $2^{-1/2}a_0$ in fcc metals. On the other hand, owing to the screening of electrical charges in metals, we may assume that it does not exceed about $3a_0/2$. Hence for fcc metals we may assume that

$$0.7 < r_0/a_0 \leq 1.5 \tag{12.75}$$

holds.

2. There are no measurements of the high-temperature *positron diffusivity* that do not make use of the trapping by thermal-equilibrium defects. In agreement with the theoretical prediction for diffusivity limitation by phonon scattering of e^+ in Bloch-wave states [60], the measurements at lower temperatures on the four cubic metals Al, Cu, Ag, and Mo [77] could be approximately described by

$$D^+(T) = D^+(T_\mathrm{D})(T_\mathrm{D}/T)^{1/2}\,, \tag{12.76}$$

where T_D denotes the Debye temperature. In Cu, which will be discussed in more detail below, (12.76) gives an acceptable description of the experimental data up to about 500 K. (For a more detailed experiment–theory comparison and small corrections to (12.76) resulting from the temperature dependence of the elastic constants, see [78,79].) However, at elevated temperatures the positrons may populate metastable states in which they diffuse by hopping according to the Arrhenius law [61]

$$D^+(T) = D^+_0 \exp(-H^+/k_\mathrm{B}T)\,. \tag{12.77}$$

Since in Cu e^+ trapping by vacancies in thermal equilibrium sets in only at about 870 K, there is a broad uncharted temperature interval in which the transition between the temperature laws (12.76) and (12.77), described in detail elsewhere [61], may take place. This means that in estimating the specific trapping rate $\sigma_{1\mathrm{V}}(T)$ at high temperatures by means of (12.47), we may well have to use (12.77) rather than the extrapolation of low-temperature determinations of $D^+(T)$ using (12.76). As will be discussed below, in the relevant temperature range (12.77) gives a lower diffusivity than predicted by (12.76).

3. If (12.76) applies, i.e. if the 'free' positrons are in Bloch states, the specific trapping rate may be limited not by the positron diffusivity but by the quantum mechanical transition rate from the Bloch states to the vacancy

bound state [80]. Then σ_{1V} becomes temperature-independent. (The intermediate case between transition-rate limitation and diffusion limitation has been treated in detail elsewhere [59].) If, however, (12.77) applies, i.e. if the 'free' positrons are in self-localized states, there will be no 'transition barrier', hence we are justified in estimating $\sigma_{1V}(T)$ using (12.47).

Because of the uncertainties just discussed, e^+ annihilation is not capable of providing us with accurate absolute values of C_V^{eq} or C_{1V}^{eq}. It is, however, an excellent tool for determining H_{1V}^F if (a) $C_V^{eq} \approx C_{1V}^{eq}$ and (b) the last term in (12.67) may be neglected. Condition (a) may be considered to be satisfied at least approximately since in almost all metals the 'concentration window' ends well below T_m. The last term in (12.67) is zero if σ_{1V} is not controlled by the positron diffusivity but by the quantum-mechanical *transition rate* from the 'free' positron states to the trapped state. The opposite limit of *diffusion limitation* of σ_{1V} has been discussed in Sect. 12.6.4. Here the possibility that at high temperatures the e^+ occupy self-localized states and move by hopping is of particular interest, since then the last term of (12.67) is negative and, in general, not negligible. In such cases, disregarding this term (as is usually done in the literature) may result in serious overestimates of H_{1V}^F. How to recognize this situation is clearly an important issue, to be discussed in the following paragraphs.

For cubic metals, the situation would be simple if $D^+(T)$ measurements at low or intermediate temperatures demonstrated that the positron diffusivity obeyed (12.77) with $H^+ > 0$ and $10^{-7}\text{m}^2\text{s}^{-1} \leq D_0^+ \leq 10^{-5}\text{m}^2\text{s}^{-1}$. (This is the order of magnitude expected for D_0^+, taking into account that in hopping not only the positron but also the surrounding ions have to move.) The only metal in which this has possibly been observed is Cd (see [81] and Footnote 8). In all other cases the experimental evidence for the high-temperature occupancy of a self-localized e^+ state moving by hopping comes entirely from the so-called prevacancy effect. The prevacancy effect is an 'anomalous' temperature dependence of the positron annihilation signal below the onset of trapping by thermal-equilibrium vacancies that goes in the same direction as the vacancy trapping but is weaker.[8] According to the interpretation of the pre-vacancy effect in terms of the occupation of a metastable self-localized

[8] The 'prevacancy effect' was discovered by P.C. Lichtenberger, C.W. Schulte, and I.K. MacKenzie [82] in Doppler-broadening measurements on Cd and subsequently interpreted [61] as due to the occupation, at elevated temperatures, of a metastable self-localized positron state that diffuses by hopping according to (12.77). Diffusivity measurements on Cd single crystals [81] indicate indeed such a behaviour for e^+ diffusion in the basal plane [83]. Since, however, in this case the e^+ diffusivity perpendicular to the basal plane appears to be normal [i.e., at most weakly temperature-dependent and much faster than predicted by (12.77)], the e^+ trapping rate in Cd is *not* expected to be *strongly* affected by the temperature dependence (12.77). Small deviations from the 'normal' behaviour, however, should exist [83]. Strong evidence for such 'mixed' positron states (localized in the basal plane but with a Bloch-wave character in the direction *perpendicular* to

state [61], in this temperature range the positron diffusivity predicted by (12.77) is much lower than that according to (12.76). Hence the modification of the annihilation signal due to the positron self-localization shows up at temperatures well below those at which $D^+(T)$ obeys (12.77). It is therefore not surprising that in fcc metals the temperature variation of the positron diffusivity according to (12.77) has so far not been observed.

Let us now assume that in the temperature range of high-temperature trapping experiments the positron diffusivity does obey (12.77). In the simplest case (cubic metal, monovacancies only) the trapping rate κ then follows the Arrhenius law

$$\kappa(T) = \kappa_0 \exp(-(H^{\mathrm{F}}_{1\mathrm{V}} + H^+)/k_{\mathrm{B}}T)\,. \tag{12.78}$$

The activation enthalpy is now the sum of two positive unknowns, $H^{\mathrm{F}}_{1\mathrm{V}}$ and H^+; therefore, its experimental determination gives us an *upper limit* to $H^{\mathrm{F}}_{1\mathrm{V}}$ only. Unfortunately, the pre-exponential factor of the trapping rate, κ_0, is not sufficiently sensitive against the two alternative forms of $D^+(T)$ to allow a clear-cut distinction between (12.76) and (12.77). We demonstrate this by considering the four cubic metals referred to in item 2, p. 387. Here the positron diffusivity at the Debye temperature, T_{D}, was approximately the same, namely $D^+(T_{\mathrm{D}}) = (1.3 \pm 0.3) \times 10^{-4}\mathrm{m}^2\mathrm{s}^{-1}$ [77]. If we characterize the temperature range in which $H^{\mathrm{F}}_{1\mathrm{V}}$ is to be determined by the temperature T^+ and approximate (12.76) in that range by an Arrhenius law, we obtain the effective activation enthalpy $-k_{\mathrm{B}}T^+/2$ and the pre-exponential factor $D^+(T_{\mathrm{D}}) = (T_{\mathrm{D}}/\mathrm{e}T^+)^{1/2}$ with $\mathrm{e} = 2.718\ldots$. Between the four metals mentioned above the ratio $(T_{\mathrm{D}}/T^+)^{1/2}$ varies by about a factor of three. A reasonable average value of the effective pre-exponential factor at T^+ is thus $D_0^+ = 4 \times 10^{-5}\mathrm{m}^2\mathrm{s}^{-1}$, with variations by a factor of three in both directions. The uncertainty in κ_0 is enhanced further if the question is open whether (12.47) is valid or not. Control of the specific trapping rate $\sigma_{1\mathrm{V}}$ by the quantum-mechanical transition rate clearly makes the two alternatives less distinguishable. Even in the most favourable case, viz. that of a very high transition rate and thus validity of (12.47) and (12.75), the lack of precise knowledge of r_0 brings in an uncertainty factor of about two. A further contribution to the uncertainty in κ_0 comes from our limited knowledge of $S^{\mathrm{F}}_{1\mathrm{V}}$ (at least for the time being); it amounts to another factor of about two. Recalling the above order-of-magnitude estimate of D_0^+ we have indeed to conclude that the accumulated uncertainties are too large for the experimental determination of a numerical value of κ_0 to permit a decision between the alternatives (12.76) and (12.77).

the basal plane) has been found not only in Cd but also in various measurements on Zn single crystals (for references see [83]). The strong anisotropy of these states has the consequence that the annihilation signals of polycrystals depend on the texture of the specimens, which is usually not stated in the publications. This may account for conflicting evidence reported in the literature, e.g. on the tetragonal metal In (see, for example, [57], Sect. 3.2.2).

The chances for a valid decision between the temperature laws (12.76) and (12.77) and thus for the correct interpretation of the high-temperature e^+ data become much better if from differential dilatometry we have quantitative information on C_V^{eq} within or at least close to the 'concentration window' of the e^+ measurements as well as a reliable estimate of S_{1V}^F. Because of its practical importance, the specific case of Cu is considered in the next paragraph.

In Cu, the differential-dilatometry data extend with reasonably good accuracy down to about $T^{-1} = 9.0 \times 10^{-4}\,K^{-1}$, with $C_V^{eq}(1111\,K) = 7 \times 10^{-5}$ [15, 23]. The 'concentration window', i.e. the temperature regime over which the positron trapping rate κ has been measured with good accuracy, extends up to about $T^{-1} = 9.5 \times 10^4\,K^{-1}$. Based on the identification of H_{1V}^F with $-d \ln \kappa / d(1/k_B T)$ as determined by several groups [84–95], the numerical value $H_{1V}^F = (1.28 \pm 0.05)\,eV$ was recommended in the Landolt-Börnstein data collection, apparently with a bias in favour of larger values ([12], Sect. 2.3.2). It is easy to see that this assignment *cannot be correct.* Since

$$\begin{aligned} C_V^{eq}(1111\,K) \exp(-1.28\,eV/1111\,k_B K) &= 7 \times 10^{-5}/1.6 \times 10^{-6} \\ &= 45 = \exp(3.8) \end{aligned} \tag{12.79}$$

and since at these low vacancy concentrations $C_{1V}^{eq} = C_V^{eq}$ should hold within experimental accuracy, the assignment [12], if correct, would require $S_{1V}^F = 3.8\,k_B$. Such a large monovacancy formation entropy for a close-packed metal such as Cu has no theoretical support whatsoever and can therefore not be accepted. In an alternative attempt to interpret their own differential-dilatometry and positron-annihilation data, Hehenkamp and his associates [15] chose the specific trapping rate σ_{1V} in such a way that the two data sets could be jointly fitted by *one* Arrhenius law. Their numerical value $H_{1V}^F = 1.19\,eV$ implies, by the same chain of arguments as before, $S_{1V}^F = 2.9\,k_B$, which is still unacceptably large.

The difficulties outlined in the preceding paragraph are resolved by recourse to the explanation of the pre-vacancy effect observed on Cu in terms of the occupation, in the temperature range of vacancy trapping, of a slowly diffusing metastable e^+ state [61]. If we treat the H_{eff} value of Cu of Table 12.1 in the same way as was done above for Au, we obtain $H_{1V}^F = 1.03\,eV$ and $0.15\,eV \leq H^+ \leq 0.25\,eV$.[9] The consistency of this interpretation may be

[9] The combination of $H_{1V}^F = 1.03\,eV$ with the Cu self-diffusion data [95] shows very clearly that the 'recommended' assignment ([12], p. 233) of the Stage-III migration energy $(0.70 \pm 0.02)\,eV$ to the monovacancy migration enthalpy, H_{1V}^M, must be wrong. This is another example of the detrimental effects of disregarding quenching experiments. There is, in fact, overwhelming evidence that the observed defect migration enthalpy of about 0.70 eV is that of the migration of $\langle 100 \rangle$ dumbbell self-interstitials (see, for example, [96]). – *Note added in proof:* G. Neumann, V. Tölle and C. Tuijn, Physica **B271**, 21 (1999), have recently reanalyzed the high-temperature differential-dilatometry and positron-annihilation

tested by inserting the numerical values at $T^{-1} = 9 \times 10^{-4}\,\mathrm{K}^{-1}$ as well as the inequality (12.75) into the right-hand side of the relationship

$$D^+(T) = a_0^2 \kappa(T)(a_0/r_0)/16\pi C_\mathrm{V}^\mathrm{eq}(T)\,, \tag{12.80}$$

following for fcc metals from (12.47) and (12.66). With $\kappa(1111\,\mathrm{K}) = 1.02 \times 10^{10}\,\mathrm{s}^{-1}$ [94] this gives us

$$2.5 \times 10^{-7}\,\mathrm{m}^2\,\mathrm{s}^{-1} < D^+(1111\,\mathrm{K}) < 5.4 \times 10^{-7}\,\mathrm{m}^2\,\mathrm{s}^{-1} \tag{12.81}$$

or, with the above inequality for H^+,

$$1.2 \times 10^{-6}\,\mathrm{m}^2\,\mathrm{s}^{-1} < D_0^+ < 8 \times 10^{-6}\,\mathrm{m}^2\,\mathrm{s}^{-1}\,. \tag{12.82}$$

The inequality (12.82) is in full agreement with the above order-of-magnitude estimate of D_0^+, hence the interpretation is indeed consistent. Similar considerations may be applied to all those metals for which quantitative information on both $C_\mathrm{V}^\mathrm{eq}(T)$ and $\kappa(T)$ is available.

The comparison between the *muon spin rotation* (μ^+SR) and the *positron annihilation* techniques is particularly interesting, since both are based on the same phenomenon, viz. the trapping of a highly mobile probe with positive electric charge by a less mobile defect with negative, strongly screened electric charge. The differences between the two probes are due to their different masses (this accounts for the higher binding energy and the slower diffusion of the positive muons) and to the huge ratio (about 10^4) between the mean lifetimes of μ^+ and e^+. A detailed comparison between the two techniques as well as with the available self-diffusion and quenching data has recently been made on bcc iron [75]. This required the solution of two specific problems.

1. In ferromagnetic materials the properties of atomic defects depend on the magnetic order, which, to a first approximation, may be described quantitatively by the magnetization. In bcc iron, the high-temperature data are obtained either in the paramagnetic regime or in the weakly magnetized regime close to the Curie temperature, T_C, whereas the defect migration after quenching, irradiation, or cold-work is investigated in the magnetically fully ordered state. The problem of how to relate these two data sets to each other may be considered solved (for details and references see [75,97,98]).

---

data on Cu using a more detailed numerical approach than the one employed above. They arrive at the same values for monovacancies, viz. $H_\mathrm{1V}^\mathrm{F} = 1.03\,\mathrm{eV}$ and $S_\mathrm{1V}^\mathrm{F} = 1.1\,k_\mathrm{B}$. For the melting-point ratio of vacant sites in divacancies to those in monovacancies they obtain $2C_\mathrm{2V}(T_\mathrm{m})/C_\mathrm{1V}(T_\mathrm{m}) = 0.68$. This value corresponds to $H_\mathrm{2V}^\mathrm{B} = 0.23\,\mathrm{eV}$ and $S_\mathrm{2V}^\mathrm{B} = -2.8k_\mathrm{B}$. By combining their results with the self-diffusion data, the authors derive $H_\mathrm{1V}^\mathrm{M} = 1.02\,\mathrm{eV}$ and $H_\mathrm{2V}^\mathrm{M} = 0.64\,\mathrm{eV}$, in agreement with the preceding conclusion on the assignment of Stage III and with the results of Seeger and Mehrer [76].

2. In a temperature range that is crucial for the positron-annihilation measurements, the fcc γ-phase interrupts the α- and δ-phases of bcc Fe. The difficulty resulting from this was solved partially by positron-lifetime spectrometry with very high statistics near the α/γ phase transition [99]. The very narrow δ-phase (on the T^{-1} scale) lies outside the 'concentration window' of positron annihilation. Differential dilatometry in the δ-phase would therefore be extremely helpful and provide us with an additional test of the conclusions drawn so far.

The most important conclusions are:

1. In bcc iron, at high temperatures the *trapping capability* (cf. Sect. 12.6.5) of positive muons is larger than that of positrons. While μ^+SR data may be obtained at temperatures below or near to T_C, the positron trapping data are confined to paramagnetic Fe, i.e. to higher temperatures.
2. The binding energy μ^+–monovacancy is so high that μ^+ detrapping need not be taken into account. However, the e^+–monovacancy binding energy amounts to only about 1.1 eV; in the analysis of the high-temperature positron-annihilation data on bcc iron it is therefore essential to allow for e^+ detrapping.
3. The monovacancy formation enthalpy in the fully ordered ferromagnetic state obeys

$$1.61\,\mathrm{eV} \leq H_{1V}^{F} \leq 1.75\,\mathrm{eV}\,. \tag{12.83}$$

4. By combining the inequality (12.83) with the 'low-temperature' activation enthalpy of self-diffusion in α-Fe, (2.91 ± 0.04) eV [95], the monovacancy migration enthalpy is found to be

$$H_{V}^{M} = (1.24 \pm 0.13)\,\mathrm{eV}\,. \tag{12.84}$$

5. Equations (12.83) and (12.84) are compatible with the quenching experiments, which in α-Fe are particularly difficult owing to the low vacancy concentrations involved (estimated to be at most 1×10^{-6} at the α/γ phase boundary) and the likelihood of the interference of fast-diffusing interstitial impurities such as C and N.

The difference between the result (12.84), deduced from thermal-equilibrium and self-diffusion data, and the 'Landolt–Börnstein H_{1V}^{M} value' (0.55 ± 0.03)eV ([12], p. 126, taken from [100]) is so large that the latter assignment cannot even be approximately correct. Thus, in bcc iron, too, the high-temperature experiments support fully the interpretation of the recovery stage III in terms of self-interstitial migration [101].

Two of the results on bcc iron are believed to be generally valid for bcc transition metals:

1. The monovacancy *migration* enthalpy is distinctly lower than the *formation* enthalpy, in contrast to the fcc metals, in which these two quantities are often not too different. This makes the determination of H_{1V}^{M} by quenching experiments in bcc metals more difficult than in fcc metals.

2. Under otherwise comparable conditions, the binding energy of positrons in vacancies is smaller than in fcc metals. For example, it is virtually certain that in high-temperature trapping experiments on vanadium, niobium, and tantalum [102] the escape of the positrons from the vacancies plays an important role. Together with the fact that so far successful differential dilatometry has not been reported on a bcc transition metal, items 1 and 2 are responsible for the much poorer state, compared with fcc and some of the hcp metals, of our knowledge about intrinsic atomic defects in most bcc metals. For the future research on intrinsic atomic defects in bcc refractory metals the reliable picture achieved for the intrinsic atomic defects in at least one high-melting-point bcc metal, viz. Fe, may serve as a guideline. The reader will have noted that we owe this exceptional situation to a large extent to the ferromagnetism of bcc iron.

The deductions from the various *calorimetric techniques* have recently been compared with the results of differential-dilatometry and positron-annihilation measurements in the review by Y. Kraftmakher [10], to which the reader is referred for detailed references.[10] For fcc metals, for which quantitative comparisons are possible, the calorimetric investigations gave considerably larger $C_V(T_m)$ values than differential dilatometry with the exception of an elaborate analysis of various high-temperature calorimetric measurements on Al by R.C. Shukla, C.A. Plint, and D.A. Ditmars [103]. Assuming $H_V^F = 0.66\,\mathrm{eV}$ (i.e., a value fully compatible with the differential-dilatometry and positron-annihilation data on Al), the authors estimate $S_V^F \leq 1.8\,k_B$. Accounting for $C_V(T_m) = 9 \times 10^{-4}$ (cf. Table 12.1) requires $S_V^F = 1.2\,k_B$. This value fits the calorimetric data reasonably well and is compatible with the theoretical expectations. For the other fcc metals approximate ratios of the $C_V(T_m)$ values furnished by high-temperature calorimetry to those of Table 12.1 are 10 (Cu), 3 (Au), and 13 (Pb), hence major discrepancies exist. For the hexagonal metal Zn the corresponding ratio is 8. For the refractory bcc metal *quantitative* comparisons are more difficult for reasons indicated above. Nevertheless, it can be stated that in metals such as Nb, Ta, Mo, and W the $C_V(T_m)$ values deduced from high-temperature calorimetry are too large and the corresponding H_{1V}^F values too small to be compatible with the positron-annihilation data. (For a collection of data see [10].) Even when fully allowing for the uncertainties mentioned above, we have to conclude that in the bcc refractory metals the discrepancies between high-temperature calorimetry and positron annihilation are even larger than in the close-packed metals.

To a large extent, the discrepancies just mentioned may be attributed to the fact that in high-temperature calorimetry the vacancy contribution is deduced by subtracting from the measured specific heat a 'background' of insufficiently known temperature dependence. (Note that in ideally performed

[10] The review [10] covers also techniques not treated in the present chapter, e.g. measurements of the high-temperature electrical resistance.

differential-dilatometry measurements the 'background problem' is avoided altogether, while the e^+ and μ^+ techniques 'decorate' vacancies by mobile probes.) As the key to a successful application of high-temperature calorimetry to the study of intrinsic defects in thermal equilibrium, the background problem will be discussed in some detail in the following paragraphs.

High-temperature calorimetry gives us the specific heat at *constant external pressure*, c_p. From a theoretical view point, a much simpler quantity is the specific heat at *constant volume* V of the system,

$$c_V = \partial F/\partial T\mid_V , \tag{12.85}$$

where F denotes the free energy per atom. For isotropic and cubic materials, the two quantities are related according to

$$c_p = c_V + V_\mathrm{A} B_T \beta^2 T . \tag{12.86}$$

(As before, V_A denotes the atomic volume.) The isothermal bulk modulus, B_T, is defined by

$$B_T^{-1} := -\partial \ln V/\partial p\mid_T \tag{12.87}$$

and the volume expansivity, β, by

$$\beta := \partial \ln V/\partial T\mid_p . \tag{12.88}$$

Since (12.86) is a thermodynamic relationship [104], it holds for any mechanism of thermal expansion. In the application to the present problem, we have to take into account that in the second term on the right-hand side of (12.86) we have, in addition to the explicit T-dependence, an implicit dependence resulting from the temperature variations of V_A, B_T, and β. This difficulty may be circumvented by making use of the empirical result [104,105] that for many metals the quantity

$$c_p^2/V_\mathrm{A} B_T \beta^2 =: k_\mathrm{B} T_\mathrm{NL} , \tag{12.89}$$

which has the dimension of energy, is approximately *temperature-independent.* Since (12.86), if rewritten as

$$c_p = c_V + (c_p^2/k_\mathrm{B}) T/T_\mathrm{NL} , \tag{12.90}$$

is known as the *Nernst–Lindemann equation* [104], we denote the temperature T_NL defined by (12.89) as the *Nernst–Lindemann temperature*. Table 12.2 lists the Nernst–Lindemann temperatures deduced from the so-called dilation term $A = (k_\mathrm{B} T_\mathrm{NL})^{-1}$ as given by K.A. Gschneidner Jr. ([105], Table XIV) as well as the ratios $T_\mathrm{m}/T_\mathrm{NL}$ of 18 cubic metals. So far, it has apparently gone unnoticed that the distributions of these ratios for the bcc and the fcc metals do not overlap and that for each group the distributions are fairly narrow if the anomalously low Li value (presumably related to the small ratio of $T_\mathrm{m}/T_\mathrm{D}$ and thus to the influence of quantum statistics) is left out. The average value

Table 12.2. Nernst–Lindemann temperatures $T_{\rm NL}$ calculated from the dilation term A as given by K.A. Gschneider Jr. [105], and ratios $T_{\rm m}/T_{\rm NL}$ of melting temperature to Nernst–Lindemann temperature

	Li	Na	K	Al	Pb	Cu	Ag	Au	Ni
$T_{\rm NL}[10^4{\rm K}]$	2.45	1.32	1.17	2.06	1.45	3.09	2.28	2.45	4.59
$10^2 T_{\rm m}/T_{\rm NL}$	1.85	2.81	2.88	4.53	4.29	4.39	5.41	5.45	3.76
	Pd	Pt	V	Nb	Ta	α-Fe	Cr	Mo	W
$T_{\rm NL}[10^4{\rm K}]$	4.25	4.4	8.75	9.05	9.26	5.09	7.48	12.03	12.22
$10^2 T_{\rm m}/T_{\rm NL}$	4.29	4.64	2.49	2.89	3.53	3.55	2.87	2.4	2.99

of the remaining nine bcc metals is $\langle T_{\rm m}/T_{\rm NL}\rangle = 2.9 \times 10^{-2}$, that of the eight fcc metals $\langle T_{\rm m}/T_{\rm NL}\rangle = 4.6 \times 10^{-2}$.

Solving (12.90) for c_p gives us

$$\begin{aligned} c_p &= (k_{\rm B}T_{\rm NL}/2T)\{1 - [1 - 4(c_V/k_{\rm B})T/T_{\rm NL}]^{1/2}\} \\ &= c_V\{1 + c_V T/k_{\rm B}T_{\rm NL} + 2(c_V T/k_{\rm B}T_{\rm NL})^2 + {\rm O}[(c_V T/k_{\rm B}T_{\rm NL})^3]\}\,. \end{aligned} \quad (12.91)$$

The specific heat at constant volume, c_V, may be subdivided into the contribution $c_{\rm ion}$ due to the thermally activated motion of the atoms and the electronic contribution $c_{\rm el}$. The first few terms of the asymptotic expansion of $c_{\rm el}$ in powers of T read [106]

$$\begin{aligned} c_{\rm el} &= (\pi^2/3)D(E_{\rm F})k_{\rm B}^2 T\{1 + k_{\rm B}^2 T^2[7D''(E_{\rm F})/60D(E_{\rm F}) \\ &\quad - D'(E_{\rm F})^2/12D(E_{\rm F})^2] + {\rm O}(k_{\rm B}^4 T^4)\}\,. \end{aligned} \quad (12.92)$$

In (12.92) $D(E_{\rm F})$ denotes the density of electronic states per atom at the Fermi energy (including the renormalization due to the interaction between phonons and electrons, which itself is temperature-dependent); $D'(E_{\rm F})$ and $D''(E_{\rm F})$ are the first and second derivatives of $D(E)$ at the Fermi energy. Equation (12.92) shows that the deviation of $c_{\rm el}$ from a linear increase with increasing T depends on the variation of the density of states with electron energy E near the Fermi energy. It is large and positive if the Fermi energy is near a minimum of $D(E)$, and large and negative if the Fermi energy lies near an inflexion point or a maximum of $D(E)$.

To illustrate the preceding, let us consider some examples. The experimental information on $c_{\rm el}$ is primarily obtained by identifying it with the linear term in the *low-temperature* T-dependence of the specific heat. Within the rows of the periodic table of the elements, $c_{\rm el}$ oscillates strongly from element to element. (For the fourth, fifth, and sixth period see [105], Fig. 18.) If we use (12.92) to express the experimental results (for references see [105]) in terms of $D(E_{\rm F})$ and use the fifth row as an example, we obtain for Rb $1.07\,{\rm eV}^{-1}$, for Zr $1.23\,{\rm eV}^{-1}$, for Nb $3.25\,{\rm eV}^{-1}$, for Mo $0.89\,{\rm eV}^{-1}$, and for Ag

$0.28\,\mathrm{eV}^{-1}$. In the present context it is important to note that high or low $D(E_\mathrm{F})$ values are characteristic of the entire groups of the periodic table (Group-V transition metals: V $3.83\,\mathrm{eV}^{-1}$, Ta $2.48\,\mathrm{eV}^{-1}$; Group-VI transition metals: Cr $0.61\,\mathrm{eV}^{-1}$, W $0.52\,\mathrm{eV}^{-1}$). Generally speaking, owing to the contribution of the d-states to the Fermi surface [107], in the transition metals $D(E_\mathrm{F})$ is larger than in non-transition metals. (The low $D(E_\mathrm{F})$ value of Ag mentioned above is characteristic of Cu, Ag, Zn, and Cd as well.) Examples of rather large electronic contributions to c_V are the Group-V transition metals. If we consider the leading term of the expansion (12.92) only (in this case admittedly giving us overestimates), we obtain $c_\mathrm{el}(T_\mathrm{m}) = 2.4\,k_\mathrm{B}$ (V), $2.5\,k_\mathrm{B}$ (Nb) and $2.3\,k_\mathrm{B}$ (Ta), i.e. contributions that are by no means small compared with the Dulong–Petit value

$$c_\mathrm{ion} = 3\,k_\mathrm{B}\,. \tag{12.93}$$

The fact that the corresponding values of the Group-VI transition metals are small by comparison, the largest being the Mo value $c_\mathrm{el}(T_\mathrm{m}) = 0.73\,k_B$, indicates (in agreement with band-structure calculations) that in this group the Fermi energies lie at or close to a *minimum* of $D(E)$. Therefore, according to (12.92), we may expect a substantial positive T^3-contribution of electronic origin to the high-temperature specific heat of the Group-VI transition metals.

Let us now turn to the contribution of the *motion of the ions* to the specific heat, $c_\mathrm{ion} = c_\mathrm{ion}(T)$. As long as this motion may be described by a set of harmonic oscillators (so-called *quasiharmonic approximation*), at temperatures above the Debye temperature the equipartition theorem gives us the Dulong–Petit value (12.93). Since the pioneering work of M. Born and E. Brody [108] and of E. Schrödinger [109] (cf. [110, 111]) many attempts have been undertaken to treat the effects of *anharmonicity* on c_ion (for references see [10]). Apart from the result that for small anharmonicities beyond the quasiharmonic approximation the deviations from the Dulong–Petit value (12.93) are proportional to T, not very many insights of general validity appear to have been obtained. The usual perturbation-theory treatment of the anharmonicities is presumably insufficient anyway, since it does not allow for modes such as the creation and collapse of short-lived Frenkel pairs.

The possible importance of *short-lived Frenkel pairs* in the present context results from several facts:

1. Frenkel pairs, whether short- or long-lived, do not show up in differential dilatometry (cf. Sect. 12.6.3). They do, however, contribute to the *high-temperature* calorimetric measurements.
2. As long as their lifetimes are short compared to the mean positron lifetime, short-lived Frenkel pairs do not show up in positron-annihilation studies either. Frenkel pairs that live only for about one tenth of the inverse Debye frequency or shorter are not *permanent imperfections* on the time scale of the positron annihilation techniques.

3. The formation enthalpy of close Frenkel pairs depends strongly on the distribution of V–I separations; it is less than the energy required to form a widely separated pair. The entropy connected with the Frenkel-pair modes is quite large because of the large number of possible configurations.
4. If in the formation of a short-lived Frenkel pair substantial displacements of several atoms are involved, the collapse path need not be the same as the generation path. Some of the atoms in the neighbourhood will end up on lattice sites other than their original ones. This means that *self-diffusion* has taken place. In several refractory bcc metals there is indeed a high-temperature contribution to the self-diffusivity that is hard to understand in terms of mono-, di-, etc. vacancies but might well be due to the reshuffling of the atoms by the creation and annihilation of short-lived Frenkel pairs.

As a first step in the quantitative comparison experiment–theory, (12.90) may be used to calculate $c_V(T)$ from the measured $c_p(T)$. From $c_p(3580\,\mathrm{K}) = 6.3\,k_\mathrm{B}$ of W follows $c_V(3580\,\mathrm{K}) = 5.1\,k_\mathrm{B}$, i.e. an excess of $2.1\,k_\mathrm{B}$ over the Dulong-Petit value, from $c_p(3200\,\mathrm{K}) = 5.14\,k_\mathrm{B}$ of Ta follows $c_V(3200\,\mathrm{K}) = 4.2\,k_\mathrm{B}$, i.e. an excess of only $1.2\,k_\mathrm{B}$ (data taken from [10], Fig. 15). The last result is highly significant since the linear extrapolation of the low-temperature electronic specific heat gives $c_\mathrm{el}(3200\,\mathrm{K}) = 2.25\,k_\mathrm{B} > 1.2\,k_\mathrm{B}$. This shows very clearly that in the representation of $c_V(T)$ by a series of increasing powers of T there must be *negative* contributions to the high-temperature specific heat of Ta. (Analogous arguments hold for V and Nb.) One of them has already been identified above: for the Group-V transition metals the T^3-term in c_el is negative, cf. (12.92). However, it is conceivable that going beyond the quasiharmonic approximation leads to negative contributions to c_ion, too. On the other hand, as indicated above, in the Group-VI transition metals we may expect substantial positive T^3-terms in the asymptotic expansion of c_el.

The majority of the published attempts to deduce vacancy formation enthalpies from high-temperature calorimetry (cf. [10]) paid insufficient attention to the complications discussed in the preceding paragraphs. In subtracting the non-defect contributions to the high-temperature specific heat most analyses relied either on rather simple models or on formal extrapolations of $c_p(T)$ as measured at intermediate temperatures. In the opinion of the present writer the failures to reach agreement with the differential-dilatometry and positron-annihilation data must be primarily attributed to inadequate treatment of the calorimetry background problem. Furthermore, it appears likely that close to the melting temperature the generation of short-lived Frenkel pairs, disregarded in the published analyses, contributes substantially to the high-temperature specific heat of the bcc metals.

Finally, we raise the question: 'What should be done in order to utilize fully the potential of high-temperature calorimetry for the study of intrinsic defects in metals?' We emphasize that high-temperature calorimetry is arguably the best method to obtain independent information on short-lived Frenkel pairs, which may play an important role in diffusion processes close to the melting temperature. This requires, in the first place, more complete theoretical treatments of $c_{\mathrm{ion}}(T)$ and $c_{\mathrm{el}}(T)$. As has recently been demonstrated, it is now possible to compute with good accuracy the free energy of metals at high temperatures by the local density approximation to the density functional theory [112, 113]. So far the comparison with experiment has concentrated on the melting curve. It should be investigated under which conditions such computations give us $c_{\mathrm{ion}}(T)$ with an accuracy that is adequate for the present purpose. In order to obtain $c_{\mathrm{el}}(T)$ with sufficient accuracy, the calculations of the electron-mass enhancement due to the electron–phonon interaction (cf., for example, [114]) must be extended to finite temperatures. Once the background problem has been solved, the best strategy for separating the contributions of short-lived close Frenkel pairs and long-lived vacancies to the high-temperature specific heat is presumably to calculate the vacancy contribution from differential-dilatometry and/or positron-annihilation data by means of (12.34) and to obtain the Frenkel-pair contribution by subtraction. This might then permit us to establish the magnitude of the Frenkel-pair contribution to the self-diffusivity.

Acknowledgement. The author is very grateful to Professor Werner Frank, Stuttgart, not only for his critical reading of this chapter and the numerous suggestions resulting from it, but also for several decades of fruitful and stimulating collaboration. Many of the views presented have been developed jointly. The author is also indebted to Professor Manfred Fähnle, Stuttgart, for in-depth discussions of ab initio calculations on defects in metals, to Professor G. Neumann, Berlin, for his critical reading of the manuscript, to Ms. Anke Geigle for her assistance in preparing the typescript, and to Ms. Adelheid Duhm of Springer Verlag for her unselfish help.

References

1. A. Seeger, E. Mann, and R. v. Jan: J. Phys. Chem. Solids **23**, 639 (1962)
2. M. Hohenstein, A. Seeger, and W. Sigle: J. Nucl. Mater. **169**, 33 (1989)
3. W. Sigle and A. Seeger: Phys. Stat. Sol. (a) **146**, 57 (1994)
4. F.G. Fumi: Phil. Mag. **46**, 1007 (1955)
5. A. Seeger and H. Bross: Z. Physik **145**, 161 (1956)
6. W. Frank, U. Breier, C. Elsässer, and M. Fähnle: Phys. Rev. B **48**, 7676 (1993)
7. W.M. Lomer, in: *Vacancies and other Point Defects in Metals and Alloys*, (Inst. of Metals, London 1958) p. 79
8. A. Seeger: Crystal Lattice Defects **4**, 221 (1973)
9. A. Seeger: *Diffusion and Point-Defect Properties.* Defect and Diffusion Forum **95–98**, 147 (1993)

10. Y. Kraftmakher: *Equilibrium Vacancies and Thermophysical Properties of Metals*. Physics Reports **299**, 79 (1998)
11. R.M.J. Cotterill, M. Doyama, J.J. Jackson, and M. Meshii (Eds): *Lattice Defects in Quenched Metals* (Academic Press, New York and London 1965)
12. Landolt-Börnstein: New Series, Group III, Vol. 25, *Atomic Defects in Metals*. H. Ullmaier (Ed.) (Springer, Berlin Heidelberg New York 1991)
13. O.M. Corbino: Lincei Rendic. (5) **19**, 133 (1909)
14. O.M. Corbino: Lincei Rendic. (1) **20**, 223 (1911)
15. T. Hehenkamp: J. Phys. Chem. Solids **55**, 907 (1994)
16. J. Beyer and C. Wagner: Z. Phys. Chem. **B** 32, 113 (1936)
17. A. Gott: Ann. Phys. (Lpzg) **41**, 520 (1942)
18. J. van Dujn and J. van Galen: Physica **23**, 375 (1957)
19. R. Feder and A.S. Nowick: Phys. Rev. **109**, 1959 (1958)
20. R.O. Simmons: *The Experimental Study of Point Defects in Pure Metals*. In: Radiation Damage in Solids, D.S. Billington (Ed.) (Academic Press, New York and London 1965) p. 568
21. R.O. Simmons and R.W. Balluffi: Phys. Rev. **129**, 1533 (1969)
22. K. Differt, A. Seeger, and W. Trost: Mater. Sci. Forum **15–18**, 99 (1987)
23. J.-E. Kluin and T. Hehenkamp: Phys. Rev. B **44**, 11597 (1991)
24. J. Eshelby: J. Appl. Phys. **24**, 1249 (1953); **25**, 255 (1954)
25. V. Schott, M. Fähnle, and P.A. Madden: J. Phys.: Condens. Matter **12**, 1171 (2000)
26. J. Wolff, J.-E. Kluin, and T. Hehenkamp: Mater. Sci. Forum **105–110**, 1329 (1992)
27. T. Hehenkamp, personal communication
28. K. Mosig, Dr.rer.nat.thesis, Universität Göttingen, (1993)
29. R.O. Simmons and R.W. Balluffi: Phys. Rev. **119**, 600 (1960)
30. R.O. Simmons and R.W. Balluffi: Phys. Rev. **125**, 862 (1962)
31. R.O. Simmons and R.W. Balluffi: Phys. Rev. **117**, 528 (1960)
32. G. Bianchi, D. Mallėjac, C. Janot, and G. Champier: Compt. Rend. Acad. Sci. (Paris) B **263**, 1404 (1966)
33. B. von Guėrard, H. Peisl, and R. Zitzmann: Appl. Phys. **3**, 37 (1974)
34. R. Feder and A.S. Nowick: Phil. Mag. **15**, 805 (1967)
35. C. Janot, D. Mallėjac, and B. George: Phys. Rev. B **2**, 3088 (1970)
36. H.M. Gilder and G.M. Wallmark: Phys. Rev. **182**, 771 (1969)
37. M. Current and H.M. Gilder: Phys. Rev. B **16**, 2386 (1977)
38. R. Balzer and H. Sigvaldason: J. Phys. F: Metal Phys. **9**, 171 (1979)
39. R. Feder and A.S. Nowick: Phys. Rev. B **5**, 1244 (1972)
40. C. Janot and B. George: Phys. Rev. B **12**, 2212 (1975)
41. A. Seeger: Phil. Mag. A **64**, 735 (1991)
42. W. Rupp, U. Ermert, and R. Sizmann: Phys. Stat. Sol. **33**, 504 (1969)
43. V.I. Gol'danskii and P.E. Prokop'ev: Sov. Phys. Solid State **6**, 2641 (1965)
44. W. Brandt. In: *Positron Annihilation*, A.T. Stewart and L.O. Roellig (Eds) (Academic Press, New York and London 1985) p. 369
45. S. Berko and J.K. Erskine: Phys. Rev. Letters **19**, 307 (1967)
46. B. Bergersen and M.J. Stott: Solid State Commun. **7**, 1203 (1969)
47. D.C. Connors and R.N. West: Phys. Letters **30A**, 24 (1969)
48. I.K. MacKenzie, T.L. Khoo, A.B. McDonald, and B.T.A. McKee: Phys. Rev. Letters **19**, 946 (1967)

49. A.N. Goland and T.M. Hall: Phys. Letters **25A**, 397 (1973)
50. A. Seeger: Appl. Phys. **4**, 183 (1975)
51. A.S. Davydov: *Quantum Mechanics* (Pergamon, Oxford 1965) Sect. 37
52. J. Eshelby: *The Continuum Theory of Lattice Defects.* In: Solid State Physics, Vol. 3. F. Seitz and D. Turnbull (Eds) (Academic Press, New York 1956) p. 79
53. G. Leibfried and N. Breuer: *Point Defects in Metals I: Introduction to the Theory* (Springer, Berlin Heidelberg 1978)
54. A. Seeger: J. Phys. F: Metal Phys. **3**, 248 (1973)
55. I.K. MacKenzie, T.W. Craig, and B.T.A. McKee: Phys. Letters **36 A**, 227 (1971)
56. A. Seeger: *Positron Annihilation: A New Tool for Studying Imperfections in Crystals.* In: Frontiers in Materials Science, L.E. Murr and C. Stein (Eds.) (Marcel Dekker, New York and Basel 1976) Chap. 5
57. R.N. West: *Positron Studies of Lattice Defects in Metals.* In: Positrons in Solids, P. Hautojärvi (Ed.) (Springer, Berlin Heidelberg 1979) Chap. 3
58. A. Seeger and F. Banhart: Helv. Phys. Acta **63**, 403 (1990)
59. W. Frank and A. Seeger: Appl. Phys. **3**, 61 (1974)
60. A. Seeger: Phys. Letters **40A**, 135 (1972); **41A**, 267 (1972)
61. A. Seeger: Appl. Phys. **7**, 85, 257 (1975)
62. H. Stoll: *MeV Positron Beams.* In: Positron Beams and their Applications, P. Coleman (Ed.) (World Scientific, Singapore 2000) Chap. 8
63. A. Seeger: Rad. Phys. Chem. **58**, 4111 (2000)
64. A. Seeger: Phys. Letters **53**, 324 (1975)
65. A. Seeger: Hyperfine Interact. **6**, 313 (1979)
66. A. Seeger: *Positive Muons and Pions in Materials Research: Perspectives and Future Developments.* In: Muons and Pions in Materials Research, J. Chappert and R.I. Grynszpan (Eds) (North-Holland, Amsterdam 1984)
67. A. Schenk: *Muon Spin Rotation Spectroscopy* (Adam Hilger, Bristol 1985)
68. R. Kadono, W. Higemoto, K. Nagamine, and F.L. Pratt: Phys. Rev. Letters **83**, 987 (1999)
69. J.H. van Vleck: Phys. Rev. **74**, 1168 (1948)
70. O. Hartmann: Phys. Rev. Letters **39**, 832 (1977)
71. K. Fürderer, H.-P. Döring, M. Gladisch, N. Haas, D. Herlach, J. Major, H.-J. Mundinger, J. Rosenkranz, W. Schäfer, L. Schimmele, M. Schmolz, W. Schwarz, and A. Seeger: Hyperfine Interact. **31**, 81 (1986)
72. K. Fürderer, H.-P. Döring, M. Gladisch, N. Haas, D. Herlach, J. Major, H.-J. Mundinger, J. Rosenkranz, W. Schäfer, L. Schimmele, M. Schmolz, W. Schwarz, and A. Seeger: Mater. Sci. Forum **15–18**, 125 (1987)
73. G. Majer, R. Messer, A. Seeger, W. Templ, K. Fürderer, M. Gladisch, and D. Herlach: Phil. Mag. Letters **57**, 57 (1988)
74. M. Hampele, D. Herlach, A. Kratzer, G. Maier, J. Major, H.P. Raich, R. Roth, C.A. Scott, A. Seeger, W. Templ, M. Blanz, S.F.J. Cox, and K. Fürderer: Hyperfine Interact. **65**, 1081 (1990)
75. A. Seeger: Phys. Stat. Sol. (a) **167**, 289 (1998)
76. A. Seeger and H. Mehrer: *Analysis of Self-Diffusion and Equilibrium Measurements.* In: Vacancies and Interstitials in Metals, A. Seeger, D. Schumacher, W. Schilling, and J. Diehl (Eds) (North-Holland, Amsterdam 1970) p. 1
77. E. Soininen, H. Huomo, P.A. Huttonen, J. Makinen, A. Vehanen, and P. Hautojärvi: Phys. Rev. B **41**, 6227 (1990)

78. R. Blank, L. Schimmele, and A. Seeger: Mater. Sci. Forum **105-110**, 603 (1992)
79. R. Blank: *Positronendiffusion in kubischen Kristallen.* Dr.rer.nat. thesis, Universität Stuttgart (1992)
80. C.H. Hodges: Phys. Rev. Letters **25**, 284 (1970)
81. P.J. Schultz, K.G. Lynn, and B. Nielsen: Phys. Rev. B **32**, 1369 (1985)
82. P.C. Lichtenberger, C.W. Schulte, and I.K. MacKenzie: Appl. Phys. **6**, 305 (1975)
83. F. Banhart and A. Seeger: Mater. Sci. Forum **105–110**, 825 (1992)
84. O. Sueoka: J. Phys. Soc. Jpn **36**, 464 (1974)
85. W. Triftshäuser and J.D. McGervey: Appl. Phys. **6**, 177 (1975)
86. H. Fukushima and M. Doyama: J. Phys. F: Metal Physics **6**, 677 (1976)
87. P. Rice-Evans, T. Hlaing, and D.B. Rees: J. Phys. F: Metal Physics **6**, 1079 (1976)
88. S. Nanao, K. Kuribayashi, S. Tanigawa, and M. Doyama: J. Phys. F: Metal Physics **7**, 1403 (1977)
89. G. Dlubek, O. Brümmer, and N. Meyendorf: Appl. Phys. **13**, 67 (1977)
90. J.L. Campbell, C.W. Schulte, and R.R. Gingerich: J. Nucl. Mater. **69–70**, 609 (1978)
91. A.S. Berger, S.T. Ockers, and R.W. Siegel: J. Phys. F: Metal Physics **9**, 1023 (1979)
92. M.J. Fluss, L.C. Smedskjaer, R.W. Siegel, D.G. Legnini, and M.K. Chason: J. Phys. F: Metal Physics **10**, 1763 (1980)
93. H. Bosse, A. Sager, W. Lühr-Tank, and T. Hehenkamp: J. Phys. F: Metal Physics **16**, 1337 (1986)
94. H.-E. Schaefer, W. Stuck, F. Banhart, and W. Bauer: Mater. Sci. Forum **15–18**, 117 (1987)
95. H. Mehrer, N. Stolica, and N.A. Stolwijk: *Self-diffusion in Solid Metallic Elements.* In: Landolt-Börnstein, New Series, Vol. III/26. H. Mehrer (Ed.) (Springer, Berlin Heidelberg New York 1990) Chap. 2
96. A. Seeger: *The Interpretation of Radiation Damage in Metals.* In: Fundamental Aspects of Radiation Damage in Metals, M.T. Robinson and F.W. Young Jr. (Eds.), Conf.-75006 P1 (National Techn. Inf. Service, US Department of Commerce, Springfield VA 1975) p. 419
97. M. Fähnle and A. Seeger: Phys. Stat Sol. (a) **91**, 609 (1985)
98. M. Fähnle and M. Khatun: Phys. Stat. Sol. (a) **126**, 109 (1991)
99. Y. Shirai, H.-E. Schaefer, and A. Seeger. In: *Positron Annihilation,* L. Dorikens-Vanpraet, M. Dorikens, and D. Segers (Eds) (World Scientific, Singapore 1989) p. 419
100. S. Takaki, J. Fuss, H. Kugler, U. Dedek, and H. Schultz: Radiat. Effects **79**, 87 (1983)
101. W. Decker, J. Diehl, A. Dunlop, W. Frank, H. Kronmüller, W. Mensch, H.-E. Schaefer, B. Schwendemann, A. Seeger, H.-P. Stark, F. Walz, and M. Weller: Phys. Stat. Sol. (a) **52**, 239 (1979)
102. K. Maier, M. Peo, B. Saile, H.-E. Schaefer, and A. Seeger: Phil. Mag. A **40**, 701 (1979)
103. R.C. Shukla, C.A. Plint, and D.A. Ditmars: Int. J. Thermophysics **6**, 517 (1985)
104. M.W. Zemansky: *Heat and Thermodynamics.* 3rd edn (McGraw-Hill, New York 1951)

105. K.A. Gschneider Jr.: *Physical Properties and Intercorrelations of Metallic and Semiconducting Elements.* In: Solid State Physics, Vol. 16. F. Seitz and D. Turnbull (Eds) (Academic Press, New York and London 1964)
106. E.C. Stoner: Proc. Roy. Soc. (London) A **154**, 656 (1936)
107. N.F. Mott: Proc. Roy. Soc. (London) A **152**, 42 (1935)
108. M. Born and E. Brody: Z. Physik **6**, 132, 140 (1921); **8**, 205 (1922)
109. E. Schrödinger: Z. Physik **11**, 170 (1922)
110. G. Leibfried: *Gittertheorie der mechanischen und thermischen Eigenschaften der Kristalle.* In: Encyclopedia of Physics, Vol. VII, Pt. 1, S. Flügge (Ed.) (Springer, Berlin Heidelberg 1955) Chap. G
111. M. Blackman: *The Specific Heat of Solids.* In: Encyclopedia of Physics, Vol. VII, Pt. 1. S. Flügge (Ed.) (Springer, Berlin Heidelberg 1955) Sect. 20
112. G.A. de Wijs, G. Kresse, and M.J. Gillan: Phys. Rev. B **57**, 8223 (1998)
113. D. Alfè, M.J. Gillan, and G.D. Price: Nature **401**, 462 (1999)
114. G.D. Mahan: *Many-Particle Physics*, 2nd edn (Plenum, New York and London 1990) Chap. 6

Index

GPSR Compliance

The European Union's (EU) General Product Safety Regulation (GPSR) is a set of rules that requires consumer products to be safe and our obligations to ensure this.

If you have any concerns about our products, you can contact us on ProductSafety@springernature.com

In case Publisher is established outside the EU, the EU authorized representative is:

Springer Nature Customer Service Center GmbH
Europaplatz 3
69115 Heidelberg, Germany

Batch number: 10704030

Printed by Printforce, the Netherlands